CRIMINAL JUSTICE ILLUMINATED

Juvenile Justice

A Social, Historical, and Legal Perspective

Second Edition

Preston Elrod, PhD
Professor
Department of Correctional
and Juvenile Justice Studies
Eastern Kentucky University
Richmond, Kentucky

R. Scott Ryder, JD
Juvenile Court Director
St. Joseph County, Michigan

JONES AND BARTLETT PUBLISHERS
Sudbury, Massachusetts
BOSTON   TORONTO   LONDON   SINGAPORE

World Headquarters
Jones and Bartlett Publishers
40 Tall Pine Drive
Sudbury, MA 01776
978-443-5000
info@jbpub.com
www.jbpub.com

Jones and Bartlett Publishers Canada
6339 Ormindale Way
Mississauga, ON L5V 1J2
Canada

Jones and Bartlett Publishers International
Barb House, Barb Mews
London W6 7PA
United Kingdom

Jones and Bartlett's books and products are available through most bookstores and online book-sellers. To contact Jones and Bartlett Publishers directly, call 800-832-0034, fax 978-443-8000, or visit our website www.jbpub.com.

Substantial discounts on bulk quantities of Jones and Bartlett's publications are available to corporations, professional associations, and other qualified organizations. For details and specific discount information, contact the special sales department at Jones and Bartlett via the above contact information or send an email to specialsales@jbpub.com.

Production Credits
Publisher, Public Safety Group: Kimberly Brophy
Acquisitions Editor: Stefanie Boucher
Production Director: Amy Rose
Associate Production Editor: Carolyn F. Rogers
Marketing Associate: David Weliver
Cover and Text Design: Anne Spencer
Photo Research: Kimberly Potvin
Cover Image: photo © Joel Gordon 1997; column © Ron Chapple/Thinkstock/Alamy Images
Chapter Opener Image: © Masterfile
Composition: Jason Miranda
Printing and Binding: Malloy, Inc.
Cover Printing: Malloy, Inc.

Library of Congress Cataloging-in-Publication Data

Elrod, Preston.
  Juvenile justice : a social, historical, and legal perspective / Preston
Elrod, R. Scott Ryder—2nd ed.
      p. cm.
  Includes bibliographical references and index.
  ISBN 0-7637-3307-5 (hardcover : alk. paper)
  1. Juvenile justice, Administration of--United States. I. Ryder, R.
Scott. II. Title.
  KF9779.E44 2005
  345.73'08--dc22

                                                            2005007399

Printed in the United States of America
09 08 07 06 05   10 9 8 7 6 5 4 3 2

To my family—Carol, Colin, Ryan, Karyn, Douglas, Linda, Herman, and Catherine
P.E.

To my wife, Denise, and my children, Joshua, Adam, and Tim, my stepdaughter Kirby Jarzeboski, and to my parents, Dick and Polly
R.S.R.

Contents

Preface

This book is intended to serve as a comprehensive introduction to the field of juvenile justice. It is designed to help readers understand the complexities of present juvenile justice practice by presenting a thorough examination of the social, historical, and legal context within which delinquency and juvenile justice practice occurs. We have worked diligently to include a broad base of information so that those without experience in the field will be able to grasp the operation of juvenile justice and the processing of cases by juvenile justice agencies. We examine the relevant literature on the effectiveness of various juvenile justice interventions, as well as a variety of practical, philosophical, ethical, and legal issues that invariably arise when juvenile justice agencies become involved in the lives of delinquent youths and their families. We also examine a variety of commonly believed myths about juvenile offenders and juvenile justice.

Our mission is to make juvenile offenders (a very diverse group of youths) and juvenile justice (a very diverse range of formal and nonformal responses to delinquent behavior) accessible to readers. We have, of course, relied heavily on the scientific literature in developing the contents of this book. However, we have also relied on our own experiences as juvenile justice practitioners (our combined years as practitioners total more than 30), as well as our experiences as teachers and instructors of juvenile justice practitioners and university students.

Despite the incredible amount of time and energy that has gone into this book, we recognize that the final product is not perfect. Indeed, there are several aspects of juvenile justice that create difficulties for anyone attempting to develop a comprehensive introductory text in this area. One significant challenge has been to present the diversity of current approaches to juvenile justice. Not only is there considerable variability in the practice of juvenile justice nationwide, but there can be considerable variability within a single juvenile justice agency. Although we were not able to capture all of the nuances that characterize juvenile justice practice, we have attempted to create a picture of juvenile justice that accurately portrays its main features. Moreover, we have endeavored to present a balanced view of juvenile justice that critically examines juvenile justice practice while acknowledging the many positive things that are done for children and families in juvenile courts and other juvenile justice agencies each day. We hope that we have succeeded.

Many people contributed to this book. We acknowledge our wives, Carol Elrod and Denise Ryder, and our children, Joshua, Adam, and Tim Ryder and Kirby Jarzeboski, and Colin and Ryan Elrod, who supported our efforts in countless ways. Special thanks go to Carol Elrod for her assistance with editing the manuscript. Her efforts have been a significant contribution to this project. Also we thank the court staff and judges of the Family Divisions of the 9th Circuit Court, Kalamazoo County, Michigan, and the 45th Circuit Court, St. Joseph County, Michigan, and the many capable and caring individuals we have come across who are dedicated to serving the needs of children and families. Their work has been an inspiration to both of us. We also thank the many individuals who were willing to spend time with us and share their experiences in juvenile justice. Without their help, this book would not have been possible. We give special thanks to Brenda Foley, Sasha Sloan, Eva Sylvester, Stephanie Sims, Janet Snow, Jennifer Fielder, Steve Parson, Jeff Cantrell, Amber Wells, and Janice Marcum, who provided various types of support for this project. Special thanks also go to several anonymous reviewers who provided helpful critiques of our work and to Professor Kevin Minor, Eastern Kentucky University, for helpful comments on the early part of the manuscript.

Finally, we thank a number of individuals at Jones and Bartlett Publishers, particularly Chambers Moore and Carolyn Rogers, who have assisted us with this project. They have been and continue to be a pleasure to work with. Thank you all!

Acknowledgments

About the Authors

Preston Elrod, PhD
Associate Professor
Department of Correctional and Juvenile Justice Studies
Eastern Kentucky University
Richmond, Kentucky

Preston Elrod graduated from Presbyterian College, Clinton, South Carolina, with a B.A. in history. He attended Western Michigan University, where he received his M.A. (sociology, honors) and Ph.D. (sociology/criminology). He has worked as the site director of a model federally funded school-based delinquency prevention project and for a juvenile court as an intake officer and a probation department supervisor. He began his university career in 1989 and is presently a professor in the Department of Correctional and Juvenile Justice Studies at Eastern Kentucky University, where he teaches courses on theory, delinquency and crime prevention, juvenile justice, and juvenile corrections, among others. He has published articles on a variety of criminal justice topics including effective interventions for juvenile offenders, adolescent jail inmates, and juvenile justice policy. His primary academic interest is in the development and evaluation of programs designed to reduce juvenile crime. When not at work, he enjoys spending time with his family, coaching youth soccer, and participating in a variety of outdoor activities.

R. Scott Ryder, JD
Juvenile Court Director
St. Joseph County, Michigan

Scott Ryder graduated magna cum laude from Wittenberg University, Springfield, Ohio, in 1971 with a B.A. in history. He attended Indiana University School of Law in Bloomington, Indiana, where he received his Juris Doctor degree in 1974. He began his involvement in juvenile justice in 1975, working as an assistant prosecuting attorney in Shiawassee County, Michigan, with primary responsibility for proceedings in the juvenile court. Since May 1979, he has served as the chief referee of the Kalamazoo County (Michigan) Juvenile Court and as a referee in the Family Division of the 9th Circuit Court, Kalamazoo County, Michigan. He retired from Kalamazoo County in May 2004 and is presently the Juvenile Court Director of the St. Joseph County, Michigan, Juvenile Court. In addition to his duties in the court, since 1981 he has been an instructor for the Michigan Judicial Institute, the training agency for the Michigan Supreme Court, where he has instructed numerous court employees on various legal and procedural issues related to family court operation. He is an Adjunct Assistant Professor at Western Michigan University teaching a course in criminal law and procedure. He has also published in the area of juvenile justice. When he is not on the bench, teaching, or working, his hobbies are golf, reading, and refereeing soccer. He is married to Denise and has three sons and a stepdaughter.

The Context of Juvenile Justice: Defining Basic Concepts and Examining Public Perceptions of Juvenile Crime

1

Chapter Objectives

After studying this chapter, you should be able to

- Define basic concepts necessary for understanding the juvenile justice process
- Explain what is meant by a legal definition of delinquency
- Describe the shortcomings of legal definitions of delinquency
- Explain why it is important to understand diversity within the population of juvenile offenders
- Explain why age is the primary criterion for defining a juvenile from a legal perspective
- Describe the problems associated with using age to define the clients of the juvenile justice process
- Describe those factors that contribute to variability and conflict in juvenile justice practice
- Explain why "juvenile justice system" may be a misleading term
- Describe the role that public perception plays in responding to the "delinquency problem"

Chapter Outline

- Introduction
- Defining Delinquency
- Defining Juveniles: The Legal Perspective
- The Juvenile Justice System
- Public Perceptions of Delinquency and the Politics of Juvenile Justice

- Chapter Summary
- Key Concepts
- Review Questions
- Additional Readings
- Notes

■ Introduction

The creation of separate institutions for the treatment of youths who engage in illegal or immoral behavior is a recent historical development. During the colonial period and up until the early 1800s, youths were subject to the same criminal justice process as adults. Consequently, children who were apprehended for crimes were tried in the same courts and, when found guilty, were often, although not always, given the same punishments as adults.[1]

Today, most Americans are aware that a separate justice process exists for **juveniles**. However, many people lack an understanding of how this process works, and others hold views that are based more on myth than reality. Moreover, most lack an understanding of the social, political, and economic contexts that shape juvenile crime and juvenile justice practice.

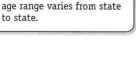

juvenile A youth who falls within an age range specified by state law. The age range varies from state to state.

For Your Information

Throughout this text, important points of information helpful for understanding juvenile justice practice are highlighted. In addition, myths about the operation of juvenile justice are explored in sidebars separated from the main text.

labeling Defining and responding to a youth as a delinquent. Labeling can lead to stigmatization, lost opportunities, and additional delinquent behavior.

Although many citizens lack an understanding of how the juvenile justice process works, they often express strong opinions about the causes of delinquent behavior and how such behaviors should be dealt with. Such opinions are found in letters to local newspapers, in radio talk show programs, in television commentaries, and in the everyday discussions that people have about current events. Some members of the public call for increased efforts to prevent delinquency and rehabilitate young offenders. Others call for harsher punishments for young "thugs" and demand that youths who commit "adult" crimes be treated like adults. Similar sentiments are found among political leaders as well. Although rhetoric designed to capture the public's attention may be seen by some as "good politics," it is problematic for several reasons. First, it contributes little to the public's understanding of juvenile crime and the development of effective methods of responding to juvenile offenders. Second, **labeling** youths as "thugs" dehumanizes them and makes it easier to respond to them in ways that protect neither their interests nor the interests of the community. Third, inflammatory rhetoric about juvenile crime leads to ineffective policies that squander precious resources and lead to increased public cynicism about our ability to deal with juvenile crime.

Young people do commit serious offenses, although most juvenile crime does not result in serious injury or property loss. The question is, what is the best response to the illegal behavior of youths? Which responses are more likely to help youths develop and make more socially productive decisions? Which responses are more likely to represent a sound investment of public funds and protect community safety? These are not easy questions to answer, but any viable response to juvenile crime must be predicated on sound knowledge of what causes delinquency as well as a clear understanding of past and present responses to the juvenile crime problem, including the effectiveness of those responses. A primary purpose of this text is to help readers gain such an understanding.

Although this text is intended to help students understand the social, historical, and legal context of juvenile crime and juvenile justice, and to understand present juvenile justice responses to juvenile crime, bear in mind that juvenile justice is ultimately a human endeavor. Indeed, what makes juvenile justice interesting, challenging, and potentially very rewarding is that many youths and their families are helped by the many committed, highly trained, and caring individuals who work in juvenile justice. However, other youths are not helped, despite the best efforts of juvenile justice professionals, and in other instances youths and their families are harmed by those who are ostensibly charged with protecting their well-being. Unfortunately, in some instances, those who work in juvenile justice lack the knowledge, training, support, and commitment necessary for effective practice.

 Juvenile Justice Practice

Ultimately, juvenile justice involves many people making decisions, sometimes very difficult decisions and often under arduous conditions, that can have a profound influence on the lives of others.

This chapter is intended to introduce the basic concepts necessary for understanding present juvenile justice practice. It begins by defining delinquency. Next, it provides a profile of juvenile offenders, examines public perceptions of delinquency (which form an important part of the contemporary context of juvenile justice), and explores the concept of a juvenile justice system.

delinquency Those behaviors that fall within the jurisdiction of the juvenile court and result in processing by official juvenile justice agents.

status offenses Acts that are considered illegal when committed by a child but not when committed by an adult (e.g., running away, school truancy, and failure to obey parents' directions).

Defining Delinquency

From a legal standpoint, <u>delinquency</u> consists of those behaviors that are prohibited by the family or juvenile code of the state and that subject minors (i.e., persons not legally adults) to the jurisdiction of the juvenile court. Behaviors prohibited by juvenile codes can be grouped into two general categories: (1) behaviors that would be defined as criminal offenses if committed by adults (malicious destruction of property, larceny, robbery, motor vehicle theft, etc.), and (2) behaviors that are prohibited only for minors, which are called <u>status offenses</u> (school truancy, running away from home, incorrigibility, etc.).

Defining delinquency as behaviors that violate the legal code of the state seems straightforward, yet it fails to completely represent the reality of delinquency and juvenile justice responses to youths' illegal behaviors. For example, police respond to only some of the actions that are legally defined as delinquent. Police often ignore some (typically minor) illegal behaviors that are prohibited by legal codes. Also, as noted in Chapter 6, how police respond to youths alleged to be engaged in illegal behavior can vary considerably from one community to another. Police in one community may arrest youths who do not go to school, whereas in another community failure to attend school receives little attention by the authorities. Moreover, the delinquent activities of some youths tend to be more visible than the activities of others, thus increasing the likelihood that certain youths will come to the attention of the police and be labeled delinquents. Defining delinquency as behavior that violates the legal code ignores these nuances in juvenile justice practice. In recognition of these realities, some have suggested that a more useful definition of delinquency would focus on those acts that official agents select for enforcement rather than on all legally prohibited behaviors.[2] From this perspective, delinquency actually represents a sample of those behaviors prohibited by state law, and delinquents are, for all practical purposes, youths who are subject to formal processing by the authorities.

Another problem with using a legal definition is that such definitions cover an extremely broad range of behaviors from incorrigibility (i.e., not obeying one's parents) to serious criminal actions (e.g., homicide). From a legal standpoint, almost all minors could be considered delinquents, because most youths engage in at least one illegal behavior at some time during their juvenile years. For example, a 2001 survey of high school seniors' drug and alcohol use revealed that 85% had used alcohol and 52% had used marijuana.[4] The percentage of youths who fail to obey their parents, also illegal in many <u>jurisdictions</u>, is likely to be even higher. According to a strict legal definition, most youths would be considered delinquents, even though most people would not consider many of these youths to be delinquents, nor would they consider their actions to be illegal.

jurisdiction A geographical area subject to a particular law or court.

The Saints and the Roughnecks

In a classic study of delinquency, "The Saints and the Roughnecks," William Chambliss pointed out that lower-class youths, who tend to be visible to the community, nonmobile, and not very adept at meeting the social expectations of authorities, are more likely to be selected for punishment than affluent youths, who are generally less visible and more mobile. By observing the Saints, eight white males from solid middle-class families, and the Roughnecks, six lower-class white males, Chambliss discovered that the Saints actually engaged in more frequent delinquent behaviors, although the Roughnecks engaged in somewhat more serious behaviors, such as fighting and property offenses. The Roughnecks, however, were more likely to be seen as delinquents destined for lives of trouble than were the Saints, who were seen as upstanding students with bright futures.

In explaining the difference in the reaction to these two groups, Chambliss stated that the Roughnecks' activities took place within the purview of the community because the Roughnecks did not have access to cars. In contrast, the Saints, who did have access to cars, were able to travel to the periphery of town or to other towns, and thus their delinquent behaviors were less visible. Furthermore, Chambliss noted that, during contact with the authorities, the demeanor of the Saints allowed them to avoid difficulty. The Saints generally responded in ways that were felt to be more acceptable by authority figures. As a result, the authorities possessed a perceptual bias that led them to see lower-class youths as more problematic—a bias reinforced by the visibility of the Roughnecks and their lack of social skills.[3]

There are still other problems with legal definitions. As just suggested, legal definitions do not make a distinction between those who are caught and those who engage in delinquent behavior. Yet, this distinction may be important, because those who are caught are subject to the loss or the restriction of their freedom. Moreover, we should not assume that our reactions to juvenile offenders will necessarily lead to a cessation of their illegal behavior. As Harold Garfinkel notes, the process of labeling a youth a delinquent may be seen as a status degradation ceremony through which the youth's identity is (possibly) transformed into a lesser form of humanity.[5] Not only may labeling cause the individual to see him- or herself differently (e.g., as a troublemaker, thief, or delinquent), but it also may cause others to respond differently to or avoid the person, leading to rejection and the restriction of law-abiding opportunities and thus increasing the likelihood of further delinquency.[6]

MYTH VS. REALITY — Punishment

Myth—Youths should always be punished when they violate the law.

Reality—Sometimes our responses to youths can increase the likelihood of further deviant behavior, or what sociologist Edwin Lemert calls "secondary deviance." Lemert claims that persons may engage in initial acts of deviance, such as delinquency, for a variety of reasons. Lemert terms such deviance "primary deviance." However, through repeated interaction between someone identified as deviant and authority figures, a process that may involve labeling and stigmatization, the individual may reorganize his or her identity around a more deviant role, thus increasing the likelihood of further acts of deviance. This secondary deviance is not a product of the original factors that produced the initial acts of deviance but an adjustment to or means of defense against societal reactions to the primary deviance.[7]

A number of studies have uncovered the problems faced by persons on whom certain negative labels have been placed, such as mental patients and those convicted of crimes. For example, Charles Frazier documented the problems experienced by "Ken," a young man who lived in a small town and was labeled a "criminal" at a public trial. Subsequent to his official labeling, people began to see previous events in Ken's life as indications of deviance. In addition, former friends and associates began to reject Ken, and their rejection led Ken to see himself as a criminal.[8] Research by Schwartz and Skolnick in the United States and by Buikhuisen and Dijksterhuis in the Netherlands highlights the problems faced by "offenders" in their efforts to seek employment.[9] In these studies, prospective employers were given job applications of fictitious persons that contained varying amounts of information regarding criminal involvement. Not surprisingly, both studies found that employers were more likely to consider prospective candidates with no criminal history than those with criminal histories. Although such studies do not address possible positive effects of labeling, such as the avoidance of negative behaviors out of fear of receiving a negative label or the potential deterrent effects of shame and embarrassment associated with criminal justice involvement, they do challenge the notion that responding to offenders by labeling them is always beneficial. Moreover, as noted earlier, sometimes juvenile justice responses actually result in youths being harmed. For example, youths are sometimes placed in correctional programs where they are exposed to more hardened offenders and are victimized by other youths and/or staff.

Another difficulty with legal definitions of delinquency is that they obscure potentially important differences between youths involved in illegal behavior. Some youths become involved in the juvenile justice process because of their involvement in status offenses, others because of criminal offending.

Distinguishing Between Status Offenders and Deliquents

Most states make a distinction between youths who engage in criminal behaviors by designating them "delinquent offenders" or "public offenders" and those who engage in status offenses, who are called **PINS** (persons in need of supervision), **CHINS** (children in need of supervision), or "undisciplined children."

PINS Persons in need of supervision; children who engage in status offenses.

CHINS Children in need of supervision; see **PINS**.

However, making distinctions between status and criminal offenders still does not take into account the many ways in which youths differ from one another. Indeed, youths who engage in illegal behavior constitute a diverse population. They vary considerably in terms of their psychological and biological characteristics and the social context within which they live. Moreover, these variations in psychological and biological characteristics, as well as social context, need to be considered when making decisions about their treatment.

(3) Finally, legal definitions of delinquency present it as an either/or phenomenon and ignore both the frequency and seriousness of delinquent conduct. In reality, most youths engage in infrequent and minor types of delinquent behavior, and only a small percentage of the juvenile population engages in repetitive and serious delinquent actions. Consequently, it would arguably be more accurate to view delinquency as a form of behavior that falls on a behavioral continuum ranging from extreme conformity to extreme nonconformity.[10]

In fact, juvenile justice practitioners spend a considerable amount of time attempting to determine how they should respond to youths who cluster around various points along this continuum and how they should respond to youths with different personal and social characteristics. How many resources should we devote to minor offenders? If we fail to devote resources to minor offenders, will they become serious offenders? How many resources should we devote to serious offenders? Which juvenile offenders represent a threat to the public and which are "good kids" who have made a mistake? These are questions that juvenile justice practitioners wrestle with daily. Moreover, how juvenile justice practitioners answer these questions can have profound effects not only on youths who come to the attention of the authorities but also on their families and the wider community. Consider the (actual) case of Marcus (see **Box 1-1**). How has the court responded to this case? From Marcus's point of view, what have been the pros and cons of court involvement? Do you think the court responded appropriately?

Box 1-1

Interview: Marcus, a Delinquent Offender

Q: How old were you when you first became involved in the juvenile court system, and how old are you now?

A: I was 15 when I was first arrested, and I am 16 years old now.

Q: What was your presenting delinquent offense? Were you alone or with someone?

A: A friend and I robbed a convenience store, and we had a pistol; the charge was armed robbery.

Q: What was your sentence or disposition?

A: I was sentenced in July of 1997 to probation and put into the day treatment program. They monitor me 24 hours a day, including before and after school; but I get to live at home.

Q: Since being sentenced, have you committed any new delinquent crimes or violated your probation?

A: I violated my probation by trying to buy some marijuana after a few months on probation. I got 24 hours in detention and in-home detention for Valentine's weekend so I couldn't spend Valentine's Day with my girlfriend. I learned my lesson and have not violated since!

Q: What are the benefits that you believe you are receiving from your involvement with the juvenile court?

A: My involvement with the court has really helped turn my life around. The referee who sentenced me made me attend football workouts all summer as part of my probation. I ended up playing on my high school's freshmen football team and was elected captain and selected as most valuable player. These good things that have happened to me have helped my confidence. I feel good about them. My day treatment worker, teachers, and football coaches have been very positive persons in my life. The structure and the pressure of having rules that I must follow helps keep me out of trouble. In addition to football, the day treatment program gives good rewards for obeying the rules, like traveling to amusement parks. We also do stuff in the community.

Q: What don't you like about your involvement with the court?

A: It's not too bad so long as you obey the rules, so long as you don't mess up. I don't like the tracking and monitoring part of day treatment because it really takes away my freedom; but I know if I violate the rules, I'll only get deeper into trouble and deeper into the system.

Q: How do you feel about committing a crime that hurt someone?

A: I feel bad about it. I know it was wrong; I know I shouldn't have done it. I got the idea from a bunch of guys.

Q: How are you doing in school?

A: My grades are improving because I know I have to be eligible to play football next fall.

Q: When do you think you will be off court supervision?

A: I am not sure of a date, but I must complete my community service hours. I have about 3 $\frac{1}{2}$ hours left. Probably not until the school year is finished, maybe longer.

> Q: What goals for your future do you have now?
> A: I want to graduate from high school with a 3.0 grade point average. I want to play football in college at the University of Michigan or North Carolina. I want to major in athletic administration and be a coach.
> Q: What have you learned from your court involvement?
> A: Even though you make a real big mistake, there's always someone out there to give you another chance. You need to take advantage of that chance or you'll get locked back up.
> Q: Has your experience with the court been enough to keep you from committing another delinquent offense?
> A: Yes!!

Which definition of delinquency is best? Should we use a legal definition and include all youths who violate juvenile laws, or should we employ a definition based on system reaction to delinquent behavior?

As noted earlier, strictly legal definitions of delinquency ignore much of its complexity. Moreover, delinquency can be defined in more than one way. Because this text focuses on the processing of youths by official agencies that make up the juvenile justice process, a legal and justice agency response definition will be used. According to this definition, delinquency consists of those behaviors that fall within the jurisdiction of the juvenile court and result in processing by official agents of juvenile justice. Keep in mind that this definition does not take into account the differences that exist among youths who engage in delinquency, nor does it take into account the differences that exist among the types of delinquency that youths commit. These differences are important, however, because they help determine how youths are processed by juvenile justice agencies.

■ Defining Juveniles: The Legal Perspective

Implicit in the definition of juvenile delinquency is an assumption about which youths are considered juveniles. As noted earlier, those persons who are subject to the jurisdiction of the juvenile court are considered minors; however, the legal definition of a minor varies from state to state.

In the majority of states, in the District of Columbia, and in the federal system, the upper age limit for juvenile court jurisdiction is 17 years. This means that once youths turn 18, they are subject to the jurisdiction of adult courts. In a number of other states (Georgia, Illinois, Louisiana, Massachusetts, Michigan, Missouri, South Carolina, and Texas), youths become adults in the eyes of the criminal justice system when they become 17. In a few states (Connecticut, New York, and North Carolina), adult court jurisdiction begins at age 16.[11]

In addition to an upper age limit, some states have a lower age limit for juvenile court jurisdiction, which varies from 6 to 10 years of age. Where a lower age limit exists, minors younger than the statutorily specified age cannot engage in delinquent behavior. These youths are felt to be too young to have criminal intent, or *mens rea*. The idea that children cannot commit crimes is based

doli incapax Not capable of criminal intention or malice; unable to formulate criminal intent (*mens rea*); not able, due to lack of maturity, to know right from wrong; not criminally responsible.

waiver A legal process that enables a juvenile to be tried as an adult.

remand A legal process whereby a juvenile is sent to adult court for trial; see **waiver**.

transfer A legal process whereby a juvenile is sent to adult court for trial; see **waiver**.

certification A legal process whereby a juvenile is remanded to adult court for trial; see **waiver**.

on the legal principle of *doli incapax*, which holds that young children do not know right from wrong. As a result, they cannot be held liable for actions that would be considered illegal if committed by an adult.

Although each state sets age limits that determine which youths are subject to juvenile court jurisdiction, some state juvenile codes allow juvenile justice agencies to continue jurisdiction over a youth for a specified period once he or she passes the maximum age for juvenile court jurisdiction. For example, in North Carolina youths become subject to adult court processing at 16 years of age. Nevertheless, once the juvenile court takes jurisdiction over a youth, it can continue that jurisdiction until the youth reaches his or her eighteenth birthday. However, a youth past the age of 16 who is under the continuing jurisdiction of the juvenile court would be tried as an adult if apprehended for a new criminal offense.

Although state juvenile codes establish the general parameters for juvenile court jurisdiction, state laws also make it possible to transfer some juveniles to criminal courts for trial. The process by which some juveniles are processed as adults is typically called "**waiver**," "**remand**," "**transfer**," or "**certification**" to adult court. An in-depth examination of waiver is found in Chapter 9.

Using age as a basis for waiver satisfies legal criteria that require clear-cut definitions for those juveniles who are eligible for trial in adult court; however, it suffers from at least two shortcomings. First, it flies in the face of existing knowledge about human psychological, social, and physical development. An adult is commonly defined as a person who has reached maturity, whereas a juvenile is commonly defined as a person who is no longer a child but not yet an adult. At what stage of life, however, does one become an adult, and what exactly is maturity? Psychologists who take a developmental approach suggest that a mature person is one who possesses certain skills that are the product of both cognitive development and the nature of the person's interactions with his or her environment. Moreover, they maintain that these skills are acquired in a developmental sequence. According to this perspective, the ability of individuals to understand their world, including how they relate to others and how their behavior affects and is affected by others, depends on their level of cognitive development and the quality of the interactions they have had with their environment. Accordingly, their ability to understand the effects of their actions on others and themselves is not static, but changes over time.[12]

Human Development

Jean Piaget argues that the ability of the individual to understand and interpret his or her world proceeds in a series of stages, beginning with the sensorimotor period, which lasts roughly from birth until age two, and ending with the formal operations period, which lasts from roughly age 11 through adulthood. During the sensorimotor period, the cognitive abilities of the child are quite different than those of an adult. During this period the young child learns to make sense out of visual, tactile, auditory, and other sensations and focuses on how the world responds to the child's actions. However, during the middle school years, the child enters the formal operations stage, at which point the child's cognitive abilities change greatly. During this period the child is able to understand and interpret the world differently because of his or her ability to engage in more abstract thought. In addition, the development of the child's cognitive abilities is to some extent influenced by the child's environment.[13]

Following Piaget, Lawrence Kohlberg argues that an individual's ability to engage in moral reasoning develops over time and is affected by his or her cognitive development, education, and social experience.[14] At the earliest level of moral reasoning, the individual is more egocentric, focusing attention on how actions affect him or her. At higher stages of moral development, the individual considers how his or her actions affect everyone.[15] Importantly, a number of studies that have examined the relationship between levels of moral reasoning and delinquency have found that low levels of moral reasoning are related to delinquency and criminal behavior. Although these studies do not prove that low moral reasoning causes delinquency, they do suggest that low moral development, along with other psychological and social factors, appears to play an important role.[16]

2. A second shortcoming of relying on age as the major determinant of adulthood is that it fails to consider variations in the social and psychological development of youths. Although developmental psychologists indicate that cognitive and moral development occur in stages that typically begin and end around specific ages, they recognize that youths vary considerably in their rates of development. Also, some research indicates that boys may develop "other-oriented" reasoning or the ability to be empathic later than girls.[17] Consequently, the fact that a youth has turned 17 or 18 and become subject to the jurisdiction of an adult criminal court does not mean that he or she is a mature adult capable of making adult decisions.

Given the lack of consensus about when youths become adults, it is not surprising that state laws often reflect considerable confusion over this very issue. Although a state law may mandate that youths who are 16 or 17 fall under the jurisdiction of adult criminal courts, other laws within that state may deny those youths full participation in adult social and political life by denying them the right to vote, possess alcohol, and enter into legal contracts until they reach 18 or 21 years of age. From a developmental perspective, this type of inconsistency highlights the problems associated with using age as a criterion for determining when one becomes an adult. Moreover, it is important to consider psychological and moral development issues because doing so makes evident the variation that exists among youths who become the clients of both the juvenile and adult justice process and because knowledge of child development is needed in order to develop more effective responses to juvenile offenders.

■ The Juvenile Justice System

In referring to juvenile justice practices, it is common to use the phrase "the juvenile justice system"; however, the extent to which juvenile justice constitutes a system is a matter of debate. Although state juvenile codes usually specify two main purposes of the juvenile court—to protect the best interests of children and to protect the community—daily juvenile justice operations typically result in considerable variability in practices as well as conflict between juvenile justice agencies and practitioners. In this section, we examine factors responsible for variability in juvenile justice practice and for conflict between agencies and practitioners.

MYTH VS. REALITY — Goals of the System

Myth—The juvenile justice system comprises police agencies, courts, and correctional institutions, which all work together to achieve the same common goals.
Reality—Although the various components of the juvenile justice process (i.e., police agencies, courts, and correctional institutions) and those who work within these different components are charged with protecting the public and serving the best interests of children, they often work at cross-purposes. Importantly, those who work in different areas of juvenile justice, such as police officers, district attorneys, defense attorneys, judges, and probation officers, have different roles and responsibilities. Moreover, those who work in juvenile justice perceive their roles differently and are influenced by a variety of factors that result in a considerable amount of conflict over how particular cases or types of cases should be handled.

One factor that leads to variability in juvenile justice practices is variability in the law. Each state has its own juvenile laws, and they vary regarding the age ranges that fall under the jurisdiction of the juvenile court. Consequently, a youth who is 16 in South Carolina is considered a juvenile, whereas 16-year-olds in North Carolina fall under the jurisdiction of adult courts. Moreover, the expressed purposes of juvenile codes also vary from state to state. Some juvenile codes emphasize prevention and treatment (e.g., those in Idaho, New Hampshire, Pennsylvania, and Tennessee), others emphasize punishment (e.g., those in Arkansas, Iowa, Oregon, and Rhode Island), and still others emphasize all three objectives (e.g., those in Alabama, Colorado, Delaware, and Washington).[18] Differences in the law exist within states as well, because states contain smaller jurisdictional units, like cities, townships, and counties, that can develop local ordinances prohibiting certain juvenile behaviors. For example, the City of Charlotte, North Carolina, has a curfew ordinance that prohibits youths under the age of 16 from being out between 11:00 p.m. and 6:00 a.m. unless they are supervised by an adult. However, this ordinance does not affect youths under 16 who live outside the city limits.

Also influencing variability in juvenile justice practices are differences in the social, political, and economic environments found in different communities. The level of juvenile crime, the range of economic opportunities for youths, the quality of the public schools, the existence of activist community groups that demand certain responses to juvenile crime, and a host of other social, political, and economic factors can influence juvenile justice practices. As a result, the response to a youth who violates the law (e.g., by shoplifting or failing to obey his or her parents) in one county may be quite different than the response to those who commit the same offense in an adjacent county in the same state.

Another factor that leads to variability in juvenile justice practice is the **discretion** that juvenile justice decision makers (police officers, district attorneys, probation officers, etc.) have when determining how to respond to youths who violate the law. Discretion is the ability to make judgments on one's own authority. However, it is important to recognize that the use of discretion itself is influenced by a number of political, social, economic, legal, and other factors. For example, juvenile justice decision makers work within a legal context that places statutory limits on their decision-making ability. Moreover, their decisions may be constrained by monetary resources and by their awareness of what the community will tolerate or support. In addition, juvenile justice decision makers have different **juvenile justice ideologies**, which is to say that they have different ideas about what causes juvenile crime and what should be done about it.

Remember also that, since their inception, juvenile courts, like other courts, have taken an individualized approach to dealing with offenders. This individualized approach has been reflected in efforts by juvenile justice decision makers to consider the characteristics of each offender (e.g., his or her age, maturity, mental capacity, and home environment) and the circumstances surrounding the offense (e.g., whether there was provocation) in determining the appropriate response. Finally, different juvenile justice practitioners have different roles and responsibilities in the juvenile justice process. These different roles and responsibilities inevitably lead to some degree of conflict between agencies and individuals as they strive to carry out what they feel their position requires of them. Because local decision makers within juvenile justice have discretion; because they have different views about what constitutes an appropriate response to various types of delinquent behavior; because they have different roles and responsibilities; and because the social, economic, political, and legal context of juvenile justice practice places limits on juvenile justice decision making, the fact that youths are often dealt with differently, even within the same juvenile justice agency, should not be surprising. Nor should it be surprising that there can be considerable conflict between the agencies and individuals responsible for responding to delinquent youths. In some instances, of course, juvenile justice practice involves relatively coordinated responses to juvenile offenders and relatively low levels of conflict. What makes the idea of a juvenile justice "system" open to question are the many instances where there are low levels of coordination and high levels of conflict between juvenile justice agencies and/or practitioners.

discretion The authority to exercise judgment.

juvenile justice ideology A set of ideas about what causes juvenile crime and what should be done about it.

■ Public Perceptions of Delinquency and the Politics of Juvenile Justice

The preceding sections have addressed some important issues related to basic concepts employed in juvenile justice. In this section, we turn our attention to understanding public perceptions of delinquency. These perceptions are important because they color much of what we think about the so-called delinquency problem and how we should respond to it.

Interestingly, public concern over delinquency is not new. Indeed, each generation of Americans seems to believe that the country is experiencing a juvenile crime wave.[19] Concern about youth crime and waywardness in Eastern cities led reformers in the early 1800s to develop the first specialized juvenile institutions. In the late 1800s, other reformers, called the "child savers," had similar concerns and developed the first juvenile courts. During the 1950s, many members of the public were alarmed at what they perceived to be the negative influences of youth culture on adolescents. Movies about young rebels, played so well by actors such as James Dean and Marlon Brando, caused many Americans to question the influence of the media on middle-class youths, who, it was felt, were being seduced by lower-class values that spawned delinquency. As a result, various efforts were undertaken to censor movies, comic books, and other media believed to spread the wrong values.[20]

Although public concern about delinquency is hardly new, each wave of concern produces its own set of solutions to the delinquency problem. Changes in the juvenile justice response to offenders have not always coincided with actual increases in the level of delinquency nor are they always in line with our knowledge of effective responses to juvenile offenders. Since the late 1970s, concerns about serious, violent, and chronic juvenile offenders, sometimes referred to as juvenile "superpredators," have led to a variety of "get tough" efforts, including legislation in a number of states intended to ensure that serious juvenile offenders receive maximum terms of confinement, often by making it easier to process juveniles in adult courts.[21] Unfortunately, these policies have been driven more by the politics of juvenile justice than by our knowledge of juvenile crime or juvenile offenders. As we demonstrate in Chapter 2 and Chapter 14, various data sources on delinquency provide no evidence of a juvenile crime wave nor do they indicate that there is a growing population of violent juvenile offenders. This highlights three important facts about juvenile justice. First, how we respond to juvenile offenders is, to a large extent, a reflection of our perception of juvenile crime. Two, changes in our *perception* of the nature and extent of juvenile offending, irrespective of the actual level of juvenile crime, can produce changes in juvenile justice policies. Three, responses to juvenile crime are the result of a political process in which particular sets of ideas about what constitutes a reasonable response to the delinquency problem win out over others. Unfortunately, this process is often driven by public perceptions based on sensational and distorted media accounts of juvenile crime that influence public beliefs about the level of juvenile crime, the etiology of delinquency, the characteristics of juvenile offenders, and the most appropriate response to the juvenile crime problem. One result is that existing policies frequently are at odds with our current knowledge about delinquency and about the best way to deal with juvenile offenders. In fact, many of these policies are harmful to youths, their families, and the communities in which we live because they are ineffective and, in some cases, counterproductive. Such policies also have a negative impact on those who work in juvenile justice because they limit the ability of these individuals to facilitate successful client outcomes.

Public Opinion

Although increasingly more punitive responses to juvenile offenders have been developed around the country,[22] there is considerable evidence that the public still favors the traditional rehabilitative focus of the juvenile court when it comes to treating juvenile offenders.[23] Thus, efforts to treat youths more punitively may not necessarily reflect public sentiment.

As noted earlier, public perception that a serious juvenile crime problem existed during the last two decades resulted in a variety of policies intended to get tough with juvenile offenders, despite a lack of evidence that the juvenile crime problem was actually getting worse. The cumulative effect of these policies was to increase reliance on incarceration as a response to delinquency, a policy that continues today in many jurisdictions, even though there is no sound evidence that incarceration has any significant effect on levels of violent juvenile crime. Nevertheless, the "get tough" approach continues to have considerable appeal because it fits well with many people's perceptions of what should be done. An important question to ask, however, is the following: What is the most appropriate response to juvenile crime? The information provided throughout this text is intended to assist you in formulating your own answer to this question (as well as answers to many other questions you will have).

■ Chapter Summary

This chapter was designed to introduce some of the basic concepts necessary for developing a clear understanding of the juvenile justice process. It began by defining two important concepts, delinquency and juvenile, and exploring a number of shortcomings of these definitions. Clearly, there are a number of ways of thinking about what constitutes delinquency and who is to be considered a delinquent (and consequently subject to the jurisdiction of juvenile justice agencies). The chapter also pointed out the broad range of behaviors that are treated as juvenile offenses and the tremendous variability in the psychological, biological, and social characteristics of the youths who engage in illegal behavior.

In addition, the chapter suggested that talk about the juvenile justice system may be best viewed as rhetorical. It examined a variety of factors that produce variability and conflict in juvenile justice practice and that make systemic responses to juvenile offenders difficult. This is not to deny that some commonalities in juvenile justice operations across the United States exist. Nevertheless, it is important to recognize that there is often considerable conflict between the agencies and individuals responsible for responding to juvenile offenders—conflict that impedes systemic responses to juvenile crime.

Finally, the chapter discussed the important role of public perceptions and politics in the development of juvenile justice policy. As noted, public opinion about juvenile justice is not always based on accurate knowledge of the juvenile justice process or juvenile offenders. Nevertheless, public perceptions and beliefs about what cause delinquency and the best way to respond to it play critical roles in the politics of juvenile justice. Unfortunately, the politics of policy development do not always produce responses to delinquency that help youths or their families, nor do they always lead to safer communities and high success rates among juvenile justice practitioners.

KEY CONCEPTS

certification A legal process whereby a juvenile is remanded to adult court for trial; see **waiver**.

CHINS Children in need of supervision; see **PINS**.

delinquency Those behaviors that fall within the jurisdiction of the juvenile court and result in processing by official juvenile justice agents.

discretion The authority to exercise judgment.

doli incapax Not capable of criminal intention or malice; unable to formulate criminal intent (*mens rea*); not able, due to lack of maturity, to know right from wrong; not criminally responsible.

jurisdiction A geographical area subject to a particular law or court.

juvenile A youth who falls within an age range specified by state law. The age range varies from state to state.

juvenile justice ideology A set of ideas about what causes juvenile crime and what should be done about it.

labeling Defining and responding to a youth as a delinquent. Labeling can lead to stigmatization, lost opportunities, and additional delinquent behavior.

PINS Persons in need of supervision; children who engage in status offenses.

remand A legal process whereby a juvenile is sent to adult court for trial; see **waiver**.

status offenses Acts that are considered illegal when committed by a child but not when committed by an adult (e.g., running away, school truancy, and failure to obey parents' directions).

transfer A legal process whereby a juvenile is sent to adult court for trial; see **waiver**.

waiver A legal process that enables a juvenile to be tried as an adult.

REVIEW QUESTIONS

1. What two broad categories of behaviors fall within legal definitions of delinquency?

_____

_____

_____

_____

2. What are the shortcomings of legal definitions of delinquency?

_____

_____

_____

_____

3. How common is delinquent behavior in the United States?

_____

_____

_____

_____

4. How might labeling a youth a delinquent affect his or her future behavior?

_____

_____

_____

_____

5. Are all juvenile offenders alike? What evidence can you cite to support your view?

_____

_____

_____

_____

6. What is the definition of a juvenile from a legal perspective?

_____

_____

_____

_____

7. What are the problems associated with defining a juvenile using an age criterion?

_____

_____

_____

_____

8. What is meant by the term _waiver_ (or _transfer_, _remand_, and _certification_)?

_____

_____

_____

_____

9. Is there a juvenile justice "system"? Provide support for your view.

_____

_____

_____

_____

10. According to state statutes, what are the two primary functions of the juvenile courts?

_____

_____

_____

_____

11. What factors lead to variability and conflict in juvenile justice practices?

_____

_____

_____

_____

12. How do public perceptions of the delinquency problem affect responses to juvenile crime?

_____

_____

_____

_____

13. What role does politics play in juvenile justice?

_____

_____

_____

_____

ADDITIONAL READINGS

Bernard, T. J. 1992. *The Cycle of Juvenile Justice*. New York: Oxford University Press.

Gilbert, J. 1986. *A Cycle of Outrage: America's Response to the Juvenile Delinquent in the 1950s*. New York: Oxford University Press.

Schwartz, I. M. 1989. *(In)Justice for Juveniles: Rethinking the Best Interests of the Child*. Lexington, MA: Lexington Books.

NOTES

1. A. M. Platt, *The Child Savers: The Invention of Delinquency* (Chicago: University of Chicago Press, 1977).

2. R. Cloward and L. Ohlin, *Delinquency and Opportunity: A Theory of Delinquent Gangs* (New York: The Free Press, 1960).

3. W. J. Chambliss, "The Saints and the Roughnecks," *Society* 11, no. 1 (1973): 341–355.

4. A. L. Pastore and K. Maguire, eds. *Sourcebook of Criminal Justice Statistics*. Retrieved October 17, 2001, from http://www.albany.edu/sourcebook/1995/pdf/t368.

5. H. Garfinkel, "Conditions of Successful Degradation Ceremonies," *American Journal of Sociology* 61 (1956): 420-424.

6. C. Frazier, *Theoretical Approaches to Deviance: An Evaluation* (Columbus, OH: Merrill, 1976).

7. E. Lemert, *Social Pathology: A Systematic Approach to the Theory of Sociopathic Behavior* (New York: McGraw-Hill, 1951).

8. Frazier, *Theoretical Approaches to Deviance*.

9. R. D. Schwartz and J. H. Skolnick, "Two Studies of Legal Stigma," in *The Other Side: Perspectives on Deviance*, ed. H. S. Becker (New York: The Free Press, 1964); W. Buikhuisen and P. H. Dijksterhuis, "Delinquency and Stigmatization," *British Journal of Criminology* 11 (1971): 186.

10. R. S. Cavan and T. N. Ferdinand, *Juvenile Delinquency*, 3rd ed. (Philadelphia: Lippincott, 1975).

11. M. W. Zawitz, ed., *Report to the Nation on Crime and Justice*, 2nd ed. (Washington, DC: U.S. Department of Justice, 1988).

12. See L. Kohlberg, *Child Psychology and Childhood Education: A Cognitive-Developmental View* (New York: Longman, 1987) for an example of the developmental perspective.

13. R. E. Mayer, *Educational Psychology: A Cognitive Approach* (Boston: Little, Brown, 1987).
14. Kohlberg, *Child Psychology and Childhood Education*.
15. C. Bartol, *Criminal Behavior: A Psychosocial Approach* (Englewood Cliffs, NJ: Prentice-Hall, 1991).
16. D. A. Andrews and J. Bonta, *The Psychology of Criminal Conduct* (Cincinnati: Anderson, 1994).
17. M. Hoffman, "Sex Differences in Empathy and Related Behaviors," *Psychological Bulletin* 84 (1977): 712–722; M. Morash, "An Explanation of Juvenile Delinquency: The Integration of Moral-Reasoning Theory and Sociological Knowledge," in *Personality, Theory, Moral Development, and Criminal Behavior*, eds. W. S. Laufer and J. M. Day (Lexington, MA: Lexington Books, 1983).
18. H. N. Snyder and M. Sickmund, *Juvenile Offenders and Victims: A National Report* (Washington, DC: Office of Juvenile Justice and Delinquency Prevention, 1995).
19. T. J. Bernard, *The Cycle of Juvenile Justice* (New York: Oxford University Press, 1992).
20. J. Gilbert, *A Cycle of Outrage: America's Reaction to the Juvenile Delinquent in the 1950s* (New York: Oxford University Press, 1986).
21. B. Krisberg et al., "The Watershed of Juvenile Justice Reform," *Crime and Delinquency* 32 (1986): 5–38.
22. P. Torbet and L. Szymanski, "State Juvenile Responses to Violent Juvenile Crime: 1996-97 Update," *Juvenile Justice Bulletin* (Washington, DC: Office of Juvenile Justice and Delinquency Prevention, 1998).
23. M. M. Moon et al., "Is Child Saving Dead? Public Support for Juvenile Rehabilitation," *Crime and Delinquency* 46 (2000): 38–60.

Measuring the Extent of Juvenile Delinquency

- Key Concepts
- Review Questions
- Additional Readings
- Notes

■ Introduction

Chapter 1 examined a variety of conceptual issues important for understanding the operation of juvenile justice in the United States. In addition, it indicated that public perceptions of juvenile crime and politics influence juvenile justice operations. However, as noted in Chapter 1, the perceptions of juvenile crime and juvenile justice held by the public and policy makers are not always based on knowledge about the actual levels of delinquency or the actual operation of juvenile justice agencies. To close the gap between perception and reality, it is important to consider several questions: How extensive is juvenile delinquency in the United States? To what degree, if at all, has the level of delinquency changed over time? How extensive is juvenile justice processing of offenders, and to what degree has such processing changed over time? Unfortunately, there are no simple answers to these. Moreover, the answers arrived at will depend on which data sources and measures are used to investigate the questions. To assist readers in developing an understanding of the extent of delinquency and the response to delinquency in the United States, this chapter explores various data sources and measures of delinquency. It also examines trends in juvenile offending and how the processing of juvenile offenders has changed over time. The data sources and measures used to understand delinquency are derived from official and other sources of data. The chapter concludes with a consideration of the different questions addressed by different data sources and measures and the suitability of various data sources for answering different questions about delinquency in the United States.

■ Official Data Sources

official data Information on juvenile delinquency collected by formal juvenile justice agencies, such as police agencies, juvenile courts, and juvenile detention and correctional facilities.

There are two types of data on juvenile delinquency: data derived from official sources and data derived from other sources. These data are collected by different agencies and individuals using different methodologies, and they often present different pictures of the extent of delinquency as well as youths who engage in delinquent behavior. **Official data** are collected by juvenile justice agencies, such as law enforcement agencies, juvenile courts, and various correctional programs that handle juveniles. These agencies collect a wealth of information on those with whom they have contact. This information helps policy makers and practitioners understand the number and characteristics of youths processed by juvenile justice agencies and the offenses they commit.

The Uniform Crime Report

The most well-publicized source of official data on delinquency is the FBI's <u>Uniform Crime Report (UCR)</u>. The UCR separates crimes into two broad categories: Part I (or Crime Index) offenses and Part II crimes. Part I crimes are further separated into Crime Index violent offenses (murder and nonnegligent manslaughter, forcible rape, robbery, and aggravated assault) and Crime Index property offenses (burglary, larceny-theft, motor vehicle theft, and arson). (Note: In this chapter, all mention of violent offenses should be taken to refer to violent offenses as defined by the UCR Crime Index, and likewise for all mention of property offenses.) Part II crimes consist of all offenses (except traffic violations) that are not Index offenses. Also, the UCR contains three basic types of data: (1) the number (frequency) and kinds of offenses that are known to the police; (2) the frequency of arrests for different kinds of offenses; and (3) data on the characteristics of offenders, such as their gender, race, age, and whether they live in large or small metropolitan areas or more rural areas.[1] Because the UCR contains information on arrests by age group, it is an important source of information on juvenile crime and the responses of law enforcement agencies to the illegal behaviors of youths.

Uniform Crime Report (UCR) The most comprehensive compilation of known crimes and arrests. The report is published each year by the Federal Bureau of Investigation.

The UCR Focuses on the Incidence of Arrests

Note that arrest data contained in the UCR refers to the incidence of arrests (the frequency of arrests) rather than the prevalence of arrests (the number of individual persons who are arrested). For instance, an individual may be arrested several or more times in one year. Each of these arrests may be counted in the UCR.

According to UCR data, juvenile crime represents a significant problem in the United States. In 2002, there were over 1.6 million arrests of persons under 18 years of age. However, as can be seen in **Table 2-1**, the great majority of those arrests (85%) were for nonviolent crimes. Indeed, over half of all arrests (64%) of persons under 18 were for nonassaultive Part II offenses. Approximately 12% of all arrests of persons under 18 years of age were for <u>status offenses</u>, such as running away from home, curfew violations, and loitering.[2]

status offense An act that is considered a crime or legal violation when committed by a juvenile but not by an adult (e.g., running away from home, incorrigible behavior, failure to attend school, and failure to obey school rules).

Table 2-1 Arrests of Persons Under 18 Years of Age by Crime Type, 2002

Crime Type N		Percentage of Total
Index offenses	415,607	25.6
Index violent offenses	66,508	4.1
Murder and nonnegligent manslaughter	973	0.1
Forcible rape	3,361	0.2
Robbery	17,893	1.1
Aggravated assault	44,281	2.7
Index property offenses	349,099	21.5
Burglary	61,843	3.8
Larceny-theft	248,861	15.3
Motor vehicle theft	32,544	2.1
Arson	5,851	0.4
Part II offenses	1,208,585	74.4
Other assaults (including offenses against family and children)	175,568	10.8
Forgery and counterfeiting, fraud, embezzlement	11,091	0.7
Buying, receiving, possessing stolen property	18,819	1.2
Vandalism	75,955	4.7
Carrying, possessing weapons	25,288	1.6
Prostitution and commercialized vice	1,095	0.1
Sex offenses (except forcible rape and prostitution)	13,877	0.9
Drug abuse violations	133,754	8.2
Gambling	1,114	0.1
Driving under the influence	15,214	0.9
Liquor law violations	106,014	6.5
Drunkenness, disorderly conduct	152,577	9.4
Curfew and loitering	103,155	6.3
Runaways	90,349	5.6
Other	284,715	17.5
Total Arrests	**1,624,192**	**100.0**

Note: Percentages may not total 100 due to rounding.

Source: Reprinted from FBI (2003), Crime in the United States 2002, *Uniform Crime Reports,* p. 250, Table 41, United States Government Printing Office, Washington, DC.

Although most arrests of persons under 18 were for nonviolent crimes, offenses against people still account for a substantial portion of all youth arrests. For example, in 2002, persons under the age of 18 accounted for 17% of all arrests (both Part I and Part II arrests). Furthermore, persons under 18 years of age accounted for 26% of all Crime Index arrests, 15% of all arrests for Index violent offenses, and 30% of all arrests for Index property offenses.[3] These percentages take on added significance when one considers that persons between the ages of 10 and 17 (inclusive) account for only 15% of the total U.S. population between the ages of 10 and 89.[4] This means that youths are disproportionately represented in arrest data for Index property offenses. They are not, however, disproportionately represented in arrest data for Index violent offenses.

MYTH VS. REALITY — Violent Juvenile Crime

Myth—Juveniles are over-represented in arrests for violent offenses.
Reality—Although people between 10 and 17 years of age accounted for a disproportionate share of arrests for Index violent offenses during some years in the past, this has not been the case in recent years.

Juvenile Arrest Trends

Although there were noticeable peaks in juvenile arrests during the late 70s and early 80s and then again in the late 80s and early 90s, the number of arrests of youths under 18 years of age has been fairly stable over the past 25 years. In fact, the number of persons under 18 years of age who were arrested during each of the first 3 years of this century was lower than the number arrested during all but 2 years between 1980 and 2000 (see **Figure 2-1(a)**). An examination of arrests of persons under 18 years of age for Index violent and property offenses presents a similar picture.[5]

Of particular concern to many policy makers, juvenile justice practitioners, researchers, and members of the public in recent years has been juveniles' involvement in violent crime. Indeed, the number of arrests of persons under 18 years of age for Index violent offenses increased 48% from 1989 to 1994. However, the number of arrests of persons under 18 for violent offenses began to decline in 1995, and this general decline was still evident in 2002.[6] Thus, arrest data provide no support for the belief that juvenile crime has steadily increased or become more serious over time.

(a)

(b)

Figure 2.1 Total arrest rates per 100,000 youth 10–17 years of age, 1975–1996. Source: Data from Bureau of the Census, *Current Population Estimates and Projections,* Series P-25, No. 917, 1982, U.S. Government Printing Office; Bureau of the Census, *Current Population Reports: Population Projections for the United States by Age, Sex, Race, and Hispanic Origin, 1992–2050,* P25-1092, 1992, U.S. Government Printing Office; Federal Bureau of Investigation, *Uniform Crime Reports,* 1975–1996.

Population Effects on Juvenile Arrest Trends

The number of arrests of persons under 18 is an indicator of the total number of juvenile crimes committed. The number of juvenile crimes committed, however, is partially determined by the size of the juvenile population. In other words, as the size of the juvenile population (i.e., potential juvenile offenders) increases, so does the absolute level of juvenile crime (assuming the rate of juvenile crime remains constant). In fact, it has been estimated that changes in the age structure of the population may account for as much as 40% of the changes in the arrest rate.[7] Although the actual strength of the relationship between levels of juvenile crime and the size of the juvenile population is a matter of debate, population size does have some effect on levels of delinquency. Consequently, another important indicator of juvenile crime is the juvenile **arrest rate**. An examination of the juvenile arrest rate is important because it takes into consideration the size of the juvenile population.

arrest rate The number of arrests adjusted for the size of the population; often reported as arrests per 100,000 youths 10–17 years of age in the population.

FYI **Juvenile Arrest Rate**

The juvenile arrest rate per 100,000 juveniles is calculated as follows:

$$\frac{\text{Number of juvenile arrests}}{\text{Total juvenile population}} \times 100{,}000 = \text{juvenile arrest rate per 100,000 juveniles}$$

Because the juvenile court jurisdiction age range varies from state to state, juvenile arrest rates are typically calculated by using the number of arrests of persons between the ages of 10 and 17 taken from the UCR (youth under 10 are rarely arrested relative to older youth) and dividing that number by population figures taken from census data.

Juvenile Arrest Rate Trends

An examination of juvenile arrest rate trends since 1980 indicates that juvenile crime has remained rather stable during this period. Indeed, a downward trend in the juvenile arrest rate was evident by the late 1990s (see **Figure 2-1(b)**).[8] A similar trend was also apparent when Index arrest trends were examined. As can be seen in **Figure 2-2**, arrest rates for Index offenses were declining by the latter half of the 1990s. However, when the Index property offense arrest rate and the violent offense arrest rate are examined separately, a somewhat different picture emerges. As is true of the overall juvenile arrest rate, the property offense arrest rate for juveniles remained quite stable between 1980 and the mid-1990s, at which time it began to decline as the figure shows. The violent offense arrest rate for juveniles, however, while remaining remarkably stable during most of the 1980s, increased by over 60% between 1988 and 1994.[9] Beginning in 1995, however, there was a significant decline in juvenile Index violent crime arrest rates, as seen in the figure.

Figure 2.2 Arrest rates per 100,000 youths 10–17 years of age (Crime Index property offenses), 1976–1996. Source: Data from Bureau of the Census, *Current Population Estimates and Projections,* Series P-25, No. 917, 1982, U.S. Government Printing Office; Bureau of the Census, *Current Population Reports: Population Projections for the United States by Age, Sex, Race, and Hispanic Origin, 1992–2050,* P25-1092, 1992, U.S. Government Printing Office; Federal Bureau of Investigation, *Uniform Crime Reports,* 1975–1996.

Juvenile Crime Trends

An examination of both arrest trends and arrest rate trends in Figures 2-1(a) and 2-1(b) reveals almost identical patterns and indicates that, overall, juvenile crime has not become significantly more problematic over time.

The Race and Gender of Arrested Juveniles

In addition to information about numbers of arrests, the UCR also provides information on the racial and gender characteristics of persons who are arrested. According to 2002 UCR data, whites accounted for approximately 72% of the arrests of persons under 18 years of age. Whites also represented approximately 70% of those under 18 who were arrested for Index property offenses and approximately 55% of those arrested for Index violent crimes. In contrast, African American youths accounted for 26% of all arrests, about 27% of Index property arrests, and almost 43% of Index violent arrests. Other racial groups accounted for the remainder of juvenile arrests.[10] However, although white youths accounted for the majority of juvenile arrests, African American youths were arrested in disproportionate numbers, because they account for only about 16% of the population between 10 and 17 years of age.[11]

The UCR data further indicate that juvenile lawbreaking is predominately the domain of males. For example, 2002 UCR data indicate that females accounted for approximately 29% of all arrests of persons under 18 years of age. Females accounted for about 18% of persons under 18 years of age arrested for Index violent offenses and approximately 32% of those under age 18 arrested for Index property crimes.[12]

Problems with Using Arrest Data as an Indicator of Juvenile Crime

Although police arrests of juveniles constitute an important indicator of the extent of juvenile crime, there are a number of problems with using official data as a measure of delinquency. First, many juvenile offenses do not come to the attention of the police. Because crimes known to the police are usually reported by citizens, the ability or willingness of citizens to make complaints influences the number of crimes included in the UCR. Moreover, a variety of factors affect the ability and willingness of citizens to report crimes to the police. For example, some offenses go undetected, which prevents them from being reported. Furthermore, the seriousness of the offense, the relationship between the offender and the victim, fear of possible reprisals by the offender, the belief that nothing will be done to the offender, embarrassment, and the involvement of the complainant in criminal behavior he or she wants to keep from the police are factors that can influence a citizen's willingness to report their victimization.[13] As a result, many offenses do not come to the attention of law enforcement agencies.

Hidden Juvenile Crime

The amount of juvenile crime that goes unreported—the **"dark figure" of juvenile crime**—is not known; however, it appears that well over half of all juvenile offenses are not reported to the authorities. For instance, findings from the National Youth Survey (NYS), a large-scale self-report study; the UCR; and other sources indicate that only about 24% of juveniles who commit offenses for which they could be arrested are actually taken into custody by the police.[14] Moreover, the probability that an individual will be arrested for even serious offenses is estimated to be only 2 in 100.[15]

dark figure of juvenile crime The amount of juvenile crime that goes unreported.

Second, a number of factors influence the accuracy and completeness of the data that are collected and published in the UCR. For example, police must make a decision as to whether an offense technically qualifies as an offense, and this decision can be influenced, for example, by political pressures to reduce or inflate crime rates, the quality of the interaction between the police and the offender (e.g., whether the offender is hostile toward the police), and the style of policing employed in a particular jurisdiction (e.g., the extent to which the police use a more legalistic and formal style of policing rather than a more preventive style, which encourages the informal processing of cases when possible).[16] These factors can lead to variability across jurisdictions regarding the extent to which illegal behaviors are actually recorded as crimes, thus making comparisons across jurisdictions suspect. In addition, factors such as political pressure and changes in the style of policing can result in changes in arrest data over time that do not accurately reflect real changes in the level of crime. For example, pressure exerted by a powerful interest group can lead to police crackdowns on particular types of behaviors (e.g., truancy, prostitution). This, in turn, will be reflected in an increase in the numbers of arrests for those behaviors even though their actual incidence did not change. Also, in the UCR data, not all offenses are counted when multiple offenses occur. Only the most serious offense is reported, except where arson is also involved (an act of arson is always counted even if it is not the most serious offense).[17]

Police Discretion

It is important to note that **police discretion** plays an important role in determining whether a particular act is recorded as a crime. Moreover, police decision making is influenced by a variety of factors (see Chapter 6).

police discretion The authority of police to make their own judgments about which crimes or delinquent acts are subject to investigation and which juveniles are subject to arrest.

Third, arrest data say as much about the actions of the police as about the extent of crime. Because police departments make decisions to focus their attention on certain types of criminal activity (i.e., the types of offenses contained in the UCR) and these types more often occur within particular communities (i.e., lower-class neighborhoods), certain actions (e.g., robbery, vandalism, larceny, curfew violations) are more likely to be responded to and the people in those communities (i.e., poor people, minorities) are more likely to be arrested and seen as criminals.

MYTH VS. REALITY — Problems with the UCR

Myth—The UCR contains a comprehensive listing of crimes that are committed in the United States.

Reality—Although the media, political leaders, and policy makers often use the UCR as their primary source of national, state, and local crime data, it suffers from a number of shortcomings as a source of data on juvenile crime. These include the following: (1) it omits a sizable amount of crime that is never reported to the police; (2) it contains only limited data on the characteristics of people arrested; and (3) a number of factors influence the accuracy of the data that is recorded.

Using Clearance Data to Understand Juvenile Crime

Although arrest data provide important insights on juvenile crime, they suffer from one important limitation—they *overestimate* juveniles' involvement in criminal activity.[18] This occurs because juvenile offending is more likely to involve other persons, often juveniles, than crimes committed by adults. As a result, juvenile offenses are more likely than adult offenses to involve multiple arrests, even though some of those arrested are not knowing or willing participants in criminal activity.

Another way to understand the extent to which juveniles are involved in crime, which provides a more accurate estimate of juvenile involvement in offending, is to examine clearance data. An offense is cleared when at least one person is arrested for the offense, charged with an offense, or turned over to the court for prosecution. In addition, "exceptional clearances" occur in situations where the offender dies, the victim refuses to cooperate in the prosecution of the case, or the offender is denied extradition and is prosecuted in another jurisdiction.[19] Of the Crime Index offenses cleared by law enforcement in 2002, 18.0% involved only juveniles. Furthermore, persons under 18 accounted for 11.9% of violent crime clearances and 20.3% of property crime clearances. These percentages are considerably lower than the arrest percentages presented earlier for Index violent and property offenses. However, it is likely that these data also overestimate juvenile involvement in crime because juveniles are more likely to be arrested for crimes like robbery than adults.[20]

Trends in the proportion of violent crimes attributed to juveniles present an overall picture that is similar to that portrayed by arrest data. Although the proportion of violent crimes attributed to persons under 18 grew from approximately 9% in the late 1980s to 14% in 1994, it had decreased to 12% in 2001 and 2002. Like arrest data, an examination of clearance data indicates that juvenile crime has remained quite stable over time and has shown some decline in recent years.

The Uses of Official Data

As noted, official data on juvenile delinquency, such as UCR arrest statistics, are used regularly by politicians, juvenile justice policy makers, and the media in their efforts to understand juvenile crime. Moreover, because much of what the public knows about juvenile crime comes from the media, politicians, and policy makers, these data play a critical role in educating the public about juvenile crime. For example, newspapers and television news departments regularly analyze and report UCR data on juvenile arrests. Of course, the media also rely on the findings of researchers who study juvenile crime, but these researchers often depend on official data as well. Researchers investigating police responses to juvenile offending rely on official arrest data to determine how many youths are arrested by the police, the basic demographic characteristics of those youths, and the types of offenses for which youths are arrested. Similarly, researchers interested in the way youths are dealt with by juvenile courts and correctional institutions employ official court and corrections data. In addition, official arrest data has been used in countless studies that have explored the **etiology** of delinquency and the effectiveness of various types of juvenile justice programs.

etiology
The study of causation.

unofficial data
Information collected by researchers not connected with formal juvenile justice agencies (i.e., police agencies, juvenile courts, and correctional institutions).

self-report study A study in which subjects are asked to report their behavior (e.g., their involvement in illegal activities).

■ Unofficial Data Sources

Besides official data sources, there are several other important **unofficial sources of data** on juvenile crime, including studies performed by researchers not connected to formal juvenile justice agencies (i.e., police, juvenile courts, and juvenile detention and corrections facilities). Among the main types are self-report studies of delinquency, cohort studies, developmental studies, observational studies, and victimization studies. Each of these is discussed in the following sections.

Self-Report Studies

As the name implies, **self-report studies** ask people to report on their own behavior. A primary advantage of the self-report method is that it can elicit information on offenses not known to the police. Consequently, it allows researchers to better understand the dark figure of juvenile crime. Another advantage is that a variety of data important to understanding delinquency (e.g., income, education, quality of life, work, family life, and peer group affiliations) can be collected. Unlike the UCR and other official sources of data, self-report studies have been conducted by a variety of researchers unaffiliated with formal juvenile justice agencies. Moreover, some long-term self-report studies have provided important information on changes in juvenile delinquency over time. A good example of an ongoing self-report study is the Monitoring the Future study conducted by researchers at the University of Michigan's Institute of Social Research (ISR). This study, which began in 1975, collects a variety of data on the attitudes and behaviors of a national sample of high school seniors.

As one might expect, self-report studies indicate that youths engage in considerably more illegal behavior than is indicated by official data. Particularly when offenses such as school truancy, alcohol consumption, using a false ID, petty larceny, and vandalism are examined, delinquency appears to be normal rather than abnormal.[21] Self-report studies generally indicate that most youth do not engage in serious criminal activity, a finding that mirrors the picture of delinquency presented in the official data sources. Nevertheless, some self-report studies show that a sizable percentage of juveniles are involved in serious offenses. For example, according to data collected in 2000 as part of the Monitoring the Future study, almost 12% of the respondents indicated that during the preceding year they had hurt someone badly enough to require a doctor's attention; approximately 10% indicated that they had stolen something worth more than $50; and approximately 14% reported that during the year they had gotten into a serious fight at work or school.[22] In another ongoing large-scale study of high school students conducted by the Centers for Disease Control in 2001, 17% of respondents indicated that they had carried a weapon at least once in the 30 days prior to the survey, and 33% reported that they had engaged in a physical fight one or more times in the previous 12 months.[23]

Although self-report studies reveal that delinquency is a pervasive phenomenon in the United States, they do not indicate that juvenile crime is getting worse. Although there are short-term fluctuations in youths' self-reported delinquency, an examination of long-term trends in delinquent behavior indicates that it has remained quite stable over time.[24] For instance, an examination of data collected as part of the Monitoring the Future Study between 1990 and 2002 indicates that for most offenses no clearly identifiable trends were evident. To the extent that trends could be discerned in the data they tended to be toward less involvement in delinquency.

Another important self-report study is the National Youth Survey (NYS), an ongoing study that began in 1976 and is conducted by researchers at the University of Colorado's Behavioral Research Institute. The NYS is a **panel study**, which means that a sample of youths (now adults) are surveyed each year over a number of years. Panel studies are valuable because they allow researchers to examine changes in young people's behaviors and attitudes over time and into adulthood.

The NYS indicates that youths engage in a range of criminal activity, although most are involved in minor offenses. It also reveals that the frequency of offending and the kinds of offenses youths commit change over time. Youths' illegal behaviors tend to increase in severity as they reach their late teens and move into their early adult years, but after that period offending decreases. Younger juveniles tend to engage mostly in status offenses, whereas older youths engage in more property and personal offenses.[25]

Analysis of the NYS has also produced some important findings regarding the relationship among race, social class, and delinquency. One finding of considerable interest, reported by Delbert Elliott and Suzanne Ageton, is that the levels of offending for African American youths and white youths are similar. Although

panel study A study that involves the examination of the same select group over time.

the researchers found that African American youths reported slightly more involvement in serious crimes, the size of the difference was not as great as indicated by arrest data.[26] This suggests that the disparity in the arrest levels of white and African American youths is, in part, a reflection of the actions of police agencies.

The relationship between social class and delinquency found in the NYS is also of considerable interest. For example, Delbert Elliott and David Huizinga found that the prevalence of delinquency did not differ among social classes when all types of offenses were considered. In other words, the proportion of middle-class and lower-class youths who engaged in delinquency was similar. However, when the researchers examined different types of offenses, they discovered significant class differences. Middle-class youths had the highest rates of involvement in such offenses as stealing from their family, cheating on tests, cutting classes, disorderly conduct, lying about their age, and drunkenness. Lower-class youths had higher rates of involvement in more serious offenses, such as felony assault and robbery.[27]

Self-report studies also have examined the relationship between gender and delinquent behavior. Overall, these studies indicate that females engage in considerably more delinquency than is indicated by arrest data, which suggests a possible gender bias within juvenile justice. Nevertheless, females engage in less delinquency than males and tend to be involved in less serious types of delinquency, although differences in delinquent behavior between males and females are much smaller when minor offenses are examined.[28]

A very important point to note regarding recent self-report research is that it produces an overall picture of delinquency that is similar to that painted by official data. Although the extent of delinquency depicted in self-report research is considerably greater than is depicted in arrest data, the pattern of delinquency is similar. However, despite improvements in recent self-report studies, they are plagued by several problems. First, there is no standard reporting format used in self-report studies. Consequently, it is difficult to compare different studies. It is particularly difficult to compare older self-report studies with more recent studies because older studies focused exclusively on less serious types of delinquent behavior and employed less sophisticated methodologies, raising questions about the accuracy of the results obtained. Second, although newer self-report studies include more serious offenses, they still do not cover all types of serious delinquency. Thus, the full range of delinquent behaviors is not covered. Third, there is some doubt as to whether present studies have been able to capture a completely representative sample of youths to study. Self-report studies that rely on school populations miss students who are not in school or who have dropped out. Even studies that do not sample from school populations may fail to include some youths. For instance, research conducted by Stephen Cernkovich, Peggy Giordano, and Meredith Pugh indicated that institutionalized chronic offenders have not been adequately represented in self-report studies.[29] Regardless of these problems, however, self-report studies are a good way to (1) estimate the dark figure of juvenile delinquency, (2) determine the extent of delinquency using more representative samples, and (3) examine factors that are believed to be related to delinquent behavior.

MYTH VS. REALITY — Self-Report Studies Provide Valuable Data

Myth—Self-report studies are not reliable because youths are unwilling to be truthful about their behavior.
Reality—Efforts to examine the validity of the self-report studies indicate that the self-report method is a valid method of measuring delinquent behavior.[30]

Cohort Studies

cohort study A study in which some specific group (e.g., those born during a certain year in a particular geographic location) are studied over a period of time.

<u>Cohort studies</u> are designed to examine specific subpopulations over a period of time. Several important cohort studies have been published since the 1970s, and these provide an important picture of delinquency within the juvenile population. Several of these studies were conducted by Marvin Wolfgang and his associates at the University of Pennsylvania. One of the best known of these studies was published in 1972 by Wolfgang, Robert Figlio, and Thorsten Sellin under the title *Delinquency in a Birth Cohort*, and it has had a profound influence on many people's thinking about delinquency.[31]

Wolfgang, Figlio, and Sellin studied a cohort of 9,945 males who were born in Philadelphia in 1945 and resided there until 1963, when they became 18 years of age. For each youth the researchers collected information on formal police contacts from police records. In addition, they collected information from public, private, and parochial (Catholic) school records on academic achievement, IQ, types of schools attended, school and residential changes, and highest grade completed, and they determined each youth's socioeconomic status based on the youth's place of residence.

socioeconomic status (SES) A person's position in a stratified society based on criteria such as place of residence, family income, educational background, and employment history.

The results of this study were important not only because of the valuable insights provided, but also because of the policy responses the results suggested. Of the 9,945 cohort subjects, 35% had at least one formal contact with the police. Among white youths, 28.6% had some official police contact, while 50.2% of nonwhite cohort members had at least one official police contact. Among higher <u>socioeconomic status (SES)</u> youths, 26.5% had at least one official police contact, compared with 46% of the lower SES youths. Indeed, the results indicated that race and SES were the two most important variables related to official police contact, and that school-associated variables were related to delinquency. Consequently, the researchers noted that a number of variables related to delinquency seemed indicative of what they referred to as a "disadvantaged position." For example, they reported that the nonwhite youth who has official police contact is "likely to belong to the lower socioeconomic group, experience a greater number of school and residential moves (that is, be subject to the disrupting forces of intra-city mobility more than the nondelinquent) and have the lowest average grade completed, the lowest achievement level, and the lowest IQ score."[32] These findings were similar to those found in other studies of official delinquents.

The findings however, that received the most attention concerned what Wolfgang, Figlio, and Sellin termed the "chronic offender." The researchers divided the cohort into three groups: nonoffenders, one-time offenders, and **recidivists.** Of those youths who had an official police contact, 46% were one-time offenders and 54% were recidivists. Moreover, the recidivists were more likely to be in a "disadvantaged position" compared to other cohort members and to commit more serious offenses. The researchers categorized the recidivists into two additional groups: (1) those who had more than one police contact but less than five, whom they labeled "nonchronic recidivists," and (2) those who had five or more contacts, whom they labeled "chronic recidivists." Of the cohort, 1,235 youths (35.6% of all delinquents or 12.4% of the entire cohort) were nonchronic recidivists. Chronic recidivists numbered 627 youth (18% of all delinquents or 6% of the total cohort). Although the chronic offenders made up only 6% of the cohort, they accounted for a disproportionate share of the offenses attributed to the entire cohort, including serious offenses. For example, chronic offenders accounted for over half (51.9%) of all offenses attributed to the cohort. Even more striking was their involvement in serious criminality. This 6% of the cohort accounted for 71% of the homicides, 73% of the rapes, 82% of the robberies, and 69% of the aggravated assaults attributed to the cohort. Yet chronic offenders were more likely to commit more serious offenses, there was little indication that the seriousness of those offenses increased over time. Indeed, the analysis revealed that, in general, the offenses that youths commit do not necessarily become more serious over time. Nevertheless, the researchers found that in those instances when youth do repeat offenses, increases in severity are common.[33] They also found that youths who had a police contact at an early age often had lengthy delinquent careers.

Finally, Wolfgang and his associates found that nonwhite and low SES youths were more likely to be treated more harshly by juvenile justice agencies than white and higher SES youths, even when controlling for offense seriousness and prior record. Moreover, they discovered that juvenile justice responses to chronic offenders had little positive effect on their recidivism. Indeed, juvenile justice responses such as fines, probation, and incarceration did not reduce recidivism.

In a subsequent study published in 1987, Wolfgang, Figlio, and Terence Thornberry examined a 10% sample ($N = 975$) of the 1945 cohort from the time the cohort members were 18 years of age until they turned 30 years of age in order to examine the criminal involvement of members beyond their juvenile years.[34] The researchers supplemented Philadelphia police records with FBI data and conducted personal interviews with 58.2% of the follow-up sample in order to collect self-report and attitudinal data. The results of this study indicated that 47% of the sample had a recorded police contact by age 30. Also, as was found in the original study, being nonwhite and being poor were strongly related to having an arrest record. The researchers also classified the offenders into three groups: (1) juvenile offenders, (2) adult offenders, and (3) persistent offenders (those who were both juvenile and adult offenders). Their analyses indicated that persistent offenders were more likely to come from disadvantaged backgrounds and have extensive delinquency histories. Similar to their original findings, most offenses committed by adults were non-Index offenses; however, unlike the earlier find-

recidivist A person who continues to commit crimes despite efforts to rehabilitate or deter him or her.

ings, adult offenses tended to become more serious over time. Indeed, the most serious offenses were more likely to be committed by persistent offenders.

Of course, it is possible that the findings of the 1945 cohort study conducted by Wolfgang and his associates were not representative of other cohorts. Consequently, replications of this study are needed to determine the extent to which its findings might be generalizable to other populations. In 1977 such an effort began, and its culmination was the publication of *Delinquency in Two Birth Cohorts* in 1985 and *Delinquency Careers in Two Birth Cohorts* in 1990.[35] In these publications, Wolfgang, Figlio, and Paul Tracy examined a cohort of all youths (males and females) who were born in Philadelphia in 1958 and resided there at least from their 10th to their 18th birthdays (a total of 27,160 youths). Again, data on police contacts were gathered as well as information from public and private school records. Finally, cohort members were assigned a socioeconomic status position derived from an analysis of census data. Analyses of the 1958 cohort data revealed that race and socioeconomic status were again related to delinquency, although the relationships were not as strong as in the 1945 cohort. Again, being in a disadvantaged position—indicated by being nonwhite, possessing low SES, experiencing residential instability, doing poorly in school, and failing to graduate from high school—was found to be related to delinquency. Again, chronic offenders were found to account for a disproportionate share of criminal involvement, including involvement in serious offending. In the 1958 cohort, chronic offenders made up 7.5% of the entire cohort and 23% of the delinquents. However, they accounted for 61% of all offenses attributed to the cohort, including 61% of the homicides, 75% of the rapes, 65% of the aggravated assaults, and 66% of the offenses that resulted in injury. Moreover, a greater percentage of serious offenses were committed by the 1958 cohort.[36]

Also, because information was collected on females, data on female contacts with the police were available. These data indicated that males were far more likely than females to have a police contact by their 18th birthday. Of cohort males, 32.8% had such contacts, compared with only 14.1% of females. Moreover, male contacts were more likely to involve serious offenses. Indeed, the overall male to female offense ratio was 4:1, but it was 9:1 for Crime Index offenses and 14:1 for violent Crime Index offenses. Finally, an analysis of the disposition of offenders found that, although differences existed in the processing of nonwhites and whites, these differences were not as strongly related to race and socioeconomic status as had been the case for youths born in 1945. Moreover, court responses to the juvenile offenders appeared to be more effective in the 1958 cohort.[37]

The disproportionate involvement in criminality of a small population of chronic offenders has been noted by other researchers as well. For instance, Lyle Shannon examined three cohorts of youths born in Racine, Wisconsin, in 1942, 1949, and 1955 and in each case uncovered the existence of a chronic offender population.[38] According to Shannon, less than 25% of each cohort's male subjects had five or more nontraffic offenses. Nevertheless, this group accounted for 77 to 83% of the police contacts involving males. Moreover, from 8 to 14% of the subjects in each cohort accounted for all of the serious felony offenses. According to Shannon, approximately 5% of each cohort was responsible for the majority of all offenses committed by the cohort and was also responsible for about three-quarters of all felonies committed by cohort members.

A cohort study conducted at Ohio State University called the Dangerous Offender Project focused on the violent juvenile offender. In this study, Donna Hamparian and her associates examined a cohort of 1,138 youths who were born in Franklin County (Columbus), Ohio, between 1956 and 1960 and who had been arrested for at least one violent offense before turning 18 years of age.[39] Rather than discovering a large number of violence-prone youth, the researchers found that only about 2% of the juveniles in Columbus had been arrested for a violent offense. Moreover, although youths in the cohort averaged about four arrests during their juvenile years, few youths were involved in repetitive acts of violence.

Two federally funded studies in the 1980s produced findings similar to those reported by Hamparian. In one of these studies, over 340,000 case records submitted to the National Juvenile Court Archive by courts in 12 states revealed that 6% of the juveniles referred to juvenile courts in those states for a law violation in 1984 were referred for a violent offense. Moreover, the researchers found that nonwhites and males were more likely to be referred to courts for violent offenses.[40] In the second study, staff at the National Center for Juvenile Justice examined the court careers of over 69,000 juveniles dealt with by the Maricopa County (Phoenix) juvenile court. Like other studies, the researchers found that only 5% of the youths referred to the juvenile court were referred for a violent offense and less than 1% had more than one violent court referral in their career. However, although violent offenders were the least common type of juvenile offender, these juveniles were the most likely to return to court charged with a violent offense, and this finding held for both males and females.[41]

Cohort studies have produced valuable information about youth crime and delinquent offenders. In fact, the attention they have focused on chronic and violent offenders has led to a number of policy proposals intended to better control this population, often through incarceration or transfer to adult courts for trial. Although important, such studies suffer from some weaknesses, notably their reliance on official data to measure delinquency. Indeed, research conducted by Dunford Franklyn and Delbert Elliott as part of the National Youth Survey found that a substantial amount of serious delinquency goes undetected by the police. In their study, only about 24% of serious career offenders reported that they had ever been arrested, which suggests that reliance on official data leads to an underestimation of the size of the chronic offender population.[42] Nevertheless, trends in self-report data provide no support for the contention that serious or violent juvenile crime has substantially increased over time.[43]

Developmental Studies

In recent years, other researchers have built upon and added to our understanding of the chronic and violent offender as well as our understanding of youths' patterns of offending over time. One such approach that aids our understanding of these phenomena is found in what is called <u>developmental criminology</u>. Essentially, developmental criminology is concerned with changes in individual and group patterns in offending over time and understanding factors that influence these patterns. One developmental approach to understanding juvenile crime is being developed by Terrie Moffitt. According to Moffitt, a small group of individuals engage in various types of antisocial behavior (e.g., biting and hitting

developmental criminology An area of criminology concerned with changes in individual and group patterns in offending over time and understanding factors that influence these patterns.

when they are 4 years of age, shoplifting at age 10, selling drugs at 16 years of age, robbery at 22, and child abuse at age 30) throughout their lives. She refers to these individuals as *life-course-persistent* offenders. Because these antisocial behaviors begin early in the life course, Moffit argues that they are influenced by various neuropsychological factors that affect children's behavioral development, temperament, cognitive abilities, or all three. In addition, she maintains that these youths are also at greater risk of being exposed to environments that increase the likelihood of antisocial behaviors.[44] Thus, from Moffitt's point of view, life-course-persistent offending is influenced by a variety of neuropsychological and environmental factors that lead to the onset of antisocial behavior and help maintain it over the life course.

According to Moffitt, the great majority of youths who engage in delinquent behavior are *adolescence-limited* offenders. These individuals' antisocial behavior is of short duration and takes place during the period of adolescence. Among this population, delinquent behavior can begin and end abruptly, the types and seriousness of offenses can change over time, and youths' behavior can vary in different situations. For example, these youths may engage in sporadic delinquent acts with friends but conform to rules at home and school. According to Moffitt, the delinquency exhibited by adolescence-limited offenders is a product of social mimicry in which adolescence-limited youths model the behaviors of their life-course-persistent peers. For adolescence-limited youths, delinquent behavior can be seen as a way of exerting autonomy, taking risks, and demonstrating their maturity. As these youths mature and they take on jobs and other social responsibilities, however, their involvement in delinquency declines because antisocial behavior becomes less rewarding.[45]

Developmental studies such as those developed by Moffitt and others, like other studies that are intended to understand delinquency, suggest that certain types of interventions for juvenile offenders are necessary. From the developmental perspective, interventions with some youths should begin early, and they should be developmentally appropriate. We will return to these two issues later as we examine correctional interventions in Chapter 11, Chapter 12, and also in Chapters 15 and 16 at the conclusion of the book.

Observational Studies

observational study A study in which the researcher observes and collects data on subjects in a field setting.

Chapter 1 discussed an important observational study done by William Chambliss that examined two groups of boys, the Saints, a group of middle-class males, and the Roughnecks, a group of lower-class males.[46] The study showed that the "outsider" status of the Roughnecks operated to reinforce a conception of them as boys headed for trouble, whereas the middle-class background of the Saints led authorities to see them as bright youths "sowing their wild oats." As a result, the Roughnecks were more likely than the Saints to be formally processed by the authorities. Such studies are important because they remind us that societal perceptions play a critical role in determining which youths become clients of the juvenile justice process.

Another important observational study was done by Herman and Julia Schwendinger and published in their highly regarded book *Adolescent Subcultures and Delinquency*.[47] The Schwendingers' work is based on over 4 years of participant observation of peer groups in both working-class and middle-class communities in southern California (not to mention the many years they spent developing a theoretical understanding of delinquency). Through their research, the Schwendingers were able to elaborate on what they argued was a critical shortcoming of many theories of delinquency—their inability to explain the substantial amount of middle-class delinquency documented by self-report studies as well as the extensive lower-class delinquency more likely to be reflected in official data.

Rather than relying on social class as a starting point for understanding delinquency, the Schwendingers focus on adolescent peer groups (see **Box 2-1**). According to the Schwendingers, the particular form of capitalist development that occurred in the United States has produced a large variety of adolescent social types, peer networks, and status groups that compose what is often referred to as "youth culture." As the Schwendingers note, youth culture actually comprises various subcultures and peer networks that cut across social class lines. These subcultures are reflected in various "social types" that have their own distinct designations, such as "intellectuals," "hodads," "greasers," "homeboys," "socialites," "preppies," "athletes," and others, as well as distinctive dress and linguistic patterns that are found in these groups. Moreover, they note that some of these adolescent peer networks encourage delinquency among their members and can be categorized as falling into different "stradoms" comprising distinctive social types. What the Schwendingers' research reveals is that youths who belong to these stradoms are more likely to engage in delinquent behaviors than youths who do not, regardless of their social class. For example, they note that "intellectuals" are the least delinquent youths in a peer society whereas members of "streetcorner" stradoms are the most delinquent. They also note that as stradoms mature, the types of delinquency committed by their members change.

Box 2-1

Interview: Herman and Julia Schwendinger

Q: Both of you have backgrounds in social work. In what ways has this influenced your study of delinquency?

A: From the very beginning of our social work careers, we worked with adolescents and delinquent gangs in high-crime areas. Herman's own childhood and adolescence involved gang activity in a poverty-stricken New York City community, and his earliest assignments as a social group worker included field work with street gangs. In her first social work position, Julia also worked with children and teens in a poor high-crime community. Our research evolved from questions raised by our social work experience. In fact, Herman, for instance, originally attended UCLA as a part-time student to acquire knowledge that he could use as an administrator of a youth program that served delinquents as well as nondelinquents.

Q: What led to your interest in studying crime and delinquency?

A: Herman's initial UCLA experience highlighted the lack of fundamental information about delinquent groups, and his interest in developing such knowledge was strongly encouraged by sympathetic faculty. Julia's doctoral dissertation on rape went hand in hand with her work in founding the first antirape group in the country.

Q: I know that Herman has a background in math and quantitative analysis, but he has also spent a lot of time observing youth groups. How important are qualitative methods in understanding youth behaviors?

A: Qualitative work has been crucial for developing some of our most important theoretical ideas. Also, our recent work indicates that ethnographic observations combined with quantitative methods (e.g., field experiments and network analysis) can produce much more rapid advances than traditional surveys and interviews for the study of delinquency.

Q: How has this observational work informed our understanding of youth behavior?

A: We believe that our work will have considerable impact when a new generation of criminologists abandons armchair theorizing and gets out in the field. Unfortunately, today too many are stuck with "social control" and "delinquency subculture" theories that emphasize personality abnormalities and ignore the variety of adolescent subcultures in our society. Subcultures composed of jocks, preppies, homeboys, metalbangers, freaks, brains, etc., exhibit different forms and degrees of misconduct. But even the most delinquent subculture cannot be attributed to pathological traits of leaders, early childhood disorders, or the social incapacities of the members at large.

Q: What are some of the problems that you encountered in doing observational research?

A: Observational research has problems that are never encountered by researchers who ride around in patrol cars or spend time devising survey questionnaires. One has to deal with boredom from hanging around at all hours listening to teen gossip and small talk; creating trusting relationships and guaranteeing trust; staving off manipulation and being used by conflicting groups or individuals; trying to deal with ethical questions due to knowledge about crimes that have occurred and that will occur; and developing an awareness of and techniques to assure personal safety.

Q: What else would you like to tell students interested in field work?

A: Despite the myriad problems related to participant observation, we had a lot of fun. We formed warm friendships (and sometimes became "role models") which have been maintained to this day with some of our "subjects" and their subsequent families.

Herman Schwendinger. Psychology, CCNY; MSW, Columbia; PhD, Sociology, UCLA.

Julia Schwendinger. Sociology, Queens College; MSW, Columbia; PhD, Criminology, University of California, Berkeley

Participant observation is a very time-consuming form of research, and participant observation studies, like other studies, can have weaknesses. It is possible, for example, that research results gleaned from observing youths in one locality may not be generalizable to youths living somewhere else or during a different period. However, because participant observers spend a considerable amount of time with their subjects, they are in a position to learn things about the subjects' behaviors and attitudes that cannot be easily learned through other research methods.

Victimization Studies

Another valuable source of information about juvenile offending is <u>victimization studies</u>, which involve the collection of data from crime victims through victim surveys or interviews. These studies allow researchers to gather a range of information about the experiences of crime victims, such as the level of victimization within the population, relationships between offenders and victims, circumstances surrounding the victimization (e.g., where the victimization occurred, the victim's possible contribution to the victimization), and characteristics of victims and offenders. Victimization studies can also help researchers understand the dark figure of crime. The most important source of data on youth victimizations comes from the National Crime Victimization Survey (NCVS), a large-scale national survey conducted by the Bureau of Justice Statistics, a branch of the U.S. Department of Justice, and the U.S. Census Bureau.

Although the NCVS examines only a narrow range of criminal offenses where estimates of the age of the offender are available, it can be used to assess offending rates for certain violent offenses. As might be expected, levels of offending based on victimization data are considerably higher than those found in UCR data. In a study of trends in violent offending based on NCVS data, James Lynch found that offending rate estimates calculated from NCVS data were 9 to 15 times greater than those based on the UCR. However, he also found that the general trends in violent offending seen in NCVS data are similar to those revealed in the UCR. Like the UCR data, NCVS data indicate that there has been a recent decline in violent offending among juveniles.[48]

Although victimization data can provide some important insights on levels of juvenile offending, they are even more valuable in aiding our understanding of youths' victimization experiences. Overall, NCVS data indicate that violent and property crime victimization has been declining since the early 1990s. However, this data also indicates that youths have high rates of victimization compared to other groups. For example, in 2002, persons ages 16 to 19 had the highest rate of violent crime victimization of any age group, followed by those 12 to 15 years of age and those 20 to 24 years of age.[49] In addition, some groups of teenagers have higher rates of violent crime victimization than others. American Indian youths, males, teens between 12 and 17, and teenagers in disadvantaged communities have the highest rates of victimization.[50]

victimization study A study that focuses on the crime victimization experiences of subjects.

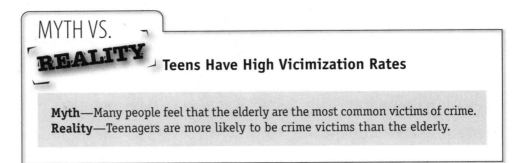

MYTH VS. **REALITY**   **Teens Have High Vicimization Rates**

Myth—Many people feel that the elderly are the most common victims of crime.
Reality—Teenagers are more likely to be crime victims than the elderly.

The NCVS also provides information about the relationship between offenders and victims as well as information on the location and time of offenses. Most teenagers are victimized by people they are at least acquainted with, and a sizable percentage of victimizations involve offenders who are well known to the victim; for instance, in 1996, among teens 12 to 17 years of age, the victim knew the perpetrator in 64% of reported victimizations. Moreover, a sizable amount of teenage victimization occurs at school. In 1995, about 15% of students reported some type of victimization at school. However, serious violent victimization in school is rare and is concentrated in a small number of schools.[51]

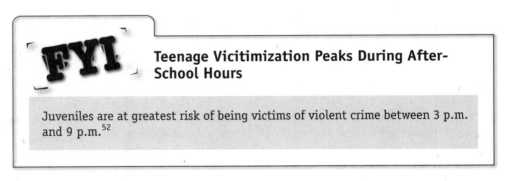

FYI   **Teenage Vicitimization Peaks During After-School Hours**

Juveniles are at greatest risk of being victims of violent crime between 3 p.m. and 9 p.m.[52]

Although victimization studies are a rich source of information about crime, they are limited in their ability to describe the extent of delinquency for several reasons. Many crimes involve no contact between victim and offender; consequently, the age of the offender is unknown. Also, when there is personal contact, the victim may be unable to provide an accurate description of the offender or even determine the offender's age. Some youths simply look older than their actual age, whereas some adults look younger than they are. Despite these limitations, victim data constitute an important source of juvenile crime data. They can be used to estimate offending rates among juveniles for some violent offenses, and they provide valuable information about youths' victimization experiences.

■ Legal Issues

The Use of Arrest versus Conviction Data

The presumption of innocence is an accepted common law principle in the United States. Any person arrested must be formally charged, arraigned, tried, and convicted beyond a reasonable doubt before he or she can be considered guilty. Therefore, arrest data, such as the information found in the UCR and other formal and informal data, may not accurately reflect the extent of adult or juvenile crime. Being arrested for an alleged offense does not mean one is guilty. Nevertheless, arrest data are often used as measures of juvenile crime because juvenile courts have varying rules of confidentiality regarding juvenile conviction records, which makes such records difficult to examine in many jurisdictions. Furthermore, police departments generally have fewer confidentiality requirements, which make arrest data more accessible and contributes to their use as a measure of juvenile crime.

Confidentiality for Juveniles

As long as there have been juvenile courts, there has been debate concerning the confidentiality of juvenile court records and proceedings. As with many legal debates, both sides of the issue have good points to make. On the side favoring the confidentiality of juvenile proceedings and records is the argument that the fundamental reason for the establishment of juvenile courts was to allow the mistakes and indiscretions of children to be dealt with away from the range of public scrutiny so that the emphasis could be on rehabilitation, not stigmatization.

Advocates on the other side of the issue contend that allowing juvenile courts to operate outside the scrutiny of the public has led to abuses that have harmed more children than confidentiality has helped. They also point out that society has a legitimate interest in protecting itself from dangerous individuals, be they juveniles or adults, and no dangerous person should be allowed to use confidentiality to avoid public scrutiny or appropriate consequences. Indeed, this argument has gained more weight as the public has grown increasingly concerned with the rise in serious juvenile crime.

Another unintended consequence of confidentiality and closed courtrooms is that it prevents the public from seeing the difficult cases that the courts handle and the societal problems represented in juvenile court cases. Not only does this lead people to downplay the seriousness and legitimacy of juvenile courts, but it also allows some people to ignore social trends and problems that are apparent to court personnel. Unfortunately, "out of sight, out of mind" seems to apply to much of what happens in the juvenile justice process. For an excellent discussion of these public policy conflicts and balance, see *People v. Smith*, 437 Mich 293, 470 NW2d 70, 78 (1991), in which it is stated, "The purpose of the court rule, and of similar rules or statutes in other jurisdictions, is to prevent a juvenile record from becoming an obstacle to educational, social, or employment opportunities. When, however, a juvenile offender appears in court again as an adult, his juvenile offense record may be considered in imposing sentence. The law contemplates a differentiation in sentencing between first-time offenders and recidivists, juvenile or adult."[53]

■ Chapter Summary

Taken together, official and unofficial sources of delinquency data provide valuable insights regarding delinquency and responses to delinquency within the United States. These data sources indicate that delinquency is widespread within our society and consists of a wide range of behaviors, from serious crimes against persons to failure to obey one's parents. They also indicate that far more males than females engage in delinquent behavior and that male dominance in delinquency increases with the seriousness of the offense. Yet most delinquency, whether committed by males or females, involves minor offenses and does not come to the attention of the authorities. Further, none of the data sources suggest that juvenile crime has gotten appreciably worse over the last 20 years. Taken together, various data sources indicate that juvenile crime, including violent juvenile crime, has remained stable or has been declining in recent years.

Clearly, official and unofficial data sources provide a more comprehensive view of juvenile crime than one gets from the media, which present a distorted view of juvenile crime. However, no one data source provides a comprehensive picture of delinquency. Moreover, each method of measuring delinquency and juvenile justice processing possesses some weaknesses. Consequently, caution needs to be exercised by those who use each of the data sources in formulating policy responses to the delinquency problem.

Which data source is best, and which method of measuring delinquency or the processing of juvenile offenders should one rely on? Perhaps the best answer to this question is to say that it depends on the question to be addressed. For example, if the object is to discover the number of juvenile arrests and the characteristics of those arrested, the UCR arrest data would be most helpful. On the other hand, someone interested in understanding the dark figure of delinquency and exploring a variety of factors related to delinquency would find self-report data to be more helpful. Indeed, the various sources of data on juvenile crime are probably best viewed as complementary—each tells something about the phenomena of juvenile crime and the processing of juvenile offenders.

KEY CONCEPTS

arrest rate The number of arrests adjusted for the size of the population; often reported as arrests per 100,000 youths 10–17 years of age in the population.

cohort study A study in which some specific group (e.g., those born during a certain year in a particular geographic location) are studied over a period of time.

dark figure of juvenile crime The amount of juvenile crime that goes unreported.

developmental criminology An area of criminology concerned with changes in individual and group patterns in offending over time and understanding factors that influence these patterns.

etiology The study of causation.

observational study A study in which the researcher observes and collects data on subjects in a field setting.

official data Information on juvenile delinquency collected by formal juvenile justice agencies, such as police agencies, juvenile courts, and juvenile detention and correctional facilities.

panel study A study that involves the examination of the same select group over time.

police discretion The authority of police to make their own judgments about which crimes or delinquent acts are subject to investigation and which juveniles are subject to arrest.

recidivist A person who continues to commit crimes despite efforts to rehabilitate or deter him or her.

self-report study A study in which subjects are asked to report their behavior (e.g., their involvement in illegal activities).

socioeconomic status (SES) A person's position in a stratified society based on criteria such as place of residence, family income, educational background, and employment history.

status offense An act that is considered a crime or legal violation when committed by a juvenile but not by an adult (e.g., running away from home, incorrigible behavior, failure to attend school, and failure to obey school rules).

Uniform Crime Report (UCR) The most comprehensive compilation of known crimes and arrests. The report is published each year by the Federal Bureau of Investigation.

unofficial data Information collected by researchers not connected with formal juvenile justice agencies (i.e., police agencies, juvenile courts, and correctional institutions).

victimization study A study that focuses on the crime victimization experiences of subjects.

REVIEW QUESTIONS

1. What is the distinction between official and unofficial sources of data on delinquency?

_____

_____

_____

2. What types of data are contained in the UCR?

_____

_____

_____

3. How much of a problem is juvenile crime according to UCR arrest data?

_____

_____

_____

4. Why is it important to consider arrest rate trends when attempting to examine juvenile crime trends?

_____

_____

_____

5. What does the examination of juvenile arrest rate trends tell us about juvenile crime over the last 20 years?

_____

_____

_____

_____

6. What problems are associated with using UCR arrest data as an indicator of juvenile crime?

_____

_____

_____

_____

7. What are the primary uses of official data?

_____

_____

_____

_____

8. What is the picture of delinquency presented by self-report studies?

_____

_____

_____

_____

9. What are potential problems associated with self-report studies?

_____

_____

_____

_____

10. What are the major findings of cohort studies of delinquency?

_____

_____

_____

_____

11. What insights have developmental criminology provided about juvenile offending and its treatment?

_____

_____

_____

_____

12. What are major findings of the observational studies of delinquency conducted by William Chambliss and by Herman and Julia Schwendinger?

_____

_____

_____

_____

13. In what ways are observational studies superior to official data sources for learning about delinquency?

_____

_____

_____

_____

14. What do victimization studies tell us about juvenile crime and teenage victimization?

_____

_____

_____

_____

ADDITIONAL READINGS

Chambliss, W. J. 1973. "The Saints and the Roughnecks." _Society_ 2, no. 1: 341–355.

Elliott, D. S. and S. S. Ageton. 1980. "Reconciling Race and Class Differences in Self-Reported and Official Estimates of Delinquency." _American Sociological Review 45_: 95–110.

Howell, J. C. 2003. _Preventing and Reducing Juvenile Delinquency, A Comprehensive Framework._ Thousand Oaks, CA: Sage.

Lynch, J. P. 2002. "Trends in Juvenile Violent Offending: An Analysis of Victim Survey Data." _Juvenile Justice Bulletin._ Washington, DC: Office of Juvenile Justice and Delinquency Prevention.

Moffitt, T. E. 1993. "Adolescent Limited and Life-Course Persistent Antisocial Behavior: A Developmental Taxonomy." *Psychological Review 100*: 674–701.

Schwendinger, H. and J. S. Schwendinger. 1985. *Adolescent Subcultures and Delinquency.* New York: Praeger.

NOTES

1. Federal Bureau of Investigation, *Crime in the United States 2002, Uniform Crime Reports* (Washington, DC: U.S. Government Printing Office, 2003).

2. Federal Bureau of Investigation, *Crime in the United States 2002, Uniform Crime Reports.*

3. Federal Bureau of Investigation, *Crime in the United States 2002, Uniform Crime Reports.*

4. U.S. Census Bureau, *Projections of the Total Resident Population by 5-Year Age Groups, and Sex with Special Age Categories: Middle Series, 2001 to 2005.* Retrieved February 16, 2004, from http://www.census.gov/population/projections/nation/summary/np-t3-b.pdf.

5. See Federal Bureau of Investigation, *Crime in the United States, Uniform Crime Reports*, 1975–2002.

6. Federal Bureau of Investigation, *Crime in the United States 2002, Uniform Crime Reports.*

7. D. Steffensmeir, "Is the Crime Rate Really Falling? An 'Aging' U.S. Population and Its Effect on the Nation's Crime Rate, 1980–1984," *Journal of Research in Crime and Delinquency 24* (1987): 23–48.

8. See P. J. Cook and J. H. Laub, "The (Surprising) Stability of Youth Crime Rates," *Journal of Quantitative Criminology 2* (1986): 265–277; D. Osgood et al., "Time Trends and Age Trends in Arrests and Self-Reported Illegal Behavior," *Criminology 27* (1989): 389–417.

9. H. Snyder, "Law Enforcement and Juvenile Crime," *Juvenile Offenders and Victims: National Report Series.* Washington, DC: Office of Juvenile Justice and Delinquency Prevention (2001).

10. Federal Bureau of Investigation, *Crime in the United States 2002, Uniform Crime Reports.*

11. Bureau of the Census, *Projections of the Total Resident Population by 5-Year Age Groups, Race, and Hispanic Origin with Special Age Categories: Middle Series, 2001 to 2005.* Retrieved February 15, 2002, from http://www.census.gov/population/projections/nation/summary/np-t4-b.pdf.

12. Federal Bureau of Investigation, *Crime in the United States 2002, Uniform Crime Reports.*

13. R.M. O'Brien, "Crime and Victimization Data," in *Criminology: A Contemporary Handbook*, ed. J. F. Sheley, 2nd ed. (Belmont, CA: Wadsworth, 1995).

14. D. Huizinga and D. S. Elliott, "Juvenile Offenders: Prevalence, Offender Incidence, and Arrest Rates by Race," *Crime and Delinquency 33* (1987): 206–223.

15. D. S. Elliott, "Lies, Damn Lies, and Arrest Statistics," Paper presented at the annual meeting of the American Society of Criminology, Boston, 1995.

16. O'Brien, "Crime and Victimization Data."

17. O'Brien, "Crime and Victimization Data."

18. Snyder, "Law Enforcement and Juvenile Crime."

19. Federal Bureau of Investigation, *Crime in the United States 2002, Uniform Crime Reports.* Retrieved February 23, 2004 from http://www.fbi.gov/ucr/cius_02/pdf/3sectionthree.pdf.

20. Snyder, "Law Enforcement and Juvenile Crime."

21. See J. Bachman et al., *Monitoring the Future* (Ann Arbor, MI: University of Michigan, 1994); M. Gold, "Undetected Delinquent Behavior," *Journal of Research in Crime and Delinquency 3* (1966): 27–46; M. Gold, *Delinquent Behavior in an American City* (Belmont, CA: Brooks/Cole, 1970); M. Erickson and L. Empey, "Court Records, Undetected Delinquency, and Decision-Making," *Journal of Criminal Law, Criminology, and Police Science 54* (1963): 446–469; J. Short and F. I. Nye, "Extent of Unrecorded Delinquency," *Journal of Criminal Law, Criminology, and Police Science 49* (1958): 296–302.

22. A. L. Pastore and K. Maguire, eds., *Sourcebook of Criminal Justice Statistics.* Retrieved February 24, 2004 from http://www.albany.edu/sourcebook/1995/pdf/t346.pdf.

23. A.L. Pastore and K. Maguire, eds., *Sourcebook of Criminal Justice Statistics.* Retrieved February 24, 2004 from http://www.albany.edu/sourcebook/1995/pdf/t358.pdf.

24. Osgood et al., "Time Trends and Age Trends in Arrests and Self-Reported Illegal Behavior."

25. D. S. Elliott et al., *The Prevalence and Incidence of Delinquent Behavior: 1976–1980, National Youth Survey Report No. 26* (Boulder, CO: Behavioral Research Institute, 1983).

26. D. S. Elliott and S. S. Ageton, "Reconciling Race and Class Differences in Self-Reported and Official Estimates of Delinquency," *American Sociological Review* 45 (1980): 95–110.

27. D.S. Elliott and D. Huizinga, "Social Class and Delinquent Behavior in a National Youth Panel," *Criminology* 21 (1983):149–177.

28. M. Chesney-Lind and R.G. Shelden, *Girls, Delinquency and Juvenile Justice*, 3rd ed. (Belmont, CA: Thompson/Wadsworth, 2004).

29. S. Cernkovich et al., "Chronic Offenders: The Missing Cases in Self Report Delinquency Research," *Journal of Criminal Law and Criminology* 76 (1985): 705–732.

30. M. Hindelang et al., *Measuring Delinquency* (Beverly Hills, CA: Sage, 1981).

31. M. Wolfgang et al., *Delinquency in a Birth Cohort* (Chicago: University of Chicago Press, 1972).

32. Wolfgang et al., *Delinquency in a Birth Cohort*, 246.

33. P. E. Tracy et al., *Delinquency Careers in Two Birth Cohorts* (New York: Plenum Press, 1990).

34. M. E. Wolfgang et al., *From Boy to Man, from Delinquency to Crime* (Chicago: University of Chicago Press, 1987).

35. P. E. Tracy et al., *Delinquency in Two Birth Cohorts*. (Washington, DC: U.S. Department of Justice, 1985); Tracy et al., *Delinquency Careers in Two Birth Cohorts*.

36. Tracy et al., *Delinquency Careers in Two Birth Cohorts*.

37. Tracy et al., *Delinquency in Two Birth Cohorts*.

38. L. Shannon, *Assessing the Relationship of Adult Criminal Careers to Juvenile Careers: A Summary* (Washington, DC: U.S. Department of Justice, 1982).

39. D. M. Hamparian et al., *The Violent Few* (Lexington, MA: Lexington, 1978).

40. Office of Juvenile Justice and Delinquency Prevention, "The Juvenile Court's Response to Violent Crime," *Office of Juvenile Justice and Delinquency Prevention Update on Statistics* (Washington, DC: U.S. Department of Justice, 1989).

41. H. Snyder, *Court Careers of Juvenile Offenders* (Washington, DC: Office of Juvenile Justice and Delinquency Prevention, 1988).

42. D. Franklyn and D. Elliott, "Identifying Career Offenders Using Self-Report Data," *Journal of Research in Crime and Delinquency* 21 (1984): 57–86.

43. For a recent review of research on this topic see J. C. Howell, *Preventing and Reducing Juvenile Delinquency, A Comprehensive Framework* (Thousand Oaks, CA: Sage, 2003).

44. T. E. Moffitt, "Adolescent Limited and Life-Course Persistent Antisocial Behavior: A Developmental Taxonomy." *Psychological Review* 100 (1993): 674–701.

45. Moffitt, "Adolescent Limited and Life-Course Persistent Antisocial Behavior: A Developmental Taxonomy."

46. W. J. Chambliss, "The Saints and the Roughnecks," *Society* 2 (1973): 341–355.

47. H. Schwendinger and J. S. Schwendinger, *Adolescent Subcultures and Delinquency* (New York: Praeger, 1985).

48. J. P. Lynch, "Trends in Juvenile Violent Offending: An Analysis of Victim Survey Data." *Juvenile Justice Bulletin* (Washington, DC: Office of Juvenile Justice and Delinquency Prevention, 2002).

49. C. M. Rennison and M.R. Rand, *Criminal Victimization, 2002, Bureau of Justice Statistics* (Washington, DC: U.S. Department of Justice, 2003).

50. J. L. Lauritsen, "How Families and Communities Influence Youth Victimization," *Juvenile Justice Bulletin* (Washington, DC: Office of Juvenile Justice and Delinquency Prevention, 2003); H. N. Snyder and M. Sickmund, *Juvenile Offenders and Victims: 1999 National Report* (Washington, DC: Office of Juvenile Justice and Delinquency Prevention, 1999).

51. Snyder and Sickmund, *Juvenile Offenders and Victims: 1999 National Report*.

52. Snyder and Sickmund, *Juvenile Offenders and Victims: 1999 National Report*.

53. *People v. Smith*, 437 Mich 293, 470 NW2d 70, 78 (1991).

The Social Context of Juvenile Delinquency and Juvenile Justice

3

Chapter Objectives

After studying this chapter, you should be able to

- Explain the connections among poverty (economic inequality), social disorganization (divorced couples and single-headed families), and juvenile delinquency

- Describe how special interest groups influence political and government decision making in a manner that adversely affects children

- Describe the key role that the family plays in the socialization of children and the relationship between child socialization and delinquency

- Explain how parental roles influence the socialization of children

- Describe the importance of parental economic or workplace power, parental criminality, and family conflict, and explain how each of these may influence delinquency

- Explain the connection between parental abuse and/or neglect of children and delinquency

- Explain the connection between school performance and delinquency

- Describe how schools themselves, as well as societal factors, contribute to student failure

- Describe the phenomenon of "social disorganization" and how such disorganization can encourage the creation of a delinquent "culture" in neighborhoods or urban areas

- Explain how youth peer groups and/or gangs differ from one another and how youths' peer associations encourage or discourage delinquency

- Describe how the development, management, and distribution of political and economic resources influence both youths and the operation of the social institutions that make up the juvenile justice process

Introduction

Delinquency and the practice of juvenile justice occur, not in a vacuum, but in a social context. This does not mean that individual factors such as biological makeup and psychological functioning do not play a role in delinquency or the operation of juvenile justice. Nor does it imply that individuals do not make choices, often conscious choices, to engage in delinquent behaviors. However, it recognizes that individuals and the choices they make cannot be adequately understood without considering the social contexts within which they live and act.

As noted in Chapter 1, the social context helps shape our views of juvenile crime and the operation of juvenile justice through the portrayal of delinquency and juvenile justice in the media. In fact, much of what most people (including the makers of juvenile crime policy) know about juvenile crime and juvenile justice comes from the media. However, the social context of juvenile delinquency and juvenile justice comprises more than the media. In the United States, juvenile delinquency and juvenile justice are influenced by a variety of factors found within the political economy of the United States and within communities, families, schools, peer groups, and other important socializing institutions. How political and economic arrangements and socializing institutions such as families, communities, and schools influence delinquency is a primary focus of theory and research within the field of criminology. Indeed, courses in criminology, juvenile delinquency, and criminological theory focus attention on how factors such as economic inequality, school failure, residence in a high-crime neighborhood, child discipline practices, child abuse, association with criminally involved peers, and many other factors are related to delinquency. Explanations of criminal behavior that refer to such factors comprise a significant body of criminological theory. Moreover, theories are important, as Stephen Pfohl has noted, because they "provide us with an image of what something is and how we might best act toward it."[1] The development of good theories of delinquency, then, could help reduce delinquency.

Although the purpose of this text is not to examine the many theoretical explanations of delinquency, it should be kept in mind that delinquency is the product of a complex set of factors and that juvenile justice exists within a complex social context that has real and profound effects on the practice of juvenile justice. Importantly, juvenile justice practitioners often spend considerable energy attempting to help their clients develop more law-abiding responses to the social context within which their clients live. Many also spend considerable energy attempting to manipulate or alter various aspects of the social context in ways that will lessen the likelihood of continued delinquent behavior.[2]

This chapter discusses some of the important components of the overall social context that influence juvenile crime and the operation of juvenile justice. It begins by considering the effects of economic and political factors on delinquency. Next, it explores a variety of important social institutions such as families, neighborhoods, schools, and peer associations and their relationship to youths' delinquent behavior. It then examines how the social context influences the development of delinquency theory. It closes by looking at the ways in which the social context influences the practice of juvenile justice.

■ The Political and Economic Context of Juvenile Delinquency and Juvenile Justice

The Political Economy of Delinquency

The term **political economy** refers to the basic economic and political organization of society. The political economy determines the ways in which economic and political resources are developed, managed, and distributed. The development, management, and distribution of these resources are important because they directly affect the ability of basic socializing institutions such as families and schools to meet people's needs. For example, the distribution of job opportunities and the pay earned by workers influences the amount of economic resources possessed by families and can influence the quality of life experienced by family members. As the National Commission on Children noted in a 1993 report, "Economic security is fundamental to children's well-being. Children need material support to have a better chance to grow up healthy, succeed in school, and become capable and caring adults."[3] Furthermore, decisions regarding the development, management, and distribution of economic resources do not occur in a social vacuum; rather, they are products of a political process in which different groups with conflicting concerns and varying degrees of power try to protect and further their interests.

political economy The basic political and economic organization of society, including the setting of political priorities and the management and distribution of economic resources. People's economic status largely determines the opportunities that are available to them, their feeling of political empowerment, their ability to access the political system, and their ability to influence political decisions.

The Political Context

The forms of political organization found in most modern societies can be separated into two main categories: democratic forms and authoritarian forms. The United States, although relatively democratic, is not a pure democracy where every person has an equal voice in decision making and not every political decision maximizes the common good. Instead, governmental decision making is regularly influenced by powerful special interests that seek to maximize benefits for themselves at the expense of others. These **special interests** consist of individuals, families, corporations, unions, and various other organizations, and they use a variety of means to obtain tax breaks, favors, subsidies, and favorable rulings from congressional committees, regulatory agencies, and executive agencies.[4]

Although the government can and does act in ways that benefit the majority of Americans, it is not always neutral.[5] When governmental entities decide to regulate certain activities and not others, when they enforce certain laws and not others, when they decide on which resources to develop and how to manage and distribute these resources, their decisions and actions invariably favor some groups and interests over others. Whether to provide government support for abortion or to withdraw it, whether to increase or reduce regulations concerning gun ownership, whether to reduce or eliminate support for summer jobs programs for youth, or whether to make it more difficult for low-income youths to receive loans for higher education while providing a range of tax breaks and subsidies to large corporations are hardly neutral decisions. In determining what should be done, government is generally biased toward policies that benefit the wealthy, especially the big business community.[6]

The ability to influence political decision making is a form of power. **Power** is the "capacity of some persons to produce intended and foreseen effects on others."[7] However, power is not equally distributed across individuals and groups in our society. Instead, it is concentrated among those who possess substantial economic resources—resources that can be used to directly and indirectly influence government decision making. The expenditure of those resources to influence government decision making is seen as a reasonable cost because the decisions reached partially determine the development, management, and distribution of resources within society and typically benefit those who possess substantial economic resources.

From a juvenile delinquency and juvenile justice standpoint, the question we must ask ourselves is this: To what extent do present political arrangements contribute to the well-being of children and to the healthy functioning of other socializing institutions, such as families and communities, that have an important impact on children's lives? We must also ask, to what extent do present political arrangements encourage delinquency, either directly or indirectly? Decisions to invest our resources in ways that encourage the healthy development of communities, schools, families, and children are political decisions. This does not mean that families, communities, schools, and other organizations do not share responsibility for the welfare of children and for the problem of delinquency. Community decision making and school decision making are themselves inherently political activities. However, it must be kept in mind that government decision making

special interest group An organization consisting of persons, corporations, unions, and so on, that have common political and often economic purposes and goals. These individuals or groups come together to use their influence, monetarily and politically, to persuade politicians and other persons in power to act favorably toward them.

power The ability to influence political and economic decisions and actions. Power is the ability to force, coerce, or influence a person's actions or thoughts.

does have important ramifications for the quality of life experienced by many Americans and for the level of delinquency that exists in our society. The following section examines one important outcome of decision making in the United States, namely, economic inequality, as well as the relationship of economic inequality to delinquency.

The Economic Context

There are two primary forms of economic organization in the modern world: <u>capitalism</u> and <u>socialism</u>. Although pure capitalism and pure socialism do not exist, the economic organization of each country tends toward one form or the other. In the United States the predominant form of economic organization is capitalism. This form of economic organization is characterized by three basic principles: private ownership of personal property, personal profit, and competition. In addition, according to those who favor pure capitalism, government interference in economic life should be kept to a minimum. The basic view of capitalism's supporters is that "the profit motive, private ownership, and competition will achieve the greatest good for the greatest number in the form of individual self-fulfillment and the general material progress of society."[8]

Clearly, many people in the United States are able to achieve considerable prosperity. Yet it is also true that many others have relatively few material resources. In 2002, it was estimated that slightly over 12% of the U.S. population (almost 35 million people) were officially poor. Most of these poor people are white, and most live in metropolitan areas. Moreover, 38% of all poor people worked during 2002, although many worked either part time or part year.[9] However, other estimates of the number of poor persons—estimates not based on the official poverty line—suggest that the real number is considerably higher than official estimates.[10] Even using official data, the number of persons considered *near* poverty is estimated to be some 47 million people.[11]

capitalism An economic system based upon three fundamental principles: (1) private ownership of property, (2) competition between economic interests, and (3) personal profit as a reward for economic risk and effort.

socialism Socialism represents a form of political and economic system that is based on democratic decision making, equality of opportunity for all, collective decision making designed to further the interests of the entire community, public ownership of the means of production, and economic and social planning.

Official Poverty

The poverty line is defined as three times the amount of income necessary for a minimally nutritional diet. In 2002, the official poverty line for a family of four was $18,392.[12] Many economists and others argue that the official poverty line underestimates the number of poor persons, since those earning slightly above the poverty line still lack access to adequate shelter, diet, housing, clothing, and medical care, yet they are not considered officially poor. Moreover, researchers indicate that in some places families would need double the poverty rate to meet their basic needs.[13]

It is important to note that poverty is not equally distributed within the population. Minorities, children, and families headed by women are disproportionately represented among the impoverished in the United States. In 2002, approximately 8% of whites had an income below the official poverty level, compared with approximately 24% of African Americans and 22% of Hispanics. Although most officially poor people in the United States are white, African Americans and Hispanics are disproportionately represented in poverty statistics. Similarly, almost 17% of persons under age 18 and 27% of families headed by women were below the poverty line in 2002.[14]

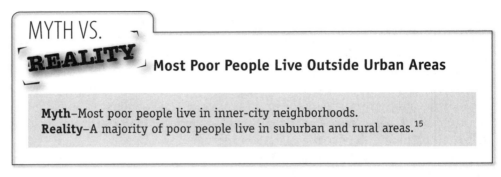

MYTH VS. REALITY   Most Poor People Live Outside Urban Areas

Myth–Most poor people live in inner-city neighborhoods.
Reality–A majority of poor people live in suburban and rural areas.[15]

The gap between the wealthiest Americans and the poorest Americans has widened in recent years. Research that has examined trends in income inequality indicates that it has increased since the 1970s. The increase in income inequality appears to be related to six factors: (1) the shift from better-paid manufacturing jobs to less well-paid service jobs; (2) technological advances that have increased the demand for more highly skilled workers at the expense of less skilled workers; (3) globalization, which forces American workers to compete with foreign workers, driving down American wages; (4) immigration, which drives down wages among less skilled workers; (5) a decline in unionization that decreases the ability of workers to bargain for higher wages; and (6) the segregation of poorer people into areas where their opportunities for economic well-being are lessened.[16]

MYTH VS. REALITY   The Working Poor

Myth—Anyone can achieve the American dream if he or she is willing to work hard.
Reality—Many poor people work. In 2000, some 4.7 million children lived in working poor families. For a single-parent family, this means members of the family worked at least 1,040 hours during the year, yet they did not earn above the poverty line. A primary reason why many working people are poor is that they earn low wages. For example, in 2000, 42% of working poor mothers earned less than $5 per hour.[17]

For most Americans, the economic resources of the families they are born into play a critical role in their future economic status. A person born into a poor family will likely remain poor or close to the poverty level, and a person born into a wealthy family will likely remain wealthy.[18] This does not mean, however, that everyone's economic circumstances remain static over time. Even individuals from middle-class families experience some unexpected change in their economic circumstances at some point in their lives, and most will experience periodic, though predictable, changes in their standard of living during their lifetimes. Divorce or separation, the death of a spouse or parent, becoming a family head or spouse, unemployment, work loss due to retirement, and a reduction in the work hours of individuals and family members can lead to financial hardship. Thus, it is not only the poor who are likely to suffer economic hardship during their lifetimes; they are merely more likely to suffer longer and more acutely due to their economic situation.

MYTH VS. REALITY — Many Find the American Dream Elusive

Myth—Only the poor who refuse to work are in jeopardy of failing to realize the American dream.

Reality—The typical American family is finding it more difficult than ever to own a home and a car and send the children to college, especially if the family lives on one income. The main reason is that the costs of housing, transportation, education, and health care have risen substantially since the 1970s.[19]

An important key to economic well-being is the ability to find and hold a job that pays a decent wage. Yet many Americans have a hard time locating such jobs. Periodic economic recessions in the 1980s, 1990s, and the early 2000s; decisions by U.S. companies to move their production operations to other countries; reductions in the workforce as a result of automation; efforts to make businesses more efficient; and a shift away from the production of goods to the provision of services not only have reduced job opportunities for the traditionally unemployed, but also have affected the wages and the future employment prospects of those who are working and those who will soon be entering the workforce. Although these trends have influenced people of all racial groups and ages, they have had a profound influence on people living in urban areas, many of whom are minorities, and children.

The shift from manufacturing jobs to service jobs; the movement of entry-level jobs in manufacturing, retail sales, and customer services from urban areas to the urban periphery or overseas; and the relocation of high-tech jobs to cities have resulted in lessened job opportunities for low-skilled urban residents. Moreover, a lack of affordable rental housing in the urban periphery coupled with racial residential segregation prevents many poor and low-skilled workers from relocating to areas that are closer to these jobs. Further, low levels of automobile ownership and poor public transportation mean that poor low-skilled urban residents find it difficult to commute to jobs.[20] Indeed, research indicates that distance from home is a major factor that influences workforce participation for both men and women, but it is particularly strong for women who are often hesitant to take jobs that are far from home because they do not want to be too far away from their children's caregivers.[21] Simultaneously, many new jobs that have been created in urban areas require a high degree of skill and are not well matched to the skills of many urban residents. The result has been a high degree of economic segregation characterized by concentrated pockets of poverty, particularly in urban areas,[22] at one extreme and areas of affluence at the other.

Economic resources are important because of the material goods and opportunities that such resources can procure. For those living in affluent areas it means having a range of public services and private resources. It means well-funded schools, recreational and cultural facilities, programs for youths, and access to a range of private resources that can be used to support and respond to family and individual problems. In poor communities it often means a lack of quality public or private resources for youths and families. It means that those who are disproportionately exposed to the stress of life have the fewest resources to help them deal with the problems that they encounter.

Adverse economic conditions, such as unemployment and declining wages, place a strain on many families. For example, research on unemployed men has found them to feel less satisfied with themselves and with their lives in general. They also tend to feel more victimized, anxious, depressed, and hostile toward others than men who are employed.[23] Financial loss appears to be related to changes in men's attitudes and parenting practices, and many fathers whose earnings decline become more tense, irritable, and explosive and more arbitrary and punitive in responding to their children.[24]

Of course, women are also affected by financial hardship. In 2002, women accounted for 58% of the individuals living in poverty, and half of all families in poverty were headed by women,[25] many of whom worked in low paying jobs.[26] Indeed, females who head families have multiple responsibilities and demands that go beyond their own immediate needs. They are responsible for providing food, shelter, and clothing for their children; seeing that they get to school; keeping them healthy; supervising and monitoring their behavior; meeting their emotional and psychological needs; managing family finances; and meeting a wide variety of other demands that emanate from employers, relatives, and others. In addition, they must deal with issues of gender discrimination, poor labor market prospects, and welfare policies and programs that provide minimal relief and are often seen as humiliating.[27]

Economic hardship also affects the young. Poor infants are much more likely to suffer low birth weight, to die during the first year of life, and to suffer hunger or abuse while growing up. They are less likely to receive immunizations or adequate medical care, and they are less likely to undergo the type of cognitive development that will allow them to do well in school.[28] Moreover, negative psychological and social effects are associated with poverty. Long-term exposure to poverty is related to increased levels of anxiety and unhappiness among children, and current exposure to poverty is associated with behaviors such as disobedience and aggression.[29]

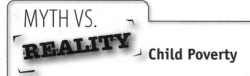

Child Poverty

Myth–Although child poverty is a problem in the United States, fewer American children live in poverty than children in other industrialized countries.

Reality–The percentage of children living in poverty in the United States during the 1990s was higher than the percentages in most other industrialized countries. For example, countries such as Sweden, Norway, Finland, France, Hungary, Germany, Spain, Poland, and Japan have lower levels of child poverty than the United States.[30]

Economic resources clearly influence the ability of people to obtain those things that are necessary for survival and that are desired in our society. Such resources influence the life chances that individuals have. As Hans Gerth and C. Wright Mills noted, life chances involve "everything from the chance to stay alive during the first year after birth to the chance to view fine art, the chance to remain healthy and grow tall, and if sick to get well again quickly, the chance to avoid becoming a juvenile delinquent—and very crucially, the chance to complete an intermediary or higher educational grade."[31]

Furthermore, economic status is an important factor from a juvenile justice perspective, for it appears to be strongly related to delinquent behavior. Indeed, many theories of crime and delinquency stress the role of economic inequality in the production of crime, and these theories have received some support in the research literature. For example, ethnographic studies of low income neighborhoods,[32] large-scale self-report studies of the link between **social class** and delinquency, and research on the relationship between persistent childhood poverty and delinquency have found that poverty appears to have an effect on delinquent behavior. In a large-scale study of social class differences in delinquency, Delbert Elliott and David Huizinga found that, although class differences were almost nonexistent when minor types of delinquency were examined, lower class youths reported significantly more involvement in serious forms of delinquency.[33] In a more recent study of the relationship between persistent poverty and delinquency, Roger Jarjoura, Ruth Triplett, and Gregory Brinker found that exposure to persistent poverty is associated with delinquency. Moreover, they also found that exposure to persistent poverty during early childhood is related to an increased likelihood of delinquency later in life.[34]

Political and economic arrangements in our society that foster inequality between Americans clearly influence the types of lives that many people, including children, experience. As noted, there is evidence that political and economic arrangements do have an important effect on delinquent behavior. In addition, they provide a context that influences the operation of other important socializing institutions, such as families, schools, communities, and peer groups, that are

social class A person's or family's economic, social, and political standing in the community. Social class can be directly related to type of employment, neighborhood, circle of contacts, and friends.

related to delinquency. The remaining sections of this chapter examine the role of these socializing institutions in the lives of children and the research that has looked at the relationship between these socializing institutions and delinquency.

■ The Family and Delinquency

family A collective body of any two or more persons living together in a common place under one head or management; a household; those persons who have a common ancestor or are from the same lineage; people who live together and who have mutual duties to support and care for one another.[151]

There appears to be widespread agreement among both social scientists and the general public that the <u>family</u> plays a key role in child development and socialization. The family can be a place where members love each other, care for one another, and provide a mutually beneficial environment for healthy human growth. On the other hand, the family can be characterized by conflict, a lack of mutual support and nurturance, and violence. Like other major socializing institutions, families are profoundly influenced by the political and economic context within which they operate. As D. Stanley Eitzen and Maxine Baca Zinn note,

> The unequal distribution of society's rewards and resources determines patterns of family living. Families are embedded in a class hierarchy. Furthermore, the family is the principal unit in the social class system. Most children will have the same life chances as their parents. The "haves" will pass on their advantages to their children, the "have-nots" will pass along their disadvantages to their offspring. A family's placement in the class system is the single most important determinant of family life.[35]

A family's placement in the political and economic structure is also important because such placement determines the family's access to and interconnection with other institutions, such as work establishments, schools, churches, and voluntary associations. These institutions can function as resources for the family and can facilitate access to other resources.[36]

Not only does the family determine the economic conditions within which children live, but it also is the primary shaper of a child's values, personality, and behavior. Not surprisingly, a variety of criminological theories suggest that the family plays a significant role in the production or prevention of delinquent behavior, and much research has been devoted to examining various aspects of the family that appear to be associated with delinquency. As Walter Gove and Robert Crutchfield note, "the evidence that the family plays a critical role in juvenile delinquency is one of the strongest and most frequently replicated findings among studies of deviance."[37] Essentially, this research has focused on two broad areas—family structure and family relations—both of which are felt to have a strong influence on juvenile delinquency and later criminality. Indeed, one important conclusion that can be drawn from this research to date is that the individual's experiences during infancy and early childhood influence behavior over the life course.[38]

Family Structure

Family structure refers to the ways in which families are constituted. For instance, are both natural parents present, is a stepparent present, is the family headed by a single parent, and how large is the family? One element appears clear: The structure of the American family is changing dramatically.

The Impact of the Single-Parent Home on Delinquency

One important fact is that children today are more likely to live in single-parent households than in the past. For example, in 1970 about 12% of all children lived with one parent, typically their mother,[39] but by 2002 the percentage of children living in single-parent homes had risen to approximately 28%, and another 4% lived in households where neither parent was present.[40] There are two primary reasons for the increase in the number of children who live in single-parent homes. First, the divorce rate has been rising over the past 20 years. The United States has the highest divorce rate in the world, with about half of all marriages ending in divorce. Moreover, the majority of all divorces involve children. A second reason why more children are living in single-parent homes is that more children are being born out of wedlock. The number of out-of-wedlock births has increased significantly since the 1960s, especially among teenagers, and the rise has contributed to the increasing number and proportion of children in single-parent homes.[41]

Of course, the high rate of marital dissolution and the growing number of out-of-wedlock births have not occurred in a social and economic vacuum. Rather, they are influenced by a complex set of social and economic factors. Although most Americans indicate they value marriage, there is considerable evidence that the importance given to marriage and two-parent families has eroded in recent years. The result has been a relaxation of social constraints on divorce, out-of-wedlock childbearing, and single parenthood.[42] At the same time, the economic prospects of many young men, particularly those who are minorities, have worsened, which has not only led to joblessness but also influenced marital dissolution and has made them less attractive as marriage partners who can support a family.[43]

The relationship between single-parent families (or what is sometimes referred to as "**broken homes**") and juvenile delinquency has been the focus of considerable debate within the fields of criminology and juvenile justice. Indeed, some research has discovered statistically significant relationships between single-parent homes and delinquency.[44] However, research also suggests that the effects of coming from a single-parent home may not be the same for all youths. For instance, some research has indicated that coming from a single-parent home is associated, not with serious delinquency, but with status offenses like running away from home and truancy.[45] Also, some research has found that white youths, young females,[46] and youths from high-income families[47] are more likely to be affected in more adverse ways by parental absence.

broken home A home or family in which at least one natural (biological) parent is absent.

Multiple Family Transitions and Deliquency

Many children experience family breakup several times during their lives. Research indicates that children who experience multiple family transitions during their lives are more likely to engage in delinquent behavior.[48]

Although there does appear to be a relationship between living in a single-parent home and involvement in delinquent behavior, the relationship is not very strong. Furthermore, the research uncovering such a relationship has often been based on official data. A potential problem with using official data is that youths from single-parent homes may be treated differently than youths from intact homes by the authorities.[49] For example, police may be more inclined to formally process youths from single-parent homes than youths from two-parent homes. The existence of such a bias is suggested by the fact that studies relying on self-report data have often failed to find a strong relationship between single-parent homes and delinquency.[50]

Research that has not discovered a relationship between single-parent homes and delinquency suggests that what is most important is not whether youths come from a single-parent or two-parent family but the quality of the relationship that exists between those parents that are present and their children.[51]

Broken Homes and Deliquency

One problem with examining the relationship between broken homes and delinquency is the vagueness of the concept of a "broken home." According to the standard definition, a broken home is a home in which at least one natural (biological) parent is absent. However, this definition ignores the potential differences in types of broken homes and variations in parental absence. For instance, a number of events might cause the absence of a parent, including divorce, separation, disability, military service, and job responsibilities. The length of parental absence can vary as well. Unfortunately, the effects of different types of parental absence on youths and their relationship to delinquency are not known.

The Effect of Family Size on Delinquency

Another aspect of family structure that may influence delinquency is family size. In his classic study of delinquency, Travis Hirschi found that, even when controlling for academic performance, parental supervision, and attachment between youths and their parents, family size was related to delinquency.[52] This finding is supported by research conducted in England, although the relationship discovered was much weaker for middle-class families than for lower-class families,[53] which suggests that it may be economic resources rather than family size that is most important. Perhaps parents of larger families who have substantial economic resources may be better able to meet their children's needs in ways that reduce the probability of delinquency. Some researchers have questioned the relationship between family size and delinquency, arguing that a more important variable is having a delinquent sibling. Their line of reasoning is that having a delinquent sibling is related to delinqency and that in larger families youths are more likely to have a delinquent sibling.[54]

Family Relations and Delinquency

The term *family relations* encompasses the quantity and quality of interactions and relationships among family members. Like family structure, however, family relations are also influenced by the larger social context within which families reside as well as the economic condition of the family itself. As noted earlier, families have changed in a variety of ways in recent years in response to political and economic developments. This section begins by examining how changes in the family may have influenced family socialization. It then discusses different aspects of family socialization that have been found to be related to delinquency.

The Influence of Women's Employment on Family Relations and Delinquency

In 1960, less than 7% of married women with children participated in the paid labor force.[55] In 2002, 67% of married women with children were employed.[56] Indeed, one of the most important changes in the American family has been the growing number of women, including mothers, who are entering the workforce. The effects of mothers' employment on children are not entirely clear, however. By working outside the home, mothers are able to improve the economic well-being of their families. Indeed, in poor families and many middle-income families, the mother must work if the children are to receive adequate care.[57] A working mother can make a significant contribution to the economic well-being of the family; however, it also means that the mother has less time to spend interacting with her children. This is a problem that single working mothers share with parents in dual-earner families, where both husband and wife work outside the home. In response, many families have come to rely increasingly on child-care or leave children to care for themselves for at least part of the day.[58]

Latchkey Children

Children who care for themselves at least part of the day are often referred to as **latchkey children**. In 1999 it was estimated that about 3.3 million 6- to 12-year-olds regularly spent time unsupervised or in the care of a young sibling.[59]

latchkey children
Children who care for themselves at least part of the day due to parental unavailability because of employment.

Although many people feel that a decline in the amount of parent–child interaction can obstruct child development, the extent of the impact is far from apparent. Some research indicates that there is virtually no developmental difference among toddlers who attend day-care programs and children taken care of at home, although there may be some negative health outcomes for infants.[60] Moreover, quality child-care programs can facilitate children's cognitive, emotional, and social development, particularly for disadvantaged children.[61] Even self-care does not always result in poor outcomes. Many children who are left unsupervised become more independent and learn to become more responsible for themselves over time. In contrast, however, other children who care for themselves experience loneliness and isolation and may be significantly disadvantaged by a lack of adult supervision.[62] According to criminologist Travis Hirschi, a lack of parental supervision is related to an increase in the likelihood of delinquent behavior.[63] Furthermore, research has shown that when children are not involved with their parents, they are more likely to report involvement in delinquency.[64] However, it appears that parental employment is not the most significant factor in determining youths' involvement in delinquency. For example, research by Thomas Vander Ven and his colleagues found that mothers' employment had very little effect on children's delinquent behavior when they were adequately supervised.[65] Consequently, mothers' employment may have little impact on delinquency. What appears to be more important is whether children receive appropriate supervision and the quality of parent–child relations.

Researchers have also examined the ways in which parents' roles and experiences in the workplace influence their relationship with their children. Criminologists Mark Colvin and John Pauly argue that parents tend to reproduce at home the authority relations they experience in the workplace. The problem is that the differences in power characteristic of a capitalist economy result in workplace experiences that are authoritarian and coercive. These experiences in turn engender coercive and authoritarian relationships within many homes—relationships that are not conducive to the establishment of intimate bonds between family members, and in turn, increase the likelihood of delinquency.[66]

The relationship between the economic sphere and the family is also the focus of John Hagan's power control theory. According to Hagan, "Work relations structure family relations, particularly relations between fathers and mothers and in turn relations between parents and their children, especially mothers and their daughters."[67] Data collected by Hagan and his colleagues in Canada tend to support this hypothesis. From Hagan's perspective, the power that parents have within the workplace is typically reproduced within the family. Hagan argues that when both parents are in positions of power in the workplace, the parents share power and the family structure is egalitarian. In such families, male and female children are socialized in similar ways, which results in comparable levels of delinquency among male and female children. However, in traditional patriarchal families, where the mother remains at home, as well as in single-parent families, daughters are more likely to become the objects of control by mothers, who socialize their daughters to avoid risk. One outcome is that in such families males tend to engage in more delinquency than females.[68]

The broader political and economic environment within which people live does appear to exert important influences on the types of relationships found within families. Moreover, a wealth of information exists indicating that various types of family relations are associated with delinquency.

Family Socialization and Delinquency

An important process through which family relations are developed is the process of socialization. Socialization refers to the ways that a child is taught cultural roles and normal adult responsibilities, and it involves a variety of interactions, such as touching, holding, hugging, kissing, and talking to the child; listening to the child; feeding and clothing the child; and taking care of the child's need for safety, security, and love. All of these interactions convey important messages to children.

One important socialization process—one that is associated with delinquent behavior—is the process by which social control is developed and implemented in the family. One form of social control consists of the bonds that children develop to the family and family members. Indeed, research indicates that youths who lack closeness to parents or caregivers or who feel there is little family cohesiveness are more likely to engage in delinquency.[69] In addition, some researchers have uncovered a relationship between family conflict, hostility, a lack of warmth and affection among family members, and delinquency,[70] and others have found a relationship between parental criminality and delinquency (the children of parents who are involved in criminal behavior are more likely to engage in delinquency than youths whose parents are not involved in criminality).[71]

Family crises and change also have an effect on family relations and appear to be related to delinquency. There is some evidence that disruptions in family life, such as moving to a new residence, separation or divorce of parents, and family conflict, can produce pressures that push youths toward acting out behaviors.[72] Overall, these studies suggest that when the quality of parent–child relations is poor, delinquency is more likely. They also suggest the converse: Positive parent–child relations act to control delinquent behavior.

Of course, parents exert other forms of social control, including the imposition of discipline. There is considerable evidence that inconsistent discipline, as well as overly harsh or lax discipline, is related to delinquent behavior.[73] Unfortunately, some parents' responses to their children's objectionable behavior are not simply lax or overly harsh but are neglectful or abusive.

<u>Child abuse</u> consists of acts of commission—things done to children. Types of child abuse include physical, sexual, psychological, and emotional abuse. **Neglect** consists of acts of omission; in other words, when a parent or guardian fails to meet the needs of his or her child (e.g., the need for food, shelter, medical care, clothing, education, or affection). Neglect can be physical, emotional, or psychological. However, although child abuse and neglect consist of different forms of behavior, they often occur simultaneously. That is, children who are abused are often neglected as well, and children who are neglected are often abused.

The sad fact is that many children are subjected to abuse and neglect by

child abuse Nonaccidental acts directed against children that result in physical, sexual, psychological, or emotional harm.

neglect The nonaccidental failure of a caregiver to meet the physical, psychological, or emotional needs of a minor or another person in the charge of the caregiver.

Actual Level of Child Maltreatment not Known

In 1999 almost 3 million cases of child abuse and neglect were reported to state agencies. Of these cases, there were 800,000 substantiated or indicated cases of child abuse and neglect reported to child protection agencies. These were cases where there was clear evidence or good reason to believe that maltreatment had occurred or was likely to occur.[74] However, because many incidents of abuse and neglect go unreported to the authorities, the actual level of child maltreatment is unknown.

their parents. Moreover, abuse and neglect have been found to be related to a variety of health, cognitive, educational, and social difficulties, including brain injuries, mental disorders, poor school performance, fear, anger, and antisocial behavior.[75] In one of the most comprehensive studies done on the relationship among child abuse, neglect, and criminality, Cathy Spatz Widom and Michael Maxfield found that youths who were abused or neglected were more likely to engage in delinquent and adult criminal behaviors than youths who were not subjected to such treatment. Youths who were abused or neglected were 59% more likely to be arrested as a juvenile and 28% more likely to be arrested as an adult than youths with no abuse or neglect history. Moreover, youths who had been abused or neglected were 30% more likely to be arrested for a violent crime.[76] These findings support the idea of a cycle of violence (a cycle in which those who experience violence as children are more likely to engage in violence as adults).

Parent or Caretaker Victimization Is a Significant Social Problem

An examination of police data from 12 states found that approximately one in five incidents (19%) of violent crime against youths reported to the police involved the maltreatment of children by their parents or caretakers.[77]

Aside from child abuse and neglect, another indicator of the quality of family life is conflict between parents. Such conflict sometimes takes the form of domestic violence, which usually, although not always, involves males abusing their female companions. Like child abuse, the actual extent of domestic violence is not known because much of this behavior is not reported to the police. Nevertheless, data collected in 1998 as part of the National Crime Victimization Survey (NCVS) indicates that domestic violence is a significant problem.

Domestic Violence Is a Significant Social Problem in the U.S.

In 1998 men were the victims of approximately 160,000 violent crimes committed against them by an intimate partner. In that same year, women were the victims of some 900,000 incidents of violence at the hands of an intimate partner, such as their current or former spouse or boyfriend.[78] Indeed, it has been estimated that a woman has a higher probability of being assaulted by her partner in her own home than a police officer has of being assaulted on the job.[79]

Like child abuse, domestic violence rarely occurs as an isolated incident, possibly because each act of violence tends to reduce the inhibition against violence.[80] Also like child abuse, domestic violence has negative effects on children. Research indicates that children who observe domestic violence tend to be more withdrawn and anxious; they are more likely to perform well below their peers in school performance, organized sports, and social activities; and they are more likely to exhibit aggressive and delinquent behaviors.[81] In addition, research suggests that domestic violence and child abuse are related. For example, research on the co-occurrence of domestic violence and child maltreatment indicate that there is a 30 to 60% overlap between violence directed at women and violence directed at children in the same families.[82]

There is an undeniable connection between violence in families and violent children. Parents who physically abuse their children teach them that it is okay to physically confront other people when angry and that physical violence is an acceptable form of interaction between people. Men who abuse their wives or girlfriends teach their male children that it is acceptable to hit women. By their behavior, they also teach female children that it is acceptable for women to be hit by men and that women should expect and accept physical abuse by men.

Many Americans Have Attitudes that Encourage Spousal Abuse

In a national survey conducted in 1980, 25% of the male respondents and 16% of the female respondents approved of husbands hitting their wives.[83]

In a similar manner, children who have been sexually abused, especially those who are victimized by a close family member or a person who has authority over them, often become perpetrators themselves. The inappropriate form of social and sexual contact inherent in sexual abuse teaches the child victim all of the wrong lessons about appropriate sexuality and, just as importantly, breaks down natural or societal taboos regarding sexual behavior.

■ Schools and Delinquency

School is another institution that has a profound influence on the lives of young people. Today, a much larger percentage of the youth population attends school than was true in the past. For example, in 1890 only 7% of the school-age population attended school. However, owing to the development of compulsory education laws, today over 95% of school-age youths are enrolled in school.[84] Moreover, the amount of time that youths spend in school has been increasing. For example, in 1940 only 38% of persons in the United States between the ages of 25 and 29 had completed high school. By 2001, the percentage of persons between 25 and 29 who had completed high school had risen to almost 88%.[85]

Clearly, school is an important institution because it gives young people the academic skills that are critical for effective participation in today's society. Yet school is important for other reasons as well. It has become the primary socializing institution "through which . . . community and adult influences enter into the lives of adolescents."[86] It is in school that youths learn values, attitudes, and skills, such as punctuality and deference to authority, that are necessary for participation in economic and social life. For many families, much of the interaction between parents and children revolves around school-related issues.[87] Indeed, school is generally acknowledged to affect "the lives of youth in ways which transcend the more obvious influences of academic knowledge acquisition."[88] It is, among other things, the place where youths develop a better sense of who they are and how they stand in relationship to others, both peers and adults.

Aside from its role as a major socializing institution, school is important because it is a primary determinant of both economic and social status. It is for many people the primary avenue to economic and social success, because it confers the credentials necessary for entry into well-paying jobs. Consequently, persons who complete high school, in general, earn significantly more than those who do not.[89]

MYTH VS. REALITY **Overall, the Benefits of Education Vary by Gender and Race**

Myth—Education is a path to social and economic equality for all Americans. **Reality**—Arguably there are many benefits to receiving an education, because income tends to improve with educational attainment. However, not everyone with the same amount of education does equally well. For example, men with professional degrees can expect to earn almost $2 million more than women with equivalent degrees over the course of their lives. Also, non-Hispanic whites, on average, earn substantially more than African Americans or Hispanics during the course of their lives at every level of educational attainment.[90]

Although clear benefits are tied to educational attainment, many youths experience both academic and social problems at school. Consider some of the following statistics cited by Richard Lerner in his book *America's Youth in Crisis*:

- About 25% of the approximately 40 million children and adolescents enrolled in America's 82,000 public elementary and secondary schools are at risk for school failure.
- About 4.5 million children 10–14 years old are one or more years behind in their modal grade level.
- Each year about 700,000 youths drop out of school.
- Unemployment rates for dropouts are more than double those of high school graduates.

- At any point in time, about 18% of all adults 18–24 years old and 30% of dropouts 23–25 years old are under the supervision of the criminal justice system. Among African Americans, the corresponding percentages are about 50% and 75%.[91]

Unfortunately, as these statistics demonstrate, for many youths school is an unpleasant experience characterized by failure and dropout. Furthermore, there is considerable evidence that failure in school and other school-related factors are related to delinquent behavior. The remainder of this section examines some of the school-related factors found to be associated with delinquency.

School Failure and Delinquency

As Gary Jensen and Dean Rojek note, "One of the most persistent findings concerning the school and delinquency is that students who are not doing well in school have higher rates of delinquency than those who are faring better."[92] After reviewing a number of studies, John Phillips and Delos Kelly found a strong relationship between school failure and delinquency. They also discovered that, opposite to the view held by some, school failure precedes delinquency and not the reverse.[93] Moreover, a more recent review of the research in this area by Eugene Maguin and Rolf Loeber found that children who were not doing well academically were almost twice as likely to engage in delinquency as youths earning good grades.[94]

Considerable research has been done on a number of other factors found to be related to school failure and delinquency. For example, research on students' feelings of belonging, attachment, and commitment to school reveal that these factors are related to school violence, vandalism, and delinquency. Studies have also found that students who do not like their teachers or school are more likely to report involvement in delinquency than those who claim to have a strong attachment to their teachers or school.[95] Similarly, in the 1978 *Safe School Study Report*, produced by the National Institute of Education and delivered to the U.S. Congress, student alienation was said to be an important factor linked to school violence and property loss.[96]

It is hardly surprising to find that students who are less committed to school, who are less attached to their teachers and schools, and who feel alienated are more likely to engage in disruptive or delinquent behaviors in and out of school. However, a lack of attachment and commitment to school should not be seen simply as a product of individual failure but as a product of the position students occupy in relation to others in school and as a product of the differences in opportunity available to students in the educational environment. One factor that determines the position that students occupy and the opportunities they may be exposed to is social class.

Social Class

The importance of social class to the school failure–delinquency relationship was first made explicit by Albert Cohen in his 1955 book, *Delinquent Boys*.[97] According to Cohen, school is the one place youths of all social classes come together and compete for status. However, working-class youths are at a disadvantage in this competition because they lack the necessary skills to be successful and because success is defined in middle-class terms. One tempting response available to working-class youths is to form a delinquent subculture (with its own status system) that adheres to nonconventional values and encourages delinquent behavior.

Although Cohen claims that the social class position of working-class youths acts as a stumbling block, other evidence indicates that school practices constitute the main barrier to achievement for many students.[98] Indeed, various practices make it difficult for some students to succeed academically or socially within the school environment. Moreover, a number of these practices have been found to encourage delinquent behavior.

Tracking

Tracking, the sorting of students according to ability or achievement, is one practice that fosters inequality between students and has been found to be associated with delinquent behavior. Common in American schools, tracking typically begins early in students' educational careers. Once assigned to a particular track, students tend to stay in that track.

Although tracking is common, a variety of negative consequences are associated with this practice, including delinquency. Research has found that students in college preparatory tracks get much higher grades than those in noncollege tracks.[99] Also, being placed in a noncollege track has been found to be related to a lack of participation in school activities, lowered self-esteem, school misbehavior, dropping out, and delinquency.[100]

Irrelevant Curricula

Today, school curricula are designed primarily for students who are planning to attend college, while noncollege and technical programs are frequently of inferior quality. As a result, many students have difficulty understanding how much of what they are taught will help them in the future roles they will occupy.[101] This appears to be particularly true of low-income, non-college-track students, who often feel that school is a waste of time. Research indicates that when students feel that school is not relevant to their future job prospects, rebelliousness, school violence, property loss, and delinquency tend to increase.[102]

School Dropouts and Delinquency

For many students, dropping out is seen as a solution to the problems they face in school. Yet dropping out has substantial negative consequences for the individual. Owing to continuing technological sophistication, the skills that people need to function effectively in society have increased. Consequently, dropouts are often woefully unprepared to compete for and maintain positions requiring even basic skills. As a result, they face diminished job prospects and often experience difficulty meeting subsistence income needs.

In addition to the economic consequences of dropping out of school, there are psychological and social consequences associated with leaving school before graduation. Research indicates that dropouts usually regret their decision to drop out,[103] and dropping out appears to be associated with further dissatisfaction with themselves and their environment.[104] In addition, dropouts typically have lower occupational aspirations than those who graduate, and they also have lower educational aspirations for their children.[105]

Not surprisingly, there appear to be important differences between youths who remain in school and those who drop out. Compared with students who stay in school, those who drop out tend to be from low socioeconomic status groups, to be members of minority groups, and to come from homes with fewer study aids and where there are fewer opportunities for non-school-related learning. Dropouts are also more likely to come from single-parent homes, to have mothers who work, and to receive less parental supervision. In addition, compared to those who stay in school, future dropouts are more likely to receive poor grades and low scores on achievement tests in school, are less likely to be involved in extracurricular activities, and are more likely to have school discipline problems.[106]

As noted earlier, dropping out of school clearly makes it more difficult for the individual to obtain a well-paying job in an economy that requires increasing technological sophistication on the part of workers. Not surprisingly, many people assume that there is a direct relationship between dropping out and involvement in delinquency. However, research on the relationship between dropping out and delinquency has produced conflicting results. For example, research conducted by Delbert Elliott and Harwin Voss found that youths who drop out of school engage in more delinquency than those youths who remain in school. Yet their research also found that the level of delinquency among youths who drop out was greatest right *before* they dropped out rather than after. Also, the reasons youths gave for dropping out were directly tied to their negative schooling experience. Specifically, dropouts tended to find school alienating, they were not successful academically, and they associated with peers who were involved in delinquency.[107] The finding that delinquency tends to decrease once youths drop out suggests that negative school experiences encourage youth to engage in delinquency and leave school. Once the condition for these negative experiences is eliminated, however, the motivation to engage in delinquency tends to decrease.

On the other hand, other studies have uncovered evidence that when youths drop out of school, their involvement in criminal activities tends to immediately *increase*. For instance, research conducted by Terrence Thornberry and his colleagues found that dropouts, compared with students who stayed in school, were more likely to engage in crime soon after leaving school. Furthermore, they found consistently higher arrest rates for dropouts until the two groups reached their mid-twenties.[108] Thornberry and his colleagues argue that delinquency immediately increases for school dropouts because leaving school severs ties with an important conventional socializing institution, namely school. Similar results have been noted in a more recent study by Marvin Krohn and his associates who analyzed data from the Rochester Youth Development Study. Their analysis revealed that school dropouts tended to engage in more delinquency and reported more drug use than youths who remained in school.[109]

The literature on school dropouts clearly has policy implications. Some have suggested that compulsory education laws be relaxed based on research findings that dropping out leads to a decrease in delinquent behavior.[110] However, others are strongly opposed to such a policy. They argue that the research is not clear on the relationship between dropping out and delinquency. Their position is based on the research that indicates that dropping out of school is associated with a variety of negative outcomes such as increased involvement in delinquency.

Consequently, they maintain that the focus of policy should be on reducing the dropout rate by improving the ability and willingness of schools to meet the educational needs of *all* students.

■ The Community and Delinquency

As Robert Bursik and Harold Grasmick note in their book *Neighborhoods and Crime*, concern about the influence of the neighborhood on delinquency and crime is hardly new.[111] Indeed, since the development of cities in the United States, considerable emphasis has been placed on the negative influences found within some areas of the urban environment—influences believed to be related to delinquency, adult crime, and a host of other social problems, such as poverty and drunkenness.[112] However, particular attention began to be devoted to some of the more negative aspects of the urban environment around the turn of the century, and it was then that sociologists such as Clifford Shaw and Henry McKay started their pioneering work designed to understand the influence of the community on delinquency.

As part of their efforts, Shaw and McKay mapped areas of Chicago where official delinquents lived. The maps indicated that the highest rates of delinquency were located in deteriorating inner-city areas characterized by decreasing population, a high percentage of foreign-born persons and African American households, little home ownership, low rental values, close proximity to industry and commercial establishments, and an absence of agencies designed to promote community well-being.[113] Also, they discovered that, despite changes in the ethnic composition of these high delinquency areas over time, the delinquency rates remained relatively constant.[114] This convinced Shaw and McKay that the high delinquency rates in certain areas could not be attributed to residents' individual pathologies but resulted from a set of conditions that added up to, in their terminology, "**social disorganization**." They believed that these transitional neighborhoods typically suffered a breakdown in social control characterized by a lack of community cohesiveness, common values, and institutions that prevented delinquency.[115] They also thought that in these neighborhoods gangs and delinquent groups formed where delinquent traditions were passed from one generation to the next (i.e., through "cultural transmission"). According to Shaw and McKay, cultural transmission accounted for high rates of delinquency in these areas, despite changes in their ethnic composition over time.[116]

social disorganization
The state of a neighborhood in which it is unable to exert control over its members. Social disorganization can result from a decrease in population, a large transient population, the existence of groups with differing values, little home ownership, low property values, proximity to industry and commercial establishments, and a lack of agencies designed to promote a sense of community.

The Chicago Area Project Attempted to Put Theory into Practice

Not only did Shaw and McKay want to understand the ways in which conditions in communities influence levels of delinquency, they also were interested in creating programs designed to help disorganized communities respond to delinquency. In 1932, Clifford Shaw initiated the Chicago Area Project in three high-delinquency neighborhoods; the project eventually grew to include more neighborhoods and lasted for approximately 30 years. The intent of the project was to get local residents to respond to local problems, including delinquency, by developing social programs, such as sports and recreation, gang intervention, counseling, discussion groups, and community improvement campaigns.[117] Unfortunately, however, no systematic evaluation of the effectiveness of the Chicago Area Project was conducted. Although some researchers claim that the project likely had little real impact on delinquency because it failed to address the underlying causes of community disorganization,[118] others have suggested that it may have increased community cohesion and led to reductions in delinquency in some areas.[119]

A number of more recent research efforts have documented the importance of community influences on delinquency. Ora Simcha-Fagan and Joseph Schwartz examined the effects of both community and individual factors on delinquency in a study of 12 New York City neighborhoods and found that communities characterized by low organizational participation by residents and the existence of a criminal subculture are likely to experience high levels of delinquency.[120] Robert Bursik examined delinquency rates, along with a variety of other variables, for 74 communities in Chicago from 1930 to 1970.[121] In many of the communities studied, the results were similar to those found by Shaw and McKay: Delinquency rates remained high even though the racial and ethnic composition of the communities changed. Some communities did not exhibit this pattern, however. In one community, an increase in delinquency rates appeared to be influenced by rapid changes in *adjacent* neighborhoods. This finding suggests the importance of examining the ways in which communities are linked together as well as possible causal factors outside the community. Still other research has found that neighborhoods go though cycles of change.[122] Although some of the changes, such as neighborhood deterioration, have been found to be related to increases in delinquency, other changes, such as neighborhood revitalization, may be associated with decreases.[123]

Research on the relationship between community characteristics and delinquency has found that communities do play an important role in the encouragement of delinquent behavior. Communities that are economically deprived appear to be particularly susceptible to high levels of criminality. Indeed, economic deprivation appears to interact with a variety of other community and family characteristics that produce increased levels of delinquency, and this appears to be true regardless of the racial makeup of community members.[124] Poor, physically deteriorating communities, where drugs are readily available to youths, where residents are uninvolved in community organizations, and where criminal subcultures exist are likely to experience high rates of delinquent behavior.[125] In such communities, there appears to be a lack of close personal ties between residents and a variety of criminal role models that result in a lessening of restraints on illegal behavior.

■ Peers and Delinquency

Concern about delinquency as a group phenomenon is hardly new. In fact, concern about youth groups that threatened citizens was a major impetus for the establishment of the first juvenile courts in the United States, which occurred around the beginning of the 20th century. Furthermore, early efforts to study delinquency using a sociological approach typically focused on the group nature of delinquent behavior. The early research of Shaw and McKay in Chicago, as well as some of the more popular theoretical work on delinquency up through the mid-1960s, focused on gangs and other types of delinquent subcultures. There is considerable evidence that youths' peers exert a strong influence on their behavior, and such influence seems apparent when groups engage in delinquent behaviors. There is some evidence that the larger a youth's **accomplice network** (the pool of potential co-offenders a youth associates with) the more likely a youth is to engage in delinquency.[126] Moreover, other research on the relationship between peers and delinquency indicates that many, though not all, youths pass through a progression from no delinquency to more serious delinquency that involves the following steps: (1) youths interact with mildly delinquent peers *before* the onset of delinquency; (2) minor delinquency results from this association; (3) involvement in minor delinquency leads to interactions with more delinquent peers; (4) this interaction leads to involvement in more serious forms of delinquency.[127] However, peer influence may also be present when an individual commits a lone act of deviance. Conversely, simply because two or more youths in the same location are engaging in delinquent acts does not mean that their illegal behavior is a product of some group dynamic.[128]

accomplice network
the pool of potential co-offenders with whom a youth interacts.

Most Deliquency Occurs in a Group Context

Studies that have examined the group context of delinquency suggest that between 62% and 93% of delinquent activity occurs in group settings. A recent study that examined data from a nationally representative sample of youths discovered that 73% of their delinquent offenses were committed in the company of others. This included 91% of all burglaries and alcohol violations, 79% of all drug violations, 71% of all assaults, 60% of all acts of vandalism, and 44% of all thefts.[129] Interestingly, official arrest data suggest a higher percentage of delinquency occurs in groups than does self-report data. The discrepancy is possibly explained by the **group hazard hypothesis**, which holds that delinquency committed in groups is especially likely to be detected and to result in a formal response. If youths who engage in delinquency in groups are more likely to be arrested than lone delinquents, then official arrest data would overestimate group delinquency rates.

group hazard hypothesis
The hypothesis that delinquency committed in groups is more likely to be detected and to result in a formal response by the authorities.

One type of peer group that has received attention in both the research literature and the popular press is the gang. However, there is little agreement among researchers regarding the proper definition of the term *gang*. Sometimes the term is used to describe any congregation of youths who have joined together to commit a delinquent act. At other times it is used to refer to more structured ongoing groups that hold or defend a particular territory. Indeed, some communities do have highly structured groups that fit the popular conception of a gang. Nevertheless, it is worth noting that research on gangs has discovered that gangs vary in a number of ways, including their involvement in delinquent behavior.[130]

Other researchers have noted that some groups actually have very little organization and cohesion. Lewis Yablonsky has argued that a more accurate description of many youth groups involved in crime would be "near group."[131] Such groups are characterized by (1) diffuse role definitions, (2) limited cohesion, (3) impermanence, (4) minimal consensus on norms, (5) shifting membership, (6) disturbed leadership, and (7) limited definition of membership expectations. Yablonsky draws our attention to the fact that not all youth groups that engage in delinquent activities are highly organized and cohesive.

This is not to deny that organized gangs are a significant problem in many communities around the country. However, although gang membership appears to be common in some areas, only a minority of youths belong to gangs at any point in time, even in neighborhoods where gangs exist.[132] Nevertheless, gang membership is linked to drug dealing, vandalism, violent crime, and a variety of other illegal activities, although different gangs may favor different types of criminal activities. Moreover, illegal gang activity appears to be increasing in many cities, as is the number of gang members,[133] despite the fact that more gang members are being arrested, prosecuted, and incarcerated than ever before. There is also considerable evidence that many gangs are becoming more oriented toward violence than in the past.[134] Early studies of gangs found that violent activity was not common in these groups. Furthermore, when gangs did engage in violence, they rarely used firearms.[135] Today, gang activity is more likely to be violent and lethal because of the availability and possession of sophisticated weapons by many gang members and the types of violent behaviors they exhibit (e.g., driveby shootings).[136]

Gang Involvement Is Associated with Criminality

Gang members have higher rates of delinquent involvement than nongang youths. Moreover, in areas where there are gang problems, gang members account for a majority of delinquent acts that are reported; this is particularly true for serious offenses. For example, one study that examined a sample of Denver youths found that 14% were gang members. However, these youths accounted for 79% of serious violent crimes, 71% of serious property offenses, and 87% of drug sales.[137]

A variety of factors are associated with youth involvement in gangs. Gang formation appears to be facilitated by a social context characterized by poverty, inequality, social disorganization, easy access to drugs, and an absence of well-paying jobs.[138] Many youths who join gangs are marginalized within their community. Such youths face a variety of stressful conditions: They have few legitimate opportunities for earning money, and they have few strong bonds to conventional institutions, such as school and family. In many instances, gang members come from destitute and dysfunctional families where parents exhibit poor parenting practices. Moreover, many gang members have family members who are involved in gangs, and they have few positive educational or vocational role models.[139] Research also indicates that learning disabilities and poor academic performance, having friends who engage in problem behaviors, early use of drugs and involvement in violence at a young age are strong predictors of gang involvement.[140] For many marginalized youths, gangs hold out the promise of economic and social opportunities.[141] Gangs also provide youths with a sense of belonging and status[142] as well as protection from other gangs and a means for dealing with a socioeconomic environment that fosters aggression and violence.[143]

Youth involvement in gangs appears to be linked to increases in delinquent behavior. For example, several studies have found that, prior to their joining gangs, gang members' involvement in delinquency was similar to that of non-gang youths. When these youths joined a gang, however, their involvement in delinquency, particularly violent delinquency and drug sales, increased. However, after youths left gangs, with the exceptions of drug sales, their involvement in delinquency decreased.[144]

Organized gangs are one type of peer group seemingly related to increased delinquency, but researchers have noted that other peer groups apparently encourage delinquent behavior as well. As noted earlier, Yablonsky argues that many youth groups involved in delinquency lack the type of organization and cohesion often associated with gangs. In studying delinquency in Flint, Michigan, in the 1960s, Martin Gold found that a considerable amount of delinquent behavior occurred spontaneously in rather loosely structured youth groups.[145] Gold concluded that delinquency of this type resembles a "pickup game" where opportunities for delinquent behavior present themselves to ordinary peer groups, leading to delinquent behavior.

The research of Herman and Julie Schwendinger has done the most to highlight the complexity of peer groups and how various peer groups contribute to delinquency.[146] Their book *Adolescent Subcultures and Delinquency* focuses on the complexity of youth culture and the variety of peer groups that exist. The Schwendingers' observational studies of youth culture in southern California communities reveal that youth culture is far more complex than many assume, a finding supported by other researchers.[147] Rather than being a monolithic entity or strictly based on social class, youth culture comprises a variety of subcultures and peer networks that cut across class lines. The various peer formations have their own designations (e.g., "intellectuals," "greasers," "homeboys," "socialites," "athletes") and have their own distinctive dress and linguistic patterns. They are also accorded differing degrees of status and prestige by their members as well as by other youths and adults.

Although the Schwendingers identified a variety of adolescent subcultures, they noted that there were three persistent types: streetcorner, socialite, and intellectual groups. Moreover, they found that delinquency tended to vary between these types. For example, delinquency is less common among intellectuals, who focus on academic or technical interests (e.g., computers, mathematics, electronics, and physics), have little interest in adolescent fashion, and often spend considerable time doing homework or participating in adult-sponsored activities. In contrast, streetcorner groups are more likely to consist of youths who are economically and politically disadvantaged and who engage in delinquency, including serious delinquency. Falling between intellectual and streetcorner groups, as regards involvement in serious delinquency, are socialite groups, which typically consist of youths from economically and politically advantaged families. Although members of these groups engage in less serious delinquency than members of streetcorner groups, they nevertheless engage in a considerable amount of garden variety delinquency, such as driving violations, vandalism, drinking, petty theft, truancy, gambling, and sexual promiscuity. The Schwendingers' studies not only help us understand the complexity of youth culture, but also indicate the necessity of carefully examining the ways in which youth subcultures encourage or inhibit delinquency among their members.

By highlighting the contextual nature of delinquent behavior and pointing out some of the ways in which economic, political, family, school, neighborhood, and peer contexts influence juvenile offending, this section has offered the reader a chance to develop a broader and more detailed picture of delinquency. The fact that delinquent behavior is affected by factors that lie outside the individual does not imply, of course, that youths are incapable of exercising some degree of free will. Clearly, youths who engage in delinquent behavior make a choice to act as they do, although the decision-making processes that they use could not be described as particularly mature or rational.

■ The Social Context of Delinquency Theory

Social, economic, and political contexts affect not only delinquent behavior, but also the explanations, or theories, of delinquency that gain prominence during particular historical periods. A **theory** of delinquency is a statement or a set of statements that is designed to explain how one or more events or factors are related to delinquency.[148] Such theories are important for two primary reasons. First, they help us make sense of delinquency and understand why it occurs. Second, they guide us in our attempts to reduce crime. Importantly, hypotheses about why delinquency occurs suggest actions we might take in order to reduce it.

Theories of delinquency, like other theories, can be in sharp conflict. Some of them are based on the assumption that political and economic conditions play a crucial role in generating delinquency within American society, whereas others treat delinquency as primarily the product of rational choices made by individual youth. Our object here is not to review these theories (an object more fitting for a book on criminological theory or juvenile delinquency) but to emphasize their role in helping us understand how and why particular factors appear to be related to juvenile crime and what their role is in guiding responses to the

theory A statement or set of statements that is designed to explain how two or more phenomena are related to one another.

problem of delinquency.

Theories of delinquency, like delinquency itself, are products of a particular historical context. Most people, including those who study delinquency, have ideas about why youths engage in delinquent behavior, ideas that are influenced by their particular life experiences.[149] For example, people in medieval Europe favored explanations that treated deviant behavior as a product of otherworldly spirits. Such explanations made perfectly good sense to people within that particular historical context. Today, influenced by ideas derived from the social and behavioral sciences and by popular notions about human behavior, people are much less inclined to explain delinquency in ways that would make sense to medieval Europeans.

This does not mean that at present there is general agreement over the causes of delinquency. Indeed, the historical context within which we live is conducive to the promulgation of a variety of theoretical perspectives on delinquency and considerable debate over its causes. Moreover, the differing theoretical perspectives on delinquency lead to differing, sometimes opposing, responses to delinquent behavior. A theory based on the idea that delinquency is the product of choices made by rational actors leads to policies that stress punishment as a logical response. In contrast, a theory based on the idea that delinquency is the product of the oppression of youths calls forth a different type of policy response,[150] as would a theory that views delinquency as the product of abnormal thinking patterns. In short, social context influences theoretical explanations of delinquency, which in turn suggest various juvenile justice responses to the delinquency problem.

■ The Social Context of Juvenile Justice

The previous sections of this chapter examined the influence of social context on delinquency and on the ways we explain delinquency. Social context also influences the practice of juvenile justice. Indeed, as noted previously, different explanations of delinquency are associated with different juvenile justice responses. Other aspects of the sociopolitical environment influence the practice of juvenile justice as well, and some of these are examined in the remainder of this chapter.

The Influence of the Political Economy on the Practice of Juvenile Justice

The development, management, and distribution of political and economic resources not only has an impact on the behavior of young people in our society, but also influences the operation of important social institutions, including those institutions that comprise the so-called juvenile justice system. The economic resources that are available to law enforcement agencies, juvenile courts, and correctional agencies have a profound effect on the level of staffing, the types of programs operated, and the support (both personnel and materiel) provided to the individuals and agencies charged with responding to delinquency. For example, an increase in the funding for a probation department might allow more staff to be hired and additional types of interventions to be developed. In turn, this may lead to better services being given to youths and their families and a reduction in juvenile offending. Similarly, reductions in funding levels can result in staff reductions, increased caseloads, additional stress for juvenile services workers, a reduction in services and monitoring of clients, and higher levels of recidivism.

Although the effects of changing levels of funding on the operation of juvenile justice agencies are easy to imagine, it should be kept in mind that such changes take place within a highly political environment. In other words, funding allocation decisions are the result of a political process in which various ideas and interests vie for supremacy. Moreover, this political process is found at the state, local, and agency level. At the state level, legislative as well executive decisions can have a profound effect on the levels of funding available to juvenile justice agencies for staff salaries, staff hiring and promotions, staff training, and correctional programs. State-level political decision making plays a particularly important role in those states where juvenile courts and correctional agencies are state operated. However, even when juvenile justice agencies are county or city run, state-level political decisions can affect their operation. For example, political decisions made at the state level can determine the level of monetary and human resources, as well as the types of institutions and programs (and their operation) available to local courts for the treatment of youths.

Local political decision making can also have a profound impact on the operation of juvenile justice agencies. For example, when juvenile justice agencies are county run, political decisions made by county governments play a major role in determining the level of monetary and human resources available to the courts as well as the types of programs operated at the local level. Similarly, political decisions made within juvenile justice agencies can also affect the allocation of monetary and human resources, the amount and types of training given to staff, as well as the types of programs that are operated by those agencies.

Political decision making at the state, local, or agency level is often characterized by considerable conflict between groups possessing different ideologies and interests. State legislatures, county and city governments, executive branch agencies at the state and local level, and local juvenile justice agencies typically encompass groups and individuals that have conflicting views about delinquency and juvenile justice. As a result, conflict is a common element of juvenile justice practice.

The Influence of the Local Community and the Media on the Practice of Juvenile Justice

The local community and the media play a substantial role in determining juvenile justice practice. Many communities possess at least one group that seeks to influence the local response to juvenile delinquency. Indeed, the practice of juvenile justice, like the practice of criminal justice in general, is often a highly political endeavor. In addition, if juvenile justice decision makers, such as prosecuting attorneys and judges, are elected officials, they can be quite sensitive to public perceptions of their performance. As a result, support for various juvenile justice programs and practices can be extended or withdrawn depending on perceptions of "what the public wants."

The decisions made by prosecuting attorneys and judges can also influence the actions of other important decision makers. For example, in some states, court personnel (e.g., court administrators) and corrections staff (e.g., detention unit, probation, and other casework personnel) are, in effect, employees of the judge. As a result, a judge in one of these states can have tremendous influence over the allocation of juvenile justice resources. Even in jurisdictions where judges do not directly control probation and other juvenile justice staff, they are often able to exert considerable influence over other components of the juvenile justice process.

Courts that deal with juveniles are also political entities. Most juvenile court judges are elected officials who not only are accountable to the electorate, but also must contend with other elected officials for coveted tax dollars. Competition for local dollars among local officials and law enforcement or law-related agencies, such as prosecutors, sheriffs, and the courts, continues to force courts to make decisions about juvenile services based on economic considerations, not the juveniles' best interests.

When juvenile courts are faced with budget limits restricting the number of probation staff or the amount of money available for out-of-home placements, certain actions by those courts are predictable. First, judges will limit out-of-home placements by increasing their tolerance for delinquent behavior and/or probation violations before removal. Second, many courts will try to develop dispositional alternatives, such as day treatment programs, expanded use of local detention facilities, and an increase in the number of foster home or group home beds. Although these strategies are fiscally responsible, they allow more serious delinquents to remain in the community. As a result, courts can be caught between the public outcry to get dangerous delinquents off the streets and budgetary restraints that prevent them from heeding the public's demand for protection.

The public influences nonelected juvenile justice decision makers as well as elected ones. Court administrators, middle management personnel such as chief probation officers, and other juvenile justice personnel are often sensitive to public perceptions and demands. For example, complaints by organized community groups or perceptions by local political leaders, including judges and prosecuting attorneys, that juvenile justice practices are at odds with community preferences can produce changes in these practices. Within the community, the media often play an important role in framing public perceptions of both delinquency and the operation of juvenile justice. Indeed, much of what the public knows about juvenile justice practices is a reflection of **media-controlled perceptions**. This is not surprising, because the juvenile justice process has historically been a closed process intended to prevent the stigmatization of children who come before the court. However, the media may not present the most balanced view of juvenile justice practices to the public, preferring instead to focus on the most sensational cases and the most obvious failings. Whether media coverage is balanced or not, the public relies on the media for information about juvenile justice practices, and the information received by the public can be used by influential groups in their efforts to influence juvenile justice practices.

media-controlled perceptions People's perceptions of some phenomenon (or area of interest) that are substantially shaped by the media's coverage of that phenomenon. The public's perceptions of juvenile justice are determined largely by the media, because the public lacks information from other sources about the daily operation of juvenile justice agencies.

◼ Chapter Summary

This chapter examined the context of juvenile justice, including juvenile justice theory and juvenile justice practice. More specifically, it looked at the political, economic, and social context within which delinquency occurs—a context that shapes our ideas about why youths engage in delinquency and influences the ways in which we respond to delinquent behavior. Recognizing and understanding the influence of the political, economic, and social context on delinquency is a requisite for developing more effective juvenile justice practices.

KEY CONCEPTS

accomplice network The pool of potential co-offenders with whom a youth interacts.

broken home A home or family in which at least one natural (biological) parent is absent.

capitalism An economic system based upon three fundamental principles: (1) private ownership of property, (2) competition between economic interests, and (3) personal profit as a reward for economic risk and effort.

child abuse Nonaccidental acts directed against children that result in physical, sexual, psychological, or emotional harm.

family A collective body of any two or more persons living together in a common place under one head or management; a household; those persons who have a common ancestor or are from the same lineage; people who live together and who have mutual duties to support and care for one another.[151]

group hazard hypothesis The hypothesis that delinquency committed in groups is more likely to be detected and to result in a formal response by the authorities.

latchkey children Children who care for themselves at least part of the day due to parental unavailability because of employment.

media-controlled perceptions People's perceptions of some phenomenon (or area of interest) that are substantially shaped by the media's coverage of that phenomenon. The public's perceptions of juvenile justice are determined largely by the media, because the public lacks information from other sources about the daily operation of juvenile justice agencies.

neglect The nonaccidental failure of a caregiver to meet the physical, psychological, or emotional needs of a minor or another person in the charge of the caregiver.

political economy The basic political and economic organization of society, including the setting of political priorities and the management and distribution of economic resources. People's economic status largely determines the opportunities that are available to them, their feeling of political empowerment, their ability to access the political system, and their ability to influence political decisions.

power The ability to influence political and economic decisions and actions. Power is the ability to force, coerce, or influence a person's actions or thoughts.

social class A person's or family's economic, social, and political standing in the community. Social class can be directly related to type of employment, neighborhood, circle of contacts, and friends.

social disorganization The state of a neighborhood in which it is unable to exert control over its members. Social disorganization can result from a decrease in population, a large transient population, the existence of groups with differing values, little home ownership, low property values, proximity to industry and commercial establishments, and a lack of agencies designed to promote a sense of community.

socialism Socialism represents a form of political and economic system that is based on democratic decision making, equality of opportunity for all, collective decision making designed to further the interests of the entire community, public ownership of the means of production, and economic and social planning.

special interest group An organization consisting of persons, corporations, unions, and so on, that have common political and often economic purposes and goals. These individuals or groups come together to use their influence, monetarily and politically, to persuade politicians and other persons in power to act favorably toward them.

theory A statement or set of statements that is designed to explain how two or more phenomena are related to one another.

REVIEW QUESTIONS

1. Why is delinquency theory important in juvenile justice?

_____

_____

_____

_____

2. In what ways does the political economy of the United States affect children?

_____

_____

_____

_____

3. What are the essential elements of a capitalist economy?

_____

_____

_____

_____

4. How do the economic circumstances of adults and children affect the quality of their lives?

_____

_____

_____

_____

5. In what ways is family structure related to delinquent behavior?

_____

_____

_____

_____

6. What types of family relationships have been found to be related to delinquency?

_____

_____

_____

_____

7. What kinds of violence are found in American families and how are they related to delinquent behavior?

_____

_____

_____

_____

8. What are the characteristics of school that encourage both school failure and delinquency?

_____

_____

_____

_____

9. What community characteristics have been found to be related to high levels of delinquent behavior?

_____

_____

_____

_____

10. What are the different types of youth peer groups that exist, and how do these groups encourage delinquency?

_____

_____

_____

_____

11. In what ways do economic conditions influence the operation of juvenile justice agencies?

_____

_____

_____

_____

12. In what ways do politics affect the operation of juvenile justice?

_____

_____

_____

_____

ADDITIONAL READINGS

Bowles, S. and H. Gintis. 1976. *Schooling in Capitalist America: Educational Reform and the Contradictions of Economic Life*. New York: Basic Books.

Edwards, R. C. et al. 1978. *The Capitalist System*, 2nd ed. Englewood Cliffs, NJ: Prentice-Hall.

Ehrenreich, B. 2001. *Nickel and Dimed*. New York: Henry Holt.

Howell, J. C. 2003. *Preventing and Reducing Juvenile Delinquency*. Thousand Oaks, CA: Sage.

McCord, J., ed. 1997. *Violence and Childhood in the Inner City*. New York: Cambridge University Press.

McWhirter, J. et al. 1993. *At Risk Youth: A Comprehensive Response*. Pacific Grove, CA: Brooks/Cole.

National Commission on Children. 1991. *Beyond Rhetoric: A New American Agenda for Children and Families*. Washington, DC: U.S. Government Printing Office.

Schwendinger, H. and J. S. Schwendinger. 1985. *Adolescent Subcultures and Delinquency*. New York: Praeger.

Shulman, B. 2003. *The Betrayal of Work*. New York: The New Press.

Wilson, W. J. 1987. *The Truly Disadvantaged*. Chicago: University of Chicago Press.

NOTES

1. S. J. Pfohl, *Images of Deviance and Social Control* (New York: McGraw-Hill, 1985), 9.
2. For an interesting account of how juvenile justice practitioners may work to manipulate the "system" in order to help clients, see M. D. Jacobs, *Screwing the System and Making it Work, Juvenile Justice in the No-Fault Society* (Chicago: University of Chicago Press, 1990).
3. National Commission on Children, *Ensuring Income Security* (Washington, DC: National Commission on Children, 1993).
4. W. G. Domhoff, *The Powers That Be: Processes of Ruling Class Domination in America* (New York: Random House, 1978).
5. M. Parenti, *Power and the Powerless*, 2nd ed. (New York: St. Martin's Press, 1978).
6. D. S. Eitzen and M. B. Zinn, *Social Problems*, 9th ed. (Boston: Allyn & Bacon, 2004).
7. D. Wrong, *Power: Its Forms, Bases, and Uses* (New York: Harper & Row, 1979), 2.
8. Wrong, *Power*, 23.
9. B. D. Proctor and J. Dalaker, *Poverty in the United States: 2002*. U.S. Census Bureau, Current Population Reports, P60-222 (Washington, DC: U.S. Government Printing Office, 2003).
10. M. Harrington, *The New American Poverty* (New York: Holt, Reinhart & Winston, 1984).
11. Proctor and Dalaker, *Poverty in the United States: 2002*.
12. Proctor and Dalaker, *Poverty in the United States: 2002*.
13. B. Shulman, *The Betrayal of Work, How Low-Wage Jobs Fail 30 Million Americans* (New York: The New Press, 2003).
14. Proctor and Dalaker, *Poverty in the United States: 2002*.
15. K. Short and M. Shea, *Beyond Poverty, Extended Measures of Well-Being, Current Population Reports* (Washington, DC: U.S. Department of Commerce, 1995).
16. T. Swanstrom, P. Dreier, and J. Molenkopf, "Economic Inequality and Public Policy," *City and Community*, 1 (2002): 349–372.
17. R. Wertheimer, et al., *Children in Working Poor Families: Update and Extensions* (Washington, DC: Child Trends, 2002).
18. M. Harrington, *The Other America: Poverty in the United States* (Baltimore: Penguin Books, 1963); Harrington, *The New American Poverty*; W. Ryan, *Equality* (New York: Vintage Books, 1981).
19. National Commission on Children, *Beyond Rhetoric, a New American Agenda for Children and Families*, Final Report of the National Commission on Children (Washington, DC: U.S. Government Printing Office, 1991).
20. C. Jaret et al., "Black-White Income Inequality and Metropolitan Socioeconomic Structure," *Journal of Urban Affairs*, 25 (2003): 305–333; T. Swanstrom et al., "Economic Inequality and Public Policy."
21. T. Swanstrom et al., "Economic Inequality and Public Policy."
22. T. Swanstrom et al., "Economic Inequality and Public Policy."
23. T. Buss and F. S. Reddman, *Mass Unemployment: Plant Closings and Community Mental Health* (Beverly Hills, CA: Sage, 1983); L. E. Gary, "Correlates of Depressive Symptoms among a Select Population of Black Men," *American Journal of Public Health* 75 (1985): 1220–1222.
24. N. Galambos and R. Silbereisen, "Income Change, Parental Life Outlook, and Adolescent Expectations for Job Success," *Journal of Marriage and the Family* 49 (1987): 141–149; J. Lempers et al., "Economic Hardship, Parenting, and Distress in Adolescence," *Child Development* 60 (1989): 25–39.
25. Proctor and Dalaker, *Poverty in the United States: 2002*.
26. R. Wertheimer, M. Long, and J. Jager, *Children in Working Poor Families: Update and Extensions*.
27. B. Shulman, *The Betrayal of Work*.
28. G. J. Duncan and J. Brooks-Gunn, *Consequences of Growing Up Poor* (New York: Russell Sage Foundation, 1997); National Commission on Children, *Beyond Rhetoric*.
29. J. D. McLeod and M. J. Shanahan, "Poverty, Parenting, and Children's Mental Health," *American Sociological Review* 58 (1993): 351–366.

30. B. Bradbury and M. Jantti, *Child Poverty Across Industrialized Nations*, Innocenti Occasional Papers, Economic and Social Policy Series no. 71 (Florence, Italy: UNICEF International Child Development Centre, 1999).

31. H. Gerth and C. W. Mills, *Character and Social Structure: The Psychology of Social Institutions* (New York: Harcourt, Brace, & World, 1953), 313.

32. See E. Anderson, *Streetwise: Race, Class, and Change in an Urban Community* (Chicago: University of Chicago Press, 1990); E. Anderson, "Violence and the Inner City Street Code," in *Violence and Childhood in the Inner City*, ed. J. McCord (New York: Cambridge University Press, 1997); M. Sanchez-Jankowski, "Ethnography, Inequality, and Crime in the Low Income Community," in *Crime and Inequality*, eds. J. Hagan and R. D. Peterson (Stanford, CA: Stanford University Press, 1995).

33. D. S. Elliott and D. Huizinga, "Social Class and Delinquent Behavior in a National Youth Panel: 1976–1980," *Criminology* 21 (1983): 149–177.

34. G. R. Jarjoura et al., "Growing Up Poor: Examining the Link between Persistent Childhood Poverty and Delinquency," *Journal of Quantitative Criminology* 18 (2002): 159–187.

35. Eitzen and Zinn, *Social Problems*, 329.

36. R. Rapp, "Family and Class in Contemporary America," in *Rethinking the Family: Some Feminist Questions*, eds. B. Thorne and M. Yalom (New York: Longman, 1982), 168–187.

37. W. Gove and R. Crutchfield, "The Family and Juvenile Delinquency," *Sociological Quarterly* 23 (1982): 301–319.

38. K. N. Wright and K. E. Wright, *Family Life, Delinquency, and Crime: A Policymaker's Guide, Research Summary* (Washington, DC: Office of Juvenile Justice and Delinquency Prevention, 1994).

39. National Commission on Children, *Beyond Rhetoric*.

40. J. Fields, *Children's Living Arrangements and Characteristics: 2002*, Current Population Reports (Washington, DC: U.S. Department of Commerce, 2003).

41. J. Fields and L. M. Casper, *America's Families and Living Arrangements, Current Population Reports* (Washington, DC: U.S. Department of Commerce, 2001); National Commission on Children, *Beyond Rhetoric*.

42. National Commission on Children, *Beyond Rhetoric*.

43. W. J. Wilson, *The Truly Disadvantaged, The Inner City, the Underclass, and Public Policy*. (Chicago: The University of Chicago Press, 1987).

44. R. Canter, "Family Correlates of Male and Female Delinquency," *Criminology* 20 (1982): 149–167; S. Glueck and E. Glueck, *Unraveling Juvenile Delinquency* (Cambridge, MA: Harvard University Press, 1950); L. E. Wells and J. Rankin, "Families and Delinquency: A Meta-Analysis of the Impact of Broken Homes," *Social Problems* 38 (1991): 71–90; J. Q. Wilson and R. Herrnstein, *Crime and Human Nature* (New York: Simon & Schuster, 1985).

45. Wells and Rankin, "Families and Delinquency."

46. J. Toby, "The Differential Impact of Family Disorganization," *American Sociological Review* 22 (1957): 505–512.

47. R. Chilton and G. Markle, "Family Disruption, Delinquent Conduct, and the Effects of Subclassification," *American Sociological Review* 37 (1972): 93–99.

48. T. P. Thornberry, C. A. Smith, C. Rivera, D. Huizinga, and M. Stouthamer-Loeber, *Family Disruption and Delinquency, Juvenile Justice Bulletin* (Washington, DC: Office of Juvenile Justice and Delinquency Prevention, 1999).

49. R. E. Johnson, "Family Structure and Delinquency: General Patterns and Gender Differences," *Criminology* 24 (1986): 65–84.

50. Gove and Crutchfield, "The Family and Juvenile Delinquency."

51. S. A. Cernkovich and P. C. Giordano, "Family Relationships and Delinquency," *Criminology* 25 (1987): 295–321; J. H. Laub and R. J. Sampson, "Unraveling Families and Delinquency: A Reanalysis of the Gluecks' Data," *Criminology* 26 (1988): 355–380.

52. T. Hirschi, *Causes of Delinquency* (Berkeley, CA: University of California Press, 1969).

53. M. Rutter and H. Giller, *Juvenile Delinquency: Trends and Perspectives* (New York: Gilford Press, 1984).

54. R. Loeber and M. Stouthammer-Loeber, "Family Factors as Correlates and Predictors of Juvenile Conduct Problems and Delinquency," in *Crime and Justice: An Annual Review of Research*, vol. 7, eds. M. Tonry and N. Morris (Chicago: University of Chicago Press, 1986).

55. D. J. Curran and C. M. Renzetti, *Social Problems, Society in Crisis*, 5th ed. (Boston: Allyn and Bacon, 2000).

56. Bureau of Labor Statistics, *Employment Characteristics of Families in 2002*, (Washington, DC: U.S. Department of Labor, 2003).

57. National Commission on Children, *Just the Facts*.

58. National Commission on Children, *Beyond Rhetoric*.

59. S. Vandivere et al., *Left Unsupervised: A Look at the Most Vulnerable Children*, Child Trends Research Brief (Washington, DC: Child Trends, 2003).

60. J. Ehrle, G. Adams, and K. Tout, *Who's Caring for Our Youngest Children? Child Care Patterns of Infants and Toddlers*, Occasional Paper No. 42 (Washington, DC: The Urban Institute); J. Galtry, "Child Health: An Underplayed Variable in Parental Leave Policy Debates?" *Community, Work and Family*, 5 (2002): 257–278.

61. S. L. Kagan and M. J. Neuman, "Defining and Implementing School Readiness: Challenges for Families, Early Care and Education, and Schools," in *Healthy Children 2010: Establishing Preventive Services*, eds. R. P. Weissberg, et al., (Thousand Oaks, CA: Sage, 1997).

62. Vandivere et al., *Left Unsupervised*.

63. T. Hirschi, "Crime and Family Policy," in *Juvenile Delinquency: A Justice Perspective*, eds. R. A.Weisheit and R. G. Culbertson (Prospect Heights, IL: Waveland Press, 1985).

64. Laub and Sampson, "Unraveling Families and Delinquency."

65. T. Vander Ven et al., "Home Alone: The Impact of Maternal Employment on Delinquency," *Social Problems* 48 (2001): 236–257.

66. M. Colvin and J. Pauly, "A Critique of Criminology: Toward an Integrated Structural-Marxist Theory of Delinquency Production," *American Journal of Sociology* 90 (1983): 513–551.

67. J. Hagan, *Structural Criminology* (New Brunswick, NJ: Rutgers University Press, 1989), 13.

68. J. Hagan et al., "The Class Structure and Delinquency: Toward a Power-Control Theory of Common Delinquent Behavior," *American Journal of Sociology* 90 (1985): 1151–1178; J. Hagan et al., "Class in the Household: A Power-Control Theory of Gender and Delinquency," *American Journal of Sociology* 92 (1987): 788-816; J. Hagan et al., "Clarifying and Extending Power Control Theory," *American Journal of Sociology* 95 (1990); 1024–1037; Also see B. S. Blackwell, "Perceived Sanction Threats, Gender, and Crime: A Test and Elaboration of Power-Control Theory," *Criminology* 38 (2000): 439–488.

69. C. Hanson et al., "Demographic, Individual, and Familial Relationship Correlates of Serious and Repeated Crime among Adolescents and Their Siblings," *Journal of Counseling and Clinical Psychology* 52 (1984): 528–538; Hirschi, *Causes of Delinquency*; Laub and Sampson, "Unraveling Families and Delinquency"; J. H. Rankin and R. Kern, "Parental Attachments and Delinquency," Criminology 32 (1994): 495–515; R. Smith and J. Walters, "Delinquent and Nondelinquent Males' Perceptions of Their Fathers," *Adolescence* 13 (1978): 21–28.

70. D. P. Farrington, "The Development of Offending and Antisocial Behaviour from Childhood: Key Findings from the Cambridge Study in Delinquent Development," *Journal of Child Psychology and Psychiatry* 360 (1995): 929–964; D. P. Farrington, "The Explanation and Prevention of Youthful Offending," in *Delinquency and Crime: Current Theories*, ed. J. D. Hawkins (Cambridge, England: Cambridge University Press, 1996); D. P. Farrington, "Criminological Psychology: Individual and Family Factors in the Explanation and Prevention of Offending," in *Working with Offenders*, ed. C. R. Hollin (Chichester, England: John Wiley, 1996); S. Henggeller, *Delinquency in Adolescence* (Newbury Park, CA: Sage, 1989); R. J. Sampson and J. H. Laub, *Crime in the Making* (Cambridge, MA: Harvard University Press, 1993); Wright and Wright, *Family Life, Delinquency, and Crime*.

71. Laub and Sampson, "Unraveling Families and Delinquency."

72. P. Amato and B. Keith, "Parental Divorce and the Well-Being of Children: A Meta-Analysis," *Psychological Bulletin* 110 (1991): 26–46; J. H. Grych and F. D. Fincham, "Marital Conflict and Children's Adjustment: A Cognitive-Contextual Framework," *Psychological Bulletin* 108 (1990): 267–290; M. Hershorn and A. Rosenbaum, "Children of Marital Violence: A Closer Look at the Unintended Victims," *American Journal of Orthopsychiatry* 55 (1985): 260–266; P. Jaffe et al., "Similarities in Behavior and Social Maladjustment among Child Victims and Witnesses to Family Violence," *American Journal of Orthopsychiatry* 56 (1986): 142–146; Wright and Wright, *Family Life, Delinquency, and Crime*.

73. R. Agnew, "Physical Punishment and Delinquency: A Research Note," Youth and Society 15 (1983): 225–236; Laub and Sampson, "Unraveling Families and Delinquency"; G. R. Patterson et al., "A Developmental Perspective on Antisocial Behavior," *American Psychologist* 44 (1989): 329–335; Wright and Wright, *Family Life, Delinquency, and Crime*.

74. R. Chalk et al., *The Multiple Dimensions of Child Abuse and Neglect: New Insights into an Old Problem*. Child Trends Research Brief. (Washington, DC: Child Trends, 2002).

75. Chalk et al., *The Multiple Dimensions of Child Abuse and Neglect*.

76. C. S. Widom and M. G. Maxfield, *An Update on the Cycle of Violence*, National Institute of Justice Research in Brief (Washington, DC: U.S. Department of Justice, 2001).

77. D. Finkelhor and R. Ormrod, *Child Abuse Reported to the Police*, Juvenile Justice Bulletin (Washington, DC: Office of Juvenile Justice and Delinquency Prevention, 2001).

78. C. M. Rennison and S. Welchans, *Intimate Partner Violence*, Bureau of Justice Statistics Special Report (Washington, DC: U.S. Department of Justice, 2000).

79. A. Browne, *When Battered Women Kill* (New York: The Free Press, 1987); C. Saline, "Bleeding in the Suburbs," *Philadelphia Magazine*, March 1984, 81–85, 144–151.

80. R. J. Gelles and C. P. Cornell, *Intimate Violence in Families* (Beverly Hills, CA: Sage, 1985).

81. J. R. Kolbo, "Risk and Resilience Among Children Exposed to Family Violence," *Violence and Victims* 11 (1996): 113–128.

82. National Clearinghouse on Child Abuse and Neglect Information, *In Harm's Way: Domestic Violence and Child Maltreatment* (Washnigton, DC: Author, 2003).

83. Straus, *Behind Closed Doors*.

84. U.S. Office of Education, *Digest of Educational Statistics* (Washington, DC: U.S. Government Printing Office, 1969); National Center for Educational Statistics, *Digest of Education Statistics, 2002* (Washington, DC: U.S. Department of Education, 2003).

85. National Center for Educational Statistics, *Digest of Education Statistics, 2002*.

86. K. Polk and W. E. Schafer, eds., *Schools and Delinquency* (Englewood Cliffs, NJ: Prentice Hall, 1972).

87. G. Johnson et al., *Delinquency Prevention: Theories and Strategies* (Washington, DC: Office of Juvenile Justice and Delinquency Prevention, 1979).

88. H. P. Elrod and P. C. Friday, "Delinquency Reduction through School Organizational Change: Some Thoughts on the Relationship between Theory, Process and Outcomes," paper presented at the American Society of Criminology, Atlanta, October 1986.

89. National Center for Education Statistics, *Annual Earnings of Young Adults, by Educational Attainment,* Indicator of the Month (Washington, DC: Author, 1999).

90. J. C. Day and E. C. Newburger, *The Big Payoff: Educational Attainment and Synthetic Estimates of Work-Life Earnings*, Current Population Reports (Washington, DC: U. S. Census Bureau, 2002).

91. R. M. Lerner, *America's Youth in Crisis* (Thousand Oaks, CA: Sage, 1995), 4–5.

92. G. Jensen and D. Rojek, *Delinquency and Youth Crime*, 3rd ed. (Prospect Heights, IL: Waveland, 1998), 273.

93. C. Phillips and D. H. Kelly, "School Failure and Delinquency: Which Causes Which?" *Criminology* 17 (1979): 194–207.

94. E. Maguin and R. Loeber, "Academic Performance and Delinquency," in *Crime and Justice: A Review of Research*, vol. 20, ed. M. Tonry (Chicago: University of Chicago Press, 1996).

95. R. Agnew, "A Revised Strain Theory of Delinquency," *Social Forces* 64 (1985):151–167; Hirschi, *Causes of Delinquency*; M. J. Hindelang, "Causes of Delinquency: A Partial Replication," *Social Problems* 21 (1973): 471–487.

96. D. Boesel et al., *Violent Schools—Safe Schools: The Safe Schools Study Report to the Congress*, vol. 1 (Washington, DC: U.S. Department of Health, Education and Welfare, 1978).

97. A. K. Cohen, *Delinquent Boys* (New York: The Free Press, 1955).

98. S. Bowles and H. Gintis, *Schooling in Capitalist America: Educational Reform and the Contradiction of Economic Life* (New York: Basic Books, 1976); M. Gold, "School Experiences, Self-esteem, and Delinquent Behavior," *Crime and Delinquency* 24 (1978): 290–308; Polk and Schafer, eds., *Schools and Delinquency*.

99. W. E. Schafer et al., "Programmed for Social Class Tracking in High School," in *School and Delinquency*, eds. K. Polk and W. E. Schafer (Englewood Cliffs, NJ: Prentice-Hall, 1972).

100. Schafer et al., "Programmed for Social Class Tracking in High School"; J. Oakes, *Keeping Track: How Schools Structure Inequality* (New Haven, CT: Yale University Press, 1985).

101. Boesel et al., *Violent Schools—Safe Schools;* Polk and Schafer, *Schools and Delinquency*; E. L. Wertleib, "Juvenile Delinquency and the Schools: A Review of the Literature" *Juvenile and Family Court Journal* 33 (1982): 15–24.

102. Boesel et al., *Violent Schools—Safe Schools*; Polk and Schafer, *Schools and Delinquency*; A. Stinchcombe, *Rebellion in a High School* (Chicago: Quadrangle Press, 1964).

103. S. S. Peng and R. T. Takai, *High School Dropouts: Descriptive Information from High School and Beyond*, ERIC No. ED 236 3666 (Washington, DC: National Center for Education Statistics, 1983).

104. R. B. Ekstrom et al., "Who Drops Out of High School and Why? Findings from a National Study," *Teacher's College Record* 87 (1986): 356–373.

105. J. McWhirter et al., *At-Risk Youth: A Comprehensive Response* (Pacific Grove, CA: Brooks/Cole, 1993).

106. Ekstrom et al., "Who Drops Out of High School and Why?"; P. Kaufman et al., *Dropout Rates in the United States: 2000* (Washington, DC: National Center for Education Statistics, 2001).

107. D. S. Elliott and H. L. Voss, *Delinquency and Dropout* (Lexington, MA: Lexington Books, 1974).

108. T. P. Thornberry et al., "The Effect of Dropping Out of High School on Subsequent Criminal Behavior," *Criminology* 23, no. 1 (1985): 3–18.

109. M. T. Krohn, T. Thornberry, L. Collins-Hall, and A. Lizorre, "School Dropout, Delinquent Behavior, and Drug Use," in *Drugs, Crime, and Other Deviant Adaptations: Longitudinal Studies*, ed. H. Kaplan (New York: Plenum Press, 1995).

110. J. Toby, *Violence in Schools* (Washington, DC: National Institute of Justice, 1983).

111. R. J. Bursik and H. G. Grasmick, *Neighborhoods and Crime: The Dimensions of Effective Community Control* (New York: Lexington Books, 1993).

112. T. J. Bernard, *The Cycle of Juvenile Justice* (New York: Oxford University Press, 1992); D. Rothman, *The Discovery of the Asylum* (Boston: Little, Brown & Co., 1971).

113. C. R. Shaw and H. D. McKay, "Social Factors in Juvenile Delinquency," in *National Commission on Law Observance and Enforcement, Report on the Causes of Crime*, vol. 2 (Washington, DC: U.S. Government Printing Office, 1931).

114. C. R. Shaw and H. D. McKay, *Juvenile Delinquency in Urban Areas*, rev. ed. (Chicago: University of Chicago Press, 1972).

115. Shaw and McKay, *Juvenile Delinquency in Urban Areas*.

116. Shaw and McKay, *Juvenile Delinquency in Urban Areas*.

117. R. J. Lundman, *Prevention and Control of Juvenile Delinquency*, 3rd ed. (New York: Oxford University Press, 1993).

118. J. Snodgrass, *The Jackroller at 70* (Chicago: University of Chicago Press, 1982).

119. S. Schlossman and M. Sedlack, *The Chicago Area Project Revisited* (Santa Monica, CA: Rand, 1983); A. Sorrentino, *Organizing Against Crime: Redeveloping the Neighborhood* (New York: Human Sciences Press, 1977).

120. O. Simcha-Fagan and J. E. Schwartz, "Neighborhood and Delinquency: An Assessment of Contextual Effects," *Criminology* 24 (1986): 667–703.

121. R. J. Bursik, Jr., "Ecological Stability and the Dynamics of Delinquency," in *Communities and Crime*, ed. A. J. Reiss, Jr., and M. Tonry (Chicago: University of Chicago Press, 1986).

122. L. Schuerman and S. Kobrin, "Community Careers in Crime," in *Communities and Crime*, eds. A. J. Reiss, Jr., and M. Tonry (Chicago: University Of Chicago Press, 1986).

123. S. C. McDonald, "Does Gentrification Affect Crime Rates?" in *Communities and Crime*, eds. A. J. Reiss, Jr., and M. Tonry (Chicago: University of Chicago Press, 1986).

124. J. D. Hawkins et al., "A Review of Predictors of Youth Violence," in *Serious and Violent Juvenile Offenders*, eds. R. Loeber and D. P. Farrington (Thousand Oaks, CA: Sage, 1998); L. J. Krivo and R. D. Petersen, "Extremely Disadvantaged Neighborhoods and Urban Crime," *Social Forces* 75 (1996): 619–650; F. Peeples and R. Loeber, "Do Individual Factors and Neighborhood Context Explain Ethnic Differences in Juvenile Delinquency?" *Journal of Quantitative Criminology* 10: 141–157.

125. K. Harries and A. Powell, "Juvenile Gun Crime and Social Stress: Baltimore, 1980–1990," *Urban Geography* 15 (1994): 45–63; J. D. Hawkins et al., "A Review of Predictors of Youth Violence"; W. Spelman, "Abandoned Buildings: Magnets for Crime?" *Journal of Criminal Justice* 21 (1993): 481–493.

126. M. Warr, "Organization and Instigation in Delinquent Groups," *Criminology* 34 (1996): 11–37; also see D. L. Haynie, "Friendship Networks and Delinquency: The Relative Nature of Peer Delinquency," *Journal of Quantitative Criminology* 18 (2002): 99–134.

127. D. Elliott and S. Menard, "Delinquent Friends and Delinquent Behavior: Temporal and Developmental Patterns," in *Delinquency and Crime: Current Theories*, ed. J. D. Hawkins (Cambridge, England: Cambridge University Press, 1996).

128. M. W. Klein, "On the Group Context of Delinquency," *Sociology and Social Research* 54 (1969): 63–71.

129. M. Warr, "Organization and Instigation in Delinquent Groups."

130. J. Fagan, "The Social Organization of Drug Use and Drug Dealing among Urban Gangs," *Criminology* 27 (1989): 633–669; C. R. Huff, "Youth Gangs and Public Policy," *Crime and Delinquency* 35 (1989): 524–537; L. Yablonsky, "The Delinquent Gang as a Near Group," *Social Problems* 7 (1959): 108–117.

131. L. Yablonsky, "The Delinquent Gang as a Near Group."
132. F. A. Esbensen et al., "Gang and Non-gang Youth: Differences in Exploratory Variables," *Journal of Contemporary Criminal Justice* 9 (1993): 94–116; K. G. Hill et al., *Early Precursors of Gang Membership: A Study of Seattle Youth*, Juvenile Justice Bulletin (Washington, DC: Office of Juvenile Justice and Delinquency Prevention, 2001); T. P. Thornberry et al., "The Role of Gangs in Facilitating Delinquent Behavior," *Journal of Research in Crime and Delinquency* 30 (1993): 55–87
133. T. P. Thornberry, "Membership in Youth Gangs and Involvement in Serious and Violent Offending," in *Serious and Violent Juvenile Offenders*, eds. R. Loeber and D. P. Farrington (Thousand Oaks, CA: Sage, 1998); D. L. Weisel, "The Evolution of Street Gangs: An Examination of Form and Variation," in *Responding to Gangs, Evaluation and Research*, eds. W. L. Reed and S. H. Decker (Washington, DC: National Institute of Justice, 2002).
134. A. Egley and A. K. Major, *Highlights of the 2001 National Youth Gang Survey*, OJJDP Fact Sheet (Washington, DC: Office of Juvenile Justice and Delinquency Prevention, 2003); W. B. Miller, *Violence by Youth Gangs* (Washington, DC: U.S. Government Printing Office, 1975).
135. W. B. Miller et al., "Aggression in a Boy's Street-Corner Group," *Psychiatry* 24 (1961): 283–298.
136. M.W. Klein et al., eds., *The Modern Gang Reader* (Los Angeles: Roxbury, 1995).
137. Thornberry, "Membership in Youth Gangs and Involvement in Serious and Violent Offending."
138. G. D. Curry and I. A. Spergel, "Gang Involvement and Delinquency among Hispanic and African-American Adolescent Males," *Journal of Research in Crime and Delinquency* 29 (1992): 273–291; P. I. Jackson, "Crime, Youth Gangs, and Urban Transition: The Social Locations of Postindustrial Development,"; J. D. Vigil, *Barrio Gangs* (Austin, TX: Texas University Press, 1988).
139. Curry and Spergel, "Gang Involvement and Delinquency Among Hispanic and African-American Adolescent Males;" M. Klein, *Street Gangs and Street Workers* (Englewood Cliffs, NJ: Prentice-Hall, 1971); Thornberry, "Membership in Youth Gangs and Involvement in Serious and Violent Offending;" Vigil, *Barrio Gangs*.
140. Hill et al., "Early Precursors of Gang Membership: A Study of Seattle Youth"; Thornberry, "Membership in Youth Gangs and Involvement in Serious and Violent Offending."
141. M. S. Jankowski, *Islands in the Street: Gangs and American Urban Society* (Berkeley, CA: University of California Press, 1991).
142. Vigil, Barrio Gangs; W. Miller, "Lower Class Culture as a Generating Milieu of Gang Delinquency," *Journal of Social Issues* 14 (1958): 5–19.
143. Jankowski, *Islands in the Street*.
144. Thornberry, "Membership in Youth Gangs and Involvement in Serious and Violent Offending."
145. M. Gold, *Delinquency in an American City* (Monterey, CA: Brooks/Cole, 1970).
146. H. Schwendinger and J. Schwendinger, *Adolescent Subcultures and Delinquency* (New York: Praeger, 1985).
147. R. J. Berger, "Adolescent Subcultures, Social Type Metaphors, and Group Delinquency," in *The Sociology of Juvenile Delinquency*, ed. R. J. Berger (Chicago: Nelson-Hall, 1991); G. Schwartz and D. Merten, "The Language of Adolescence: An Anthropological Approach to the Youth Culture," *American Journal of Sociology* 72 (1967): 453–468. Also see Warr, "Organization and Instigation in Delinquent Groups" and Haynie, "Friendship Networks and Delinquency: The Relative Nature of Peer Delinquency" for complimentary studies.
148. Adapted from D. J. Curran and C. M. Renzetti, *Theories of Crime* (Boston: Allyn & Bacon, 2001).
149. J. R. Lilly et al., *Criminological Theory: Context and Consequences* (Newbury Park, CA: Sage, 1989).
150. For an explication of a theory focusing on child oppression, see R. M. Regoli and J. D. Hewitt, *Delinquency in Society* (New York: McGraw-Hill, 2003).
151. H. C. Black, *Black's Law Dictionary*, 4th ed. (St. Paul, MN: West Publishing Co., 1968).

Early Juvenile Justice: Before the Juvenile Court

4

Chapter Objectives

After studying this chapter, you should be able to

- Describe how the young have historically been viewed in Western societies
- Describe the characteristics of colonial villages that encouraged conformity among children and adults
- Explain how the social context of the early 1800s led to the development of the first specialized correctional institutions for youths in the United States
- Describe the original goals of the houses of refuge and the mechanisms they used to reform youths
- Describe the shortcomings of the houses of refuge
- Describe the development and operation of placing out
- Describe the shortcomings of placing out
- Describe the role that John Augustus played in the development of probation
- Describe the focus of early juvenile probation programs
- Describe the rationale for the development of reform schools, industrial schools, and training schools
- Describe the operation of early reform schools, industrial schools, and training schools
- Describe the development and operation of early institutions for females

Chapter Outline

■ Introduction

From a historical perspective, the notion of juvenile delinquency and the establishment of separate formal social control institutions for youths are recent developments. Throughout most of recorded history, childhood did not enjoy the special status we now accord it. Our modern view that childhood and adolescence are special times during which the young need nurturing and guidance for their healthy development did not exist until the later part of the Middle Ages. Before this, the young were seen either as property or as miniature adults who were expected to assume the same responsibilities as other adults by the age of 5 or 6.[1] Because people did not recognize childhood as a distinct period in human development, they did not see a need to create a separate legal process to deal with youths who violated community norms or laws.

Throughout History Youth Have Sometimes Been Spared the Harshest Punishments

Although a separate legal process for the young did not exist at this time, there is evidence that younger people were sometimes spared the harshest punishments because of their age.

As the ideas of childhood and adolescence developed, however, the development of institutions designed for children began to be seen as a logical response to youth problems. As a result, during the 1800s a number of institutions designed to control problem youths and rectify their behavior were established in the United

States. However, the development of these early institutions, like the development of contemporary correctional responses to youths, is best understood by examining the historical context within which these developments occurred. This chapter begins by examining the position of youths in early European societies and the role that families played in social control. Next, it examines the colonial period and the types of social control mechanisms that pervaded colonial life. The final section looks at the social context that led to the development of the first specialized correctional institutions for youths in the United States.

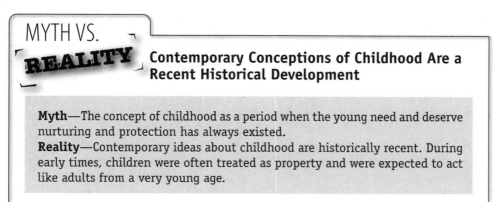

MYTH VS. REALITY — Contemporary Conceptions of Childhood Are a Recent Historical Development

Myth—The concept of childhood as a period when the young need and deserve nurturing and protection has always existed.

Reality—Contemporary ideas about childhood are historically recent. During early times, children were often treated as property and were expected to act like adults from a very young age.

■ Families and Children in Developing European Societies

The Medieval View of Childhood

How adults have responded to children throughout history has depended on existing conceptions of the young—conceptions that are best understood within their historical context. During the Middle Ages life was difficult for most people. During the fifth century, the collapse of the Western Roman Empire, which had dominated Western civilization for centuries, created many uncertainties for people in western Europe. Although our knowledge of children's lives during this time is limited, there are some indications that the young often fared poorly. For example, there is evidence that infanticide (i.e., the practice of killing children), which had been common since antiquity, continued to be practiced during the Middle Ages and beyond.[2] Historical evidence from this period indicates that mothers sometimes deliberately suffocated their offspring or abandoned them in the streets or latrines. Infants who were born deformed or were felt to be too burdensome were particularly vulnerable. According to one priest in 1527, "The latrines resounded with the cries of children who have been plunged into them."[3] In other instances, destitute parents abandoned their unwanted children or took them to monasteries to be raised by monks.[4]

Life in Europe during the Middle Ages was very different than it is today. Although the nobility, some members of the clergy, and skilled craftsmen were relatively well off, many people led a precarious existence. There were wars, rebellions, disagreements between citizens, and fires (most structures were made of wood), all of which posed regular threats to personal safety. Homelessness, drunkenness, and crime were problems in the cities. In addition, sanitation practices represented a serious threat to public health. Refuse was discarded in the streets, which contained open sewers. The refuse heaps attracted rats and other rodents and served as a breeding ground for disease.[5]

Life tended to be short, and the young were particularly at risk from various threats, including plagues and famine. Moreover, common child-rearing practices led to the premature death of many children. Swaddling (i.e., wrapping a child tightly in a long bandage) was a common practice. This prevented the child from wandering away, crawling into a sewer, or knocking over candles, which was a major concern, not only because homes and buildings were usually constructed of wood, but because they also contained a variety of highly flammable materials.[6] Swaddling also made the child easier to handle and protected it from air, sunlight, and soap, which were felt to present a threat to healthy growth.[7] Other common practices included feeding children from a horn and wrapping food in rags for them to suck. Such practices seem appalling by today's standards, but they were accepted methods of caring for the young during this time.[8]

Although the Middle Ages is often seen as a period of extreme hardship during which little social and intellectual progress occurred, by the later part of the Middle Ages Europe experienced a revival in trade and town life.[9] Moreover, by this time the concepts of childhood and adolescence had emerged, at least among some of the wealthier members of medieval society. Although historians are not in agreement as to the exact period during which childhood and adolescence began to be recognized as distinct stages of life, there is some evidence that many people began to see the young differently by the 1400s. For example, accounts of village life in southern France during the 14th century indicate that parents displayed love for their children and many grieved when their children died.[10] In addition, medical and scientific texts, literary works, and folk terminology used during the later Middle Ages in England indicate that many adults were concerned with helping children and youths make the transition to adulthood.[11] Less frequently were the young seen as miniature adults or as simply the property of their parents. Childhood was beginning to be accepted as a unique period in an individual's development, a period during which the individual should be molded and guided in order to become a moral and productive member of the community.

Changes in the conception of childhood coincided with a variety of other changes in social, political, and economic life. **Feudalism**, a form of social organization characterized by relationships between nobles (i.e., knights and greater lords), the exploitation of the peasantry by the nobility, and subsistence farming, dominated political and economic life in Europe for centuries. However, substantial increases in the size of the population, as well as the use of agricultural methods that depleted the land, strained the ability of the feudal agriculture system to feed everyone, particularly when harvests were poor.[12] Furthermore, famine and disease claimed many lives, particularly the lives of children. It is estimated that as much as two-thirds of the population died before 20 years of age, and the average life expectancy was about 30 years.[13]

Despite high mortality rates, however, the number of people in Europe grew during the 1500s and 1600s, which placed considerable pressure on both towns and rural areas. Simultaneously, political and economic power began to shift away from the church to a few monarchs who were molding strong centralized states

feudalism A system of land tenure governed by relations between nobles and lesser nobles whereby land (fief) was given by an overlord to a lesser noble (vassal) to hold (not to own) in return for military service. The land could then be given to still lesser nobles. This system depended on the agricultural production of the peasants (serfs) who worked the land.

and to a growing merchant and capitalist class. As a result, feudal forms of political and economic organization, which had been based on a subsistence system of agriculture, began to give way to a trade-oriented capitalistic system based on cash crops and the consolidation of land.[14]

Because of these broad changes in the economic and political organization of European states, large numbers of peasants were displaced from the land as landowners developed a capitalistic pasturage system geared more toward the production of cash crops. A result was the migration of peasants into the cities that strained the cities' resources and produced conflict between established workers and artisans. This conflict occurred because the influx of peasants threatened to drive down city wages and disrupt the status quo. The result was rising levels of public disorder and crime.[15]

In order to control the threat of the "dangerous classes," towns and cities passed legislation intended to restrict or discourage the immigration of peasants. For example, "poor laws" were enacted that were intended to prevent peasants from obtaining citizenship, restricted their ability to join guilds, and, in some cases, barred them from entering the cities. In England, the Statute of Artificers was passed in 1562, which restricted access to certain trades and compelled many of the rural young to remain in the countryside.[16] Despite the efforts of city dwellers and the wealthy to restrict movement into the cities, however, there was no real alternative for many of the rural poor. As a result, cities continued to grow, and urban institutions, such as the guilds and the courts, came under increasing pressure.[17]

Early Social Control: The Role of the Family

During the Middle Ages and up through the 1500s and 1600s, the family served as the primary unit of social control. Within the family, power was typically vested in the father, and both the mother and the children were expected to obey his commands.[18] However, the nuclear family of today was relatively rare. Rather, a typical family consisted of an extended group of related people and others. The father, particularly if he was a merchant, might spend considerable time away from the home. High mortality rates meant that either parent might have died and been replaced by a stepparent. In addition, servants, apprentices, and journeymen sometimes lived in the household, and neighbors were always visiting.[19] Consequently, children were exposed to a large number of adults who played a role in their socialization and control.

Because of the binding out and apprenticeship systems, a family unit often was responsible for the socialization and control not only of those born into the family, but also other youths. The practice of binding out young people consisted of placing them in the service of others for a specific period of time, from a week to many years. Although some adults became servants when they were elderly and some children when they were as young as 7 years of age, most who entered servitude did so in their teens. Binding out presented a way for poor youths to earn and save money, obtain room and board, learn a skill, prepare for adulthood, and meet a potential wife or husband.[20]

Although binding out was common among the lower class, apprenticeships were usually reserved for the wealthy. Typically, an apprenticeship began when a youth was 14 to 18 years of age and lasted 10 to 12 years. A distinguishing char-

family A social unit responsible for, among other things, the raising and socialization of children. Historically, the family in its various forms has been society's tool for teaching children how to act, what to believe, and what values to hold. (See also Chapter 3.)

binding out The placing of a youth or adult in the service of others for a specific period of time, from a week to many years. Although children as young as 7 years old and elderly adults could be placed as servants, the practice was mostly restricted to children in their teens. Binding out was a way for a poor youth to earn and save money, obtain room and board, learn a skill, prepare for adulthood, and meet a potential wife or husband.

apprenticeship A period set by contract during which a novice worked for a master in return for instruction in the master's trade. Apprenticeships were typically reserved for middle-class or wealthy youths They typically began when youths were 14 to 18 years old and lasted 10 to 12 years.

acteristic of apprenticeships was that the master was obligated to teach the apprentice a trade as well as control their behavior.[21]

Despite the existence of the binding out and apprenticeship systems and the recognized sovereignty of adults—particularly the sovereignty of adult males over their children, apprentices, and servants—undisciplined and destitute youths, social and political unrest, and crime were common problems. Indeed, the number of poor and destitute people (many of whom were children) engaged in begging, prostitution, and petty crime in London had reached such a level that in 1556 an institution specifically designed for these people, the Bridewell, was established.[22] However, youth crime and beggary were not confined to London. Consequently, in 1576 Parliament passed a law that created similar institutions in each county in England.[23]

The Bridewell Was an Early Correctional Institution for Youths and Adults

The first Bridewell was opened by private businessmen under a contractual agreement with the local government. It was designed to control a range of people both young and old who were poor, idle, and involved in petty crimes, prostitution, and lewd conduct. These institutions stressed work and, in some instances, children were committed to the Bridewell by their parents who believed these institutions would help the child learn the value of labor.

Institutions built along the Bridewell model also spread to the continent. In 1595, the best known of these, the Amsterdam House of Corrections, was established to deal, in part, with the youth crime problem of the day. It and other similar institutions were intended to instill discipline and teach youths to be hard-working and productive members of society.[24]

Early youth correctional institutions accepted a wide range of children—the destitute, the infirm, the needy, and serious offenders. Some youths were placed in these institutions by parents who believed that through institutionalization their children would learn to be more obedient. Whether youths benefited from such placement is debatable. What is not debatable is that the founders of these correctional institutions saw them source of cheap labor for local industries.[25]

■ Children in the New World

The Place of Children in Colonial Life

Population pressures, high mortality rates, the desire for land, political unrest, religious persecution, and the desire for greater wealth and influence led to a period of exploration and colonization after 1500. Children were counted on to play an important role in the expansion of European influence and became an integral part of plans to colonize the New World. Indeed, for some people the colonization of the New World was viewed as an opportunity for children to be involved in productive work. Merchants saw children as potential providers of the labor that would be needed to produce the many goods necessary for survival in the colonies. Some people felt that work was good for children, because it kept them occupied and insulated them from the temptations of the street. For others, such as the Puritans, the New World represented an environment where their religion might prosper free of religious persecution and where the souls of their children might be saved.[26]

Initially, the mid-Atlantic colonies were settled primarily by individuals— farmers, artisans, and indentured servants—instead of families. However, the need for child labor in the colonies, along with worsening conditions in Europe, soon led to the development of a variety of mechanisms by which youths could make their way to the New World. In Europe, "spirits" (commission agents of shipowners and merchants) were responsible for signing up young men, young women, and even children for the voyage to America. Sometimes the spirits bribed, tricked, or coerced people into making the voyage. In other cases, they found desperate people who desired a better life. In exchange for their passage to the colonies, those who signed with the spirits agreed to a period of indenture (contract), usually 4 years, as a way of paying for their passage.[27] To ensure that they did not renege on their contract, people were often imprisoned until they could be transported to the colonies. Furthermore, laws were passed in the colonies that allowed for the recording and enforcement of these contracts.[28] Indeed, it is believed that the majority of people who colonized the Chesapeake region during the 17th century came over as indentured servants.[29]

Conditions in Europe Encouraged Immigration to the New World

It has been estimated that about 80,000 people, or 2% of the population, left England between 1620 and 1642. Of the total who left, about 58,000 are believed to have gone to North America.[30]

Although many youths came to the New World as servants or apprentices, others came as part of a family. This was particularly true in New England, because members of Puritan families typically sailed together.[31] In other instances, poor, destitute, and wayward children were transported from European countries in order to reduce the costs of providing relief to the poor or incarcerating them and trying to correct their behavior. In addition, the colonies needed workers, and children were seen as a cheap source of labor.[32]

The great demand for labor in the colonies led not only to an influx of European children, but also to the forced transportation of large numbers of African children. In 1619, the same year that the colony of Virginia obtained an agreement for the regular shipment of orphans and destitute children from England, the first African slaves arrived in that colony. Most of these slaves were children. Slave traders thought that children would bring higher prices, and more of them could be transported in the limited cargo space of ships. Furthermore, the slave traders encouraged childbearing in order to increase their capital. As Barry Krisberg and James Austin note, "African babies were a commodity to be exploited just as one might exploit the land or the natural resources of a plantation, and young slave women were often used strictly for breeding."[33]

The Social Control of Children in the Colonies

Those who settled in the New World brought a variety of ideas with them about childhood as well as European mechanisms for responding to those who violated social and legal rules. In the New World as in the Old, the discipline of children was stern. Parents believed that corporal punishment was the appropriate method of teaching children an appreciation for correct behavior, sound judgment, and respect for their elders. The maxim of parents in colonial America was "spare the rod and spoil the child."[34]

MYTH VS. REALITY A Number of Factors Encouraged Conformity Among Colonial Youths

Myth—The stern child-rearing practices found in colonial towns and villages prevented youths from engaging in deviant behavior.
Reality—Low levels of serious misbehavior on the part of youths were most likely the product of a number of features of colonial life. Nevertheless, youthful misconduct was a concern of many adults. In writing about the youths of his town in the 1700s, religious reformer Jonathan Edwards wrote that "licentiousness for some years greatly prevailed among the youth of the town," many of whom were "very much addicted to night walking, and frequenting the tavern, and lewd practices, wherein some, by their example exceedingly corrupt others." The youths often gathered "in conventions of both sexes, for mirth and jollity, which they called frolics" and in which they spent "the greater part of the night . . . indeed family government did too much fail in the town."[35]

In addition to stern child-rearing practices, several other features of colonial life encouraged conformity among young and old. Even though the colonists came from varied social class backgrounds, they generally shared beliefs and values.[36] Also, most of them resided in small towns, where there was regular interaction between people intimately familiar with one another. Because the necessities for survival were produced locally, children spent much of their time involved in economic production and under the watchful eyes of adult members of the community.[37] In short, the organization of community life was a major factor in the establishment of conformity among community members.

Two other institutions dominated colonial social life: the family and the church.[38] During the colonial period, the family was the basic unit of economic production[39] and the primary mechanism through which social control was exerted.[40] Survival of the family and the community depended on the ability of all able-bodied persons to contribute to the production of needed commodities. Consequently, a primary responsibility of the family was to oversee the moral training and discipline of the young and perpetuate the values that supported existing social institutions and economic production. This responsibility was so important that parents who failed to instill in their young respect for community values were subject to punishment by the authorities.[41] According to the Laws and Liberties of Massachusetts (1648 edition),

> *The selectmen of every town are required to keep a vigilant eye on the inhabitants to the end that the fathers shall teach their children knowledge of the English tongue and of the capital laws, and knowledge of the catechism, and shall instruct them in some honest lawful calling, labor, or employment. If parents do not do this, the children shall be taken away and placed (boys until twenty-one, girls until eighteen) with masters who will teach and instruct them.*[42]

The family, owing to the use of binding out and apprenticeship, played a critical role in the training and control of children generally. The binding out and apprenticeship systems were responsible for the immigration of many youths and served as a primary means by which colonial youths learned a skill, earned and saved money, and prepared themselves for adulthood. They were also mechanisms of social control. Masters were responsible for the discipline of those within the household, including servants and apprentices. Children who were difficult to handle or needed supervision were often bound over to masters for care. As in Europe, apprenticeships were typically reserved for wealthy youths, and masters were obligated to teach their apprentices a trade. Under the binding out system, masters were not required to teach servants a trade, and boys were often given farming tasks while girls were assigned domestic duties.[43]

Religion, particularly in New England, was another powerful force shaping social life in the colonies. Contemporary concerns about the separation of church and state were nonexistent. Regular church attendance was expected,[44] and religious beliefs dominated ideas about appropriate behavior. Little differentiation was made between sin and crime.[45] What was believed to be immoral was also unlawful and subject to punishment by the authorities.

In colonial towns, children had few incentives or opportunities to act in ways that deviated from family and community expectations.[46] Gossip; ridicule; stern discipline by parents; work; and regular supervision by parents, masters, and others in the community ensured that most children did not stray too far from community norms. When children committed minor rule violations, their parents or masters were expected to punish them. In other instances, youths who violated community rules might be sent to the town minister for a stern lecture and a warning to avoid further infractions. However, rule violations that by today's standards would be considered minor were seen as serious by people in colonial times. For example, in some colonies a child who rebelled against his or her parents could be put to death.[47] Colonial codes contained a long list of capital offenses, such as murder, horse stealing, arson, robbery, burglary, and sodomy. According to the Massachusetts Bay Colony Laws of 1660, for sodomy, a capital crime, children under 14 were to be "severely punished" but not executed; for cursing and smiting parents, a capital crime, only those "above sixteen years old, and of sufficient understanding" could be put to death. Those above 16 years of age could be executed for arson, being a stubborn or rebellious son, and "denying the Scriptures to be the infallible word of God," all of which were capital crimes.[48]

Banishment was another method used to permanently eliminate offenders from the community.[49] On the frontier, however, banishment could be tantamount to a death sentence. In addition, punishments such as fines, whipping, branding, and the placement in stocks or a pillory served as reminders to both young and old that violations of community norms were held to be serious matters.[50] Incarceration was also used during this period. Some offenders were placed in the small jails that existed in many towns, but incarceration was used primarily to hold debtors and other offenders awaiting trial. The use of incarceration as a popular punishment for offenders did not come until later.[51]

Despite the relative homogeneity of the population of the colonies and the variety of social control mechanisms in place, social unrest still occurred,[52] including servant revolts, slave revolts, strikes, demands for political representation, and discontent among the poor. As the colonies grew, the antagonism between the poor and wealthy intensified. Although some servants became landowners after their period of servitude was over, most continued to be poor, often becoming tenants on large plantations and providing the owners with a cheap source of labor. Moreover, the towns were typically run by wealthy elites who maintained their power through intermarriage and other forms of alliance between families.

In colonial America, the majority of people were poor, and many of the poor resented their treatment by the wealthy. Even during the Revolutionary War, those who supported independence were concerned about the possibility of mutiny and the lack of support for their cause. This was particularly true in the South, where many poor people felt that a victory by the colonies would simply mean a change of one master for another. Indeed, the problems of poverty and social unrest prompted the growing cities to establish poorhouses to provide for and control the elderly, widows, the physically challenged, orphans, the unemployed, war veterans, and new immigrants. A letter to Peter Zinger's *New York Journal* in 1737 described poor street children in New York during this period: "an Object in Human Shape, half starv'd with Cold, with Cloathes out at the Elbows, Knees through the Breeches, Hair standing on end . . . . From the age about four to Fourteen they spend their Days in the Streets . . . then they are put out as Apprentices, perhaps four, five, or six years."[53]

■ Families and Children in the 19th Century

By the early 1800s, colonial social organization was changing as a result of continuing immigration and economic and social developments occurring throughout the new country. The diversity of the population had increased, and English settlers had been joined by Scotch-Irish, German, Irish, and French immigrants.[54] Family-based production, which had characterized colonial social life, was giving way to a factory-based system of production in the growing towns. Moreover, the factory-based system was beginning to supplant the binding out system, which had been the primary means by which children entered the labor force.[55] The factory and the factory boss gradually took the place of the master.

As more parents, particularly fathers, and children began to leave the home for work in factories, fundamental changes occurred in the relationships between family members and in the role of the family in controlling the behavior of children. Before the Industrial Revolution, parents were involved in making occupational choices for their children, typically when the children were 10 to 12 years of age. Moreover, there were important class differences between children. Poor children were bound out or placed in apprenticeships, whereas wealthier children were sent to schools or began preparing for careers in the military, government, and medicine by their late teens. Even within the lower classes, there were various distinctions between youths, because those who were apprenticed in the better trades would be assured of higher incomes. And, of course, there were tremendous differences in the occupational opportunities available to males and females. As a result, children developed specific identities based on their gender and class position. The advent of industrialization, along with increases in the population, led to the erosion of these traditions, however. The professions became more difficult to get into because of the greater number of people vying for entry. More often youths became employees rather than apprentices. Finally, young people began to remain at home longer, and the period of transition into full adulthood lengthened.[56]

Another change that accompanied industrialization was the decline of the large extended family. A typical family of the colonial period comprised parents, children, relatives, and perhaps servants and apprentices, but this type of family began to be replaced by the nuclear family. Also, the work that people performed for pay began to be carried out in factories and other job sites away from the home. Not only did men work outside the home, but in many cases women and children began to work outside the home as well, due to the development of industrial machinery. For example, Samuel Slater, the "father of American manufacturers," initially employed a workforce of nine boys in his Rhode Island factory. By 1801, this work-force had expanded to 100 youths between the ages of 4 and 10.[57] As apprenticeships became less important as a way of learning a trade and earning money and as the extended family declined in importance, child-care became much less of a collective responsibility (shared by family members and the community) and more of a responsibility of the mother.

Although industrialization brought prosperity to some, it was accompanied by growing concern about social unrest and crime. There were increasing numbers of poor people as well as people suffering from mental illness. In the smaller, more homogeneous towns and villages of colonial times, people had accepted these problems and attempted to devise a community response. Although some offenders and those considered rogues or vagabonds were banished or barred from communities, no systematic attempts were made to isolate the deviant or the dependent. By the early 1800s, however, notions of dependency and deviancy had changed dramatically. Deviant children were more often seen as products of pauperism and many believed that they could be transformed into productive hard-working adults.[58] It was within this context that new institutions for the control of children were developed.[59]

The Development of the First Juvenile Institutions

The United States grew significantly during the late 1700s and early 1800s. Along with the growth in the population came an increase in industry and the development of larger towns and cities. The population also became more diverse, and the conflict that had characterized colonial times continued to be a feature of American life. There was antagonism not only between the political and social elite and the poor, but also between the citizens of the new country, Native Americans, and African Americans.[60]

The population of the United States grew rapidly after the revolution. In 1790, there were fewer than 4 million Americans, and the majority of them lived within 50 miles of the Atlantic Ocean.[61] Colonial America was rural; even the largest cities were small. When Washington became president, "only two hundred thousand Americans lived in towns with more than twenty-five hundred people . . . . In 1790, no American city had more than fifty thousand residents."[62] However, by 1820 the population of New York City was about 120,000 and was growing rapidly due to immigration.[63] Indeed, by 1830, the United States was home to some 13 million people, and more than 4.5 million lived west of the Appalachian Mountains. Moreover, immigration and slavery changed the ethnic composition of many communities, making them less homogeneous than during colonial times. In contrast to the influx of immigrants, the population of Native Americans living east of the Mississippi was reduced from 120,000 to approximately 30,000 in 1820. They were either killed or forced to leave their land.[64]

Although the size of the United States grew considerably between 1790 and 1820, after 1820 the population increased even more dramatically. By 1860, the population of the United States had grown to some 31 million people. Moreover, the diversity of the population continued to expand. Increasingly, newer arrivals came from the Scandinavian countries and from Ireland and Germany.[65]

The United States presented tremendous opportunities for people and many prospered. For others, however, the American dream was elusive. The New World had always had its share of poor people. Many of those who arrived during the colonial period came as servants, and many of the immigrants to the United States were poor people seeking a better life. Both before the American Revolution and afterward, the poor outnumbered the wealthy. Moreover, poor and affluent alike faced a variety of insecurities. The affluent were worried about the threat of crime, protests, riots, periodic slave revolts, and various forms of political resistance to their leadership.[66] In addition, periodic economic downturns threatened the livelihood of many people and left many out of work.[67] Furthermore, there was no safety net to rely on in difficult times. There was no minimum wage, health care plan, social security, or pension program that might protect workers' interests.

MYTH VS. REALITY The Problem Behaviors of Youths Have Been an Ongoing Concern

Myth—In the good old days, people did not worry about youth crime because it was not a problem.

Reality—The criminal behavior of young people has been a recurring concern in larger American cities since at least the 1800s.

The Houses of Refuge

Accompanying the changes in the social and economic life of the growing cities were a host of social problems such as vagrancy, drunkenness, and crime, including crimes committed by children. It was in response to the growth in juvenile crime that the first correctional institution specifically for youths in the United States was developed. This institution, the House of Refuge, was established in New York City in 1825. Soon other **houses of refuge** were founded in other cities, such as Boston (1826) and Philadelphia (1828).[68] In the 1840s, houses of refuge opened in Rochester, Cincinnati, and New Orleans; in the 1850s, they opened in Providence, Baltimore, Pittsburgh, Chicago, and Saint Louis. By 1857, the refuge movement had expanded to the point that a national convention of refuge superintendents was held in New York. According to its committee on statistics, 17 institutions for youths were in operation in the United States. These institutions had land and buildings worth almost $2 million and total annual expenditures of some $330,000.[69]

The development of the houses of refuge represented a new approach to problem children. This new approach relied on formal child-care institutions as opposed to families, the church, and informal community controls.[70] The houses of refuge were championed by wealthy reformers who saw youth crime and waywardness as a natural outgrowth of the pauperism prevalent in cities like New York. These reformers were mostly men from established families, and their intent was to oversee the moral well-being of the community and to develop policies that would protect their way of life from the threat posed by the poor.[71] These reformers decried the unwillingness of the criminal courts to deal with children who committed minor offenses. The incarceration of youths who committed any type of criminal offense was possible because children fell under the jurisdiction of the criminal courts, but the reformers realized that adult correctional institutions did nothing to reform youth.[72] According to one of these reformers, Cadwallader Colden, who was both the mayor of New York and the presiding judge of the city's municipal court, "The penitentiary cannot but be a fruitful source of pauperism, a nursery of new vices and crimes, a college for the perfection of adepts in guilt."[73] What was needed, according to the reformers, were separate institutions for youths that could shield them from the corrupting influences of adult institutions, control them, and transform them into hard-working citizens.

house of refuge
A privately operated child-care institution developed in the United States for children who engaged in illegal behavior or were felt to be at risk of such behavior. Although the initial goal was to educate and train youths for productive lives, houses of refuge became little more than prisons characterized by harsh discipline and hard work.

In order to accomplish the reformers' mission, youths were placed in the houses of refuge for indeterminate periods of time or until their 18th or 21st birthday. Placement did not require a court hearing,[74] and a child could be committed to a refuge by a constable or a parent or on the order of a city alderman.[75]

Despite their name and the lofty goals touted by the reformers, the houses of refuge were run more like prisons and were built for the secure confinement of their inmates.[76] Indeed, the daily operation of these institutions focused more on control than reform.[77] While there, children were exposed to a daily regimen that included hard work, military drill, enforced silence, and both religious and academic training.[78] Another common practice was to operate shops within the refuge that were run by outside contractors. In these shops, children labored 8 hours a day producing goods such as shoes, brass nails, and furniture.[79] In return, the refuge was paid 10 to 15 cents per youth per day. This arrangement allowed the refuge to cover a substantial percentage of its daily operating expenses[80] and supplement the private and public funds that it received.[81] However, the child workers were often punished when they failed to meet production quotas.[82]

When children rebelled against their treatment, discipline was often severe. Minor rule violations were typically dealt with by loss of supper and an early bedtime. More serious infractions resulted in a variety of punishments, including bread and water diets, placement in solitary confinement (sometimes in a "sweat box"), manacling with a ball and chain, placement in a straightjacket, hanging by the thumbs, or whipping with a cat-o'-nine-tails.[83] When "reformed," boys were often indentured to farmers or tradesmen,[84] and girls were placed in domestic service.[85] Boys who were more difficult to manage were sometimes sent on extended whaling voyages.[86]

Despite the initial optimism concerning the ability of the houses of refuge to control and reform youths, a less sanguine view of these institutions soon developed among some reformers. Indeed, the houses of refuge faced several problems that contributed to their failure to achieve their main goals. For one thing, they admitted a range of youths, many of whom had committed no criminal offense. Thus, children who had no means of support, those who were neglected, and those who had committed criminal offenses were confined together.[87] Indeed, the diversity of youths placed in the houses of refuge made it impossible for them to meet the needs of their residents.

The length of placements also created problems. Because many children were placed for long periods of time, the houses of refuge soon became overcrowded. The number of children requiring placement always surpassed the number of available beds. Moreover, the harsh treatment inflicted on children by staff in these institutions and by the masters to whom they were indentured led many children to run away.[88] Indeed, the treatment youths received was hardly conducive to reform and was thought by some reformers to compound the problems that the houses of refuge were initially intended to solve.

Placing Out

Although institutional responses to juvenile waywardness and delinquency spread during the last half of the 1800s, many reformers recognized the inability of the houses of refuge to accommodate the large numbers of children needing placement. For example, Charles Loring Brace, director of the Children's Aid Society in New York, noted that the harsh punishments and military regimen found in these institutions had failed to transform their residents into law-abiding citizens. According to Brace, the longer a youth remained in a house of refuge, "the less likely he is to do well in outside life."[89]

One early attempted remedy for the problems of the houses of refuge was **placing out** which involved the placement of children on farms in the West and Midwest (see **Exhibit 4-1**). Placing out was first used to assist youths in finding employment once they were ready to leave the houses of refuge.[90] However, placing out picked up momentum by the mid-1800s. Faced with continuing juvenile delinquency problems exacerbated by population growth, the dislocations produced by the Civil War, and what many thought was the failure of the houses of refuge to achieve their goals, some reformers came to view placing out as a solution to the growing number of problem children.[91] Agencies such as the Children's Aid Society of New York, the Boston Children's Aid Society, and other child relief organizations began to promote placing out as the ideal way to deal with problem children, and they hired agents to help place these children with farm and ranch families in the West.[92]

Placing Out Was One Response to Problem Children

The Children's Aid Society of New York placed some 60,000 youths in western homes in the 30 years after its founding in 1853.[93]

placing out The placement of children on farms and ranches in the Midwest and West. Placing out, which was popularized by private charitable organizations in the mid-1800s (and which lasted until 1929), was an attempt to remove vagrant and wayward children from the corrupting influences of the cities and locate them in what was believed to be a more wholesome environment. The idea of using the frontier as a geographical "cure" for societal ills was consistent with the views of Frederick Jackson Turner, a 19th-century historian who believed that the American experience of the frontier helped shape the American character by cultivating individualism, industry, and self-reliance—the very values that the social reformers wanted all children to acquire.

Advocates of placing out claimed that it had several advantages over the houses of refuge. In their view, it removed children from the corrupting influences of the cities, which they saw as breeding grounds for idleness and crime. In addition, they pointed to the failure of the houses of refuge to control or rehabilitate youths[94] and suggested the rural countryside was an ideal environment for instilling in children the values the reformers cherished—discipline, hard work, and piety.

Yet, the wholesome characteristics of the country life depicted by the advocates of placing out was a far cry from the realities faced by many of the children who were sent to the West. Although some children were put into caring homes and were treated as members of the family, others were not so lucky. Considerable evidence exists that many of the children placed out were required to work hard for their keep, were abused, or were never fully accepted into the family.[95] Moreover, there was little systematic follow-up to determine if the children were treated humanely by those responsible for their care.[96]

Even when it was popular, placing out was not without its critics. Some pointed to the fact that children who were placed out engaged in delinquency after their placement.[97] Others, including supporters of juvenile institutions, made the accusation that the practice was contaminating the West with unfit youths, an accusation that struck a chord with many people in the areas to which the children were being shipped.[98] Despite these concerns, however, placing out continued to be used until after the turn of the century.

HOMELESS CHILDREN.—A company of homeless boys and girls from the Children's Aid Society of New York, will arrive at the Hight House, Dowagiac, January 31st. All interested in this work of charity are invited to meet the Agent as above, or apply to H. M. Farnam, Dowagiac.

Trains full of hope carried children away to new lives in frontier America. The lucky found loving homes and better futures, but the unknown past still draws them.

Maurice de Leleu, third from left in the back row, is shown in 1921 before boarding an Orphan Train.

BY ELAINE SCHAD
Special to the Star-Telegram

The trains rolled into hundreds of small towns across the rural American West and Midwest.

Among the passengers: thousands of orphaned or otherwise homeless children, primarily from the slums of big eastern cities, severed from family roots and facing an unknown future with families they had never met in places they had never seen.

The fortunate ones were picked early. The others re-boarded the trains and went on to the next stop, and the next, their stories eventually becoming lost in the haze of history.

They rode the Orphan Trains.

For Rachel Duesman, the ride ended July 3, 1922, when she was placed with the

Rachel Duesman visits the Gainesville depot she first saw from a "baby train."

family of Edward and Angelica Blumberg near Pilot Point.

A photocopy of a long-yellowed baptismal certificate is the only key to her roots she has found in years of searching,

but she still hopes to find a relative someday.

"I prayed to the Lord that he would help me find my brother, but I guess maybe I'm not supposed to see him," said Duesman, 73. "But God has blessed me all the way through. I've just had a good life."

Duesman was among 150,000 children who rode the trains to new lives between 1854 and 1929 in one of the largest social experiments in American history.

It is estimated that between 500 and 600 Orphan Train riders remain alive, and many of them and their descendants are trying to find their lost roots.

Marion Strittmatter of Fort Worth, a friend of Duesman's, was 3 when she became the 13th mouth to feed for a family from Orange, Texas. Strittmatter came West on one of the so-called baby trains

Exhibit 4-1 Article on Orphan Trains.
Source: Reprinted with permission from E. Schad, *Lost Ties,* 9/29/91, Star Telegram.

Probation

Another effort to deal with troubled children was initiated by a Boston shoemaker, John Augustus. Augustus spent considerable time observing the operation of the Boston Police Court and became convinced that many minor offenders could be reformed. In 1841, as a result of his concern and willingness to work with offenders, Augustus was permitted to provide bail for his first **probation** client, a drunkard who showed remarkable improvement during the period he was supervised. The court was impressed with Augustus's work and permitted him to stand bail for other minor offenders, including children. During Augustus's lifetime, he provided supervision to over 2,000 people, very few of whom violated the terms of their release.[99]

probation The supervised release of an individual by a court. Probation was first promoted by John Augustus, a Boston shoemaker. Today children are usually returned to their parents by courts on conditions of release called "rules of probation," and court or state employees are the supervisors or monitors of the persons on probation. In the 19th century, the monitoring of youths on probation was largely done by private citizen volunteers or volunteer societies.

that carried younger children.

"I'm still searching," said Stritt-matter, 69, a grandmother of 28 children, two of whom are adopted. "Mother brought me there [a New York orphanage] when she was 24 and I was 18 months old. She was Catholic, and that's all I know."

What Strittmatter has held on to these many years is a tidbit of knowledge: that her mother's name was also Marion. After decades of trying to come to grips with the reasons that would prompt a mother to leave her baby, she has reconciled herself to believing that she was loved.

"I decided that when I got older and held my first baby in my arms," she said. "Giving me her name — Marion. She tried, and she cared."

The children who rode the Orphan Trains were being sent to what social reformers expected would be better lives. Many were children of the 16 million European immigrants who passed through Ellis Island and quickly found that New York's streets were not paved with gold.

Poor sanitation was universal; cholera and tuberculosis were rampant. Many parents were sick, destitute and unable to care for themselves, much less their starving children.

By 1853, thousands of children of all ages were roaming the streets of the big East Coast cities, which had no social mechanism to provide for their care. Moved by their plight, a young minister named Charles Loring Brace founded the Children's Aid Society and devised the placing-out program, which began in New York City in 1854.

The first Orphan Train left New York City for Dowagiac, Mich., in 1854, carrying 46 boys aged 7 to 15. By 1910, Texas had received 2,100 children, according to records of the Children's Aid Society. The children usually went to farming areas, where they could work and be absorbed into rural families.

"The trains were the gimmick. It was the placing-out system that was the key," said Alice Ayler, 72, of Oklahoma City. Along with two other children, Ayler rode what she said was the last Orphan Train.

"Regardless of how our lives turned out, good or bad, it was for a purpose we were put in those circumstances," she said of the trains.

Ads placed in local newspapers attracted families — or farmers looking for free help.

Exhibit 4-1 (Continued)

After Augustus died, the Boston Children's Aid Society and other volunteers carried on his work.[100] Eventually, in 1869, the state of Massachusetts formalized the existing volunteer probation system by authorizing visiting probation agents to work with adults and child offenders who showed promise. Under a probation arrangement, youths were allowed to return home to their parents, provided they obeyed the law.[101] In addition to the supervision of youths, the visiting probation agents were charged with investigating children's cases, making recommendations to the criminal courts, and arranging for the placement of children. In 1878, a law was passed that provided for paid probation officers in Boston.[102] Subsequently, several other states authorized the appointment of probation officers. However, it was not until after the turn of the century and the development of the first juvenile court that probation gained widespread acceptance.[103]

Reform Schools, Industrial Schools, and Training Schools

Although reformers like Charles Loring Brace were skeptical of the ability of institutions to reform or control youths, the role of institutions in providing social control continued to grow during the latter half of the 1800s. The dislocations produced by the Civil War placed tremendous strain on the existing houses of refuge and the placing out system, and the number of problem youths continued to grow. In response, <u>reform schools</u>, industrial schools, and training schools proliferated during the postwar period.

The development of these schools was made possible by the willingness of state and city governments to play a larger role in the administration of institutions for wayward and criminal youths. In some instances state and city governments took over the administration of existing institutions; in others they built new institutions.[104] Although called reform schools, industrial schools, and training schools,[105] these institutions were mainly of two types: <u>cottage reformatories</u> and <u>institutional reformatories</u>.

The cottage reformatories were usually located in rural areas in order to avoid the negative influences of the urban environment. In addition, they were intended to closely parallel family life. Typical of these institutions was the Ohio Reform School, which was established in 1857. Youths in this institution were divided into separate cottage families. According to a report of the commissioners of the Ohio Reform School, the boys lived on a farm with "elder brothers" who were selected because of their ability to inspire good behavior and a love of the country in others.[106] The reformatory cottages typically contained 20 to 40 youths[107] supervised by cottage parents charged with the task of overseeing their training and education.[108]

reform school An institution for the control and reform of youths that was funded by local or state government (unlike the privately funded or self-sustaining houses of refuge). Reform schools (also known as industrial schools and training schools) were often built in rural areas to remove youths from the corrupting influences of the city. Like in the houses of refuge, however, children were often subjected to inhumane treatment in these institutions.

cottage reformatory A type of juvenile institution that was developed in the mid-1800s and intended to simulate family life. In a cottage reformatory, youths were separated into smaller groups than usual, were supervised by cottage parents, and had mentors or "big brothers" who served as normative models.

institutional reformatory A large juvenile correctional institution with hundreds of beds that eventually was characterized, like the houses of refuge, by hard work, long hours, and exploitive conditions.

Juvenile Justice Involves People Making Decisions

Ultimately, juvenile justice involves many people making decisions, sometimes very difficult decisions, and often under arduous conditions, that can have a profound influence on the lives of others.

In addition to cottage reformatories, more institutional reformatories were also developed in many states. Like the cottage reformatories, these institutional reformatories were usually located in rural areas in an effort to remove youths from the criminogenic influences of city life. However, the institutions were often large and overcrowded. For example, the Elmira Reformatory in New York was intended to be a model reformatory; it contained 500 cells, but by the late 1890s it held some 1,500 residents.[109]

Another development that occurred in the mid-1800s was the establishment of separate institutions for females, such as the Massachusetts State Industrial School for Girls, which was established in 1856.[110] Prior to this time, girls had been committed to the same institutions as boys, although there was strict gender segregation. More often than boys, girls were committed by parents or relatives for moral as opposed to criminal offenses. These moral offenses included "vagrancy, beggary, stubbornness, deceitfulness, idle and vicious behavior, wanton and lewd conduct, and running away."[111]

The expressed goal of the specialized institutions for females was to prepare them to be good housewives and mothers. Yet these institutions, like those for boys, differed little from the prisons of the day. The daily regimen revolved around the teaching of basic domestic skills, the methods of treatment were based on coercion, and life in the institution was essentially a form of punitive custody.[112]

The reform, industrial, and training schools, although they placed more emphasis on formal education, were in many respects similar to the houses of refuge. Indeed, they confronted many of the same problems, and the conditions were generally no better than those that existed in the houses of refuge, and in many cases were actually worse.

Institutional Care: A Legacy of Poor and Often Inhumane Treatment

Although the terms used to refer to institutions developed to control and reform children changed periodically,[113] the treatment of children in these institutions did not. Children in these institutions continued to be seen as a source of cheap labor that could be exploited by businesses and could help defray the costs of operations. Beatings, whippings, and other forms of corporal punishment; bread and water rations; solitary confinement; and straitjackets continued to be the standard tools to respond to recalcitrant youths or to deter would-be trouble-makers.[114] Furthermore, youths continued to be placed into apprenticeships without careful follow-up to determine if their treatment was humane. Both boys and girls were subjected to beatings and other mistreatment by their masters, and the sexual abuse of females was not unheard of.[115]

Juvenile correctional institutions continued to grow in popularity as a way of controlling problem youths throughout the latter half of the 1800s despite the economic hardships that resulted from the Civil War. The amount of state funds going to these institutions during the war had declined, which forced them to rely more heavily on contract labor to meet operating expenses. The use of child labor led to charges that the institutions were more concerned with the economic exploitation of children than their reformation. Indeed, reports of the exploitation of youths in the institutions surfaced periodically. For example, in 1871 a New York Commission on Prison Labor found that refuge youths were paid 30 cents per day for work that would command $4 per day in the community. In addition, workingmen's associations complained about the unfair competition of child inmate labor.[116]

By the end of the 19th century, a variety of institutions had been developed in response to problem children. Yet the many problems presented by children who were believed to be in need of correctional treatment—problems such as homelessness, neglect, abuse, waywardness, and criminal behavior—proved difficult to solve. Nevertheless, during the late 1800s a new group of reformers, the Child Savers, began to advocate for a new institution for dealing with problem children: the juvenile court.

■ Legal Issues

To the extent that adults in the past viewed children as property, they failed to recognize that children had legal rights as well. During colonial times and throughout the early 1800s, children were put into houses of refuge and reformatories and placed out to rural families without any kind of due process or legal recourse. As bad as it must have been for some families and children, however, slave families and children suffered more and had even less ability to protect themselves. It has only been within the last half century that the concept of children as legal entities with legal rights has been accepted by large segments of society.

■ Chapter Summary

Historically, childhood is a recent invention. For much of recorded history, the young have been regarded as "miniature adults" and expected to participate in family and community life from a very early age. On the one hand, such participation helped youths earn a place in the community and helped ensure their survival. On the other hand, it placed the young in situations where they were subject to exploitation and abuse by those who were older.

Youths were seen as both a blessing and a curse. They were seen as a blessing because they helped ensure the continuation of the family and the survival of the community. They were seen as a curse when they were too young to contribute to economic survival. Children who were physically or mentally incapable of contributing economically were viewed as a burden, and many of these were killed or abandoned.

In the past, life was often precarious for young and old alike. Disease, famine, natural disasters, and wars claimed the lives of many. Although there is clear evidence that many parents in Europe and elsewhere demonstrated care and concern for their children, there is also evidence that very little thought was given to the needs of many others. Large numbers of youths lived harsh lives characterized by abuse and exploitation, perhaps because the difficult and precarious nature of life produced a culture that downplayed the importance of children. Whatever the reason, children's lives were very different than those experienced by many children in the United States today.

Although there were many controls on the behaviors of the young, the idea that the young were always obedient in the "good old days" is a myth. Indeed, each generation of youths appears to have presented a number of challenges to their elders, and there is clear evidence that juvenile crime has long been seen as a problem, particularly by urban dwellers in Europe and the United States. However, unlike today, only one criminal justice system existed for handling illegal behavior. Consequently, youths who violated the law were frequently treated like adults, although on other occasions they were spared the harshest punishments because of their age.

By the 1800s, a number of factors made the creation of separate institutions for problem youths possible, including the growing acceptance by many that childhood and adolescence were distinct developmental periods during which the young needed guidance, protection, and correction. This view of childhood, along with a number of broad economic, social, and cultural changes, provided the foundation for juvenile correctional institutions and juvenile courts.

The 19th century was a time of enormous change in the United States. Many eastern towns were growing into crowded urban areas, the country was becoming more culturally diverse, and industrialization was well under way. Although many people benefited from these changes and the new opportunities that became available, many others were not so fortunate. Poverty, homelessness, mental health problems, disease, alcoholism, racism, a lack of social services, and crime (including juvenile crime) were reminders that not everything was perfect. The problem of juvenile crime, together with the idea that children were impressionable, led to the development of the first specialized correctional institutions for youths in the United States. These institutions, the houses of refuge, were touted as havens for youths who needed protection and guidance and as places where they could become hardworking, productive, God-fearing citizens. Yet, like other correctional institutions, they were designed to control those who were seen as problems, and eventually control became their primary focus. The result was that many of these early correctional institutions for youths were no better than adult jails and prisons.

Because of the failure of the houses of refuge, the growing number of problem youths needing control and reform, and increasing worry that the cities were breeding grounds for crime, a number of additional correctional responses aimed at youths were developed during the 1800s. These included placing out; probation; and publicly operated reform, industrial, and training schools. Each faced a number of obstacles that prevented it from effectively controlling or reforming youths. In the case of placing out, little effort was made to determine the appropriateness of placements, and some Westerners began to complain that Easterners were simply attempting to dump their problems on Western communities. Early probation officers were typically volunteers (often former police officers) who had little training and lacked the personal and community resources needed to provide assistance to many of their clients. The newer correctional institutions faced many of the same problems that had plagued the houses of refuge. Overcrowding, mixing of youths of various ages and delinquent histories, inhumane treatment, and a failure to provide follow-up services to youths upon their release prevented these institutions from achieving their correctional goals.

By the end of the 1800s, despite the existence of varied correctional responses, the problem of juvenile crime had, if anything, grown worse. As a result, a new group of reformers, the Child Savers, began to advocate for the development of a new institution to coordinate efforts to control and reform youth. This new institution, the juvenile court, is the topic of the following chapter.

KEY CONCEPTS

apprenticeship A period set by contract during which a novice worked for a master in return for instruction in the master's trade. Apprenticeships were typically reserved for middle-class or wealthy youths. They typically began when youths were 14 to 18 years old and lasted 10 to 12 years.

binding out The placing of a youth or adult in the service of others for a specific period of time, from a week to many years. Although children as young as 7 years old and elderly adults could be placed as servants, the practice was mostly restricted to children in their teens. Binding out was a way for a poor youth to earn and save money, obtain room and board, learn a skill, prepare for adulthood, and meet a potential wife or husband.

cottage reformatory A type of juvenile institution that was developed in the mid-1800s and intended to simulate family life. In a cottage reformatory, youths were separated into smaller groups than usual, were supervised by cottage parents, and had mentors or "big brothers" who served as normative models.

family A social unit responsible for, among other things, the raising and socialization of children. Historically, the family in its various forms has been society's tool for teaching children how to act, what to believe, and what values to hold. (See also Chapter 3.)

feudalism A system of land tenure governed by relations between nobles and lesser nobles whereby land (fief) was given by an overlord to a lesser noble (vassal) to hold (not to own) in return for military service. The land could then be given to still lesser nobles. This system depended on the agricultural production of the peasants (serfs) who worked the land.

house of refuge A privately operated child-care institution developed in the United States for children who engaged in illegal behavior or were felt to be at risk of such behavior. Although the initial goal was to educate and train youths for productive lives, houses of refuge became little more than prisons characterized by harsh discipline and hard work.

industrial school See **reform school**.

institutional reformatory A large juvenile correctional institution with hundreds of beds that eventually was characterized, like the houses of refuge, by hard work, long hours, and exploitive conditions.

placing out The placement of children on farms and ranches in the Midwest and West. Placing out, which was popularized by private charitable organizations in the mid-1800s (and which lasted until 1929), was an attempt to remove vagrant and wayward children from the corrupting influences of the cities and locate them in what was believed to be a more wholesome environment. The idea of using the frontier as a geographical "cure" for societal ills was consistent with the views of Frederick Jackson Turner,[117] a 19th-century historian who believed that the American experience of the frontier helped shape the American character by cultivating individualism, industry, and self-reliance—the very values that the social reformers wanted all children to acquire.

probation The supervised release of an individual by a court. Probation was first promoted by John Augustus, a Boston shoemaker. Today children are usually returned to their parents by courts on conditions of release called "rules of probation," and court or state employees are the supervisors or monitors of the persons on probation. In the 19th century, the monitoring of youths on probation was largely done by private citizen volunteers or volunteer societies.

reform school An institution for the control and reform of youths that was funded by local or state government (unlike the privately funded or self-sustaining houses of refuge). Reform schools (also known as industrial schools and training schools) were often built in rural areas to remove youths from the corrupting influences of the city. Like in the houses of refuge, however, children were often subjected to inhumane treatment in these institutions.

training school See **reform school**.

REVIEW QUESTIONS

1. What were the social conditions faced by children in the Middle Ages?

_____

_____

_____

_____

2. What role did the family play in the socialization and control of children before the 1800s?

_____

_____

_____

_____

3. What role did children play in the colonization of the New World?

_____

_____

_____

_____

4. What characteristics of colonial life encouraged conformity among both young and old?

_____

_____

_____

_____

5. What social conditions during the early 1800s contributed to the development of the houses of refuge?

_____

_____

_____

_____

6. In what ways did the houses of refuge fall short of their initial goals?

_____

_____

_____

_____

7. What was placing out?

_____

_____

_____

_____

8. What were the shortcomings of placing out?

_____

_____

_____

_____

9. Why is John Augustus called the "father of probation"?

_____

_____

_____

10. What were the basic features of the reform, industrial, and training schools that were developed during the last half of the 1800s?

_____

_____

_____

11. What were the distinguishing characteristics of the specialized female training schools?

_____

_____

_____

_____

ADDITIONAL READINGS

Bernard, T. J. 1992. *The Cycle of Juvenile Justice*. New York: Oxford.

Bremner, R. H., ed. 1970. *Children and Youth in America: A Documentary History*, 3 vols. Cambridge, MA: Harvard University Press.

Rothman, D. J. 1971. *The Discovery of the Asylum: Social Order and Disorder in the New Republic*. Boston: Little, Brown & Co.

Schlossman, S. and S. Wallach. 1998. "The Crime of Precocious Sexuality: Female Delinquency in the Progressive Era," in *Juvenile Delinquency: Historical, Theoretical, and Societal Reactions to Youth*, eds. P. M. Sharp and B. W. Hancock. Upper Saddle River, NJ: Prentice Hall.

Sommerville, C. J. 1990. *The Rise and Fall of Childhood*. New York: Vintage Books

Zinn, H. 1995. *A People's History of the United States 1492–Present*. New York: HarperCollins.

NOTES

1. P. Aries, *Centuries of Childhood: A Social History of Family Life*, trans. R. Baldick (New York: Random House, 1962).

2. C. J. Sommerville, *The Rise and Fall of Childhood* (New York: Vintage Books, 1990).

3. L. deMause, "The Evolution of Childhood," in *The History of Childhood*, ed. L. deMause (New York: Psychohistory Press, 1974), 29.

Chapter Resources

4. J. E. Boswell, "Exposito and Oblatio: The Abandonment of Children and the Ancient and Medieval Family," *American Historical Review* 89 (1984): 10–33.

5. B. A. Hanawalt, *Growing Up in Medieval London: The Experience of Childhood in History* (New York: Oxford University Press, 1993).

6. Hanawalt, *Growing Up in Medieval London.*

7. P. Robertson, "Home as a Nest: Middle-Class Childhood in Nineteenth Century Europe," in *The History of Childhood*, ed. L. deMause (New York: Psychohistory Press, 1974).

8. Hanawalt, *Growing Up in Medieval London.*

9. Sommerville, *The Rise and Fall of Childhood.*

10. E. L. R. Ladurie, "Parents and Children," in *The Middle Ages*, vol. 2, *Readings in Medieval History*, 3rd ed., ed. B. Tierney (New York: Knopf, 1983).

11. Hanawalt, *Growing Up in Medieval London.*

12. M. M. Postan, "Land and Population," in *The Middle Ages*, vol. 2, *Readings in Medieval History*, 3rd ed., ed. B. Tierney (New York: Knopf, 1983).

13. J. R. Gillis, *Youth and History* (New York: Academic Press, 1974).

14. B. Krisberg and J. F. Austin, *Reinventing Juvenile Justice* (Newbury Park, CA: Sage, 1993), 8.

15. Krisberg and Austin, *Reinventing Juvenile Justice.*

16. Krisberg and Austin, *Reinventing Juvenile Justice.*

17. Krisberg and Austin, *Reinventing Juvenile Justice.*

18. Krisberg and Austin, *Reinventing Juvenile Justice.*

19. Hanawalt, *Growing Up in Medieval London.*

20. Hanawalt, *Growing Up in Medieval London.*

21. Hanawalt, *Growing Up in Medieval London.*

22. I. J. Silverman, *Corrections: A Comprehensive View* (Belmont, CA: Wadsworth, 2001).

23. Krisberg and Austin, *Reinventing Juvenile Justice.*

24. Krisberg and Austin, *Reinventing Juvenile Justice.*

25. Krisberg and Austin, *Reinventing Juvenile Justice.*

26. R. H. Bremner, ed., *Children and Youth in America: A Documentary History*, vol. 1 (Cambridge, MA: Harvard University Press, 1970), 3.

27. Bremner, *Children and Youth in America.*

28. H. Zinn, *A People's History of the United States 1492–Present* (New York: HarperCollins, 1995).

29. Bremner, *Children and Youth in America.*

30. C. Bridenbaugh, *Vexed and Troubled Englishmen* (1968), cited in Bremner, *Children and Youth in America*, 6.

31. Bremner, *Children and Youth in America.*

32. Bremner, *Children and Youth in America*; Sommerville, *The Rise and Fall of Childhood.*

33. Krisberg and Austin, *Reinventing Juvenile Justice.*

34. Hanawalt, *Growing Up in Medieval London.*

35. R. W. Beales, "In Search of the Historical Child: Miniature Adulthood and Youth in Colonial New England," in *Growing up in America: Children in Historical Perspective*, eds. R. Hiner and J. M. Hawes (Chicago: University of Illinois Press, 1985), 21.

36. R. Michalowski, *Order, Law and Crime* (New York: Random House, 1985).

37. W. B. Waegel, *Delinquency and Juvenile Control: A Sociological Perspective* (Englewood Cliffs, NJ: Prentice-Hall, 1989), 4.

38. Waegel, *Delinquency and Juvenile Control*, 4.

39. E. Zaretsky, "Capitalism, the Family and Personal Life," in *The Capitalist System*, ed. R. C. Edwards et al. (Englewood Cliffs, NJ: Prentice-Hall, 1976).

40. Krisberg and Austin, *Reinventing Juvenile Justice.*

41. J. Demos, *A Little Commonwealth: Family Life in Plymouth Colony* (New York: Oxford University Press, 1970); see also B. Farber, *Guardians of Virtue: Salem Families in 1800* (New York: Basic Books, 1972).

42. Cited in C. E. Simonsen and M. S. Gordon, *Juvenile Justice in America* (New York: Macmillan, 1982), 19.

43. Krisberg and Austin, *Reinventing Juvenile Justice.*

44. Waegel, *Delinquency and Juvenile Control*, 4.

45. D. J. Rothman, *The Discovery of the Asylum: Social Order and Disorder in the New Republic* (Boston: Little, Brown & Co., 1971).

46. Waegel, *Delinquency and Juvenile Control*.
47. Bremner, *Children and Youth in America*.
48. E. Powers, *Crime and Punishment in Early Massachusetts* (Boston: Beacon Press, 1966), 442.
49. Rothman, *The Discovery of the Asylum*.
50. H. E. Barnes, *The Story of Punishment*, 2nd ed. (Montclair, NJ: Patterson Smith, 1972).
51. Rothman, *The Discovery of the Asylum*.
52. This section based primarily on material from Zinn, *A People's History of the United States*.
53. Zinn, *A People's History of the United States*, 49.
54. Zinn, *A People's History of the United States*.
55. Krisberg and Austin, *Reinventing Juvenile Justice*.
56. Sommerville, *The Rise and Fall of Childhood*.
57. Bremner, *Children and Youth in America*.
58. T. J. Bernard, *The Cycle of Juvenile Justice* (New York: Oxford University Press, 1992).
59. Rothman, *The Discovery of the Asylum*.
60. Zinn, *A People's History of the United States*.
61. Zinn, *A People's History of the United States*.
62. Rothman, *The Discovery of the Asylum*, 57.
63. Bernard, *The Cycle of Juvenile Justice*.
64. Zinn, *A People's History of the United States*.
65. Zinn, *A People's History of the United States*.
66. Zinn, *A People's History of the United States*.
67. Krisberg and Austin, *Reinventing Juvenile Justice*.
68. Bremner, *Children and Youth in America*.
69. Rothman, *The Discovery of the Asylum*.
70. Rothman, *The Discovery of the Asylum*.
71. Krisberg and Austin, *Reinventing Juvenile Justice*.
72. Bernard, *The Cycle of Juvenile Justice*.
73. Peirce, *A Half Century with Juvenile Delinquents*, 40.
74. Bernard, *The Cycle of Juvenile Justice*.
75. Bernard, *The Cycle of Juvenile Justice*; Rothman, *The Discovery of the Asylum*.
76. A. R. Roberts, *Juvenile Justice: Policies, Programs, and Services* (Chicago: Nelson-Hall, 1998); Rothman, *The Discovery of the Asylum*.
77. Rothman, *The Discovery of the Asylum*.
78. Rothman, *The Discovery of the Asylum*.
79. Bremner, *Children and Youth in America*.
80. R. M. Mennel, *Thorns and Thistles* (Hanover, NH: University Press of New England, 1973).
81. Bremner, *Children and Youth in America*.
82. S. L. Schlossman, *Love and the American Delinquent* (Chicago: University of Chicago Press, 1977).
83. Rothman, *The Discovery of the Asylum*; A. W. Pisciotta, "Saving the Children: The Promise and Practice of Parens Patriae, 1838–1898," *Crime and Delinquency* 28 (1982): 410–425.
84. A. Pisciotta, "Treatment on Trial: The Rhetoric and Reality of the New York House of Refuge, 1857–1935," *American Journal of Legal History* 29 (1985): 151–181; Rothman, *The Discovery of the Asylum*, 231.
85. Krisberg and Austin, *Reinventing Juvenile Justice*.
86. Roberts, *Juvenile Justice*.
87. Krisberg and Austin, *Reinventing Juvenile Justice*.
88. Mennel, *Thorns and Thistles*.
89. Cited in Roberts, *Juvenile Justice*, 98.
90. Bernard, *The Cycle of Juvenile Justice*, 65.
91. Mennel, *Thorns and Thistles*; Rothman, *The Discovery of the Asylum*.
92. Krisberg and Austin, *Reinventing Juvenile Justice*.
93. Sommerville, *The Rise and Fall of Childhood*.
94. Rothman, *The Discovery of the Asylum*.
95. Bremner, *Children and Youth in America*.
96. Krisberg and Austin, *Reinventing Juvenile Justice*.

Chapter Resources

97. Sommerville, *The Rise and Fall of Childhood*.
98. Krisberg and Austin, *Reinventing Juvenile Justice*.
99. B. R. McCarthy and B. J. McCarthy, *Community-Based Corrections*, 2nd ed. (Pacific Grove, CA: Brooks/Cole, 1991), 98.
100. L. T. Empey and M. C. Stafford, *American Delinquency: Its Meaning and Construction*, 3rd ed. (Belmont, CA: Wadsworth, 1991).
101. C. Bartollas and S. J. Miller, *Juvenile Justice in America* (Englewood Cliffs, NJ: Regents/ Prentice Hall, 1994).
102. Empey and Stafford, *American Delinquency*.
103. McCarthy and McCarthy, *Community-Based Corrections*.
104. Krisberg and Austin, *Reinventing Juvenile Justice*.
105. H. Hart, *Preventive Treatment of Neglected Children* (New York: Russell Sage, 1910).
106. Bremner, *Children and Youth in America*.
107. Bartollas and Miller, *Juvenile Justice in America*.
108. J. T. Whitehead and S. P. Lab, *Juvenile Justice: An Introduction* (Cincinnati, OH: Anderson, 1990).
109. Roberts, *Juvenile Justice*.
110. B. Brenzel, *Daughters of the State* (Cambridge, MA: MIT Press, 1983).
111. Brenzel, *Daughters of the State*, 81.
112. M. Chesney-Lind and R. G. Shelden, *Girls, Delinquency and Juvenile Justice* (Belmont, CA: Thompson/Wadsworth, 2004).
113. Hart, *Preventive Treatment of Neglected Children*.
114. Pisciotta, "Saving the Children."
115. Pisciotta, "Saving the Children."
116. Krisberg and Austin, *Reinventing Juvenile Justice*.
117. F. J. Turner, *The Frontier in American History* (New York: Holt, Rinehart, & Winston, 1920).

The Development of the Juvenile Court

Chapter Objectives

After studying this chapter, you should be able to

- Describe how the middle and upper classes in the late 19th century viewed the poor and the working class
- Describe how the elite's view of the poor and working class influenced the types of social control mechanisms that were developed
- Describe who the "child savers" were and why they were interested in poor, wayward youths
- Explain how the inability or unwillingness of criminal courts to deal with young offenders encouraged the development of the juvenile courts
- Describe the ways in which the social context of the late 1800s and early 1900s contributed to the development of the juvenile courts
- Describe the legal doctrine of *parens patriae*, and indicate how it was used as a justification for state intervention into family life
- Explain why concern about the lack of due process in juvenile courts grew during the middle of the 20th century
- Describe the significance of *Kent v. United States*, *In re Gault*, *In re Winship*, and *McKeiver v. Pennsylvania* and how these cases influenced juvenile justice practice

Chapter Outline

- Introduction
- The Social Context of the Juvenile Court
- The Legal Context of the Juvenile Court
- The Operation of the Early Juvenile Courts

- The Legal Reform Years
- Legal Issues
- Chapter Summary
- Key Concepts
- Review Questions
- Additional Readings
- Cases Cited
- Notes

■ Introduction

This chapter examines the development of the juvenile court in the United States. It describes the social context influencing the creation of the first juvenile court, established in Chicago in 1899, and the rapid spread of juvenile courts to other jurisdictions. It explores the legal developments that formed the underlying basis of juvenile court practice. It also looks at early juvenile court operation and important landmark cases that have affected contemporary practice. The chapter closes with an examination of the extent to which important legal cases actually influence the practice of juvenile justice.

■ The Social Context of the Juvenile Court

The period from 1880 to 1920, referred to as the **Progressive Era** by historians, was a time of major change in the United States. The pace of industrialization, urbanization, and immigration quickened during this period, and the population became more diverse. The stream of Irish and German immigrants trailed off at the end of the 19th century, but they were replaced by Italians, Russians, Jews, Greeks, and other immigrants from Eastern and Southern Europe-people who came from cultural backgrounds that were alien to native-born Americans.[1]

Progressive Era A period in American history, lasting roughly from 1890 to 1924, during which a variety of economic, social, and political reforms and reform movements occurred, including women's suffrage; trust busting; Prohibition; working-hour reduction; elimination of child labor; adoption of social welfare benefits; and the popular voting measures of initiative, recall, and referendum. The reforms were primarily concerned with responding to popular unrest and problems uncovered in the operation of economic, social, and political institutions.

Urban Growth Was Just One Important Change During the Progressive Era

The city of Chicago provides a good example of the changes experienced during the Progressive Era. Between 1890 and 1910, the population of Chicago doubled, growing in those two decades from 1 million to 2 million people. Between 1880 and 1890, the number of factories nearly tripled. Furthermore, by 1889 nearly 70% of the inhabitants of the city were immigrants.[2]

The dream of a better life drew many people to American cities during the Progressive Era. This dream, however, frequently remained unrealized. Wealth was becoming increasingly concentrated in the hands of business elites who sought to dominate economic life, workers were forced to work long hours in unsafe work settings for low wages and no benefits, and housing was often inadequate. Periodic economic depressions ruined many people financially and made life particularly difficult for the poor. Not surprisingly, urban areas faced a number of social problems, including poverty, labor unrest, alcoholism, disease, racial and ethnic prejudice and discrimination, and crime.

American History Is One of Conflicting Interests

The United States has seen a number of conflicts between opposing interests, oftentimes the wealthy and the poor. One of the earliest of these was Shay's Rebellion, which began in western Massachusetts during the summer of 1786. The uprising represented an attempt by debt-ridden farmers to seize court-houses in order to forestall foreclosure proceedings against them. In 1839, an anti-rent movement spread throughout the Hudson Valley, and thousands of renters attempted to make their voices heard and to resist the efforts of wealthy landlords to collect rents and taxes from poor farmers. All through the later 19th century, workers engaged in strikes and other efforts in order to win better wages and improve working conditions. The conflict between workers and employers sometimes turned violent. For example, the great railroad strikes of 1877, which involved some 100,000 strikers, resulted in 1,000 arrests and 100 deaths.[3]

In response, members of the middle and upper classes attempted to implement reforms to alleviate the worst conditions. Of course, poverty, labor unrest, alcoholism, and racial and ethnic prejudice were hardly new. Yet these problems began to take on a new tenor by the end of the 1800s. Furthermore, workers' movements had become more national in scope and were clearly seen as more threatening to political and economic elites. As a result, these elites sought to devise better mechanisms of control through which their interests might be protected.[4]

The Progressive Era Was a Time of Change

The Progressive Era was characterized by a political revolt against the social and economic evils of the Industrial Revolution and a belief that government intervention, even on a national scale, was necessary to remedy these evils. Politically, it began with a series of urban and state reform movements directed against corrupt and boss-ridden local governments.

The Progressive Era was also characterized by the belief that American ingenuity and spirit could solve social problems and make life better for people. Tremendous technological advances were taking place, including the development of the automobile, the airplane, moving pictures, the phonograph, and the electric light bulb, just to name a few. Many Americans believed that advances could also be made in social, political, and economic life.

"Like the Jeffersonians of a century before, the Progressives had an abiding faith in the intelligence and good will of the American people. Fundamentally, their remedy for the failures of democracy was more democracy. Let the will of the people really reach into the Congress, the courts, the state legislatures, and America could then solve its problems."[5]

child savers A group of reformers, mostly well-educated, middle-class Protestant women of Anglo-Saxon descent, who had the time and resources to fight for improved conditions for delinquent youths in jails and reformatories and who eventually played an important role in the creation of the juvenile courts. On the positive side, their work led to wider acceptance of adolescence as an important developmental stage of life, and they correctly believed that the adult criminal justice system was harmful to children or at best ineffective. On the negative side, their conservative ideas on the family and their "noblesse oblige" attitude toward the poor resulted in reforms that focused more on the social control of children and the protection of the interests of the wealthy than on attacking the underlying causes of problem youth behaviors.

One problem that continued to cause concern in the developing urban areas was youth waywardness and crime. Thousands of indigent children roamed the streets in the larger cities,[6] and many of these children engaged in immoral or illegal behavior that threatened the tranquility of city life. Of course, criminal laws could be used to prosecute youths who committed crimes, but if the offenses were minor, the courts were often reluctant to do more than give the culprits a stern lecture. Furthermore, the laws were ineffective against those who had not committed a criminal offense. As a result, reformers sought to develop better ways of controlling youths.

Those who led the reform effort, known as **child savers,** were primarily well-educated Protestant, middle-class women of Anglo-Saxon descent, but some were the daughters or wives of the most influential men within their community.[7] They also tended to be conservative in their thinking. For example, they believed in traditional roles for women, and like other affluent people, they took for granted their natural superiority over the poor. From the child savers' perspective, homemaking and child-care were women's primary responsibilities. However, because these traditional duties were performed in their own homes by servants, they had few outlets for their energies. Child saving provided a mechanism for these women to perform an important public service-the control and care of less fortunate children.[8]

The child savers were informed by the latest thinking on adolescence and youth crime. Adolescence was seen as a problem period in the child's development,[9] a time when youths were subject to a variety of negative influences. Indeed, youths who engaged in crime were felt to suffer from poor parenting and weak morals and to be unusually susceptible to the temptations of the streets.[10]

Initially, the child savers sought to improve jail and reformatory conditions.[11] However, their attention soon turned toward the development of more effective means of controlling the growing population of problem youths. Indeed, the primary focus of the child-saving movement was on extending government control over children through stricter supervision and improved legal mechanisms designed to regulate their behavior.[12] In short, the child savers felt that intrusion into the lives of children was necessary in order to prevent them from leading immoral and criminal lives—and, equally important, to protect the state and the interests of the wealthy.

Unlike earlier reformers, the child savers thought that the interventions needed to control and uplift children should be performed by government agencies, such as the police and courts, with the assistance of local charitable organizations. According to Julia Lathrop, an influential child saver in Chicago who later became head of the Federal Children's Bureau,

> *There are at the present moment in the State of Illinois, especially in the city of Chicago, thousands of children in need of active intervention for their preservation from physical, mental and moral destruction. Such intervention is demanded, not only by sympathetic consideration for their well-being, but also in the name of the commonwealth, for the preservation of the State. If the child is the material out of which men and women are made, the neglected child is the material out of which paupers and criminals are made.*[13]

■ The Legal Context of the Juvenile Court

Although the social conditions that existed in urban areas led reformers to advocate for changes in how communities responded to youth crime and waywardness, the reformers also recognized that existing legal mechanisms for controlling and protecting children were inadequate. One immediate concern of the child savers was the placement of children in jail. For example, a study of the county jail system in Illinois in 1869 discovered 98 children under the age of 16 in 40 different jails. In Michigan, 377 boys and 100 girls under the age of 18 were placed in county jails in 1873; 182 boys and 29 girls were placed in jails in Ohio in 1871; and over 2,000 youths, 231 of whom were under the age of 15, were placed in jails in Massachusetts in 1870. The jailing of children was a common practice.[14] However, as the Board of State Charities in Illinois noted, the county jails in the state contained cells that were "filthy and full of vermin," and they were claimed to be "moral plague spots" where children were turned into "great criminals."[15]

The child savers, concerned as they were about the conditions children were exposed to in adult jails and lock-ups, condemned the fact that adult courts failed to mete out punishment for many of the transgressions of the young. Unless their offenses were severe, criminal court judges often treated youths leniently. From the child savers' perspective, children either were treated too severely, by being placed in adult facilities, where they were corrupted by older and more hardened offenders, or were let go without receiving any assistance.[16] Thus, according to the child savers, children were neither controlled nor helped.

By the late 19th century, legal mechanisms for treating children separately from adults had been in existence for some time. For example, laws establishing the minimum age at which a child could be considered legally responsible for criminal behavior as well as age limits for placement in adult penitentiaries were enacted during the first half of the century.[17] As noted in Chapter 4, the first special institution dealing with youths, the House of Refuge, was established in New York City in 1825.

The legal justification for state intervention into the lives of children was based on the doctrine of ***parens patriae*** (the state as parent), which was given legal standing in an important case, *Ex parte Crouse*, in 1838. Mary Ann Crouse was committed to the Philadelphia House of Refuge by her mother but against her father's wishes. The father questioned Mary Ann's placement, arguing that she was being punished without having committed a criminal offense. The Pennsylvania Supreme Court ruled that Mary Ann's placement was legal because the purpose of the house of refuge was to reform youths, not punish them; that formal **due process** protections afforded to adults in criminal trials were not necessary because Mary Ann was not being punished; and that when parents were unwilling or unable to protect their children, the state had a legal obligation to do so.[18]

The right of the state to intervene in the lives of children did not go unchallenged, however. In another important case, *People v. Turner* (1870), the Illinois Supreme Court ruled that Daniel O'Connell, who was committed to the Chicago House of Refuge against his parents' wishes, was being punished, not helped by his placement. In some respects this case was similar to *Crouse*. Daniel was institutionalized even though he had committed no criminal offense, and he was perceived to be in danger of becoming a pauper or criminal. In other respects, however, the two cases were decidedly different. Both parents had objected to Daniel's placement, and, even more important, the court ruled that his placement was harmful, not helpful. Furthermore, the court decided that because placement in the house of refuge was actually a punishment, due process protections were necessary.[19]

People v. Turner was seen as an obstacle to the efforts of reformers to help and control youths. The Illinois Supreme Court's ruling, along with the increasing concern over the unwillingness of the criminal courts to sentence youths, led reformers in Chicago to consider other mechanisms by which their aims might be achieved. What they finally did was create the first juvenile court, which was established in Cook County (Chicago) in 1899.[20]

The juvenile court allowed reformers to achieve their goals of assisting and controlling children's behavior without undue interference from the adult courts and without undue concern for the due process protections afforded adults. The court was set up as a civil or **chancery court** intended to serve the "best interests" of children (as opposed to a criminal court, which focuses on the punishment of offenders). Because the new court was not a criminal court and its goal was not to punish but to help children, the need for formal due process protections was obviated.[21]

parens patriae The obligation of the government to take responsibility for the welfare of children. The doctrine of *parens patriae* recognizes the moral obligation of the government to take care of children when there is no family available or if the family is not suitable.

due process A course established for legal proceedings intended to ensure the protection of the private rights of the litigants. The essential elements of due process are (1) proper notice as to the nature of the legal proceedings, (2) a meaningful hearing in which the individual has an opportunity to be heard and/or defend him- or herself, and (3) the objectivity of the tribunal before which the proceedings take place.

chancery court English court whose primary concern is with property rights. Chancery courts, first established during the Middle Ages, were partially based on the idea that children, particularly children who owned or were in a position to inherit property, should fall under the protective control of the king. Indeed, it was in these courts that the concept of *parens patriae* was developed. Over time, the chancery courts came to be more involved in the general welfare of families and children. Because of their role in protecting the interests of children, they served as a model for the juvenile court.

■ The Operation of the Early Juvenile Courts

The juvenile court act of 1899 gave the new juvenile court in Illinois broad jurisdictional powers over people under the age of 16 who were delinquent, **dependent**, or **neglected**.[22] In addition, it required that the court be overseen by a special judge, that hearings be held in a separate courtroom, and that separate records be kept of juvenile hearings.[23] It also made probation a major component of the juvenile court's response to offenders and emphasized the use of informal procedures at each stage of the juvenile court process.[24]

dependent children
Children who are dependent on others for financial support or who do not have the support of a parent or guardian.

neglected children
Children who are not given appropriate care by the parent(s) or a guardian.

Informality Has Been a Hallmark of Juvenile Court

Indeed, procedural informality has been a hallmark of juvenile courts throughout their history.

In practice, the informality of the juvenile court meant that complaints against children could be made by almost anyone in the community. It meant that juvenile court hearings were held in offices, as opposed to traditional courtrooms, and were closed to the public, unlike adult trials, which were open. In a typical juvenile court hearing, the only people present were the judge, the parents, the child, and the probation officer. The informality also meant that few if any records were kept of hearings, that proof of guilt was not necessary for the court to intervene in children's lives, and that little or no concern for due process protections existed. Finally, it meant that judges exercised wide discretion regarding the actions they took, which could include anything from a stern warning to placement of a child in an institution.[25]

The idea of using juvenile courts to deal with youth crime spread rapidly after the passage of the Illinois legislation. Within 10 years, 10 states had established special courts for children, and by 1925 all but 2 states had juvenile courts.[26] These courts followed closely the model developed in Chicago: They were procedurally informal and intended to serve the best interests of children.

The juvenile courts were successfully implemented, in large measure, because they served a variety of interests. They served the interests of reformers who sought to help children on humanitarian grounds and the interests of those who were concerned primarily with the control of lower-class, immigrant children whose behaviors threatened urban tranquility.[27] They served the interests of the criminal courts, because they removed children from criminal court jurisdiction, freeing up time for the trying of adult offenders. Finally, they served the interests of the economically and politically powerful, because they did not require the alteration of existing political and economic arrangements.[28] They were designed to be not instruments of change, but instruments of containment.

Despite the growing popularity of the juvenile courts, they did not go unchallenged. In another important court case, *Commonwealth v. Fisher* (1905), the juvenile court's mission, its right to intervene in family life, and the lack of due process protections afforded children were examined by the Pennsylvania Supreme Court. In this case, Frank Fisher, a 14-year-old male, was indicted for larceny and committed to a house of refuge (the same house of refuge that Mary Ann Crouse had been committed to over 60 years earlier) until his 21st birthday. Frank's father objected to his placement and filed a suit that argued that Frank's 7-year sentence for a minor offense was more severe than he would have received in a criminal court.[29]

In its ruling, the Pennsylvania Supreme Court upheld the idea of a juvenile court and in many respects repeated the arguments made by the court in the *Crouse* decision. The court found that the state may intervene in families when the parents are unable or unwilling to prevent their children from engaging in crime and that Frank was being helped by his placement in the house of refuge. Further, it ruled that due process protections were unnecessary when the state acts under its *parens patriae* powers.[30]

Fisher set the legal tone for juvenile courts from their beginnings until the mid-1960s, when new legal challenges to the courts began to be mounted. These legal challenges primarily concerned expansion of juveniles' due process protections. Critics of the juvenile courts recognized that, despite their expressed goal of serving the best interests of children, the established institutions of juvenile justice often did the opposite. Although the courts were intended to help children, they did not always act as wise and benevolent parents. Indeed, their use of coercive powers to deal with a wide range of behaviors, many of which were not criminal in nature, in an informal setting without due process protections created the potential for abuse. Moreover, while reformers had looked to the juvenile courts as a mechanism for both helping and controlling delinquent and wayward children, little attention was devoted to improving other juvenile justice institutions. Various mechanisms intended to assist the court in its mission, such as probation, relied heavily on untrained volunteers.[31] Further, the institutions used by the juvenile courts for the placement of children were the same institutions used before the establishment of the courts, and children continued to be subjected to inhumane treatment.[32]

■ The Legal Reform Years

The 1960s and 1970s were a time when American institutions, including juvenile justice institutions, came under intense scrutiny, and indeed during the middle of this period the U.S. Supreme Court heard a number of cases that altered the operation of the juvenile courts. The most important of these was *In re Gault* (1967), which expanded the number of due process protections afforded juveniles. This case is discussed below, as are several others, including the first of these important cases, *Kent v. United States* (1966).

Morris Kent was accused of committing a number of break-ins and robberies in the District of Columbia. One robbery victim was raped, but the principal evidence against Kent was a latent fingerprint left at the scene of the robbery and rape. Kent was on probation at the time of these crimes, and after his apprehension he was interrogated over a 7-hour period and confessed to several house break-ins. Without a hearing or any formal notice, Kent's case was transferred to the local criminal court. His attorney tried to get the case dismissed from the adult court and moved for a psychiatric evaluation and for receipt of all social reports in the juvenile court's possession. All motions were denied, Kent was convicted by a jury in the adult court of robbery and housebreaking, and he was sentenced to 30–90 years in prison.

The matter was appealed on the jurisdictional issue of the waiver from juvenile court to adult court. It was contended that the waiver was defective on the following grounds:

- No waiver hearing was held.
- No indication was given as to why the waiver was ordered.
- Counsel was denied access to the social file and social reports reportedly used by the judge to determine the waiver.

In this case, the Supreme Court ruled that for Kent's waiver to be valid certain due process requirements were necessary. Specifically, the court held that Kent was entitled to representation by an attorney; to a meaningful hearing, even if informal; to access to any social reports, records, reports of probation, and so on, that would be considered by the court in deciding the waiver; and to be apprised of the reasons for the waiver decision.

Kent was an important case for several reasons. It resulted in the first major ruling by the U.S. Supreme Court that scrutinized the operation of the juvenile courts. After over 60 years of informal *parens patriae* procedures, the appropriateness of these procedures was being questioned, even if only narrowly, in the limited area of waivers to adult court. *Kent* also made explicit the need for due process protections for juveniles who were being transferred to adult courts for trial. The court noted that, even though a hearing to consider transfer to adult court is far less formal than a trial, juveniles are still entitled to some due process protections.

In its decision, the court made numerous references to the need for due process protections, stating that in a juvenile court a child may receive "the worst of both worlds: that he gets neither the protections afforded to adults nor the solicitous care and regenerative treatment postulated for children."[33] Once the court began to look at juvenile court processes and procedures, subsequent cases like *Gault, McKeiver,* and *Winship* became inevitable.

Having given notice that it would review the operation of the juvenile courts, within a year of the *Kent* decision the Supreme Court heard another landmark case. This case, *In re Gault* (1967), went far beyond *Kent* in its examination of juvenile court practice and extended a variety of due process protections to juveniles. The facts of the case clearly demonstrate the potential for abuse found in the informal procedures of the traditional juvenile court, and consequently this case is discussed in detail.

Gerald Gault was 15 years old when he and a friend were taken into custody by the Gila County (Arizona) Sheriff's Department for allegedly making an obscene phone call to a neighbor, Ms. Cook. At the time of his arrest, Gerald was on 6 months' probation—the result of his being with another friend who had stolen

Mother wasn't notified, plaintiff wasn't present

a wallet from a purse. Gerald was taken into custody on the verbal complaint of Ms. Cook and was taken to the local detention unit. His mother was not notified of this by the police but learned about it later that day, when she returned home and, not finding Gerald present, sent a sibling to search for him.

Upon learning that Gerald was in custody, Ms. Gault went to the detention facility and was told by the superintendent that a juvenile court hearing would be held the next day. On the following day, Gerald's mother, the police officer who had taken Gerald into custody and filed a petition alleging that Gerald was delinquent, and Gerald appeared before the juvenile court judge in chambers. Ms. Cook, the complainant, was not present. Gerald was questioned about the telephone call and was sent back to detention. No record was made of this hearing, no one was sworn to tell the truth, nor was any specific charge made, other than an allegation that Gerald was delinquent. At the conclusion of the hearing, the judge said he would "think about it." Gerald was released a few days later, although no reasons were given for his detention or release.

On the day of Gerald's release, Ms. Gault received a letter indicating another hearing would be held regarding Gerald's delinquency a few days later. A hearing was held, and again the complainant was not present, and no transcript or recording was made of the proceedings (later, what was said was disputed by the parties). Neither Gerald nor his mother was advised of any right to remain silent, of Gerald's right to be represented by counsel, or of any other constitutional rights. At the conclusion of the hearing, Gerald was found to be a delinquent and was committed to the state industrial school until age 21, unless released earlier by the court. This meant that Gerald received a 6-year sentence for an offense that, if committed by an adult, could be punished by no more than 2 months in jail and a $50 fine.[34]

In *Gault*, the Court ruled that the special circumstances that gave rise to the informal process of juvenile courts and the broad discretion of juvenile court judges did not justify the denial of fundamental due process rights for juveniles. The majority opinion stated, "As we shall discuss, the observance of due process standards, intelligently and ruthlessly administered, will not compel the states to abandon or displace any of the substantive benefits of the juvenile process." The justices went on to argue,

> *The constitutional and theoretical basis for this particular system is—to say the least—debatable. And in practice, as we remarked in the Kent case, supra, the results have not been entirely satisfactory. Juvenile Court history has again demonstrated that unbridled discretion, however benevolently motivated, is frequently a poor substitute for principle and procedures. . . . The absence of substantive standards had not necessarily meant that children receive careful, compassionate, individualized treatment. The absence of procedural rules based upon constitutional principle has not always produced fair, efficient, and effective procedures. Departures from established principles of due process have frequently resulted not in enlightened procedure, but in arbitrariness. . . . Failure to observe the fundamental requirements of due process has resulted in instances. . . of unfairness to individuals and inadequate or inaccurate findings of fact*

and unfortunate prescriptions of remedy. Due process of law is the primary and indispensable foundation of individual freedom. . . . Under our Constitution, the condition of being a boy does not justify a kangaroo court.[35]

The court made it clear that juveniles are to be afforded fundamental due process rights that measure up to the essentials of fair treatment. These include

- the right to reasonable notice of the charges
- the right to counsel (either retained or appointed)
- the right to confrontation and cross-examination of witnesses
- the right against self-incrimination

However, the Court limited the application of these rights to proceedings involving a determination of delinquency that may result in commitment to an institution.

A number of fundamental changes were required in juvenile courts after *Gault*. Many of them carried high costs. For example, the right to appointed counsel has become a major budget concern for juvenile courts, especially because this right has been expanded several times in subsequent court decisions. The right to confrontation requires the processing of subpoenas and the costs of service. Finally, due process requirements have resulted in more adversarial hearings, which take longer to complete and thus push up costs.

Gault Raised Important Questions About Juveniles' Rights

Gault raised a number of questions at the time of its release. Would procedural rules, such as the exclusionary rule, be applicable to juveniles? Do juveniles have the right to Miranda warnings before being interrogated by the police? Do juveniles have the right to jury trials? Do juveniles have the right to **bail**? At a hearing to determine a finding of delinquency, does a juvenile have to be found "guilty beyond a reasonable doubt"? Since *Gault*, many, but not all, of these questions have been answered in the affirmative, further prolonging juvenile proceedings and increasing court costs.

bail The money or bond put up to secure the release of a person charged with a crime.

Although *Gault* was a landmark juvenile law case, it was not the last Supreme Court decision that influenced juvenile court procedures. The court further expanded protections for juveniles 3 years after *Gault*. In a 1970 case, *In re Winship* (1970), it addressed the level of proof needed for a conviction of delinquency. The case involved a 12-year-old male who was found guilty of stealing $112 from a woman's purse. As a result, he was placed in a New York training school for a minimum period of 18 months, although the juvenile court indicated that the term of the sentence could be extended to the youth's 18th birthday. The judge who heard the case admitted that proof "beyond a reasonable doubt" was not established at trial but held that this level of proof was not required.

A majority of the Supreme Court justices held that proof of guilt beyond a reasonable doubt was an essential element of due process in delinquency cases. The court indicated that to allow for a lesser standard would seriously harm the confidence of the community in the fairness of the adjudicative process. The court went on to talk about the "moral force" of the criminal law and how this force would be diluted if the standard of proof was lowered. According to the court, a lower standard would lead the community to wonder whether innocent juveniles were being convicted and incarcerated.

The *Winship* decision added due process protections to those established by *Kent* and *Gault*. The result was that the concept of "proof beyond a reasonable doubt," which had long been accepted in adult criminal cases, was now applied to adjudications where juveniles were at risk of institutional placement.

MYTH VS. REALITY — Juveniles Do Not Have All of the Legal Protections Available to Adults

Myth—Today, juveniles have all of the due process protections afforded adults in criminal courts.

Reality—In some states, juveniles still lack some of the due process rights given adults.

The Supreme Court's willingness to extend due process protections to juveniles came into question the following year, however. It heard two cases jointly, one from Pennsylvania and one from North Carolina, that concerned whether juveniles should be entitled to jury trials at the trial or adjudicative stage of the juvenile court process. In the Pennsylvania case, *McKeiver v. Pennsylvania* (1971), the court procedures were similar to those used in adult criminal courts (plea bargaining, motions to suppress evidence, and so on). Moreover, one possible outcome was incarceration in a prison-like facility until the juvenile's age of majority. The North Carolina case revolved around much less formalized procedures than those in Pennsylvania, but the argument was made that the supposed benefits of the juvenile court system—discretionary intake, diversion, flexible sentencing, and a focus on rehabilitation—would not be hindered by the use of juries.

Justice Blackmun, writing for the majority, reviewed the historical reasons for a separate and distinct juvenile court and concluded by stating,

> *The arguments necessarily equate the juvenile proceeding—or at least the adjudicative phase of it—with the criminal trial, whether they should be so equated is our issue. . . . If the formalities of the criminal adjudicative process are to be superimposed upon the juvenile court system, there is little need for its separate existence. Perhaps that ultimate disillusionment will come one day; but for the moment, we are disinclined to give impetus to it."*[36]

Consequently, the Supreme Court declined to extend the constitutional right to jury trial to the juvenile system.

McKeiver is important because it made clear that the Supreme Court was unwilling to give juveniles all of the due process protections available to adults. Nevertheless, many states have extended the right to a jury trial to youths at the adjudicative phase of the juvenile justice process. The *McKeiver* decision was not unanimous, however. Justice Brennan, in his dissent in *McKeiver*, focused on the individual state procedure and whether it affords sufficient protections to the juvenile from any government overreaching and from the "biased or eccentric" judge. He thought that one crucial factor that needed to be taken into account in deciding whether to mandate a jury trial was the public "openness" of trial proceedings and the ability of the jury to act as both a finder of fact and the community's conscience, preventing the state from accusing juveniles for political purposes.[37]

Although one of the objections to jury trials in juvenile proceedings was that they would cause a backlog of cases and hamper the functioning of the juvenile court, experience has not shown such trials to be seriously obstructive. The other issue related to efficiency is whether juries can consist of less than 12 citizens. In many states, a jury of 6 is allowed for delinquency adjudications.

A collateral benefit of allowing jury trials is that they may aide rehabilitation. A juvenile who believes that "the system" has treated him or her unfairly may be less defensive if tried by an objective jury. A perception of fair treatment may go a long way toward fostering acceptance of responsibility on the part of the juvenile, to say nothing of its promotion of feelings of self-esteem. Juveniles who believe that the system treats them fairly and accepts them as significant people whose rights must be protected may overcome feelings that they were treated unjustly.

Another point that is sometimes made about the use of juries in the juvenile system is that, unlike in adult courts, the jurors are not peers of those on trial. In some instances, the juvenile may benefit from this fact, because the adults on the jury may remember their own youthful mistakes and indiscretions and feel sympathy. They may also realize, if they are themselves parents, that under other circumstances one of their children could be at the defense table.

In reality, jury trials in delinquency proceedings have not proved to be docket cumbersome, inefficient, or exceedingly expensive. Moreover, they make a statement to the community and the juvenile regarding the juvenile court's concern for fundamental due process fairness.

Juveniles vs. Criminal Courts

Looked at broadly, *Kent, Gault, McKeiver,* and *Winship* concern the degree to which adult criminal practice should be extended into the juvenile courts. *Kent, Gault,* and *Winship* made clear that certain due process protections that have long been part of adult court practice should also be given to youths being tried in juvenile courts. *McKeiver* indicated that there were limits to how far the Supreme Court was willing to go in this area. It also showed that the court was unwilling to completely dismantle the juvenile justice apparatus and that it wished to maintain some of the traditional informality of juvenile

pretrial discovery Efforts of a party to a lawsuit to obtain information prior to trial. The theory behind discovery is that all parties should go to trial with as much information as possible.

bills of particulars A written itemization of claims or charges that the defendant in a lawsuit can demand of the plaintiff that can serve as the factual basis of the allegations against the defendant.

courts.[38] These cases left unresolved, however, whether many other adult court practices should apply to the juvenile courts such as a defendant's right to bail, **pretrial discovery**, and **bills of particulars**.

Despite the due process protections extended to juveniles through the *Kent*, *Gault*, and *Winship* decisions, in practice much of the informality of the juvenile court remains intact. Indeed, one complaint is that changes in the legal procedures that supposedly govern the juvenile courts have not always resulted in fundamental changes in the daily operation of juvenile justice.[39] Many critics contend that juveniles are often denied basic protections within the juvenile justice process and that the continued informality of the juvenile courts fails to serve either the juveniles' best interests or the best interests of the community.

 Some States Allow Jury Trials

Many states have required more due process protections for juveniles than have been mandated by the Supreme Court. For example, many state laws specify that juveniles have a right to a jury trial (adjudication). In fact, at least one state, Texas, requires that all adjudications be heard by a jury. Nevertheless, jury trials are rare in states that have this right, including Texas. This is because the right is not often exercised (or, as in the case of Texas, it can be and is often waived). In practice, jury trials are frequently discouraged because they are felt to be time-consuming and costly or because they impinge on the power of the juvenile court judge.

It is one thing to be afforded various rights through Supreme Court or state court rulings and state statutes. It is another thing to ensure that those involved in the juvenile justice process know all of their rights and that they feel comfortable exercising those rights. However, a number of studies raise serious doubts about youths' access to counsel in delinquency proceedings and the quality of representation that occurs. A study by the American Bar Association Juvenile Justice Center, the Youth Law Center, and the Juvenile Law Center raised serious concerns about the quality of legal representation given youths in many juvenile courts.[40] This research found that, in many instances, juveniles are not represented by attorneys when they appear in juvenile court, attorneys who represent youths often have high caseloads and in some instances lack the proper training and experience in juvenile court to provide effective representation, and youths and their parents or guardians often fail to have a clear understanding of the legal process. As a result, youths are frequently in a vulnerable position, particularly in the early stages of the juvenile justice process, and this can lead to more severe dispositions than are warranted. These findings mirror those of other studies that have examined attorney representation in juvenile court proceedings.[41] For example, a study that examined statewide data in six states (California, Minnesota, Nebraska, New York, North Dakota, and Pennsylvania) found that in three of those states (Minnesota, Nebraska, and North Dakota) in only about half of the cases where a petition was filed was the defendant represented by counsel. In addition, this study found that youths who received assistance from counsel were more likely to receive a more severe disposition, even when the seriousness of the charges and the youths' delinquent history were taken into account.[42] Such findings indicate that not only is representation absent in many instances, but also the quality of the representation provided to juveniles is often inadequate. Moreover, the failure to receive effective representation can result in harm to juveniles involved in the juvenile court.

The continued informality of the juvenile court may also explain why very few youths contest the charges against them.[43] Moreover, many others who are not petitioned and a sizable number who go to adjudications but are not found guilty are still placed on some form of probation. Data collected by the National Center for Juvenile Justice indicates that in 1999, 34% of youths who were referred to juvenile court but were not petitioned were still placed on some form of probation, while 12% of youths who went to an adjudication but were not adjudicated were placed on probation and 19% were given some other sanction.[44]

Today, juveniles have been granted many, but not all, of the due process protections given adults in criminal trials. However, the extent to which court-mandated changes in juvenile justice procedures have actually influenced the traditional informality of the juvenile courts is open to question. Juvenile court procedures in many jurisdictions are still characterized by an informality that would be considered unacceptable for adults brought before a criminal court.[45] Is the informality necessary for the courts to carry out their mandate to serve the best interests of children and protect the community, as supporters of the traditional court procedures argue? Or does the courts' failure to adhere to more strict due process standards lead to abuses and harm many children, as critics assert? These issues continue to be the focus of an important policy debate within the field of juvenile justice.

■ Legal Issues

Kent, Gault, and *Winship* extended due process protections to juveniles in instances where juveniles are at risk of commitment. These cases also raised the issue of the rights of juveniles who are involuntarily committed by their parents to mental health facilities, drug treatment centers, and other types of treatment programs. Should children have due process rights in these circumstances? Also, what about those instances where the state or the federal government seeks treatment or commitment?

■ Chapter Summary

The juvenile courts were developed by progressive reformers (the child savers) in the late 1800s as a new means of controlling wayward and problem youths. They were characterized by a focus on rehabilitation, procedural informality, an individualized approach to cases, and the separation of juveniles from adult offenders.[46] The presumption of the child savers was that juvenile delinquents could be, and should be, controlled and induced to abandon their youthful waywardness. Although the child savers were concerned about the increasing numbers of problem children who posed a threat to community life, they also viewed "delinquents" as children who had lost their way and who needed control, guidance, and nurturing. They believed that children should feel "protected" by the state and that procedural formalities such as those found in adult criminal courts would only serve to intimidate children. They had in mind an image of a judge with his arm around a child, not a judge behind a bench with a gavel in his hand looking down on a child from on high. Their hope was that the juvenile court would form an understanding of each child's social and family history and use it to develop an appropriate "treatment."

Although many child savers were concerned about the well-being of children and sought to create institutions capable of serving their best interests, others were threatened by the many poor and immigrant children who lived in the rapidly growing urban areas and represented a threat to traditional institutions. From the point of view of the latter reformers, the main goal was simply to prevent delinquent and other problem behaviors, and the possibility that prevention could be achieved through the juvenile courts and correctional institutions was reason enough to support them.

By the early 1900s, juvenile courts had spread throughout the country. Clearly, some children were helped by their existence, but the juvenile courts did not always act in the kind and benevolent manner their supporters had envisioned. Nevertheless, juvenile court practice was not closely scrutinized until the 1960s. During the 1960s, however, some of the shortcomings of existing juvenile court practice began to receive attention by the U.S. Supreme Court. In a series of important cases, the Supreme Court extended due process protections to juveniles. The essential thrust of the rulings was that juveniles had a right to some protection from the awesome coercive powers of the juvenile court. Whether all jurisdictions adhere firmly to the principle that juveniles are entitled to due process protections is in doubt. Indeed, there is considerable evidence that many juvenile courts still operate in a very informal manner, which in effect circumvents the due process protections supposedly available to juveniles.

KEY CONCEPTS

bail The money or bond put up to secure the release of a person charged with a crime.

bills of particulars A written itemization of claims or charges that the defendant in a lawsuit can demand of the plaintiff that can serve as the factual basis of the allegations against the defendant.

chancery court English court whose primary concern is with property rights. Chancery courts, first established during the Middle Ages, were partially based on the idea that children, particularly children who owned or were in a position to inherit property, should fall under the protective control of the king. Indeed, it was in these courts that the concept of *parens patriae* was developed. Over time, the chancery courts came to be more involved in the general welfare of families and children. Because of their role in protecting the interests of children, they served as a model for the juvenile court.

child savers A group of reformers, mostly well-educated, middle-class Protestant women of Anglo-Saxon descent, who had the time and resources to fight for improved conditions for delinquent youths in jails and reformatories and who eventually played an important role in the creation of the juvenile courts. On the positive side, their work led to wider acceptance of adolescence as an important developmental stage of life, and they correctly believed that the adult criminal justice system was harmful to children or at best ineffective. On the negative side, their conservative ideas on the family and their "noblesse oblige" attitude toward the poor resulted in reforms that focused more on the social control of children and the protection of the interests of the wealthy than on attacking the underlying causes of problem youth behaviors.

dependent children Children who are dependent on others for financial support or who do not have the support of a parent or guardian.

due process A course established for legal proceedings intended to ensure the protection of the private rights of the litigants. The essential elements of due process are (1) proper notice as to the nature of the legal proceedings, (2) a meaningful hearing in which the individual has an opportunity to be heard and/or defend him- or herself, and (3) the objectivity of the tribunal before which the proceedings take place.[47]

neglected children Children who are not given appropriate care by the parent(s) or a guardian.

parens patriae The obligation of the government to take responsibility for the welfare of children. The doctrine of *parens patriae* recognizes the moral obligation of the government to take care of children when there is no family available or if the family is not suitable.

pretrial discovery Efforts of a party to a lawsuit to obtain information prior to trial. The theory behind discovery is that all parties should go to trial with as much information as possible.

Progressive Era A period in American history, lasting roughly from 1890 to 1924, during which a variety of economic, social, and political reforms and reform movements occurred, including women's suffrage; trust busting; Prohibition; working-hour reduction; elimination of child labor; adoption of social welfare benefits; and the popular voting measures of initiative, recall, and referendum. The reforms were primarily concerned with responding to popular unrest and problems uncovered in the operation of economic, social, and political institutions.

REVIEW QUESTIONS

1. In what ways did the social context of the 1800s influence the development of the "child saving" movement?

_____

_____

_____

_____

2. Who were the child savers, and what was their approach to saving children?

_____

_____

_____

_____

3. According to the child savers, what caused youth crime and waywardness?

_____

_____

_____

_____

4. What legal mechanisms existed during the early and mid-1800s to respond to children who engaged in crime or other problem behaviors?

_____

_____

_____

_____

_____

5. What were the facts of the _Crouse_ case, what were the findings of the Pennsylvania Supreme Court, and why is this case significant?

_____

_____

_____

_____

_____

6. What were the facts of the _People v Turner_ case, what were the findings of the Illinois Supreme Court, and in what ways did this case influence the development of the juvenile court in Chicago?

_____

_____

_____

_____

_____

7. What social and legal factors contributed to the development of the first statutorily recognized juvenile court?

_____

_____

_____

_____

_____

8. What were the essential characteristics of the first juvenile court?

_____

_____

_____

_____

_____

9. What were the facts of the *Fisher* case, what were the findings of the Pennsylvania Supreme Court, and why is this case significant?

_____

_____

_____

_____

_____

10. What were the essential features of early juvenile court practice?

_____

_____

_____

_____

_____

11. What issues were addressed by the U.S. Supreme Court in the *Kent*, *Gault*, *Winship*, and *McKeiver* cases? What were the important Supreme Court rulings in these cases?

_____

_____

_____

_____

_____

12. In what ways did the *Kent, Gault, Winship,* and *McKeiver* cases influence the operation of the juvenile courts?

_____

_____

_____

_____

_____

13. Explain how the continuing informality of many juvenile courts circumvents the due process protections extended to juveniles by the Supreme Court.

_____

_____

_____

_____

_____

ADDITIONAL READINGS

Bernard, T. J. 1992. *The Cycle of Juvenile Justice.* New York: Oxford University Press.
Feld, B. C. 1999. *Readings in Juvenile Justice Administration.* New York: Oxford University Press.
Hemmens, C., B. Steiner and D. Mueller. 2004. *Significant Cases in Juvenile Justice.* Los Angeles, CA: Roxbury Publishing Company.
Krisberg, B. and J. F. Austin. 1993. *Reinventing Juvenile Justice.* Newbury Park, CA: Sage.
Platt, A. M. 1977. *The Child Savers: The Invention of Delinquency,* 2nd ed. Chicago: University of Chicago Press.

CASES CITED

Commonwealth v. Fisher, 213 P. 48 (1905).
Ex parte Crouse, 4 Wharton (P.) 9 (1838).
In re Gault, 387 US 1, 87 S Ct 1428, 18 L Ed 2nd 527 (1967).
In re Winship, 397 US 358, 90 SCt 1068, 25 L Ed 2d 368 (1970).
Kent v. United States, 383 US 541 86 SCt 1045, 16 L Ed 2nd 84 (1966).
McKeiver v. Pennsylvania, 403 US 528, 91 SCt 1976, 29 L Ed 2d 647 (1971).
People v. Turner, 55 Illinois 280 (1870).

NOTES

1. H. Zinn, *A People's History of the United States 1492–Present* (New York: HarperPerennial, 1995).
2. H. Finestone, *Victims of Change* (Westport, CT: Greenwood Press, 1976).

Chapter Resources

3. Zinn, *A People's History of the United States 1492–Present.*

4. Zinn, *A People's History of the United States 1492–Present.*

5. A. H. Kelly and W. A. Harbison, *The American Constitution: Its Origin and Development* (New York: W.W. Norton, 1955), 612.

6. P .F. Clement, "Families and Foster Care: Philadelphia in the Late Nineteenth Century," in *Growing Up in America: Children in Historical Perspective,* eds. R. Hiner and J. M. Hawes (Chicago: University of Illinois Press, 1985).

7. A. M. Platt, *The Child Savers: The Invention of Delinquency,* 2nd ed. (Chicago: University of Chicago Press, 1977).

8. T. J. Bernard, *The Cycle of Juvenile Justice* (New York: Oxford University Press, 1992); Platt, *The Child Savers.*

9. C. J. Sommerville, *The Rise and Fall of Childhood* (New York: Vintage Books, 1990).

10. Bernard, *The Cycle of Juvenile Justice.*

11. Platt, *The Child Savers.*

12. Platt, *The Child Savers.*

13. R. M. Mennel, *Thorns and Thistles: Juvenile Delinquents in the United States, 1825–1940* (Hanover, NH: University Press of New England, 1973), 129.

14. Platt, *The Child Savers.*

15. Quoted in Platt, *The Child Savers,* 119.

16. Bernard, *The Cycle of Juvenile Justice.*

17. Platt, *The Child Savers.*

18. Bernard, *The Cycle of Juvenile Justice.*

19. Bernard, *The Cycle of Juvenile Justice.*

20. Platt, *The Child Savers;* Bernard, *The Cycle of Juvenile Justice.*

21. Platt, *The Child Savers;* Bernard, *The Cycle of Juvenile Justice.*

22. W. B. Waegel, *Delinquency and Juvenile Control: A Sociological Perspective* (Englewood Cliffs, NJ: Prentice-Hall, 1989).

23. L.T. Empey et al., *American Delinquency: Its Meaning and Construction,* 4th ed. (Belmont, CA: Wadsworth, 1999).

24. A. R. Roberts, "The Emergence of the Juvenile Court and Probation Services," in *Juvenile Justice: Policies, Programs, and Services,* ed. A. R. Roberts (Chicago: Nelson-Hall, 1998); Empey et al., *American Delinquency.*

25. C. Bartollas and S. J. Miller, *Juvenile Justice in America,* 2nd ed. (Englewood Cliffs, NJ: Prentice Hall, 1998); Empey et al., *American Delinquency.*

26. B. Krisberg and J. F. Austin, *Reinventing Juvenile Justice* (Newbury Park, CA: Sage, 1993).

27. Platt, *The Child Savers.*

28. Bernard, *The Cycle of Juvenile Justice.*

29. Bernard, *The Cycle of Juvenile Justice.*

30. Bernard, *The Cycle of Juvenile Justice.*

31. C. Lindner and M. R. Savarese, "The Evolution of Probation," *Federal Probation* 48 (1984): 3–10; Roberts, "The Emergence of the Juvenile Court and Probation Services."

32. Platt, *The Child Savers.*

33. Bernard, *The Cycle of Juvenile Justice,* 113.

34. J. J. Senna and L. J. Siegel, *Juvenile Law: Cases and Comments,* 2nd ed. (St. Paul, MN: West Publishing Co., 1992).

35. M. A. Bortner, *Delinquency and Justice: An Age of Crisis* (New York: McGraw-Hill, 1988), 62.

36. *McKeiver v. Pennsylvania,* 403 US 528, 91 SCt 1976, 29 L Ed 2d 647 (1971).

37. *McKeiver v. Pennsylvania.*

38. Bortner, *Delinquency and Justice;* Bernard, *The Cycle of Juvenile Justice.*

39. Bernard, *The Cycle of Juvenile Justice.*

40. P. Puritz et al., *A Call for Justice: An Assessment of Access to Counsel and Quality of Representation in Delinquency Proceedings*. (Washington, DC: Office of Juvenile Justice and Delinquency Prevention, 1995).

41. B. C. Feld, "Justice by Geography: Urban, Suburban, and Rural Variations in Juvenile Justice Administration," *Journal of Criminal Law and Criminology*, 82 (1991): 156–210; B. C. Feld, "The Right to Counsel in Juvenile Court: An Empirical Study of When Lawyers Appear and the Difference They Make," *Journal of Criminal Law and Criminology*, 79 (1989): 1185–1346; B. C. Feld, "In re Gault Revisited: A Cross State Comparison of the Right to Counsel in Juvenile Court," *Crime and Delinquency* 34 (1988): 393–424.

42. B. C. Feld, "*In re Gault* Revisited: A Cross State Comparison of the Right to Counsel in Juvenile Court."

43. Bernard, *The Cycle of Juvenile Justice*.

44. C. Puzzanchera, *Juvenile Court Statistics 1999*. (Washington, DC: Office of Juvenile Justice and Delinquency Prevention, 2003).

45. M. D. Jacobs, *Screwing the System and Making It Work: Juvenile Justice in the No-Fault Society* (Chicago: University of Chicago Press, 1990).

46. R. H. Mnookin and D. K. Weisberg, *Child, Family, and State: Problems and Materials on Children and the Law*, 2nd ed. (Boston: Little, Brown & Co., 1989).

47. H. C. Black, *Black's Law Dictionary*, 4th ed. (St. Paul, MN: West, 1968); *Herman v. Chrysler Corp*, 106 Mich App 709, 308 NW2d 616 (1981).

Public and Police Responses to Juvenile Offenders

6

Chapter Objectives

After studying this chapter, you should be able to

- Describe the role that individual citizens play in both the informal and formal juvenile justice process
- Describe the evolution of police responess to youths from colonial America to the present
- Explain the three fundamental duties of the police and how these duties influence police responses to delinquent behavior
- Explain why conflict is an inherent feature of policing and how this conflict influences how the police interact with juveniles and their families
- Describe the factors that influence police decisions to refer youths to the juvenile courts
- Define the concept of police diversion of juveniles, describe its history, and indicate how it is presently used in juvenile justice
- Explain the impact of court decisions such as *Miranda v. Arizona* and *Schall v. Martin* on police interaction with juveniles, their families, and the courts
- Describe recent trends in police responses to juvenile offending
- Describe the focus and effectiveness of policing programs and strategies such as DARE, community policing, and youth-oriented community policing in dealing with delinquent behavior

Chapter Outline

- Introduction
- Youths and the Public: The Informal Juvenile Justice Process
- Police Responses to Juvenile Offenders
- Legal Issues
- Chapter Summary
- Key Concepts
- Review Questions
- Additional Readings
- Cases Cited
- Notes

■ Introduction

This chapter examines citizen and police responses to juvenile offenders. It first explores the role that the public plays in dealing with juvenile crime by describing the operation of an informal juvenile justice process—one that consists of efforts by citizens to handle instances of juvenile crime that do not involve the police or other formal agents of juvenile justice. Next, it examines police responses to juvenile offenders and presents a general overview of the history of youth-oriented policing. It then examines contemporary issues regarding the policing of juveniles and concludes with a discussion of recent trends in youth-oriented policing.

■ Youths and The Public: The Informal Juvenile Justice Process

Examinations of the juvenile justice process usually begin with the police. The police do play a critical role in juvenile justice; in fact, they are the primary gatekeepers of the formal juvenile justice process. The public is also involved, however, because police contact with juveniles is typically the result of a citizen complaint or request for assistance. Thus, an examination of the juvenile justice system should arguably begin with the public.

As noted in Chapter 2, most juveniles, at some point, engage in illegal behaviors. Although some juvenile delinquency consists of serious offenses committed against people, the great majority of delinquent acts are nonviolent. Moreover, people respond to this delinquency in a variety of ways. Some of these responses are informal, which means that they consist of the actions taken by members of the public without reliance on official juvenile justice agencies. Formal responses, in contrast, involve official agencies, such as the police and the juvenile courts.

When citizens, whether adults or other juveniles, encounter or observe youths who they believe are engaging in illegal behavior, they may exercise any of a variety of options. One option is to ignore the offender. Another is to confront the offender. For example, a store clerk who observes a youth taking merchandise from a store shelf and attempting to leave the store may stop the youth and demand that the goods be returned, may lecture the youth and ban him or her from the store, may contact the youth's parents and request that the parents take some action, may call the police, or may decide to take some other action. Importantly, citizens use **discretion**. Discretion is the power to act on one's own authority.

Citizens regularly encounter juveniles involved in alleged violations of the law. Yet in many instances they do not contact the police. As just noted, an individual may choose to ignore the youth's actions or deal with the youth in some way that does not involve the police or other formal agents of the juvenile justice process. The actions taken by parents, neighbors, business owners, teachers, and others constitute an **informal juvenile justice process**. Citizens regularly encounter youths engaged in illegal behaviors and decide to handle those situations themselves or seek the assistance of others outside the formal juvenile justice process. This informal juvenile justice process is one, potentially powerful, mechanism that operates to control youths' behavior. Moreover, the more that citizens rely on informal control mechanisms and the more effective they are, the less necessary the formal processing of juveniles becomes.

Although many delinquent actions are handled informally, in other instances members of the public seek the assistance of formal juvenile justice agencies, such as the police or the juvenile courts, and most youths become involved in the formal juvenile justice process when citizens decide on this option. In short, the public and the police each play a major role in determining who the clientele of the formal juvenile justice process will be.

discretion The power to act on one's own authority.

informal juvenile justice process Actions to combat juvenile delinquency taken by parents, neighbors, business owners, teachers, and others who are not part of the formal juvenile justice apparatus.

■ Police Responses to Juvenile Offenders

The police are clearly an important component of the juvenile justice process. Indeed, decisions made by the police determine who the vast majority of formal juvenile justice clients will be. This is because the police are the most visible symbol of the juvenile justice process within the community. When citizens believe that juveniles are engaging in illegal or problem behaviors and seek the assistance of a formal juvenile justice agency, it is to the police that they usually turn. Consequently, how the police respond to juvenile offenders influences how many people view the juvenile justice process.

Like the responses of other institutions to juvenile offenders, police responses are best understood by examining the historical development of policing, particularly as it relates to young people. Consequently, this section begins with a brief historical overview of youth-oriented policing, then examines more contemporary approaches to policing youths.

The History of Policing Youths

The British colonists who settled the New World brought with them a variety of criminal justice institutions and practices, including English common law, a court system, punishments for crimes, and policing practices. However, colonial law enforcement had a number of shortcomings including inefficiency, corruption, and political interference.[1] Those who had formal policing responsibility, such as constables, sheriffs, and members of the night watch, engaged in a variety of activities besides law enforcement. For example, sheriffs' responsibilities included collecting taxes, conducting elections, and maintaining bridges and roads. In fact, law enforcement typically ranked among their less important areas of responsibility.[2] This is because early law enforcement officials were paid through a system of fees collected for the performance of particular duties.[3] This system encouraged early law enforcement officials to devote more time to civil functions that were safer and offered more regular income than law enforcement activities.

Policing during the colonial period was primarily reactive and often ineffective. Those who had formal law enforcement duties responded to complaints brought to their attention, but they had limited resources to investigate offenses and engaged in few efforts to prevent crime. Indeed, most criminal cases in colonial America were initiated by citizens who brought complaints before a judge; rare was the case in which a citizen contacted the police in order to initiate an investigation.[4] Rather than relying on formal mechanisms to control youth and adult behaviors, residents in colonial settlements relied more heavily on the family and the church.

MYTH VS. REALITY

During the Early Years in the U.S. Informal Controls Were the Primary Means of Social Control

Myth—Early colonial settlements established full-time police forces to maintain order.

Reality—Full-time police forces did not exist in colonial times. Indeed, the first full-time police forces in the United States were not developed until the early 1800s.[5]

By the early 1800s, industrialization, urbanization, population growth, and associated social problems, such as alcoholism, homelessness, riots, ethnic clashes, and lawlessness, led to the development of more formal methods of policing. Full-time police forces were created in larger cities during the 1830s and 1840s. By the 1870s, all the major cities had full-time forces and many smaller ones employed part-time police officers.

In these early days, policing in the United States was community based. Police officers came from the communities in which they worked and were paid more than many other workers. They walked the beat in the neighborhoods they patrolled, usually spending from 2 to 6 hours of a 12-hour shift on foot patrol. However, police forces were small, beat officers had little contact with supervisors while on patrol, and turnover was high.[6]

A Lack of Respect by Many Citizens Was a Problem for Early Police Officers

Myth—During the late 19th century and early 20th century, most city residents had a great deal of respect for the friendly neighborhood patrol officer. **Reality**—Many people lacked respect for the police, whom they saw as political hacks willing to use brutality to protect their bosses' interests.[7]

A major problem that faced early policing organizations was corruption. Because the police were paid fees for the jobs they performed, they often saw themselves as private entrepreneurs rather than public servants. Also, policing was closely tied to local politics. Police jobs were a primary form of political patronage and were controlled by local political bosses. Police officers owed their allegiance to these political bosses and the police officials who awarded them their positions rather than to the public.[8] As a consequence, they were careful to act in ways that protected the politically powerful, sometimes by ignoring illegal activities, such as gambling, prostitution, and after-hours liquor sales, and sometimes by participating in such activities themselves.[9]

Although some officers were undoubtedly respected by people in their communities, many citizens saw the police as servants of corrupt political machines.[10] It was common to view the police as primarily engaged in rigging elections, controlling ethnic and racial minorities, breaking strikes, and protecting the interests of wealthy business owners.[11] Police officers often faced considerable hostility from both young and old, and juvenile gangs often made a sport of throwing rocks at the police and taunting them.[12]

The feeling was mutual. Early police officers exhibited contempt for problem youths as well as concern about those youths who engaged in illegal behavior or who were felt to be potential offenders. For example, in his 1849 semiannual report to the mayor of New York, the Chief of Police George W. Matsell noted the following:

> I deem it to be my duty to call the attention of your Honor to a deplorable and growing evil which exists amid this community. . . . I allude to the constantly increasing numbers of vagrant, idle, vicious children of both sexes, who infest our public thoroughfares, hotels, docks. . . . Children who are growing up in ignorance and profligacy, only destined to a life of misery, shame and crime, and ultimately to a felon's doom. . . . Left, in many instances to roam day and night . . . a large portion of these juvenile vagrants are in the daily practice of pilfering wherever opportunity offers, and begging where they cannot steal. In addition to which, the female portion of the youngest class, those who have only seen some eight or twelve summers, are

addicted to immoralities of the most loathsome description. Each year makes fearful additions to the ranks of these prospective recruits of infamy and sin, and from this corrupt and festering fountain flows on a ceaseless stream to our lowest brothels—to the Penitentiary and the State Prison![13]

In dealing with youth crime and waywardness, the police employed a variety of tactics. Some of these tactics involved formal responses to youth crime, such as **arrest**. Indeed, police officers during this era had almost unlimited authority. They could place a youth in a house of refuge, seek the detention of a youth in a jail, or refer a youth to a criminal court (which could itself place the youth in a jail or prison). However, police officers also employed a variety of informal, not always legal, tactics in dealing with the young. Some officers developed close working relationships with both youths and adults in their community in order to aid them in responding to problem children. In other cases, officers warned and counseled children. On still other occasions police dispensed street justice by harassing, verbally abusing, or beating up problem youths.[14]

American policing underwent important and fundamental changes during the 20th century. Essentially these changes resulted from two factors: the development of professional policing standards and the improvement of communications technology, which led to changes in the nature of police work and police administration.[15]

The push toward police professionalism grew during the Progressive Era (roughly 1880 to 1920). This period saw a variety of social movements that focused attention on problems such as economic abuses, social welfare problems, and political and police corruption. Two of the primary architects of reform were Richard Sylvester, superintendent of the District of Columbia force from 1898 to 1915 and president of the International Association of Chiefs of Police (IACP) from 1901 to 1915, and August Vollmer, chief of police in Berkeley, California, from 1905 to 1932. Sylvester was instrumental in turning the IACP into an effective national police association that advocated for progressive police reform. Vollmer was an important innovator who developed modern police administrative methods and championed increased education for police officers. But Vollmer did more than just advocate for these reforms. He hired graduates of the University of California to work on the Berkeley force and organized the first police science courses at the university in 1916.[16]

The **reform agenda** pushed by Vollmer, Sylvester, and others had a number of goals: "eliminating political influence, appointing qualified chief executives, establishing a mission of nonpartisan public service, raising personnel standards, introducing principles of scientific management, emphasizing military-style discipline, and developing specialized units."[17] The development of specialized units in police departments was an important event in youth-oriented policing in the United States. In the early 1900s, one type of specialized police unit that began to appear was the juvenile unit (sometimes called a youth aid bureau, juvenile bureau, juvenile control bureau, juvenile division, or crime prevention bureau). The development of specialized juvenile units spread during the early 1900s as professionally oriented departments began to think more about delinquency and crime prevention.[18]

arrest Legally authorized deprivation of a person's liberty. As a general rule, a person is under arrest when he or she is not free to walk away.

reform agenda for police Agenda consisting of items such as the following: eliminating political influence and corruption, appointing qualified leaders, establishing the ideal of nonpartisan public service, raising personnel standards, implementing scientific management principles, instituting military-style discipline, and creating "specialized" units.

One significant aspect of early delinquency prevention efforts was that they became the mechanism by which women first entered policing. The first woman to be hired as a police officer was Mary Owens, who was given a job by the Chicago Police Department in 1893 after her husband, a Chicago police officer, was killed in the line of duty. Owens made visits to the juvenile court and helped detectives on cases involving women and girls.[19] In 1905 Lola Baldwin was hired by the Portland, Oregon, Police Department. Baldwin had a background in social work, and her primary responsibility was to protect young girls and women and to prevent their involvement in crime. Indeed, women were felt to be particularly suited to dealing with the needs of children, which led several departments around the United States to follow the lead of Portland and hire women to act as police social workers.[20]

The concept of delinquency prevention spread to a number of larger police departments during the early 1900s, and by 1924 approximately 90% of the nation's largest cities had specialized juvenile programs that placed welfare officers in high-crime neighborhoods, and assigned police officers to juvenile courts. Other innovative police programs, such as specialized juvenile crime units, relief programs that gave toys to children at Christmas, speakers' bureaus that gave presentations to youth organizations (e.g., the Boy Scouts and Camp Fire Girls), and police athletic leagues, were also developed.[21]

In the 1950s two additional events highlighted the growing importance of juvenile officers in police departments around the country. In 1955, the Central States Juvenile Officers Association was formed, followed by the International Juvenile Officers Association in 1957. Both of these organizations worked toward the development of professional standards and procedures for handling juveniles.[22]

During the 1960s and 1970s, the importance of the police role in the handling of juvenile offenders was given additional support through the Law Enforcement Assistance Administration (LEAA). LEAA provided federal funding for the development of hundreds of new police programs, a number of which were targeted at youths.[23] One of these programs, police diversion, became popular in a number of jurisdictions. These programs were intended to steer youths away from the formal juvenile justice process by providing various services to youths and their families. Although the services varied from jurisdiction to jurisdiction, they often included individual counseling, recreational programs, social skills training, and the development of parenting skills.[24] Other programs developed during this period included leadership and moral training programs, job assistance programs, and programs intended to reduce school truancy.[25]

In the late 1970s and 1980s, many police departments, particularly smaller ones, faced budget cuts and rethought the operation of specialized youth divisions. These departments did away with or reduced the size of specialized juvenile units.[26] Nevertheless, others developed still more specialized units that focused on family violence and gangs (issues that were becoming more problematic for police, especially in larger jurisdictions). Many departments began to focus increased attention on juvenile crime during the late 1980s and the early 1990s due to concerns about increases in juvenile violence during that period. Today, a diverse range of police programs and responses to juvenile crime can be found throughout the country.

The Role of the Police in Modern Society: Implications for Policing Juveniles

In order to understand the role of police in modern society, it is necessary to consider the "rights and duties" as well as "the normatively approved patterns of behavior" for those in law enforcement.[27] Roles have both sociological and psychological dimensions. Sociologically, the role of police consists of the position that policing has within the social structure and the activities of those who perform policing functions. Obviously, police have considerable power and responsibility and perform a variety of duties as part of their job. Psychologically, the police role is composed of the attitudes and beliefs that people (including the police themselves) have about those in policing. Importantly, the public and the police have certain expectations for those who work in policing, and these expectations are related to police performance.[28]

One thing is quite clear: The role of police in contemporary society is complex and often characterized by conflict.[29] Some scholars who have studied the police argue that the police role is composed of three basic functions: law enforcement, service, and order maintenance.[30] The law enforcement role of the police is directed toward the detection, apprehension, and prevention of illegal behavior, and the collection of evidence that can be used in the prosecution of cases in court. Police efforts to enforce legal statutes, including traffic, juvenile, and criminal codes as well as investigating crimes, chasing and arresting suspects, transporting suspects to jail, enforcing traffic laws, engaging in routine patrol, and appearing in court are all examples of the law enforcement role performed by the police.

The service role of the police encompasses efforts by the police to assist citizens in a variety of ways. Indeed, police play a central role in providing a range of services to members of the community who, because of personal, financial, economic, social, or other circumstances, need assistance.[31] Service activities include assisting motorists with stalled vehicles, giving directions to motorists and others, providing various types of information to citizens, giving first aid to injured or ill people, escorting ambulances or fire trucks, responding to the needs of people with mental health problems, and providing a variety of other services to members of the public.

The order maintenance role of the police encompasses their efforts to intervene in situations that threaten to disturb the peace or that involve face-to-face conflicts between two or more individuals. Examples of order maintenance include resolving a dispute between a tenant and a landlord or between a store clerk and a customer, dealing with a noisy drunk, and responding to a group of rowdy college students at a fraternity party. Drunkenness and rowdiness are behaviors disapproved of by some members of the public, and those who disapprove of these behaviors often ask the police to control or stop them.[32]

The order maintenance role of the police is important for several reasons. First, order maintenance activities account for as much as 80% of police activity.[33] Second, actions that disrupt the public order have the potential to erupt into violence. Third, police have considerable discretion in handling public order situations.[34] Fourth, these situations often involve juveniles and their families. Police regularly encounter individual youths as well as youths in large and small groups who may be perceived by individual police officers and some members of the community as potentially threatening to the public order.

The most appropriate police response to these situations is not always clear, however. Police feel obligated to assist people who have complaints about the behavior of juveniles, yet behaviors that members of the public find threatening or annoying are not necessarily illegal. Police officers may feel that a group of youths on a street corner represents a potentially problematic situation that demands intervention, but the youths have a constitutional right to congregate and express themselves, a right that the police are obligated to protect. As a result, the police may find themselves in a no-win situation. If they fail to disperse a rowdy group of teenagers, citizens may complain that the police are ineffectual or do not care. If they break up the group, they open themselves up to a charge of police harassment. The attitudes and behaviors exhibited by both police and juveniles in these encounters can play a large role in shaping the attitudes that police and juveniles have of one another.

As this example indicates, police often experience **role conflict** owing to their different responsibilities. They are expected to prevent crimes and respond to the illegal behaviors of individuals, including juveniles, but they are also expected to protect the rights of all citizens. In addition, the police recognize that the public does not want them to enforce all laws and that full enforcement of the law is actually counterproductive. For example, many police recognize that the arrest of a juvenile for a minor illegal activity, such as pushing another child down on a community playground (an assault), might result in more harm than good and could be better dealt with in a more informal manner.

police role conflict Clash between the duty of police to investigate crimes and prevent individuals from committing crimes and the duty of police to protect the rights of offenders, crime victims, and other citizens.

MYTH VS. REALITY — Policing Is Characterized by Role Conflict

Myth—The public expects the police to enforce all laws.
Reality—The public expects the police to engage in selective law enforcement. For example, few citizens want the police to strictly enforce all traffic laws. Indeed, the police recognize that strict enforcement of all traffic laws would result in considerable conflict between themselves and citizens.

Because police encounters with juveniles are frequent, they often occur under stressful conditions, and they play an important role in shaping the attitudes that police and juveniles have toward one another. Therefore, it is imperative that police be carefully trained in dealing with youths. Police departments have used two basic training approaches. One approach involves training individual officers in how to approach juveniles. This approach is more appropriate for smaller departments where there is little specialization; however, officers in small departments may receive little training that focuses specifically on dealing with juveniles. The second approach involves developing specialized units to deal with more serious juvenile cases and is more likely to be found in larger departments with specialized units. However, the quality of training given individual officers in handling juvenile matters, and in some cases the training of specialized juvenile officers, varies considerably across departments.[35] Some rural and smaller police departments have no officers who are specially trained to deal with juveniles.[36] In contrast, other departments have given extensive training to officers to help them to respond to juvenile crime more effectively.[37]

Police: The Gatekeepers of the Formal Juvenile Justice Process

Clearly, police work is characterized by complexity and conflict. Nevertheless, police play a crucial role in juvenile justice because they act as primary gatekeepers to the formal juvenile justice process. For example, in 1999, 84% of the delinquency cases referred to the juvenile courts came from the police. In addition, police agencies were the primary referral source for 40% of court referrals involving running away from home and 92% of the referrals involving liquor law violations.[38] Like citizens, however, the police also exercise discretion in the handling of juvenile cases. Among other options, police officers may

- warn and release the juvenile
- refer the juvenile to his or her parents
- refer the juvenile to a diversionary program operated by the police or a community agency
- refer the juvenile to the local juvenile court

Describing the typical police response to juveniles is difficult, because there is considerable variability in the options available within different jurisdictions and in how individual police officers approach juvenile offenders. The factors that influence police decision making include (1) the seriousness of the offense; (2) the police organization; (3) the community, including the resources available within the community for responding to the juvenile offender; (4) the wishes of complainants; (5) the demeanor of the offender; (6) the gender of the offender; and (7) the race and social class of the offender.

Offense Seriousness

The most important factor influencing police–juvenile encounters is the seriousness of the offense. Regardless of the other factors, as the seriousness of the offense increases, so does the likelihood of arrest. Indeed, most police–juvenile encounters involving felony offenses result in an arrest.[39] Most police–juvenile encounters, however, involve minor offenses. Only 5 to 10% involve felonies.[40] Consequently, in the majority of police–juvenile interactions, other factors play an important role in determining how police react.

The Police Organization

Police departments, like other types of organizations, develop their own particular styles of operation. They also develop a variety of formal and informal policies and procedures for handling juveniles involved in various types of illegal behaviors.[41] As a result, police departments vary in how they respond to juveniles. For example, one study that categorized police departments according to the extent to which they employed a legalistic style of policing (characterized by a high degree of professionalism and bureaucratic structure) found that the more legalistic departments were more likely to arrest juvenile suspects than less legalistic departments.[42] A study of four Pennsylvania communities found that the percentages of juveniles referred to court varied considerably from one community to another. In one community, only 9% of the juveniles who had police contacts were referred to court, whereas in another community 71% were referred.[43] Similarly, a study of 48 departments in southern California found that virtually all juveniles arrested by one department were referred to court but that in another department the great majority of juveniles were counseled and released.[44]

Clearly, there is considerable variability in how police departments respond to juvenile offenders, and this variability is partially a product of differences in department organizational characteristics and department policies and procedures. However, police departments do not operate in a political and social vacuum. They are influenced by the communities within which they operate. Consequently, community characteristics and attitudes toward the police can also influence police responses to juvenile offenders.

The Community Influence on Policing

One factor that can influence how police handle juvenile suspects is whether community programs for handling juvenile offenders exist. In some communities, officers may have a variety of options, including programs intended to provide individual, group, or family counseling; assist with conflict resolution; or provide restitution to victims. In other communities, available options are more limited.

Communities also influence policing in a variety of other ways. Through their interaction with the local community, police develop assumptions about the community, the people who reside and work there, and the ability and willingness of these people to assist in responding to crime and delinquency. When a police officer encounters a member of the community—whether a suspect, a complainant, or a witness to an alleged offense—the officer's interaction with that person is colored by his or her attitude toward the community and the subgroups that compose it. Consequently, in order to understand how communities influence policing, it is necessary to consider the attitudes police and community members have toward one another.

Research suggests that police departments operate differently in lower-class communities than in wealthier communities. The behavior of police in a given community seems to be a reflection of what police expect as well as the realities of policing that community. Police expect lower-class communities to have higher crime rates, they are aware that more arrests occur in those communities than in wealthier communities, and they are aware that lower-class communities have few resources for informally responding to the wide array of problems experienced in those communities. Consequently, when police come into contact with juveniles in lower-class areas, formal responses are likely, because informal responses are felt to be unrealistic. Indeed, research on the effect of neighborhood socioeconomic status has found that as the socioeconomic status of the neighborhood increases, the likelihood of a police–juvenile encounter ending in arrest declines.[45]

Community attitudes toward the police also exert some influence on police actions. Although the police often feel that the public is hostile toward them, the reality is that most citizens have a positive attitude toward the police. For example, citizens have more confidence in the police than a number of other American institutions including the presidency, public schools, television news, the U.S. Supreme Court, big business, banks, organized labor, church and organized religion, and the medical system.[46]

MYTH VS. REALITY — A Majority of Americans Respect the Police

Myth—The public lacks respect for police officers.
Reality—Gallup polls consistently report that the majority of citizens have a great deal of respect for the police. For example, in a 2003 Gallup Poll, 61% of the respondents indicated they had a "great deal" or "quite a lot" of confidence in the police. In that same poll, 55% indicated that they had a "great deal" or "quite a lot" of confidence in the presidency, 22% reported that they had confidence in big business, and 50% indicated that they had confidence in church or organized religion.[47]

It is true, however, that certain subgroups have less favorable views of the police than people in general. In a 2003 Gallup poll, 43% of African Americans, compared with 65% of whites, reported a great deal of confidence in the police; and 27% of African Americans, compared with 8% of whites, indicated they had little or no confidence in the police.[48] Also, some research suggests that adolescents' views of the police become more negative as they become older.[49] An early study that examined the attitudes of almost 1,000 junior high students in Cincinnati found that hostility toward the police increased as the students moved through their junior high years. The same study found that lower-class youths were less likely to have positive attitudes toward the police than upper-class youths.[50] Similar findings have been seen in several more recent studies as well as in public opinion polls that examine public confidence in American institutions, including the police.[51] Such attitudes are important in a police–juvenile encounter because they can affect how the encounter proceeds and the extent to which the parties are satisfied with the outcome.

The Wishes of the Complainants

As noted earlier, many police–juvenile interactions result from citizen complaints. Furthermore, what police do in a complaint situation typically depends on whether the complainant is present and what the complainant would like for the police to do. For example, one study of police–juvenile encounters found that when the complainant requested informal action on the part of the police, police officers always did what the complainant wanted. If the complainant asked that an arrest be made, officers complied with this request about 60% of the time.[52] In a replication of this study done 8 years later, other researchers reached similar conclusions. The results of this study indicated that when citizens and suspects were both present and when citizens indicated clear preferences regarding how they wanted the police to respond, police usually complied with these requests. The researchers concluded, based on these findings as well as on the findings of the earlier research, that citizens "largely determine official delinquency rates" because their wishes are accorded considerable weight by the police.[53]

The Demeanor of the Offender

As one might expect, how youths behave toward the police can affect whether police officers decide to make an arrest. Interestingly, youths who are unusually antagonistic or unusually polite are more likely to be arrested. In contrast, youths who are moderately respectful are less likely to be arrested—as long as the offense is not serious.[54] Apparently, police officers have basic expectations regarding the behavior of those they have interaction with. They expect suspects to be neither overly hostile nor unusually polite. When suspects' behaviors fall outside the boundaries of those expectations, then the likelihood of arrest increases.[55]

Gender of the Offender

Gender also appears to influence arrest decisions, particularly when **status offenses** are involved. However, the research in this area has produced mixed results. Some studies that examined the relationship between gender and the decision to arrest found that girls are less likely to be arrested for criminal offenses than boys, even when prior criminal record and offense seriousness are taken into account.[56] However, other studies indicate that any gender bias that existed in the past has diminished or disappeared over time,[57] although other research indicates that gender continues to influence arrest decisions in some locations.[58] Still other research indicates that the relationship between gender and arrest decisions is more complex than it first appears. For example, one study that examined police–suspect encounters in 24 police departments found that arrest rates for males and females (both juveniles and adults) were similar. However, the researchers also discovered that police officers used different arrest criteria for males and females, with the result that younger females received harsher treatment than older females. No differences between younger and older males were found. This suggests that, at least in those departments, police took a more paternalistic stance toward younger females, increasing the likelihood of formal processing.[59] Overall, this research suggests that although gender may not play a significant role in police decision making in some jurisdictions, in combination with other factors such as the type and seriousness of delinquent offense and girls' demeanor, it continues to influence practices in other locations.

Unlike the research on the relationship between gender and arrests for criminal offenses, the research on the relationship between gender and the processing of status offense cases is generally consistent. A number of studies have found that female status offenders are more likely to be formally processed than male status offenders.[60] Even studies that found that differentials in the processing of criminal offenses have declined over time report gender differentials in the processing of status offenses.[61]

status offense An act considered to be an offense partly because of the status of the person who performed the act. For example, juveniles (i.e., individuals who have a juvenile status) must obey the reasonable rules of their parents, attend school, and live at home. If they fail to do any of these things, they could be arrested and brought into the court system. Adults do not have to do any of those things and cannot be arrested for refusing to do them.

The Race and Social Class of the Offender

Strong evidence exists that the race and the social class of juveniles influence police decision making. As noted in Chapter 2, minority and poor youth are disproportionately represented in arrest statistics. However, the relationship between race or social class and police decision making is complex, which makes straightforward descriptions of this relationship difficult.

After reading the following true story, list those factors that you believe contributed to the behavior of the police. Do you feel the police handled this situation appropriately? How do incidents such as this influence police community relations?

> *These events took place in 1975. A friend and I had just completed playing a high school basketball game. We had just left our high school, Pontiac Northern, on our way to meet with other members of the team. I was driving, and my friend was in the passenger seat. As we passed a major intersection we saw police lights flashing, although they were still some distance away. As we approached the next major intersection, we saw four police cars, one at each corner of the intersection. As I entered the intersection, each of the police cars pulled into the intersection and cut off the car I was driving. The officers exited their cars with their guns drawn and demanded that we get out of the car. We asked what had we done. Did we do anything wrong? They told us to shut up and get out of the car. As we got out of the car, they grabbed us and threw us to the ground with their guns pushed against the backs of our necks. While on the ground they asked us questions about a robbery that had taken place. We said we didn't know anything about it. We told them that we had just finished playing a high school basketball game and were on our way to a party. They told us we were lying. They said a gas station in the neighborhood had been held up, and we fit the description. After we were on the ground for a short time, they said we could go. We asked, "What's going on?" Their response was that they had a line on who the perpetrators were. We argued with them about how we had been treated and asked why they had stopped us. They said we fit the description. What description was that, we asked? Their response was, "Two African American males, about 5' 11", short hair." I remember laughing about it at the time because it fit the description of just about every other black male I knew. They didn't like us questioning their treatment of us. They then searched our car and told us that if we weren't quiet they would take us to the station. The way I saw it was that just being black makes you a suspect in the eyes of many police. Prior to this, I had no personal experiences with the police, although I had witnessed a number of negative encounters between the police and other people in my neighborhood. After this incident, my trust and faith in the police was greatly diminished. (Felix Brooks, Probation Officer, MA, Political Science)*

Research on the relationship among race, social class, and delinquency has found that lower-class minority youths are especially likely to be arrested even when taking into account both prior criminal records and offense seriousness.[62] Such findings indicate that police decisions to arrest are biased against minority and poor youths. They may be even more biased against minority and poor youths who act in an unusual manner or appear to be unconventional in some other way.

As criminologist James Q. Wilson notes,

> *The patrolman believes with considerable justification that teenagers, Negroes, and lower-income people commit a disproportionate share of all reported crimes; being in those population categories at all makes one, statistically, more suspect than other people; but to be in those categories and to behave unconventionally is to make oneself a prime suspect. Patrolmen believe they would be derelict in their duty if they did not treat such people with suspicion, routinely question them on the street, and detain them for longer questioning if a crime has occurred in the area.*[63]

In contrast, a number of other studies have failed to find strong evidence of police bias,[64] while admitting that a number of factors make interpreting the relationship among race, class, and arrest difficult. For example, the effect of race on the decision to arrest could be complicated by other factors, such as police decisions to focus their attention on poor and minority areas, the wishes of the complainants, and the demeanor of the offender.

Based upon the available evidence, it appears that there are differences in how police departments treat poor and minority youths in some communities.[65] It also appears that in urban areas, where minority populations tend to be large, police are more likely to engage in discriminatory practices.[66]

The discriminatory treatment of minority and poor youths should not be interpreted to mean that police in large urban areas are blatantly racist or hold biased attitudes toward the poor. Although some police officers, like people in other occupations, are racist or strongly biased against the poor, the disproportionately high arrest rate for lower-class and minority youths may be better understood as an outgrowth of <u>**institutional racism**</u> and bias—that is, the ways in which basic social institutions operate to keep minorities and poor people in subordinate positions.

Police decisions to concentrate their surveillance activities in lower-class or minority communities, for example, increase the likelihood that they will uncover delinquent behaviors in those communities. High rates of delinquency in such communities may themselves be partial products of long-term patterns of discrimination. Further, when police encounter youths in a lower-class or minority community, some youths will present a hostile demeanor, because of past unpleasant encounters with the police, which, in turn, increases the probability that the police will respond in a more forceful way. Such interactions reinforce police perceptions that more forceful responses are justified while at the same time they encourage more negative community perceptions and responses toward the police. The result is a cycle of action and reaction that results in negative police–juvenile encounters and increased numbers of arrests.

institutional racism The ways that basic social institutions operate to keep minorities in subordinate positions. If police believe that minorities have fewer resources to respond to youth problems, believe that minority communities foster more crime, and believe that minority groups members are more hostile to the police, it increases the likelihood that police will focus more crime-fighting resources in minority communities, resulting in increases in crime, police-citizen hostility, and neighborhood problems.

Police Processing of Juvenile Offenders

Two types of police units have the most contact with juveniles: patrol units and specialized juvenile units (although not all departments have specialized juvenile units). Of these two, patrol units have the most contact with juveniles.[67] Specialized juvenile units and youth divisions typically serve as referral units that accept juvenile cases from other departmental divisions.[68] Juvenile units are usually small,[69] but they often conduct juvenile investigations on their own, particularly investigations of serious juvenile crimes. There is no agreement among police officials regarding the best response to juveniles who engage in minor types of delinquent behavior. Some maintain that police should adopt a strict law enforcement approach that focuses on enforcing laws and making juvenile arrests when law violations occur. In contrast, others argue that a crime prevention approach that favors diversion, except in the case of serious offenses, is preferable.[70] Many patrol officers and their supervisors support the crime prevention approach for nonserious juvenile offenses. This approach is also supported by labeling theorists, who call for radical nonintervention. Radical nonintervention involves avoiding formal action whenever possible in order to avoid the labeling and stigmatization of youths that may lead to additional delinquent behavior. However, as **Table 6-1** indicates (page 165), overall, police have become more formal in their processing of juvenile cases by referring a greater percentage of juvenile cases to the courts.

When a police officer encounters a juvenile who has committed an illegal act, the officer must make a decision regarding the best way to handle the case. As noted earlier, the decision the officer makes can be influenced by a variety of factors. One option, of course, is to make an arrest (in some jurisdictions this is referred to as "taking into custody"). For practical purposes, an arrest is considered to have taken place if the youth is not free to walk away.

As a general rule, the basis for arresting a juvenile is the same as for arresting an adult. The officer needs to have **probable cause**, which means that the officer has reason to believe that an offense has been committed and that the youth to be arrested committed the offense. The U.S. Supreme Court rulings in *Gault* and *Kent* as well as subsequent court cases raise the question of whether special standards should apply to the arrest of a juvenile and whether the police can deal with juveniles in the same way they deal with adults.

One of the key issues surrounding the arrest of juveniles is the interrogation of juvenile suspects. However, before determining whether special standards apply to the interrogation of juveniles, we need to look at the standards that apply to the interrogation of adults. The most important case regarding this issue is *Miranda v. Arizona*.[71] Ernesto A. Miranda had been arrested for kidnapping and rape and was in the custody of the Phoenix police. He was identified by the complaining witness and was interrogated for 2 hours without being notified of his right to have an attorney present. Finally, the police produced a signed confession, on which was typed a statement indicating that the confession was voluntary and "with full knowledge of my legal rights."[72]

probable cause Grounds sufficient to convince a reasonably competent person that a crime was committed and that the suspect committed it.

In its decision the U.S. Supreme Court indicated the following:

> *We hold that when an individual is taken into custody or otherwise deprived of his freedom by the authorities in any significant way and is subject to questioning, the privilege against self-incrimination is jeopardized. Procedural safeguards must be employed to protect the privilege . . . He must be warned prior to any questioning that he has the right to remain silent, that anything he says can be used against him in a court of law, that he has the right to the presence of an attorney, and that if he cannot afford an attorney one will be appointed for him prior to any questioning if he so desires . . . unless and until such warnings and waiver are demonstrated by the prosecution at trial. No evidence obtained as a result of interrogation can be used against him.*[73]

In its decision, the court clearly stated that adults cannot be held in custody for long periods of time and cannot be questioned without specific advice of rights to an attorney in order to protect the Fifth Amendment right to be free from self-incrimination. If these circumstances that Ernest Miranda found himself in were so coercive as to make his confession inadmissible, what types of circumstances would make the confession of a juvenile inadmissible?

The U.S. Supreme Court addressed this issue in 1979, in the case of *Fare v. Michael C.*[74] Michael C. was 16 years old at the time of his arrest on suspicion of murder. He was taken to the police station and before questioning was fully advised of his rights per *Miranda*. He was on probation to the juvenile court, and he had asked to see his probation officer. The police denied this request. He proceeded to talk to the police without an attorney, making statements and drawing sketches that implicated him in the murder.

In this case the court indicated the following: "Thus the determination whether statements obtained during custodial interrogation are admissible against the accused is to be made upon an inquiry into the *totality of the circumstances* surrounding the interrogation, to ascertain whether the accused, in fact, knowingly and voluntarily decided to forgo his rights to remain silent and to have assistance of counsel" (emphasis added).[75] The court cited age, past contacts and experience with the police, whether the juvenile was involved with the juvenile court, and the youth's intelligence as factors that were to be weighed as part of the "totality of circumstances." The court also mentioned whether the juvenile was "worn down" by improper interrogation tactics employed by the police or tricked by them as additional factors to review. In *Fare v. Michael C.*, the confession was determined to be admissible by the court.

The "totality of the circumstances" test was not a new one but had been articulated by the Supreme Court of California in *People v. Lava.*[76] In this case, the court held that the factors to be reviewed in order to determine the voluntariness of a confession by a juvenile included the juvenile's age, intelligence, education, experience, and ability to comprehend the meaning and effect of the statements made. This holding is interesting, because in *Fare v. Michael C.*, a California case, the Supreme Court of California ruled the juvenile's confession inadmissible, but the U.S. Supreme Court, adopting a similar test, ruled it admissible. As this example makes evident, courts looking at the same set of facts can interpret or construe them differently.

Fare v. Michael C. is important because it required courts to examine a variety of factors in determining whether information obtained from juveniles by the police could be used in court. It is also important because it indicated that parents, guardians, custodians, other significant adults, or attorneys do *not* have to be present when juveniles are interrogated by the police. Although the "totality of circumstances" test may sound reasonable, courts look at police conduct with hindsight that sometimes falls below 20/20 acuity. In many areas of society, children are recognized as being fundamentally different from adults. For example, minors cannot own property, sign business contracts, or engage in other adult behaviors because they are considered minors. They can, however, by themselves, confess to a murder. A strong argument can be made that this is illogical.

The issue of self-incrimination by juveniles was also given some attention by the Supreme Court of New Hampshire in the case of *State v. Benoit*. Here the court stated, "Courts employing the totality of the circumstances test do so under the belief that juvenile courts are equipped with the expertise and experience to make competent evaluations of the special circumstances surrounding the waiver of rights by juveniles."[77] The court went on to list 15 circumstances that juvenile courts should consider. Juvenile courts, however, may not routinely perform the appropriate evaluations. The expertise and experience of the court that is charged with the responsibility of protecting a juvenile's rights is a key factor determining whether those rights are, in fact, protected. Unfortunately, the experience and expertise of many juvenile courts in this area are questionable.

The *Benoit* decision also articulated the main alternative to the "totality of the circumstances" rule, namely, the "interested adult" rule. According to this rule, adopted by a few jurisdictions, no juvenile can waive the privilege against self-incrimination without having had the opportunity to consult with and have present an adult who is "friendly" to the juvenile and who understands the juvenile's rights.

The effect of the *Benoit* holding was to increase the totality factors by adding three more:

1. The juvenile must be informed of his or her rights in language understandable to him or her.

2. The juvenile must be made aware of the possibility of adult criminal prosecution or waiver, if applicable.

3. When the juvenile is arrested, the officer in charge must immediately secure from the juvenile the name of a "friendly adult" the juvenile could consult.[78]

Thus, the New Hampshire Supreme Court attempted to blend the "interested adult" test into the "totality of the circumstances" test by making the securing of a friendly adult one of the voluntary factors to be considered.

Another difference in the handling of juveniles and adults at the arrest stage is that juveniles are more likely to be detained pending trial than adults who have committed similar offenses. The U.S. Supreme Court, in *Schall v. Martin*, ruled that preventive detention of juveniles is permissible.[79] This case concerned Gregory Martin, age 14, who was arrested on December 13, 1977, and charged with first-degree robbery, second-degree assault, and criminal possession of a weapon. At 11:30 p.m., he and two other juveniles had hit a youth on the head with a loaded gun and stolen his jacket and sneakers. When arrested, he had the gun in his possession and lied to the police about where and with whom he lived. He appeared in New York Family Court on December 14 with his grandmother and was ordered detained due to the possession of the loaded gun and his lying to the police about his home address as well as the late hour of the incident. A probable cause hearing was held on December 19, and probable cause was found on all charges. At a fact-finding hearing held December 27 through December 29, he was found guilty. He had been detained a total of 15 days.

In this case the court ruled that "Children, by definition, are not assumed to have the capacity to take care of themselves. They are assumed to be subject to the control of their parents; and, if parental control falters, the state must play its part as parens patriae . . . In this respect, the juvenile's liberty interest may, in appropriate circumstances, be subordinated to the state's parens patriae interest in preserving and promoting the welfare of the child."[80] The Court went on to point out that every state and the District of Columbia permit the preventive detention of juveniles:

> *The fact that a practice is followed by a large number of states is not conclusive . . . as to whether that practice accords with due process, but it is plainly worth considering in determining whether the practice "offends some principle of justice so rooted in the traditions and conscience of our people as to be ranked fundamental" . . . we conclude that the practice serves a legitimate regulatory purpose compatible with the "fundamental fairness" demanded by the Due Process Clause in juvenile proceedings.*[81]

There are several ramifications of the *Schall* decision. *Parens patriae*, an idea associated with the juvenile court, played an important role in this decision. In fact, the idea that a juvenile should and must be under someone's control is a long-standing principle in U.S. law. As long as juvenile courts have time guidelines for the setting of meaningful hearings so that juveniles do not languish in detention, the constitutional right to be free from "excessive bail," found in the Eighth Amendment, takes a back seat to society's interest in the supervision and control of juveniles. Nevertheless, the punitive nature of some detention facilities, the possibility of a child's victimization in detention, and the widespread use of detention in some jurisdictions raise serious questions about the practice of detaining juveniles.

MYTH VS. REALITY — Many Youths Are Arrested and Detained for Minor Offenses

Myth—Most youths detained by police have committed serious offenses and pose a threat to community safety.

Reality—Most youths detained in many jurisdictions have not committed serious offenses against people. Moreover, in many jurisdictions the great majority of youths who are detained pending a court hearing are released at the hearing because they are felt to pose no threat to community safety.

The exact procedures that police must follow when taking a juvenile into custody vary from state to state and are spelled out in state juvenile codes and in police department policies. However, these codes and policies typically require officers to notify the juvenile's parents that the juvenile is in custody when an arrest is made. Often police request that parents come to the police station or the officer transports the youth home prior to any questioning. In cases where the officer feels that detention of the juvenile is appropriate, the juvenile is transported to a juvenile detention facility or, in some jurisdictions, to an adult jail if a juvenile detention facility is unavailable. In cases where the juvenile is released to his or her parents, the juvenile and parents are informed that they will be contacted by the court at a later date about their case. After the release of the juvenile, the officer completes the complaint by collecting any additional information needed to establish the offense. The completed complaint is then forwarded to the next stage of the juvenile justice process for further action. In cases where the juvenile is detained in a juvenile detention facility or adult jail, the processing of the complaint is expedited, because juvenile codes require that a detention hearing be held to determine the appropriateness of detention and the complaint will be needed at the hearing.

The interrogation and detention of juvenile suspects are sources of considerable controversy. At the heart of the interrogation controversy is a debate over the age at which juveniles are mature enough to fully understand the importance of their rights and the potential consequences associated with their waiver. Research indicates that many youths do not understand their rights when arrested. In one study, about one-third of a sample of institutionalized youths incorrectly believed that they *had* to talk with the police. The study also found that about one-third of the parents of these youths would advise them to confess to the police.[82] In another study, over half of the youths examined lacked a full understanding of the Miranda warnings.[83] Moreover, the evidence from developmental psychology indicates that youths are psychologically less mature than adults.[84] Such findings raise concerns about the vulnerability of many youths when they are questioned by the police and the reality that youths receive fewer protections than those given to adults. Concern over the detention of juveniles revolves around the fact that preventive detention amounts to punishment of someone before he or she has been found guilty of an offense. Moreover, the conditions youths are sometimes exposed to in detention units and adult jails raise additional worries. (See Chapter 12 for a more complete discussion of the placement of juveniles in detention facilities and jails.)

 Juveniles Are Sometimes Placed in Adult Jails

At midyear 2002, it was estimated that there were 7,248 juveniles detained in adult jails in the United States.[85] This figure, however, may underestimate the numbers of juveniles confined in such facilities.[86]

Trends in Police Processing of Juveniles

In 1972, approximately 51% of all juveniles taken into police custody were referred to juvenile court, approximately 45% were handled within the police department and released, and slightly more than 1% were referred to adult or criminal courts. However, the percentage of youths taken into custody and referred to juvenile courts has been increasing over time. By 2001, approximately 72% of youths taken into custody by the police were referred to juvenile court, 19% were handled within the police department and released, and 6.5% were referred to adult or criminal courts (see Table 6-1).[87] These data indicate the existence of three important trends in police processing of juveniles: (1) the referral of more youths to juvenile court, (2) the handling of fewer cases within police departments, and (3) the referral of more youths to adult courts.

Table 6-1	Percent Distribution of Juveniles Taken into Police Custody, by Method of Disposition, 1972–2002				
	Referrred to juvenile court jurisdiction	Handled within department and released	Referred to criminal or adult court	Referred to other police agency	Referred to welfare agency
1972	50.8%	45.0%	1.3%	1.6%	1.3%
1973	49.5	45.2	1.5	2.3	1.4
1974	47.0	44.4	3.7	2.4	2.5
1975	52.7	41.6	2.3	1.9	1.4
1976	53.4	39.0	4.4	1.7	1.6
1977	53.2	38.1	3.9	1.8	3.0
1978	55.9	36.6	3.8	1.8	1.9
1979	57.3	34.6	4.8	1.7	1.6
1980	58.1	33.8	4.8	1.7	1.6
1981	58.0	33.8	5.1	1.6	1.5
1982	58.9	32.5	5.4	1.5	1.6
1983	57.5	32.8	4.8	1.7	3.1
1984	60.0	31.5	5.2	1.3	2.0
1985	61.8	30.7	4.4	1.2	1.9
1986	61.7	29.9	5.5	1.1	1.8
1987	62.0	30.3	5.2	1.0	1.4
1988	63.1	29.1	4.7	1.1	1.9
1989	63.9	28.7	4.5	1.2	1.7
1990	64.5	28.3	4.5	1.1	1.6
1991	64.2	28.1	5.0	1.0	1.7
1992	62.5	30.1	4.7	1.1	1.7
1993	67.3	25.6	4.8	0.9	1.5
1994	63.2	29.5	4.7	1.0	1.7
1995	65.7	28.4	3.3	0.9	1.7
1996	68.6	23.3	6.2	0.9	0.9
1997	66.9	24.6	6.6	0.8	1.1
1998	69.2	22.2	6.8	0.9	1.0
1999	69.2	22.5	6.4	1.0	0.8
2000	70.8	20.3	7.0	1.1	0.8
2001	72.4	19.0	6.5	1.4	0.7
2002	72.8	18.1	7.0	1.4	0.7

Note: Percentages may not total 100 due to rounding.

Source: Reprinted from K. Maguire and A. L. Pastore, eds., *Sourcebook of Criminal Justice Statistics 2002*, p. 381, Table 4.26

Police and Delinquency Prevention

Although police are becoming more formal in their processing of juvenile cases, many police departments also devote resources to delinquency prevention. Indeed, such efforts have a number of advantages. For example, by preventing youths from engaging in delinquent activities, police spare youths the stigma associated with formal juvenile justice processing. Delinquency prevention also spares juvenile courts the time and costs associated with processing cases and allows the police and the courts to focus more time on serious juvenile offenses and adult crime.

One of the most widely used prevention programs is the DARE program. (DARE is an acronym for drug abuse resistance education.) This program, which involves cooperation between schools and police departments, typically targets upper elementary school children and attempts to provide them with the skills necessary for resisting peer pressure to take drugs. However, DARE curriculum materials have been developed for youths in kindergarten through the 12th grade and for parents. The program is unique because it employs uniformed police officers, who present a structured curriculum designed to

- provide students with accurate information about drugs, including alcohol and tobacco, and the consequences of taking drugs
- teach students specific skills that will allow them to resist peer pressure to experiment with drugs
- help students develop respect for police officers
- give students alternatives to drug use
- help build students' self-esteem[88]

More recently, the DARE curriculum has been updated and renamed "DARE to Resist Drugs and Violence." The new curriculum, which includes information about the use of tobacco and inhalants, attempts to teach youths conflict resolution and violence prevention skills and calls for closer cooperation between DARE officers and classroom teachers.[89] A large-scale evaluation of the DARE program conducted by researchers from the Research Triangle Institute and the University of Kentucky found that it produces some positive results. It has been found to increase students' knowledge about drugs and enhance their social skills. In addition, it appears to influence students' attitudes about drugs, their perceptions of the police, and their self-esteem. However, with the exception of tobacco, DARE's ability to influence drug use has not been demonstrated. Importantly, the finding that DARE has little if any influence on durg use has been replicated in several studies, although some evaluators note that this finding may be due to low drug usage among elementary students, who are DARE's primary targets.[90] In response to these evaluations, the DARE program has once again undergone revisions to make it more high tech and interactive. In addition, the program is undergoing another large evaluation that should shed additional light on the effectiveness of the program.[91]

Another popular prevention program initially developed by the Phoenix Police Department and sponsored by the Bureau of Alcohol, Tobacco, and Firearms is GREAT (Gang Resistance Education and Training). Like DARE, GREAT consists of a curriculum that is taught by police officers in elementary and middle schools and is found in almost every state. GREAT also underwent a curriculum revision that led to a revised curriculum in 2000. The curriculum revision was designed to make the program more interactive and to help youths better understand the relationship between gangs and violence, learn what can be done about gangs, set goals, and develop a variety of social competencies including communication and conflict resolution skills.[92] Importantly, evaluations of the GREAT program have produced some positive results, although a major goal of the program, to prevent youths from participating in gangs, may not be achieved. Nevertheless, GREAT has been associated with the development of more pro-social attitudes (including more favorable attitudes toward the police) and less favorable attitudes toward gangs. Moreover, the evidence suggests that police officers can have a positive impact on youths.[93]

Youth-Oriented Community Policing

Another significant trend in policing in recent years has been the development of problem-oriented and __community policing__ strategies. In problem-oriented policing, police attempt to identify problems that encourage criminal behavior and take steps to remedy these problems.[94] In community policing, police rely on the community to identify problems that can be cooperatively addressed by them and the community.[95] Both of these approaches are based on the realization that, in order to reduce crime, police need to take a more proactive approach in dealing with crime-generating community problems instead of simply reacting to citizen complaints after crimes occur.

Community policing employs a number of strategies, including these:

- moving officers from their patrol cars to positions of direct contact with community residents (intended to give officers information about problems that exist in the community and ideas for solutions)
- changing the officers' main mode of behavior from reactive (responding to crimes committed) to proactive (preventing crimes)
- making police operations more visible to the public and increasing police accountability
- decentralizing police operations, thereby allowing individual officers to develop greater familiarity with the needs of the communities they police and to better respond to those needs
- encouraging officers to view citizens as partners in dealing with community problems related to crime, thereby improving relations between police and citizens
- placing more decision-making power in the hands of community policing officers, who best know the community, its problems, and its expectations
- developing relationships between police and citizens that foster public initiatives aimed at preventing and solving crimes[96]

A recently developed variant of community policing is what is known as youth-oriented community policing. Youth-oriented community policing programs have multiple goals: providing multi-agency responses to children and their families in order to prevent future delinquent behavior and adult criminality; working with youths who are already involved with the juvenile courts in order to reduce the likelihood of recidivism; involving neighborhood residents in efforts to improve their communities; and developing closer relations between community residents

community policing A policing strategy in which police attempt to identify and understand the social context of delinquent and criminal behavior and then to work with the local community to rectify the problems causing crime rather than simply reacting to crime by arresting and incarcerating offenders.

and public officials, including police officers. The strategies employed to achieve these goals, although varied, often include the provision of specialized training to police officers and the development of cooperative multi-agency teams to address community problems. The specialized training given to police officers might focus on topics such as child development, the psychological impact of family violence on children, assisting victims of crime, fostering effective collaboration between citizens and police, and handling mental health issues. The multi-agency teams are typically staffed by representatives of mental health agencies, schools, social service agencies, health departments, universities, and neighborhood associations, among others. These teams focus on prevention by identifying family and community problems that negatively affect youths and lead to delinquent behavior.[97]

One thing that makes community policing stand out from other policing strategies is its focus on the social context within which delinquency takes place. Rather than simply reacting to juvenile crime, youth-oriented community policing requires law enforcement agencies to identify societal factors related to delinquency and to take positive steps to address those factors in order to prevent delinquency. As a result, youth-oriented community policing is more theory and research oriented than traditional policing strategies. By having a clear understanding of factors that encourage delinquency within the community, those using a youth-oriented community policing approach are in a better position to prevent as well as respond to delinquent behavior.

■ Legal Issues

Although the relationship between police and citizens is of long-standing concern in U.S. law, the relationship between children and police has also received attention. Many contend that children, owing to their lack of maturity, should have special protections when faced with arrest or interrogation by police. Exactly how far the police should go in protecting juveniles and what specific protections should be given to juveniles are, however, matters of debate. What is your perspective on this issue?

■ Chapter Summary

This chapter examined police and public responses to juvenile crime. As noted at the beginning of the chapter, citizens make up an informal juvenile justice process that responds in a variety of ways to youthful misbehavior. The ability of this informal juvenile justice process to handle juvenile misbehavior determines the extent to which formal components of the juvenile justice process become involved in the lives of youths.

Although many youth offenses are handled by citizens without official assistance, citizens frequently choose to involve the police, and in some cases the police initiate contact with youths on their own. The police play a critical role in juvenile justice because they serve as the primary gatekeepers of the formal juvenile justice process. Of particular significance is the fact that the police exercise discretion in how they handle alleged instances of juvenile delinquency. How the police decide to handle a juvenile case can have important ramifications for community safety as well as for the child and his or her family. In addition, how the police handle juveniles raises several legal issues regarding the relations between juveniles and the police and how far the police should go in protecting juveniles' rights. These issues have received some attention by the courts in cases such as *Fare v. Michael C.*, *People v. Lava*, *State v. Benoit*, and *Schall v. Martin*. However, many people are still concerned that youths may be denied important legal protections when confronted by the police, and this lack of protection can result in youths being harmed because of their involvement with juvenile justice agencies.

Data on the police processing of juveniles indicate that the police are becoming more formal in their responses to youths. Nevertheless, some police departments have also developed a variety of new strategies to deal with youths, such as youth-oriented community policing. To some extent, these strategies reflect the recognition by the police that their effectiveness as crime fighters requires good relationships with all members of the community, including youths. In addition, they appear to be tied to the increased emphasis on crime prevention and the growing popularity of community policing found in some police departments. However, the extent to which police departments will embrace these strategies and their effectiveness remains to be seen.

KEY CONCEPTS

arrest Legally authorized deprivation of a person's liberty. As a general rule, a person is under arrest when he or she is not free to walk away.

community policing A policing strategy in which police attempt to identify and understand the social context of delinquent and criminal behavior and then to work with the local community to rectify the problems causing crime rather than simply reacting to crime by arresting and incarcerating offenders.

discretion The power to act on one's own authority.

informal juvenile justice process Actions to combat juvenile delinquency taken by parents, neighbors, business owners, teachers, and others who are not part of the formal juvenile justice apparatus.

institutional racism The ways that basic social institutions operate to keep minorities in subordinate positions. If police believe that minorities have fewer resources to respond to youth problems, believe that minority communities foster more crime, and believe that minority groups members are more hostile to the police, it increases the likelihood that police will focus more crime-fighting resources in minority communities, resulting in increases in crime, police-citizen hostility, and neighborhood problems.

police role conflict Clash between the duty of police to investigate crimes and prevent individuals from committing crimes and the duty of police to protect the rights of offenders, crime victims, and other citizens.

probable cause Grounds sufficient to convince a reasonably competent person that a crime was committed and that the suspect committed it.

reform agenda for police Agenda consisting of items such as the following: eliminating political influence and corruption, appointing qualified leaders, establishing the ideal of nonpartisan public service, raising personnel standards, implementing scientific management principles, instituting military-style discipline, and creating "specialized" units.

status offense An act considered to be an offense partly because of the status of the person who performed the act. For example, juveniles (i.e., individuals who have a juvenile status) must obey the reasonable rules of their parents, attend school, and live at home. If they fail to do any of these things, they could be arrested and brought into the court system. Adults do not have to do any of those things and cannot be arrested for refusing to do them.

REVIEW QUESTIONS

1. What is the informal juvenile justice process, and how is it related to formal juvenile justice agencies?

_____

_____

_____

_____

_____

2. What are the typical responses available to the police who come into contact with juvenile offenders?

_____

_____

_____

_____

3. What were the basic characteristics of policing during the colonial period?

_____

_____

_____

_____

4. What problems in policing did the progressive reforms of the late 1800s and early 1900s attempt to address?

_____

_____

_____

_____

5. What innovations in youth-oriented policing occurred between the early 1900s and the 1970s?

_____

_____

_____

_____

_____

6. What are the basic roles played by the police and how do these roles influence police juvenile interactions?

_____

_____

_____

_____

7. What options do police officers have when dealing with youths who have allegedly broken the law?

_____

_____

_____

_____

8. What factors influence police discretion?

_____

_____

_____

_____

_____

9. What procedures do the police follow when taking a youth into custody?

_____

_____

_____

_____

_____

10. What are the important court cases concerning the interrogation of juveniles by the police, and what do these cases require the police to do?

_____

_____

_____

_____

_____

11. In what ways have trends in the police processing of juveniles changed over the last 20 years?

_____

_____

_____

_____

_____

12. What recent strategies or programs have been developed by the police to help them prevent or more effectively respond to delinquency? What do these strategies or programs attempt to do?

_____

_____

_____

_____

_____

ADDITIONAL READINGS

Bazemore, G. and S. Senjo. 1997. "Police Encounters with Juveniles Revisited: An Exploratory Study of Themes and Styles in Community Policing." *Policing: An International Journal of Police Strategy and Management* 20, 60–82.

Chesney-Lind, M. and R. G. Shelden. 2004. *Girls, Delinquency and Juvenile Justice.* 3rd ed. Belmont, CA: Wadsworth.

Krisberg, B. and J. F. Austin. 1993. *Reinventing Juvenile Justice.* Newbury Park, CA: Sage.

Lundman, R. J. et al. 1978. "Police Control of Juveniles: A Replication." *Journal of Research in Crime and Delinquency*, January, 74–91.

Senna, J. J. and L. J. Siegel. 1992. *Juvenile Law: Cases and Comments.* 2nd ed. St. Paul, MN: West.

Walker, S. 1992. *The Police in America: An Introduction.* New York: McGraw-Hill.

CASES CITED

Fare v. Michael C., 442 US 707, 99 S Ct 2560, 61 L Ed 2d 197 (1979).

Miranda v. Arizona, 384 US 436, 86 S Ct 1602, 16 L Ed 2d 694 (1966).

People v. Lava, 62 Cal Rptr 586, 432 P2d 202 (1967).

Schall v. Martin, 467 US 253, 104 S Ct 2403, 81 L Ed 2d 207 (1984).

State v. Benoit, 490 A2d 295 (NH 1985).

NOTES

1. S. Walker, *The Police in America: An Introduction* (New York: McGraw-Hill, 1992).
2. Walker, *The Police in America*; B. L. Berg, *Law Enforcement: An Introduction to Police in Society* (Boston: Allyn & Bacon, 1992).
3. Walker, *The Police in America*.
4. Walker, *The Police in America*.
5. C. B. Klockars, *The Idea of Police* (Newbury Park, CA: Sage, 1985).
6. C. Uchida, "The Development of the American Police: An Historical Overview," in *Critical Issues in Policing: Contemporary Readings*, eds. R. G. Dunham and G. P. Alpert (Prospect Heights, IL: Waveland Press, 1989); Walker, *The Police in America*.
7. Walker, *The Police in America*.
8. Uchida, "The Development of the American Police."
9. Walker, *The Police in America*.
10. Walker, *The Police in America*.
11. R. Trojanowicz and B. Bucqueroux, *Community Policing: A Contemporary Perspective* (Cincinnati, OH: Anderson, 1990); Walker, *The Police in America*.
12. Walker, *The Police in America*.
13. G. W. Matsell, "Semi-Annual Report, May 31–October 31, 1849, New York City Police Department," cited in R. H. Bremner, ed., *Children and Youth in America: A Documentary History*, vol. 1 (Cambridge, MA: Harvard University Press, 1970), 755.
14. C. Bartollas and S. J. Miller, *Juvenile Justice in America* (Upper Saddle River, NJ: Prentice Hall, 1998).
15. Walker, *The Police in America*.
16. Walker, *The Police in America*.
17. Walker, *The Police in America*, 13.
18. R. M. Fogelson, *Big-City Police* (Cambridge, MA: Harvard University Press, 1977).
19. C. Owings, *Women Police: A Study of the Development and Status of the Women Police Movement* (Montclair, NJ: Patterson Smith, 1969). D. M. Schultz, *From Social Worker to Crime Fighter: Women in United States Municipal Policing* (Westport, CT: Praeger, 1995).
20. S. A. Walker, *A Critical History of Police Reform* (Lexington, MA: Lexington Books, 1977).

21. Fogelson, *Big-City Police*.

22. Bartollas and Miller, *Juvenile Justice in America*.

23. L. J. Siegel et al., *Juvenile Delinquency: Theory, Practice and Law*, 8th ed. (Belmont, CA: Wadsworth, 2003).

24. A. R. Roberts, "The Emergence and Proliferation of Juvenile Diversion Programs," in *Juvenile Justice: Policies, Programs, and Services*, ed. A. R. Roberts (Chicago: Nelson-Hall, 1998).

25. Bartollas and Miller, *Juvenile Justice in America*.

26. Bartollas and Miller, *Juvenile Justice in America*.

27. M. Yinger, *Toward a Field Theory of Behavior* (New York: McGraw-Hill, 1965), 99–100, cited in Walker, *The Police in America*, 63.

28. Walker, *The Police in America*.

29. Walker, *The Police in America*.

30. J. Q. Wilson, *Varieties of Police Behavior* (New York: Atheneum, 1973).

31. A. J. Reiss, *The Police and the Public* (New Haven, CT: Yale University Press, 1971).

32. J. Q. Wilson, *Varieties of Police Behavior: The Management of Law and Order in Eight Communities* (Cambridge, MA: Harvard University Press, 1968).

33. S. P. Lab, "Police Productivity: The Other Eighty Percent," *Journal of Police Science and Administration* 12 (1984): 297–302; Wilson, *Varieties of Police Behavior*.

34. J. Whitehead and S. Lab, *Juvenile Justice: An Introduction* (Cincinnati, OH: Anderson, 1990).

35. R. Muraskin, "Police Work and Juveniles," in *Juvenile Justice: Policies, Programs, and Services*, ed. A. R. Roberts (Chicago: Nelson-Hall, 1998).

36. S. M. Cox et al., *Juvenile Justice: A Guide to Theory and Practice* (New York: McGraw-Hill, 2003).

37. Cox et al., *Juvenile Justice*; "Participant's Resource Packet for the Youth-Oriented Community Policing National Satellite Teleconference," Eastern Kentucky University Training Resource Center, Richmond, Kentucky, 1995.

38. C. Puzzanchera et al., *Juvenile Court Statistics 1999* (Pittsburgh, PA: National Center for Juvenile Justice, 2003).

39. D. Black and A. J. Reiss, "Police Control of Juveniles," *American Sociological Review* 35 (1970): 63–77; R. J. Lundman et al., "Police Control of Juveniles: A Replication," in *Juvenile Delinquency: A Justice Perspective*, eds. R. Weisheit and R. G. Culbertson (Prospect Heights, IL: Waveland Press, 1990); I. Piliavan and S. Briar, "Police Encounters with Juveniles," *American Journal of Sociology* 70 (1964): 206–214.

40. Black and Reiss, "Police Control of Juveniles"; Lundman et al., "Police Control of Juveniles: A Replication"; Piliavan and Briar, "Police Encounters with Juveniles."

41. See Lundman et al., "Police Control of Juveniles: A Replication"; Piliavan and Briar, "Police Encounters with Juveniles."

42. D. A. Smith, "The Organizational Context of Legal Control," *Criminology* 22 (1984): 19–38.

43. N. Goldman, "The Differential Selection of Juvenile Offenders for Court Appearance," in *Crime and the Legal Process*, ed. W. J. Chambliss (New York: McGraw-Hill, 1969).

44. M. W. Klein, *Police Processing of Juvenile Offenders: Toward the Development of Juvenile System Rates*, part 3 (Los Angeles: Los Angeles County Subregional Board, California Council on Juvenile Justice, 1970).

45. R. J. Sampson, "Effects of Socioeconomic Context on Official Reaction to Juvenile Delinquency," *American Sociological Review* 51 (1986): 876–885; see also A. Cicourel, *The Social Organization of Juvenile Justice* (New York: Wiley, 1968); Piliavan and Briar, "Police Encounters with Juveniles."

46. K. Maguire and A. L. Pastore, eds., *Sourcebook of Criminal Justice Statistics 2002*. Retrieved June 3, 2004, from http://www.albany.edu/sourcebook/, 115, Table 2.11.

47. Maguire and Pastore, eds. *Sourcebook of Criminal Justice Statistics 2002*.

48. Maguire and Pastore, eds. *Sourcebook of Criminal Justice Statistics 2002*, 116, Table 2.13.

49. S. H. Decker, "Citizen Attitudes Toward the Police: A Review of Past Findings and Suggestions for Future Policy," *Journal of Police Science and Administration* 9 (1981): 80–87.

50. R. Portune, *Changing Adolescent Attitudes Toward the Police* (Cincinnati, OH: Anderson, 1971).

51. D. H. Bouma, *Kids and Cops* (Grand Rapids, MI: William E. Eerdman, 1969); Decker, "Citizen Attitudes Toward the Police"; Maguire and Pastore, eds., *Sourcebook of Criminal Justice Statistics 2002*, 116, Table 2.13; K. S. Murty et al., "The Image of Police in Black Atlanta Communities," *Journal of Police Science and Administration* 17 (1990): 250–257.

52. Black and Reiss, "Police Control of Juveniles."

Chapter Resources

53. Lundman et al., "Police Control of Juveniles: A Replication," 114.

54. Black and Reiss, "Police Control of Juveniles"; Lundman et al., "Police Control of Juveniles: A Replication."

55. Lundman et al., "Police Control of Juveniles: A Replication."

56. M. D. Krohn et al., "Is Chivalry Dead? An Analysis of Changes in Police Dispositions of Males and Females," *Criminology* 21 (1983): 417–437; see also D. S. Elliott and H. L. Voss, *Delinquency and Dropout* (Lexington, MA: Lexington Books, 1974); M. Morash, "Establishment of a Juvenile Police Record: The Influence of Individual and Peer Group Characteristics," *Criminology* 22 (1984): 97–111.

57. D. Dannefer and R. K. Schutt, "Race and Juvenile Justice Processing in Court and Police Agencies," *American Journal of Sociology* 87 (1982): 1113–1132; R. J. Sampson, "Sex Differences in Self-Reported Delinquency and Official Records: A Multiple-Group Structural Modeling Approach," *Journal of Quantitative Criminology* 1 (1985): 345–367.

58. For a review see M. Chesney-Lind and R. G. Shelden, *Girls, Delinquency, and Juvenile Justice* (Belmont, CA: Wadsworth, 2004).

59. C. Visher, "Gender, Police Arrest Decisions, and Notions of Chivalry," *Criminology* 21 (1983): 5–28.

60. Krohn et al., "Is Chivalry Dead?"; K. Teilmann and P. H. Landry, "Gender Bias in Juvenile Justice," *Journal of Research in Crime and Delinquency* 18 (1981): 47-80; see also M. Chesney-Lind, "Judicial Paternalism and the Female Status Offender: Training Women To Know Their Place," *Crime and Delinquency* 23 (1977): 121–130; W. G. Staples, "Law and Social Control in Juvenile Justice Dispositions," *Journal of Research in Crime and Delinquency* 24 (1987): 7–22.

61. Krohn et al., "Is Chivalry Dead?"

62. Dannefer and Schutt, "Race and Juvenile Justice Processing in Court and Police Agencies"; Piliavan and Briar, "Police Encounters with Juveniles"; J. Fagan et al., "Blind Justice? The Impact of Race on the Juvenile Justice Process," *Crime and Delinquency* 33 (1987): 224–258; R. J. Sampson, "Effects of Socioeconomic Context on Official Reaction to Juvenile Delinquency"; Staples, "Law and Social Control in Juvenile Justice Dispositions"; National Council on Crime and Delinquency, *The Over-Representation of Minority Youth in the California Juvenile Justice System* (San Francisco: Arthur, 1992); T. P. Thornberry, "Race, Socioeconomic Status and Sentencing in the Juvenile Justice System," *Journal of Criminal Law and Criminology* 64 (1973): 90–98; T. P. Thornberry, "Sentencing Disparities in the Juvenile Justice System," *Journal of Criminal Law and Criminology* 70 (1979): 164–171.

63. Wilson, *Varieties of Police Behavior*, 40–41.

64. Black and Reiss, "Police Control of Juveniles"; Lundman et al., "Police Control of Juveniles: A Replication"; Morash, "Establishment of a Juvenile Police Record"; R. M. Terry, "Discrimination in the Handling of Juvenile Offenders by Social-Control Agencies," *Journal of Research in Crime and Delinquency* 4 (1967): 218–230; R. M. Terry, "The Screening of Juvenile Offenders," *Journal of Criminal Law, Criminology, and Police Science* 58 (1967): 173–181; R. Weitzer, "Racial Discrimination in the Criminal Justice System: Findings and Problems in the Literature," *Journal of Criminal Justice* 24 (1996): 309–322.

65. D. C. Gibbons and M. D. Krohn, *Delinquent Behavior*, 5th ed. (Englewood Cliffs, NJ: Prentice Hall, 1991).

66. Dannefer and Schutt, "Race and Juvenile Justice Processing in Court and Police Agencies."

67. Walker, *The Police in America*.

68. Muraskin, "Police Work and Juveniles."

69. Walker, *The Police in America*.

70. National Institute for Juvenile Justice and Delinquency Prevention, *Police–Juvenile Operations: A Comparative Analysis of Standards and Practices*, vol. 2 (Washington, DC: U. S. Government Printing Office, n.d.).

71. *Miranda v. Arizona*, 384 US 436, 86 S Ct 1602, 16 L Ed 2d 694 (1966).

72. *Miranda v. Arizona*, 384 US 492.

73. *Miranda v. Arizona*, 384 US 479.

74. *Fare v. Michael C.*, 442 US 707, 99 S Ct 2560, 61 L Ed 2d 197 (1979).

75. *Fare v. Michael C.*, 442 US 724, 725.

76. *People v. Lava*, 62 Cal Rptr 586, 432 P2d 202 (1967).

77. *State v. Benoit*, 490 A2d 295 (NH 1985).

78. *State v. Benoit*, 304.

79. *Schall v. Martin*, 467 US 253, 104 S Ct 2403, 81 L Ed 2d 207 (1984).

80. *Schall v. Martin*, 467 US 265.

81. *Schall v. Martin*, 467 US 268.

82. G. D. Robin, "Juvenile Interrogations and Confessions," *Journal of Police Science and Administration* 10 (1982): 224–228.

83. L. E. Holtz, "Miranda in a Juvenile Setting: A Child's Right to Silence," *Journal of Criminal Law and Criminology* 78 (1987): 534–556.

84. E. S. Scott and T. Grisso, "The Evolution of Adolescence: A Developmental Perspective on Juvenile Justice Reform," *The Journal of Criminal Law and Criminology* 88 (1998): 137–189.

85. P. M. Harrison and J. C. Karberg, "Prison and Jail Inmates at Midyear 2002," *Bureau of Justice Statistics Bulletin* (Washington, DC: U.S. Department of Justice, 2003).

86. I. M. Schwartz, *(In)Justice for Juveniles: Rethinking the Best Interests of the Child* (Lexington, MA: Lexington Books, 1989), 77.

87. Maguire and Pastore, eds. *Sourcebook of Criminal Justice Statistics 2002*, 381, Table 4.26.

88. C. Dunn, cited in C. L. Ringwalt et al., *Past and Future Directions of the D. A. R. E. Program: An Evaluation Review*, draft final report (Research Triangle Park, NC: Research Triangle Institute, 1994).

89. Ringwalt et al., *Past and Future Directions of the D. A. R. E. Program.*

90. R. R. Clayton et al., "Sensation Seeking as a Potential Mediating Variable for School-Based Prevention Interventions: A Two-Year Follow-Up of DARE," *Journal of Health Communications* 3 (1991): 229–239; Ringwalt et al., *Past and Future Directions of the D. A. R. E. Program*; D. P. Rosenbaum et al., "Cops in the Classroom: A Longitudinal Evaluation of Drug Abuse Resistance Education," *Journal of Research in Crime and Delinquency* 31 (1994): 3–31.

91. For updates on the DARE curriculum and program evaluations, see www.dare.com.

92. For information on the GREAT curriculum, see www.atf.treas.gov/great/greatbrochure.

93. F. Esbensen et al., *The National Evaluation of the Gang Resistance Education and Training (GREAT) Program, Final Report* (Washington, DC: U.S. Department of Justice, 2002).

94. H. Goldstein, *Problem-Oriented Policing* (New York: McGraw-Hill, 1990).

95. Trojanowicz and Bucqueroux, *Community Policing.*

96. National Institute of Justice, "Community Policing in the 1990's," *National Institute of Justice Journal* 225 (1992): 2–8.

97. "Participant's Resource Packet for the Youth-Oriented Community Policing National Satellite Teleconference."

Chapter Resources

Juvenile Diversion

Chapter Objectives

After studying this chapter, you should be able to

- Define the concept of diversion and describe how it is used in dealing with youthful offenders.
- Describe the social context of the 1960s and 1970s and indicate how it encouraged the development of diversion programs.
- Explain why juvenile justice might be seen as a type of diversion from the criminal justice process.
- Describe the theoretical underpinnings of the concept of diversion.
- Describe specific diversion programs and whether or not they are effective.
- Describe common problems with diversion programs and how these problems impact youths and families.

Chapter Outline

- Introduction
- Early Efforts at Diversion
- The Theoretical Foundation of Diversion
- The Policy Implications of Societal Reaction Theory
- Other Rationales for Diversion Programs
- The Spread of Diversion Programs Since the 1960s
- Contemporary Diversion Strategies and Programs
- The Effectiveness of Diversion
- Legal Issues

- Chapter Summary
- Key Concepts
- Review Questions
- Additional Readings
- Notes

diversion The processing of juvenile offenders in ways that avoid the formal juvenile justice process.

■ Introduction

Although there is a lack of agreement on how to define **diversion**, the intent of diversion is to respond to delinquent youths in ways that keep them out of the formal juvenile justice process.[1] Diversion is based on the fact that formal responses to youths who violate the law, such as arrest and adjudication, do not always protect the best interests of children, nor do such responses necessarily protect the community. Indeed, some formal responses may be harmful to many youths and increase the likelihood of future delinquent behavior. This is because formal processing may cause a youth to develop a negative or delinquent self-image,[2] may stigmatize the youth in the eyes of significant others,[3] or may subject the youth to inhumane treatment. In addition, formal processing may restrict the youth's opportunities to associate with law-abiding peers or to engage in conventional activities, thereby increasing the chances of future delinquent activity. Consequently, efforts to divert youths from the juvenile justice process (e.g., by warning and releasing) as well as efforts to divert youth to specific diversionary programs (e.g., counseling and community service programs) have long been a part of juvenile justice practice. More recently, however, diversion has been touted as a way to reduce the problem of minority overrepresentation in juvenile justice.[4] This chapter is intended to provide an overview of the history of diversion programming. In addition, it examines the theoretical rationale for diversion and explores various types of diversion programs and their effectiveness. The chapter concludes by discussing a number of potential problems that confront diversion programs as well as issues that should be considered prior to the development and implementation of these interventions.

■ Early Efforts at Diversion

Efforts to divert children from normal criminal justice processing have a long history. As noted in Chapter 2, prior to the 1800s and the development of specialized correctional institutions for youths, children were subject to the same laws and the same criminal justice process as adults. Nevertheless, there is considerable evidence that they were often spared the harshest penalties due to their age. Police officers, sheriffs, constables, and others responsible for law enforcement have at times decided to ignore the illegal actions of youths. Law enforcement officers have also occasionally handled matters on their own, by warning youths, returning them to their parents for punishment, or meting out punishment themselves. To some extent, efforts to spare youths from the most severe punishments or to let them escape punishment altogether were based on the recognition that the young were, in important ways, different from adults and that formal or severe punishments would rarely benefit the offender or the community.

The development of the houses of refuge in the early 1800s and the juvenile court movement in the late 1800s are often cited as early examples of efforts to divert youths.[5] Indeed, the houses of refuge were established, in part, to divert children from the harsh conditions of adult correctional institutions, which were seen by some reformers as harmful to youths and likely to produce more dangerous offenders.[6] The child savers of the late 1800s established juvenile courts in an effort to divert children from adult jails and prisons and from adult courts, which were often reluctant to sanction and control wayward children.[7]

Juvenile Courts Were Created to Divert Children from Criminal Courts

The creation of specialized juvenile correctional institutions as well as the juvenile courts was based on the recognized need to respond to children and adults differently. It is the height of irony that today, in legislatures across the country, the age for the transfer of juvenile cases to adult courts is being lowered and the automatic waiver to adult courts for certain offenses is becoming law. Nowadays, more and more children are being "diverted" to the adult system—a system that is not particularly effective in dealing with many adults and is hardly capable of meeting the needs of youths.

Although diversion has a long history, the development of routine diversionary strategies and specialized diversion programs has substantially increased since the late 1960s. The decade of the 1960s was a time of widespread social unrest, increases in juvenile arrests, and critical scrutiny of basic institutions, including juvenile justice. Concern about the effectiveness of formal responses to juvenile offenses was clearly reflected in recommendations made by the 1967 President's Commission on Law Enforcement and the Administration of Justice. These recommendations, formulated for the purpose of improving juvenile justice practice, served as a catalyst for the development of diversion programs during the late 1960s and 1970s. According to the commission,

> The formal sanctioning system and pronouncement of delinquency should be used only as a last resort. In place of the formal system, dispositional alternatives to adjudication must be developed for dealing with juveniles, including agencies to provide and coordinate services and procedures to achieve necessary control without unnecessary stigma. . . . The range of conduct for which court intervention is authorized should be narrowed, with greater emphasis upon consensual and informal means of meeting the problems of difficult children.[8]

In order to facilitate the number of diversionary responses available to communities, the commission called for the establishment of youth services bureaus. These bureaus were intended to supplement existing community agencies that dealt with children and to coordinate programs and services for both delinquent and nondelinquent youths. They were also intended to serve as an alternative to juvenile court processing, allowing, it was hoped, hundreds of thousands of youths to be diverted from the formal juvenile justice process each year.[9] Through federal grants, primarily from the Law Enforcement Assistance Administration (LEAA) and the Office of Youth Development and Delinquency Prevention, and smaller matching grants provided by local and state governments, **youth services bureaus** were established in communities around the country.[10] Typically, they had five basic goals:

youth services bureau An agency that provides counseling, tutoring, mentoring, advocacy, and job referrals for youths in order to keep them out of the formal juvenile justice process.

1. Divert juveniles from the juvenile justice system.
2. Fill gaps in service by advocating for and developing services for youths and their families.
3. Provide case coordination and program coordination.
4. Provide modification of systems of youth services.
5. Involve youths in the decision-making process.[11]

Evaluation studies found that the typical youth services bureau was staffed by 5 or 6 full-time staff members and from 1 to 50 volunteers. Types of services offered included tutoring; recreational programming; individual, group, and family counseling; and job referral. The typical bureau operated in a low socioeconomic status urban neighborhood where there were high rates of crime and unemployment and limited resources for youths. Each year it served approximately 350 youths who were self-referred or referred by the police, schools, and parents, among other sources.[12]

Although youth services bureaus were popular in a number of jurisdictions, they experienced some problems. One difficult issue they had to deal with was the question of voluntary participation. The ideal was to provide services to youths and possibly their families on a voluntary basis. However, if a youth or a family member failed to follow through on treatment plans developed by a bureau, a referral to the appropriate juvenile court was typically made, even if the youth avoided further delinquent behavior. In addition, if a bureau determined that a youth was not likely to benefit from its services, the youth could be referred back to the original referral source.[13] Such practices, of course, could result in formal processing—hardly what diversion programs are designed to accomplish. In addition to funding youth services bureaus, LEAA helped fund a variety of other diversion programs in communities around the country. These programs included "alternative schools, job development and training programs, police social work programs, and family counseling programs for youths referred by the police, schools, and court intake personnel."[14]

In the late 1970s, however, LEAA began to alter its priorities and shift funds away from prevention and diversion programs and into law enforcement and rehabilitation programs. The initial LEAA funding for the youth services bureaus was intended to be seed money to help establish the programs, but after a certain period, usually 2 years, communities were expected to continue funding these programs on their own. In some cases, communities did assume full support of local youth services bureaus, but by 1982 a number of bureaus developed, at least in part, with LEAA seed money were discontinued due to a lack of funds.[15] Nevertheless, a perceived need for diversion programs continued to motivate the development of such programs in communities around the country.

◼ The Theoretical Foundation of Diversion

Like other programs for youths, diversion programs are supported by beliefs and theoretical concepts. As noted earlier, for a long time many law enforcement officials, juvenile justice reformers, and students of crime and punishment have doubted the reformative potential of adult courts, jails, and prisons and have sought to spare youths from criminal processing. Questions regarding the appropriateness of criminal processing for juveniles were heightened during the political and social unrest that characterized much of the 1960s and 1970s. In addition, a well-developed theoretical rationale for diversion became popularized during this time.

The Social Context of the 1960s and the Popularization of Diversion

The 1960s and early 1970s proved to be a fruitful period for the development and proliferation of diversionary responses to juvenile offenders. For one thing, it was marked by considerable social unrest, and a substantial portion of the protests that occurred during this period were initiated and led by youths. Furthermore, many Americans, both young and old, seriously questioned the operation of basic institutions, including criminal and juvenile justice agencies. In addition, there was a marked increase in the juvenile arrest rates during this period. Not only were increasing numbers of young persons being arrested and processed by juvenile courts for violent and property offenses, but more youths were being arrested and processed for status offenses as well. Indeed, according to the President's Commission on Law Enforcement and the Administration of Justice, over 25% of the youths in juvenile courts and institutions were status offenders.[16]

It was also during the 1960s and 1970s that the societal reaction (or labeling) perspective began to play a more important role in criminological thinking.[17] This perspective focused on three interrelated topics: "(1) the historical development of deviant labels, (2) the process by which labels are applied, and (3) the consequences of being labeled."[18]

The first topic encompasses two subtopics: how deviant categories, such as delinquent, are produced, and how social control mechanisms, such as the juvenile court, are established. A good example of a work focused on this topic is Anthony Platt's classic book, *The Child Savers*, which examines the "invention of delinquency" and the development of the first juvenile court in Chicago in 1899, a court that was primarily designed to control problem youths.[19]

The second topic includes as subtopics the way in which control agents apply deviant labels to others and the factors that influence individuals' efforts to resist or accept these labels.[20] Societal reaction researchers in the field of juve-

societal reaction perspective The point of view according to which social responses, particularly formal social responses, can contribute to subsequent delinquent behavior. This perspective favors the diversion and deinstitutionalization of juvenile offenders when possible and questions the wisdom of responding through formal means to each act of youth misbehavior (e.g., by arresting and trying the offender).

labeling The process by which a derogatory or otherwise negative term comes to be associated with a person. Being labeled with a term like *delinquent* or *chronic offender* may cause a person to develop negative self-perceptions and cause others to respond to the person based on the label. Furthermore, labeling can limit the opportunities available to the person and can lead him or her to develop a deviant identity, increasing the likelihood of subsequent offending.

nile justice might examine how police, judges, prosecutors, probation officers, and other control agents attempt to apply labels such as "delinquent," "status offender," and "chronic offender" to youths. In addition, they might explore a range of factors—such as youths' perceptions of themselves, family influences, peer reactions, and interactions between youths and others—that tend to reinforce a deviant label or help youths resist such a label.

The topic that has received the most attention within juvenile justice circles is the third—the consequences of being labeled. According to societal reaction theorists, labels such as "delinquent," "chronic offender," "thief," "doper," and the like can have a variety of negative consequences for those who are given such labels. For example, a label can lead others to assume things about the individual labeled that may not be true. A youth who has gotten into trouble in the past and has been labeled as a delinquent may be incorrectly treated as untrustworthy or may be treated more punitively as a result of being labeled. Indeed, the perceptions that social control agents have of others appear to play an important role in how youths are treated. If, through being labeled, youths are believed to possess undesirable characteristics associated in the public mind with criminality or potential criminality, they are more likely to be formally processed by police and other control agents and to be avoided by law-abiding individuals.

In addition, youths saddled with a negative label often have fewer opportunities for involvement in normal law-abiding activities. Without such opportunities, they are more likely to associate with those in similar circumstances, thus increasing the likelihood of further deviance.[21] Furthermore, responding to youths as if they were a lesser form of human being may lead them to see themselves in a more negative light.[22]

In an interesting study, Charles Frazier described the case of a young man, named Ken, who was tried and "branded" a criminal in a small town. Frazier noted that labeling Ken as a criminal led his former friends and associates to see him differently. Rejected by his former friends, Ken had fewer opportunities for engaging in conventional activities and began to see himself as a criminal.[23] According to societal reaction theorists, being treated as different and denied opportunities to associate with law-abiding persons may cause an individual to adopt a delinquent or criminal identity that becomes a *master status*—one that becomes the person's primary public identity and that overrides other statuses the individual may enjoy.[24]

A critical issue raised by societal reaction theorists is whether responses to deviant behavior such as delinquency can increase the likelihood of additional deviance. This possibility was spelled out by sociologist Edwin Lemert in 1951, when he distinguished between *primary* and *secondary deviance*. According to Lemert

> *Primary deviance is assumed to arise in a wide variety of social, cultural, and psychological contexts, and at best has only marginal implications for the psychic structure of the individual; it does not lead to symbolic reorganization at the level of self-regulating attitudes and social roles. Secondary deviation is deviant behavior, or social roles based upon it, which becomes a means of defense, attack, or adaptation to overt and covert problems created by the societal*

reaction to primary deviation. In effect, the original "causes" of the deviation recede and give way to the central importance of the disapproving, degradational, and isolating reactions of society.[25]

■ The Policy Implications of Societal Reaction Theory

Societal reaction theory has had a profound impact on social policy.[26] Yet its policy implications are rather different than those of other theoretical approaches. In contrast to those who argue that punishment of some sort needs to be imposed in order to deter youths from further delinquent activity, societal reaction theorists call for minimal or no intervention whenever possible in order to avoid the negative consequences of labeling. For example, Edwin Lemert made a case for the use of "judicious non-intervention,"[27] arguing that much delinquent behavior is normal. Similarly, sociologist Edwin Schur argued for a policy of <u>radical nonintervention</u> (i.e., leaving children alone whenever possible) in order to avoid the negative consequences associated with involvement in the juvenile justice process.[28]

Societal reaction theory had an important influence on those who looked critically at the juvenile justice process in the 1960s and 1970s, and it played a significant role in the development of three juvenile justice reforms implemented during this period: decriminalization, increased emphasis on due process, and deinstitutionalization. Each of these reforms stressed keeping juveniles out of the formal juvenile justice process whenever possible.[29]

radical nonintervention The strategy of leaving youth offenders alone (i.e., not placing them in the juvenile justice process) if at all possible.

Decriminalization

Advocates of the societal reaction perspective pointed out that the criminalization of some behaviors often produces more harm than good. For example, behaviors like running away from home and not going to school are undesirable in many instances, but treating them as crimes does not necessarily benefit the youths who engage in them. On the contrary, treating truants and runaways as offenders may actually have a number of undesirable consequences. In addition, formal responses to status offenses, according to critics, were expensive and ineffective. As a result, societal reaction theorists favored the <u>decriminalization</u> (i.e., redefining status offenses as social problems to dealt with by welfare agencies as opposed to criminal actions to be handled by juvenile courts) of status offenses, and this policy was adopted in a number of jurisdictions around the country.

decriminalization The act of redefining status offenses as social problems to be dealt with be welfare agencies as opposed to criminal actions to be handled by juvenile courts.

Increased Emphasis on Due Process

Societal reaction theorists were also wary of the discretionary powers available to juvenile justice officials, including police and judges. Many of these theorists claimed that juvenile justice officials often abused their authority and acted in ways that were harmful to those under their care. As a solution, Edwin Schur argued for "a return to the rule of law."[30] According to this approach, constitutional safeguards should be extended to youths to protect them from the power of the state. Moreover, punishments should be spelled out in law and should be determinate, preventing capricious actions on the part of officials inclined to extend punishments indefinitely.

Societal reaction theory has been cited as an important contributor to the movement to extend due process protections to juveniles. Today juveniles enjoy more due process protections than they did before the mid-1960s, but the extent to which the juvenile justice process has become more humane in its treatment of youths is not clear.[31] Simply extending legal protections to youths does not mean that those protections will be implemented in practice, and there is considerable evidence that the extension of due process protections to juveniles has had less of an impact on juvenile court proceedings than some have claimed.[32]

Deinstitutionalization

Another reform supported by those influenced by the societal reaction perspective was deinstitutionalization—the removal of juveniles from correctional facilities. Like diversion, deinstitutionalization was intended to protect youths from the harmful effects of incarceration. Indeed, criminologists from diverse theoretical positions argued that the reform of delinquents required efforts to "improve attachments to family and school, increase academic skills, open up legitimate opportunities, and reduce association with delinquent peers."[33] The incarceration of youths accomplish none of these objectives. As the President's Commission stated in a 1967 report, *The Challenge of Crime in a Free Society*, "Institutions tend to isolate offenders from society, both physically and psychologically, cutting them off from schools, jobs, families, and other supportive influences and increasing the probability that the label of criminal will be indelibly impressed upon them."[34]

Several states, including Massachusetts, Maryland, Missouri, Pennsylvania, Kansas, Minnesota, North Dakota, Oregon, and Utah, have made efforts to implement community-based correctional strategies,[35] and others have articulated the importance of community-based programs. The most ambitious of these efforts was led by Jerome Miller, the Commissioner of the Department of Youth Services in Massachusetts during the 1970s. Under Miller's direction, Massachusetts dramatically reduced its institutional population by closing training schools and placing youths in community-based programs.[36] In addition, a number of states have made considerable progress in removing status offenders from correctional facilities.[37]

■ Other Rationales for Diversion Programs

One argument offered in support of diversion is that it can reduce the stigma and other negative consequences of being arrested for delinquent behavior and thus reduce the probability of recidivism. Other supporting reasons and arguments have also been advanced. For example, diversion programs are believed to allow some youths to receive assistance who would not otherwise be helped, because juvenile corrections programs often fail to provide needed services to youths. Advocates also point out that diversion programs allow juvenile justice decision makers to use discretion and be more flexible than many courts in responding to youths.[38] Another reason people favor diversion is that it reduces the burden placed on juvenile court resources by decreasing the number of youths referred to the courts.

Furthermore, diversion programs promise the added benefit of freeing up scarce resources, which could then be devoted to dealing with more serious offenders.[39]

A final argument offered in support of diversion is based on the fact that many youths who engage in delinquent behavior are not identified and punished and cease offending without formal intervention. This fact suggests that formal intervention is in many cases unnecessary. Given the potential negative consequences of formal processing (according to societal reaction theorists), subjecting youths to formal processing can cause them great harm (and the community great harm, to the extent that these youths become more inclined to engage in further delinquency).

MYTH VS. REALITY — Punishment Is Not Always Needed to Correct Youth

Myth—If youths are not punished for their delinquent behavior, they will not only continue to engage in illegal behavior, but also engage in more serious types of delinquency.

Reality—Although self-report studies indicate that most youths engage in at least minor types of delinquent behavior,[40] most youths are not apprehended and processed by formal juvenile justice agencies.[41] Despite this, most of these youths will not commit serious offenses or live lives of crime when they become adults.[42]

■ The Spread of Diversion Programs Since the 1960s

As already noted, a number of diversion programs have been developed since the late 1960s. Among the more popular have been Scared Straight programs; family crisis intervention programs for status offenders; limited individual, family, and group counseling programs for status and criminal offenders; runaway shelters; individual and family counseling programs coupled with educational, employment, and recreational services; basic casework and counseling programs; dispute resolution programs involving restitution and community service; and, more recently, teen court and restorative justice programs. Unfortunately, many of these programs have not been carefully evaluated, and their effectiveness is thus unknown. The following sections examine some of the programs that have been evaluated.

Scared Straight Programs   # 5

Scared Straight programs, first developed in the 1970s, were popularized by two films about the **Juvenile Awareness Project**, which began at Rahway State Prison in New Jersey, a maximum-security institution, in September 1976. The first film,

Juvenile Awareness Project A program in which long-term prison inmates meet with juveniles in counseling, educational, or confrontational sessions for the purpose of exposing the youths to the harsh realities of prison life. *Scared Straight!* is an example of this type of progam.

Scared Straight!, received considerable publicity and presented testimonial evidence that suggested the program was a tremendous success. Indeed, the film won both an Emmy and an Academy Award, it was shown to countless youth groups and school classes, and it aired on national television in 1979. The second film, essentially a 10-year follow-up of the initial documentary, made similar claims regarding the effectiveness of the Juvenile Awareness Project. In California, "scaring kids straight" became so popular that legislation was introduced that required the busing of 15,000 juveniles to state prisons for confrontation sessions similar to those at Rahway. In many other communities, tours of prisons and jails and confrontations with inmates were arranged in order to scare kids straight.[43]

The Juvenile Awareness Project was developed by a group of inmates called the lifers who were serving long prison terms at Rahway. The goal of the lifers was to expose youths to the harsh realities of prison life. In order to accomplish their goal, the lifers did more than just describe conditions in Rahway. The basic intervention employed consisted of a confrontation in which verbal, and at times, physical abuse and verbal and physical intimidation were used to convey the brutality and human indignities characteristic of life in a prison such as Rahway. The inmates who developed the program believed that once youths understood where the consequences their delinquent behavior might lead, they would be deterred from subsequent delinquency.

MYTH VS. REALITY — Evaluations of Scared Straight Programs Fail to Produce Positive Results

Myth—Once youths understand what prison is like, they will think twice about committing delinquent acts.

Reality—Evaluations of scared straight programs fail to demonstrate that these programs are effective.[44]

Although the scared straight films indicated that these programs were successful, an evaluation of the Juvenile Awareness Project at Rahway by James Finckenauer was far less encouraging.[45] Finckenauer found that the youths who typically attended the Juvenile Awareness Project at Rahway were less delinquent than the films suggested. More disturbing, his research indicated that youths who went through the Rahway Project were more likely to engage in subsequent delinquency than youths who had not attended the program.[46] There were, however, some problems with Finckenauer's evaluation. The evaluation design had called for the random assignment of youths to the experimental condition (the Rahway program) or to a control condition (no treatment). However, some of the agencies who selected youths did not select them on a completely random basis. Consequently, it is possible that the two groups may not be comparable. Even so, evaluations of programs similar to the Rahway program have produced results much like those reported by Finckenauer.

In a review of the evaluations of two similar programs, Richard Lundman noted that in one of the programs, youths who were exposed to the prison program, although one that was not based on confrontation like the one at Rahway, were *more* likely to engage in subsequent delinquency than youths in a control group (who did not participate in the program). This finding mirrors the one from the *Scared Straight!* evaluation. The second program appeared to be more promising, but the evaluation of this program was seriously flawed, making the results questionable. In a review of a number of other programs that involve exposing youths to prison life as well as an examination of **meta-analyses** (a data analysis technique that makes possible the examination of treatment effects across different studies) of these types of programs, James Finckenauer and Patricia Gavin also concluded that these programs are ineffective.[47] In summing up the evaluation literature on prison-based confrontation programs, Lundman stated that "At best, taking juveniles to prison for intensive confrontation sessions has no effect. At worst, . . . it makes children more delinquent rather than less."[48]

Efforts to "Scare Kids Straight" Continue to Be Found Around the Country

Despite evidence that the Scared Straight program at Rahway may be harmful to youths, it is still in existence, and similar programs continue to operate around the country. There is even an MTV version that is shown periodically, and there is a Scared Straight Web site.

Family Crisis Intervention Programs

Another type of diversion program involves providing crisis intervention services to children and families in hopes of preventing subsequent problems and formal court involvement. A good example of this type of diversion program was developed in Sacramento, California, in 1969. This program, known as the Sacramento 601 Diversion Project, was intended to divert status offenders from detention and formal court processing by having specially trained probation officers provide short-term crisis intervention and treatment services to the youths and their families.[49]

This program focused on status offenders who were referred to the juvenile court by police, parents, and school officials. The youths referred to this program were typically white, poor, female, and under 15 years of age. Also, three-quarters had no previous contact with the court. The usual reason for referral was the existence of family problems.[50] When a youth was referred to the court, a specially trained probation officer would contact the youth's parents and request that the parents and youth come to Juvenile Hall for an immediate counseling session. When they arrived at Juvenile Hall, the probation officer would read the Miranda warning and explain that participation was voluntary. The officer indicated, however, that if they did not wish to participate in an immediate counseling session, their case would be referred to the court. The youth and parents were also told that if they agreed to participate in a counseling session, they could return for a limited number of sessions (the limit was five) if they chose to do so. Most parents and their children agreed to participate.[51] Once a juvenile waived his or her rights, a counseling session began in which the probation officer used family intervention techniques. Essentially these techniques were intended to get the family to look at the problem as a family problem and one that should be addressed by the entire family rather than by locking up the youth.[52]

An evaluation of the Sacramento 601 Diversion program, which employed an **experimental design**, indicated that it produced some promising results. Only about 14% of those youths handled by the specially trained project probation officers were detained and only about 4% were petitioned to the court. Among youths handled by regular probation officers, 55% were detained and about 20% were petitioned to court.[53]

In addition, the program evaluation found that the project reduced recidivism. A follow-up study of youths 12 months after their involvement in the program found that 46% of those who received the specialized diversion services had another court referral and 22% were referred to the court for a criminal offense. In comparison, 54% of those handled by regular probation officers were subsequently referred and about 30% were referred for a criminal offense.[54]

Diversion Projects That Employ Individual, Family, and Group Counseling Services

Given the apparent success of the Sacramento County 601 Diversion Project, 11 other jurisdictions in California developed similar programs and began to conduct evaluations of those projects in the mid-1970s. These projects were developed in Compton, El Centro, Fremont, Irvine, La Colonia, Mendocino, Simi Valley, Stockton, Vacaville, and Vallejo, California, and were similar to the Sacramento 601 Diversion Program in a number of ways. They substituted short-term family and individual counseling for detention and referral to juvenile court.[55]

There were, however, important differences between these projects and the Sacramento project. For example, more of the youths involved in these projects were male, Hispanic, and African American, and a number were diverted after committing criminal offenses. In these projects, juveniles were referred by either the police or by probation officers. Youths received individual, family, and group counseling for 4 to 6 weeks on average, and the counseling included nearly 6 hours of contact services.[56]

The evaluators of these projects employed a **quasi-experimental design**. That is, they did not use random assignment of youth to experimental and control conditions but rather attempted to approximate equivalence of the groups by comparing youths serviced by the diversion programs with youths who had similar race, age, sex, ethnicity, prior arrest, and referral source characteristics but were not serviced by the programs. Arrests of the individuals in the two groups were then compared six months after the arrest that brought them to the attention of the authorities.[57]

experimental design Research design where subjects are randomly assigned to experimental and control groups.

quasi-experimental design Research designs that can be used to explore possible cause-and-effect relationships when true experimental designs are not possible. What distinguishes these designs from experimental designs is that they do not involve random assignment of subjects to experimental and control groups.

A comparison of the two groups revealed that youths serviced by the diversion programs were arrested slightly less often during the 6-month follow-up period than youths in the other group (about 25% compared with about 31%).[58] These results suggest that diversion programs may produce modest reductions in recidivism. The researchers also found that diversion seems to have some benefit for males and females and for both criminal and status offenders.

Programs That Employ Individual and Family Counseling in Conjunction with Employment, Educational, and Recreational Services

As the number of diversion programs increased in the United States, researchers raised questions about their effectiveness. Among other things, they looked at the types of evaluations that had been done. In most instances, the diversion research in California was based on quasi-experimental designs. Consequently, it was difficult to tell if it was the programs themselves that had reduced recidivism or some unknown factor. Furthermore, it was unclear if diversion could be used effectively with criminal offenders, although there was some indication that it could. Finally, none of the evaluations of diversion programs had employed a "no treatment" group. Consequently, it was difficult to determine what the benefits of diversion programs were compared with doing nothing—a radical nonintervention approach.[59] In response to these concerns and others, the Office of Juvenile Justice and Delinquency Prevention (OJJDP) made $10 million available for the National Evaluation of Diversion Programs in the 1970s. As a result, four jurisdictions—Kansas City, Missouri; Memphis, Tennessee; Orange County, Florida; and New York City—developed experimental diversion projects. Each project excluded status offenders and included radical nonintervention as one of the experimental conditions. Experimental conditions consisted of (1) assigning youths to the diversion project, (2) doing nothing, and (3) referring the youths back to the original referral source for normal handling. This meant that some, but not all, of these youths penetrated the juvenile justice system.[60]

Diverted juveniles in each of the four projects received individual and family counseling, and educational services were available to youths. However, each of the projects tended to emphasize different services. For example, in Orange County, recreational services were emphasized, whereas in Kansas City, some diverted juveniles were assigned adult counselors who acted as advocates in dealings with public agencies and organizations, such as schools.[61]

An evaluation of these projects indicated that there was no difference in re-arrests between the three groups 6 and 12 months following the intervention. At the 6-month point, 22% of the youths who had been enrolled in a diversion program, 22% of those released (radical nonintervention), and 22% of those handled in the usual way had experienced at least one re-arrest. After 12 months, 31% of those diverted, 30% of those released, and 32% of those handled in the usual way had experienced at least one re-arrest.[62] The finding of "no difference" is important for at least two reasons: It suggests that criminal offenders may benefit from diversion, and it suggests that doing less may be just as effective as doing more—that placement in the juvenile justice process may not be the best way of dealing with *some* offenders.

Programs That Use Basic Counseling and Casework Services

Other diversion programs provide basic counseling and casework services to youths. One example is the Adolescent Diversion Project in Michigan, evaluated between 1976 and 1981. The subjects involved in this project consisted of both status and criminal offenders, but youths charged with Crime Index offenses against persons and those already on probation were not included. The two most common offenses committed by the subjects were larceny-theft (34%) and breaking and entering (24%). The average age of the subjects was 14, 84% of the subjects were male, and slightly less than 30% were African American or Hispanic.[63] The longer of the two studies of the Adolescent Diversion Project began in 1976 and lasted until 1981. It examined the effectiveness of six different interventions provided by university students working for a course grade. Each student received specialized training in the type of intervention to which he or she had been assigned, and each worked with clients for 6 to 8 hours a week for 18 weeks. The students were monitored and evaluated by graduate students, by their clients, and by the professors responsible for the study. The interventions included family-focused behavioral contracting and client advocacy for youths provided by trained students; behavioral contracting and client advocacy for youths provided by trained students and a juvenile court employee; individually tailored client interventions designed by students given minimal training; relationship building led by specially trained university students; and standard court processing.[64]

A second study, which began in 1979 and lasted until 1981, limited diversion interventions to behavioral interventions and child advocacy. As in the other study, some youths were returned to the referral source for normal processing. However, the types of people who provided the services expanded to include graduate students and community volunteers. Each service provider was again trained to provide diversion services to one client, each spent 6 to 8 hours a week with the client, and each was carefully monitored and evaluated.[65]

The two studies of the Adolescent Diversion Project indicate that a number of the interventions appear to have been effective in reducing subsequent delinquent activity, at least when compared with normal court processing. The interventions that were the most effective were those that provided behavioral contracting and child advocacy services and those that focused on relationship building. Particularly effective were those interventions led by community volunteers. The least effective involved the provision of behavioral contracting and child advocacy services (by students and a juvenile court employee) and standard court processing. These findings support the view of societal reaction theorists, who claim that formal intervention can have a negative effect on youths.[66]

■ Contemporary Diversion Strategies and Programs

Today, as in the past, diversion is frequently used in juvenile justice. Also as in the past, diversion strategies are of two basic types: radical nonintervention and involvement in a diversion program. A police policy that entails the warning and release of some juvenile offenders would be an example of radical nonintervention. A policy that encourages police officers to refer youths (and possibly their parents) to a diversion program would be an example of the second type of strategy.

Contemporary diversion programs are operated by various types of juvenile justice and community agencies. Many communities have diversion programs operated by the local juvenile court or probation agency, others have diversion programs run by the police department, and still others have diversion programs operated by separate public or private agencies. Whether operated by juvenile justice or community agencies, diversion programs may employ a variety of interventions intended to reduce the continued involvement of youths in delinquency. The interventions include providing basic casework services to youths; truancy courts; crisis intervention to assist youths and their families; individual, family, and group counseling; conflict resolution; participation in restitution and community service programs; and teen courts, among others. Following are three examples of more contemporary diversion programs that have received some attention in the evaluation literature.

Youth Enhancement Services (YES)

Youth Enhancement Services (YES) was intended to provide a variety of services to primarily minority youths in Daupin County, Pennsylvania (Harrisburg), who were felt to be at risk of delinquency as well as those who had contact with the police. Its major goals were to improve school performance and behavior and reduce rates of arrest and recidivism among minority youths. Program referrals came from a variety of community sources including the police, and services were provided to clients by a range of community agencies including Girls Inc., the Boys Club, and the YMCA. These services consisted of some combination of client needs assessments, mentoring and adult support, peer group discussions, family support services, neighborhood and community projects, educational assistance, and job readiness training.[67]

An evaluation of YES compared outcomes for youths who were referred to the program but who chose not to participate (44% of the youths referred to the program), those who participated and received less than 30 hours of service over the course of a year (24% of referrals), and those who received at least 30 hours of service (32% of the referrals). As might be expected, youths who received more than 30 hours of service had characteristics that were likely to increase the odds of program success. They were less likely than other youths to have prior arrest, to be a self-referral, or to be referred by their family, and on average they did better in school. The evaluation revealed no significant differences between the three groups on school performance or behavior. The program did, however, appear to reduce minority youths' likelihood of contact with the police. Over a 2-year follow-up period, 50.6% of those who were referred to the program but who did not participate had at least one new arrest compared to 41.3% in the low attendance group, and 25.8% in the high attendance group.[68]

Teen Court

Another type of diversion program that has grown in popularity in recent years is teen court. A distinguishing feature of these courts is that juveniles, as opposed to adults, determine the case disposition from a range of sentencing options that are available under the guidance of adults that oversee the program. Typically these courts handle relatively young offenders with no prior arrests, and in most instances they deal primarily with cases involving theft, minor assaults, disorderly conduct, possession and use of alcohol, and vandalism.[69]

Advocates of teen court programs maintain that these programs make use of youths' desire to be accepted by their peers to encourage them to take responsibility for their behavior and engage in more socially acceptable behavior in the future. Moreover, supporters argue that not only does teen court help youths who are the defendants in cases, but also those youths who staff the court as well as the community as a whole. By involving a number of youths and adults in the teen court process, advocates believe that teen courts engender greater community involvement in the legal system, encourage community cohesion, promote law-abiding behavior and do this in a cost effective manner.[70]

At present, four basic teen court models are found around the country:

- *Adult judge.* An adult judge presides over the court and oversees court operations. Youths act as attorneys, jurors, clerks, bailiffs, and other court officers and staff. This is the most popular model around the country.
- *Youth judge.* Like the previous model, but here a youth serves as judge.
- *Tribunal.* Youth attorneys present their cases to a panel of three youth judges who decide on the appropriate disposition. A jury composed of youths is not used in this model.
- *Peer jury.* This model does not use youth attorneys. Instead, the case is presented to the peer jury by a youth or an adult. In this model the peer jury questions the defendant directly.[71]

Because the growth in teen courts is relatively recent, there is not yet an extensive body of research documenting their effectiveness. Some program evaluations have reported recidivism rates of 3 to 8% between 6 and 12 months of program completion. In addition, there is good evidence that at least some of these programs produce improved perceptions of authority, justice, and the legal system. Other studies, however, have reported recidivism rates of 24 to 32%.[72] Although some of the research conducted to date has produced promising results, until more comprehensive studies of these programs are completed, more definitive statements about the viability of teen court programs is premature.

Restorative Justice/Conflict Resolution/Mediation Programs

restorative justice A process designed to repair the harm done to the victim and the community that entails bringing the interested parties together in a nonconfrontational setting where an agreeable resolution to the youth's illegal behavior can be reached.

Restorative justice programs consist of a variety of interventions designed to mediate or resolve conflicts between juvenile offenders and their victims. Proponents of restorative justice argue that simply focusing on punishment of the offender hinders efforts to achieve justice. In contrast to present efforts that focus solely on the offender, **restorative justice** is concerned with achieving justice for both the offender and the victim. As a result, these programs are intended to provide rehabilitative services to offenders and to deliver some therapeutic benefit for victims. Moreover, they are believed to be cost effective, and their focus on offender accountability and victim restitution and/or community service are appealing to the public.

One diversion program being used with young offenders is *restorative justice conferencing*. In a restorative justice conference, the youth who committed the offense, the victim, and supporters of the offender and victim meet with a trained facilitator to discuss the offense and examine the harm that it caused. Supporters have an opportunity to discuss how they have been affected by the offense, and at the conclusion of the conference, a written agreement is completed specifying how the offender can make amends to the victim and others harmed by the incident. Typically, the agreement involves an apology and restitution to the victim, but it can also include other elements such as community service or requirements that youths improve their school performance or behavior at home.[73]

Although there is limited research on restorative justice conferencing with young offenders, research on programs using some form of restorative justice have produced generally favorable results.[74] One evaluation that employed an experimental design and looked at restorative justice conferencing for first-time young offenders (median age 13 years old) who had committed nonserious offenses noted several positive outcomes. An analysis of interviews with victims revealed that those who participated in conferencing expressed much higher levels of satisfaction with the way their cases were handled compared to those in other diversion programs. Levels of satisfaction were similar for youths and parents involved in conferences and other types of diversion programs. Those randomly assigned to participate in conferences compared to those assigned to other diversion programs, however, had significantly higher program completion rates than those in other diversion programs and significantly lower levels of rearrest 6 and 12 months after their initial arrest. These findings indicate that restorative justice conferencing is a promising approach for diverting young offenders from the formal justice process and can produce high levels of satisfaction among all conference participants.[75]

■ The Effectiveness of Diversion

Supporters of diversion strategies and programs maintain that such programs decrease the number of youths involved in the formal juvenile justice process, reduce offending among youths who receive diversionary treatment, minimize the stigma associated with formal intervention, are more cost-effective than formal processing, reduce the level of coercion employed in juvenile justice, and reduce minority overrepresentation in juvenile justice.[76] As noted earlier, a number of evaluation studies of diversionary programs have found them to be superior to formal court processing in a number of respects,[77] but some evaluation studies have found that they often fall short of their goals. In fact, although some evaluation studies indicate that diversion programs can reduce recidivism or are at least as effective as formal processing at reducing recidivism,[78] other studies have found that some diversion programs are associated with higher levels of subsequent offending.[79]

In addition to possible increased recidivism, other problems have been found to plague some diversion programs. Although these programs are often touted as a way to reduce the number of youths involved in the juvenile justice process,

net-widening Increasing the availability of programs to handle youth offenders outside of the formal juvenile justice process and thereby causing youths who would otherwise have been left alone to be placed into programs.

some diversion strategies may actually lead to **net-widening**.[80] Indeed, research on diversion programs has found that some programs increase the number of youths involved in the juvenile justice process by 33% to 49%.[81]

A rationale for the development of diversion programs is the belief that such programs are less stigmatizing than formal court processing. However, research suggests that involvement in a diversionary program can be as stigmatizing as involvement in formal court programs. For example, one study that compared the self-concepts of incarcerated youth to those of youths in diversion programs found no significant differences between the two groups.[82]

The coercive nature of many diversion programs is another concern. In recommending diversion, the 1967 President's Commission on Law Enforcement and the Administration of Justice called for voluntary diversion programs. However, in some jurisdictions, potential clients are given no real choice.[83] For example, a juvenile offender might be told that unless he or she agrees to participate in the diversion program, the case will be referred for court action. Under such circumstances, most potential clients will feel compelled to participate. (Recall the choices given to youths and parents in the Sacramento 601 Diversion Program.)

Still another concern is the cost. As the cost of institutional placement as well as other formal responses has risen, diversion has been seen by many as a cost effective alternative to formal processing. However, research that has examined the costs associated with diversion indicates that, whereas some diversion programs are less expensive than many formal programs, others are actually more costly.[84] Like other programs, diversion programs often require staff and support, which means that resources that might be devoted to other needs have to be allocated to diversion. Furthermore, if net-widening occurs, even though the per-client cost of diversion may be lower than the per-client cost of a formal program, the increased volume of cases can make the diversion program more expensive.

A final concern is the possibility that diversion programs may deny youths due process.[85] Diversion may be coercive and may consist of intrusive interventions, such as a requirement that a youth, and possibly the youth's family, participate in counseling or some other type of treatment. Coercion is problematic at any level of the juvenile justice process, but it is particularly problematic at the pre-adjudicatory stage of the juvenile justice process, where youths have not been proven guilty of an offense.

In sum, the research on diversion has produced mixed results. A number of problems have been associated with diversion programs, but despite these problems diversion appears to have some merit. As diversion programs are developed and implemented, however, problems that plague many diversion programs should be addressed, and diversion programs should be carefully evaluated after they are implemented to ensure that they are accomplishing their objectives.

■ Legal Issues

Due process, which is intended to assure an individual certain protections against government intervention, is in some ways in conflict with the idea of diversion. First, in many jurisdictions, eligibility for diversion may be decided by police officers who are unfamiliar with the proper criteria for diversion or how to apply them. Thus, they are ill equipped to make diversion referrals. As a result, some youths who commit the same offenses will be diverted and others arrested. Second, the availability of diversion programs may increase the likelihood that a youth will become formally involved with juvenile justice agencies as a result of net-widening. Third, if coercion is used to get youths to participate in diversion programs, they are being denied their rightful due process protections. Fourth, involvement by diversion staff in the lives of family members on a "voluntary" basis may be just as intrusive as if it was done by court order. Fifth, many diversion programs make referrals to the juvenile court when "clients" are felt to be uncooperative, which is another way these programs can be coercive and widen the net of court control.

◼ Chapter Summary

This chapter examined diversion, which encompasses responses to juvenile offenders that are intended to avoid formal court processing. The use of diversion is typically motivated by the belief that formal responses to youth offenders do not always serve the best interests of the youths (nor those of the community) and that such responses often increase rather than decrease the likelihood of recidivism. Consequently, the immediate purpose of diversion is to minimize or avoid contact between youths and formal criminal justice agencies, such as the police or the juvenile courts.

Although diversion has a long history, it became more popular in the late 1960s and early 1970s. During this period, the rising level of juvenile crime, the willingness of many Americans to question the operation of basic American institutions, and the popularization of societal reaction theory (which raised serious questions about the potential detrimental effects of formal processing) served as the impetus for the development of a range of diversionary responses to juvenile offenders.

Diversion currently encompasses a wide range of responses, from radical nonintervention to the referral of youths and possibly their parents to a diversionary program intended to reduce the chance of future delinquent behavior. Despite the widespread use of diversion, its effectiveness is unclear. Although studies of a number of diversion programs have produced considerable support for the concept of diversion, particularly when it is compared with formal juvenile justice processing, research has also produced evidence that some diversion programs have little effect on recidivism. Indeed, some studies indicate that diversion programs may increase the likelihood of subsequent delinquent activity. In addition, some of the supposed benefits of diversion, such as cost-efficiency and the lack of stigma associated with these programs, may be illusory. Nevertheless, diversion has proven to be an effective and cost-efficient response to many juvenile offenders and will continue to receive considerable attention as a juvenile justice strategy.

KEY CONCEPTS

decriminalization The act of redefining status offenses as social problems to be dealt with be welfare agencies as opposed to criminal actions to be handled by juvenile courts.

diversion The processing of juvenile offenders in ways that avoid the formal juvenile justice process.

experimental design Research design where subjects are randomly assigned to experimental and control groups.

Juvenile Awareness Project A program in which long-term prison inmates meet with juveniles in counseling, educational, or confrontational sessions for the purpose of exposing the youths to the harsh realities of prison life. *Scared Straight!* is an example of this type of program.

labeling The process by which a derogatory or otherwise negative term comes to be associated with a person. Being labeled with a term like *delinquent* or *chronic offender* may cause a person to develop negative self-perceptions and cause others to respond to the person based on the label. Furthermore, labeling can limit the opportunities available to the person and can lead him or her to develop a deviant identity, increasing the likelihood of subsequent offending.

net-widening Increasing the availability of programs to handle youth offenders outside of the formal juvenile justice process and thereby causing youths who would otherwise have been left alone to be placed into programs.

quasi-experimental design Research designs that can be used to explore possible cause-and-effect relationships when true experimental designs are not possible. What distinguishes these designs from experimental designs is that they do not involve random assignment of subjects to experimental and control groups.

radical nonintervention The strategy of leaving youth offenders alone (i.e., not placing them in the juvenile justice process) if at all possible.

restorative justice A process designed to repair the harm done to the victim and the community that entails bringing the interested parties together in a nonconfrontational setting where an agreeable resolution to the youth's illegal behavior can be reached.

societal reaction perspective The point of view according to which social responses, particularly formal social responses, can contribute to subsequent delinquent behavior. This perspective favors the diversion and deinstitutionalization of juvenile offenders when possible and questions the wisdom of responding through formal means to each act of youth misbehavior (e.g., by arresting and trying the offender).

youth services bureau An agency that provides counseling, tutoring, mentoring, advocacy, and job referrals for youths in order to keep them out of the formal juvenile justice process.

REVIEW QUESTIONS

1. What is the definition of diversion, and what are the goals of diversion?

_____

_____

_____

_____

2. Explain the principle of "radical nonintervention" as it relates to youth delinquency and diversion.

_____

_____

_____

_____

_____

3. What are youth services bureaus, and what role do they play in the diversion of youths from the formal juvenile justice process?

_____

_____

_____

_____

_____

4. How did the social climate in the 1960s and 1970s contribute to the movement toward diversion programs?

_____

_____

_____

_____

_____

5. Why do sociologists such as Edwin Lemert support some types of diversion programs?

_____

_____

_____

_____

_____

6. What criminological theory supports the use of diversion?

_____

_____

_____

_____

_____

7. What is a "scared straight" program, and do such programs work effectively as diversion programs?

_____

_____

_____

_____

_____

8. What is the Sacramento 601 Diversion Project, and did it work?

_____

_____

_____

_____

_____

9. How effective is diversion as a mechanism for reducing or preventing recidivism? Support your position.

_____

_____

_____

_____

_____

10. What are five drawbacks to using diversion for youth offenders?

_____

_____

_____

_____

_____

ADDITIONAL READINGS

Bullington, B. et al. 1978. "A Critique of Diversionary Juvenile Justice." _Crime and Delinquency_ 24, no. 1: 59–71.

Cressey, D. R. and R. A. McDermott. 1974. _Diversion from the Juvenile Justice System._ Washington, DC: U.S. Department of Justice.

Lundman, R. J. 1993, 2001. _Prevention and Control of Juvenile Delinquency._ 2nd and 3rd eds. New York: Oxford University Press, see chapters on diversion.

Schur, E.M. 1973. _Radical NonIntervention: Rethinking the Delinquency Problem._ Englewood Cliffs, NJ: Prentice-Hall.

NOTES

1. President's Commission on Law Enforcement and Administration of Justice, _Task Force Report: Juvenile Delinquency and Youth Crime_ (Washington, DC: U.S. Government Printing Office, 1967).

2. R. M. Carter and M. Klein, eds., _Back on the Street: The Diversion of Juvenile Offenders_ (Englewood Cliffs, NJ: Prentice-Hall, 1976).

3. E. M. Schur, _Radical Nonintervention: Rethinking the Delinquency Problem_ (Englewood Cliffs, NJ: Prentice-Hall, 1973).

4. R. L. Lundman, _Prevention and Control of Juvenile Delinquency_, 3rd ed. (New York: Oxford University Press, 2001).

5. J. T. Whitehead and S. P. Lab, _Juvenile Justice: An Introduction_, 2nd ed. (Cincinnati, OH: Anderson, 1996).

6. B. Krisberg and J. F. Austin, _Reinventing Juvenile Justice_ (Newbury Park, CA: Sage, 1993).

7. A. M. Platt, _The Child Savers: The Invention of Delinquency,_ 2nd ed. (Chicago: University of Chicago Press, 1977).

8. President's Commission on Law Enforcement and Administration of Justice, _Task Force Report_, 2.

Chapter Resources

9. H. T. Rubin, *Juvenile Justice: Policy, Practice, and Law*, 2nd ed. (New York: Random House, 1985).

10. A. R. Roberts, "The Emergence and Proliferation of Juvenile Diversion Programs," in *Juvenile Justice: Policies, Programs, and Services*, 2nd ed., ed. A. R. Roberts (Chicago: Nelson-Hall, 1998).

11. A. R. Roberts, "Community Strategies with Juvenile Offenders," in *Juvenile Justice: Policies, Programs, and Services*., 2nd ed., ed. A. R. Roberts (Chicago: Nelson-Hall, 1998), 117.

12. Roberts, "Community Strategies with Juvenile Offenders."

13. Rubin, *Juvenile Justice*.

14. Roberts, "Emergence and Proliferation of Juvenile Diversion Programs," 81.

15. Roberts, "Community Strategies with Juvenile Offenders."

16. President's Commission on Law Enforcement and Administration of Justice, *Task Force Report*.

17. J. R. Lilly et al., *Criminological Theory: Context and Consequences*, 2nd ed. (Thousand Oaks, CA: Sage, 1995).

18. S. Pfohl, *Images of Deviance and Social Control: A Sociological History*, 2nd ed. (New York: McGraw-Hill, 1994), 370.

19. Platt, *The Child Savers*.

20. Pfohl, *Images of Deviance and Social Control*.

21. Lilly, *Criminological Theory*.

22. H. Garfinkel, "Conditions of Successful Degradation Ceremonies," *American Journal of Sociology*, 61 (1956): 420–424; F. Tannenbaum, *Crime and the Community* (New York: Columbia University Press, 1938).

23. C. Frazier, *Theoretical Approaches to Deviance: An Evaluation* (Columbus, OH: Merrill, 1976).

24. H. S. Becker, *Outsiders: Studies in the Sociology of Deviance* (New York: The Free Press, 1963).

25. E. M. Lemert, *Social Pathology* (New York: McGraw-Hill, 1951), 48.

26. L. T. Empey et al., *American Delinquency: Its Meaning and Construction*, 4th ed. (Belmont, CA: Wadsworth, 1999).

27. E. M. Lemert, "The Juvenile Court: Quest and Realities," in *President's Commission on Law Enforcement and Administration of Justice, Task Force Report: Juvenile Delinquency and Youth Crime* (Washington, DC: U.S. Government Printing Office, 1967).

28. E. M. Schur, *Radical Nonintervention*.

29. Empey and Stafford, *American Delinquency*.

30. Schur, *Radical Nonintervention*, 169.

31. F. T. Cullen and K. Gilbert, *Reaffirming Rehabilitation* (Cincinnati, OH: Anderson, 1982).

32. T. J. Bernard, *The Cycle of Juvenile Justice* (New York: Oxford University Press, 1992).

33. L. T. Empey and M. C. Stafford, *American Delinquency: Its Meaning and Construction* (Belmont, CA: Wadsworth, 1991), 404.

34. President's Commission on Law Enforcement and Administration of Justice, *The Challenge of Crime in a Free Society* (Washington, DC: U. S. Government Printing Office, 1967), 165.

35. C. Bartollas and S. J. Miller, *Juvenile Justice in America*, 4th ed. (Upper Saddle River, NJ: Pearson, 2005); Krisberg and Austin, *Reinventing Juvenile Justice*.

36. J. G. Miller, *Last One over the Wall: The Massachusetts Experiment in Closing Reform Schools* (Columbus, OH: Ohio State University Press, 1991).

37. H. N. Snyder and M. Sickmund, *Juvenile Offenders and Victims: 1999 National Report* (Washington, DC: Office of Juvenile Justice and Delinquency Prevention, 1999); U. S. Department of Justice, "Assessing the Effects of the Deinstitutionalization of Status Offenders," *Juvenile Justice Bulletin* (Washington, DC: U. S. Department of Justice), 1993.

38. G. F. Vito and D. G. Wilson, *The American Juvenile Justice System* (Beverly Hills, CA: Sage, 1985).

39. Vito and Wilson, *The American Juvenile Justice System*.

40. R. Agnew, *Juvenile Delinquency: Causes and Control* (Los Angeles: Roxbury, 2001); M. Gold, "Undetected Delinquent Behavior," *Journal of Research in Crime and Delinquency* 3 (1966): 27–46; J. Short and F. I. Nye, "Extent of Unrecorded Delinquency: Tentative Conclusions," *Journal of Criminal Law, Criminology, and Police Science* 49 (1958): 296–302.

41. D. Huizinga and D. S. Elliott, "Juvenile Offender Prevalence, Incidence, and Arrest Rates by Race," *Crime and Delinquency* 33 (1987): 206–233.

42. S. P. Lab, "The Identification of Juveniles for Non-Intervention," (Ph.D. diss., Florida State University, 1982), cited in Whitehead and Lab, *Juvenile Justice*, 2nd ed.

43. This section is largely based on J. O. Finckenauer and P. W. Gavin, *Scared Straight, The Panacea*

Phenomenon Revisited (Prospect Heights, IL: Waveland, 1999) and Lundman, *Prevention and Control of Juvenile Delinquency*, 3rd ed.

44. Finckenauer and Gavin, *Scared Straight, The Panacea Phenomenon Revisited*; Lundman, *Prevention and Control of Juvenile Delinquency.*

45. Finckenauer and Gavin, *Scared Straight, The Panacea Phenomenon Revisited.*

46. Finckenauer and Gavin, *Scared Straight, The Panacea Phenomenon Revisited.*

47. Finckenauer and Gavin, *Scared Straight, The Panacea Phenomenon Revisited.*

48. Lundman, *Prevention and Control of Juvenile Delinquency*, 3rd ed., 202.

49. R. Baron and F. Feeney, *An Exemplary Project: Juvenile Diversion through Family Counseling* (Washington, DC: U. S. Department of Justice, 1976).

50. Lundman, *Prevention and Control of Juvenile Delinquency,* 3rd ed.

51. Baron and Feeney, *An Exemplary Project.*

52. Lundman, *Prevention and Control of Juvenile Delinquency*, 3rd ed.

53. Lundman, *Prevention and Control of Juvenile Delinquency*, 3rd ed.

54. Lundman, *Prevention and Control of Juvenile Delinquency*, 3rd ed..

55. Lundman, *Prevention and Control of Juvenile Delinquency*, 3rd ed.

56. T. Palmer et al., *The Evaluation of Juvenile Diversion Projects: Final Report* (Sacramento, CA: California Youth Authority, 1978).

57. Palmer et al., *The Evaluation of Juvenile Diversion Projects.*

58. This section is largely based on Lundman, *Prevention and Control of Juvenile Delinquency*, 3rd ed.

59. Lundman, *Prevention and Control of Juvenile Delinquency*, 3rd ed.

60. Lundman, *Prevention and Control of Juvenile Delinquency*, 3rd ed.

61. F. Dunford et al., National Evaluation of Diversion Projects: Final Report (Washington, DC: Office of Juvenile Justice and Delinquency Prevention, 1981).

62. Lundman, *Prevention and Control of Juvenile Delinquency*, 3rd ed.

63. R. J. Lundman, *Prevention and Control of Juvenile Delinquency*, 2nd ed. (New York: Oxford University Press, 1993).

64. W. S. Davidson et al., *Alternative Treatments for Troubled Youth: The Case of Diversion from the Juvenile Justice System* (New York: Plenum, 1990).

65. Davidson et al., *Alternative Treatments for Troubled Youth.*

66. Davidson et al., *Alternative Treatments for Troubled Youth*; Lundman, *Prevention and Control of Juvenile Delinquency*, 2nd ed..

67. W. N. Welsh et al., "Reducing Minority Overrepresentation in Juvenile Justice: Results of Community-Based Delinquency Prevention in Harrisburg," *Journal of Research in Crime and Delinquency* 36 (1999): 87-110; W. N. Welsh et al., "Reducing Overrepresentation in Juvenile Justice: Development of Community-Based Programs in Pennsylvania," *Crime and Delinquency* 42 (1996): 76–98.

68. W. N. Welsh et al., "Reducing Minority Overrepresentation in Juvenile Justice: Results of Community-Based Delinquency Prevention in Harrisburg."

69. J. A. Butts and J. Buck, "Teen Courts: A Focus on Research," Juvenile Justice Bulletin (Washington, DC: Office of Juvenile Justice and Delinquency Prevention, 2000).

70. Butts and Buck, "Teen Courts: A Focus on Research."

71. P. A. Nessel, "Youth Court: A National Movement," *Technical Assistance Bulletin* 17 (Chicago, IL: American Bar Association, 2000).

72. Butts and Buck, "Teen Courts: A Focus on Research."

73. E. F. McGarrell, "Restorative Justice Conferences as an Early Response to Young Offenders," Juvenile Justice Bulletin (Washington, DC: Office of Juvenile Justice and Delinquency Prevention, 2001).

74. B. Galaway, "Crime Victim and Offender Mediation as a Social Work Strategy," Social Service Review 62 (1988): 668–683; T. Marshall and S. Merry, *Crime and Accountability: Victim-Offender Mediation in Practice* (London: HMSO, 1990); A. L. Schneider, "Restitution and Recidivism Rates of Juvenile Offenders: Results from Four Experimental Studies," *Criminology* 24 (1986): 533–552; M. Umbreit and R. Coates, "Cross-Site Analysis of Victim-Offender Mediation in Four States," *Crime and Delinquency* 39 (1993): 565–585.

75. McGarrell, "Restorative Justice Conferences as an Early Response to Young Offenders."

76. Lundman, *Prevention and Control of Juvenile Delinquency*, 3rd ed; Whitehead and Lab, *Juvenile Justice*, 2nd ed.

Chapter Resources

77. Lundman, *Prevention and Control of Juvenile Delinquency*, 3rd ed; R. Regoli et al., "Using an Alternative Evaluation Measure for Assessing Juvenile Diversion Programs," *Children and Youth Services Review* 7 (1985): 21–38; H. C. Quay and C. T. Love, "The Effect of a Juvenile Diversion Program on Rearrests," *Criminal Justice and Behavior* 4 (1977): 377–396.

78. R. Baron et al., "Preventing Delinquency through Diversion," *Federal Probation* 37 (1973): 13–18; R. Baron and F. Feeney, *Juvenile Diversion through Family Counseling* (Washington, DC: U.S. Government Printing Office, 1976); Davidson et al., *Alternative Treatments for Troubled Youth*; F. Dunford et al., *National Evaluation of Diversion Projects*; Palmer et al., *The Evaluation of Juvenile Diversion Projects*; Lundman, *Prevention and Control of Juvenile Delinquency*, 3rd ed.; Quay and Love, "The Effect of a Juvenile Diversion Program on Rearrests"; Regoli et al., "Using an Alternative Evaluation Measure for Assessing Juvenile Diversion Programs."

79. Davidson et al., *Alternative Treatments for Troubled Youth*; D. S. Elliott et al., *Diversion: A Study of Alternative Processing Practices* (Boulder, CO: Behavioral Research Institute, 1978); Finckenauer, *Scared Straight! and the Panacea Phenomenon*; Finckenauer and Gavin, *Scared Straight, The Panacea Phenomenon Revisited*; S. B. Lincoln, "Juvenile Referral and Recidivism," in *Back on the Street: Diversion of Juvenile Offenders*, eds. R. M. Carter and M. W. Klein (Englewood Cliffs, NJ: Prentice-Hall, 1976); Lundman, *Prevention and Control of Juvenile Delinquency*, 3rd ed.

80. Whitehead and Lab, *Juvenile Justice*, 2nd ed., 320.

81. T. G. Bloomberg, "Diversion from Juvenile Court: A Review of the Evidence," in *Juvenile Justice Philosophy: Readings, Cases and Comments*, 2nd ed., eds. F. L. Faust and P. J. Brantingham (St. Paul, MN: West, 1979); T. Palmer and R. V. Lewis, *Evaluation of Juvenile Diversion* (Cambridge, MA: Oelgeschlager, Gunn and Hain, 1980).

82. R. Paternoster et al., "The Stigma of Diversion: Labeling in the Juvenile Justice System," in *Courts and Diversion: Policy and Operations Studies*, eds. P. L. Brantingham and T. G. Bloomberg (Beverly Hills, CA: Sage, 1979).

83. Bloomberg, "Diversion from Juvenile Court."

84. Dunford et al., *National Evaluation of Diversion Projects*; Palmer and Lewis, *Evaluation of Juvenile Diversion*.

85. Whitehead and Lab, *Juvenile Justice*, 2nd ed.

Preadjudication Processes in Juvenile Justice

8

Chapter Objectives

After studying this chapter, you should be able to

- Describe the intake process in juvenile court
- Describe the role of discretion in the intake process
- Describe recent trends in intake decision making
- List the screening criteria commonly used in making intake decisions and explain the uses of case conferences in intake screening
- Explain the use of and need for the detention or preliminary hearing and describe the due process issues surrounding the detention of juveniles
- Explain the functions that plea bargaining and sentence bargaining play in the intake and preadjudication process

Chapter Outline

- Introduction
- The Juvenile Justice Intake Process
- Additional Juvenile Justice Preadjudication Processes
- The Use of Bail in Juvenile Justice
- Plea Bargaining in Juvenile Justice
- Legal Issues
- Chapter Summary
- Key Concepts
- Review Questions
- Additional Readings
- Notes

■ Introduction

This chapter examines a variety of pretrial or preadjudication processes that occur in juvenile courts. These pretrial processes include a variety of hearings, conferences, and decision-making events where determinations about the best methods of handling juvenile court referrals are made. Like the decision to **arrest** a juvenile, decisions made at this stage of the juvenile court process are critical for a number of reasons. First, they play a major role in determining which youths penetrate further into the juvenile court process. Second, they have a direct bearing on the freedom of juvenile offenders, because they often focus on youths' placement pending further court action. Third, they can lead to formal court processing and assignment of formal labels, such as "delinquent" or "undisciplined child." Finally, they can halt or set in motion a variety of responses to juvenile offenders that have the potential to exacerbate or reduce the likelihood of subsequent delinquent behavior. The preadjudication processes come between the police or social agency referral to the court and formal court action. They operate to screen out cases that may be better handled outside of court, meaning youths and their families (and even the victims of their misbehavior) will be spared any harm associated with court involvement. Conversely, if they do not effectively screen cases, more youths and their families will be subject to possibly unnecessary and expensive court intervention.

arrest The deprivation of a person's freedom by legal authority; the apprehension of the juvenile by the police; the physical seizing of a person by the authorities for the purpose of making that person answer for a criminal charge.[48]

MYTH VS. REALITY

Formal Court Action Is Not Necessary in Every Juvenile Case

Myth—Juveniles should always receive some type of formal sanction when they engage in delinquent behavior.

Reality—Delinquency encompasses many types of offenses, many of which do not represent a substantial threat to public or individual well-being. Moreover, most youths engage in delinquency and are not caught, yet they manage to grow up to become healthy adults. Also, there is an informal juvenile justice process, as well as diversion programs, that effectively handle many cases each year. It should be remembered that, although formal court action can and does serve the best interests of children and the community in many cases, there are other instances where the effects of formal court actions are far from positive.

■ The Juvenile Justice Intake Process

When the decision to arrest a youth is made or when a social agency or the public schools refer a youth to the juvenile court, several things occur. Although most people, when thinking of arrest, probably picture a suspect being handcuffed, taken to the station to be booked, and placed in a lockup, in a typical juvenile arrest, the youth is released into the custody of his or her parents or guardian. The police officer then completes the juvenile arrest report, which is forwarded to the **intake** unit of the juvenile court for screening. Similarly, when social agencies or schools make a referral to the juvenile court, it is the intake unit that first receives the referral.

A Variety of Factors Discourage Police from Making Arrests in Juvenile Cases

There are a number of reasons police might decide not to treat juveniles like adults in the arrest process. First, many police believe that juvenile courts are ineffective in dealing with delinquents and that all of the work required to complete a case (i.e., writing up a police report for prosecutorial review, contacting witnesses, and testifying in court) is not worth the likely outcome. Second, police are not obligated to make arrests in every case where they come into contact with a juvenile who has engaged in illegal conduct, and although they are likely to make an arrest if a serious offense has been committed, they often make "informal adjustments" in instances involving minor delinquent activities. Third, the lack of secure detention beds for juveniles in some jurisdictions makes it impossible to detain many offenders. A major problem confronting juvenile detention facilities, like adult jails, is chronic overcrowding. As a result, the majority of youths placed in juvenile detention are detained in overcrowded facilities.[1]

intake The process of introducing youths who come to the police and court's attention into the juvenile justice system and determining how far they go into the system.

intake screening The process of determining whether youths should be diverted from the juvenile system or be referred for formal or informal services. Screening criteria usually include seriousness of the offense(s), family history, prior delinquency record or court contacts, legal mandates regarding diversion, attitude, age, school performance, parental cooperation, employment history, and restitution issues. Unfortunately, screening criteria may also include, despite their inappropriateness, race, gender, and socioeconomic status.

legal sufficiency The requirement that an alleged criminal offense and the alleged perpetrator meet certain criteria before the alleged perpetrator can be charged with a criminal act. A prosecuting attorney, district attorney, or state's attorney assesses the criminal charges and the characteristics of the alleged perpetrator to ensure the criteria are met.

The purpose of the intake unit is to determine the most appropriate way to handle cases that are referred for possible juvenile court action. As noted earlier, juvenile codes are constructed so broadly that a wide range of juvenile behaviors are potentially subject to juvenile court jurisdiction. As a result, complaints alleging that juveniles have thrown snowballs at a school bus, broken a window during a baseball game, engaged in a petty property dispute, disobeyed their parents, stayed out past curfew, been involved in a minor altercation in the neighborhood, as well as many more serious matters such as serious assaults, burglaries, and property damage cases, are regularly referred to juvenile courts for action. It is the intake unit of the court that has the responsibility of determining which of these many types of complaints warrant formal court action and which might be better handled in some other way.

Traditionally, **intake screening** was performed by probation officers employed by the juvenile court or separate state agencies responsible for performing probation and intake functions, although it is common for a member of the prosecuting attorney's office to review most complaints for **legal sufficiency**. The purpose of this review is to determine if the elements of a chargeable offense are present. Essentially, the prosecuting attorney makes a determination that there are legal grounds for the court to proceed and that the prosecuting attorney's office is prepared to represent the state's interests at the adjudication. This model is still employed in most states. After the complaint is reviewed for legal sufficiency, an intake officer (typically a probation officer performing the intake role) screens the complaint in order to make a decision about how the case should be processed. This screening often involves a consideration of the type and seriousness of the offense alleged in the complaint (e.g., harm to the victim) as well as factors such as the age of the youth and his or her offense and court history (if available). In addition, information in the complaint regarding parental response to the offense, the reaction of the youth, and the behavior of the youth in the community, at school, and at home may be considered. To better understand one intake officer's perspective of his job, see **Box 8-1**.

Box 8-1

Interview: Frank Weichlein, Intake Supervisor, Ninth Judical Circuit Court, Family Division, Kalamazoo, MI

Q: What is your educational and employment background?

A: I received a BA in sociology from Western Michigan University in 1978, and an MA in public administration from WMU in 1992. I started out my work in juvenile justice by interning at the Kalamazoo County Juvenile home and upon graduation was hired to work there as a child guidance worker. I moved to probation after two years and did that from 1980-86. I have been the intake supervisor since 1986.

Q: What are your duties and responsibilities as intake supervisor?

A: First of all, I am responsible for overseeing the entire intake process, which includes accepting and screening of all complaints and petitions that come to the court. I either make or oversee the initial decision as to whether a child or family receives diversion or gets referred to the formal docket. Second, I directly supervise 11 intake staff. Third, I am responsible for the development, implementation, and supervision of diversion programs run by the court. Also, with the unification of courts in Michigan we now have added responsibilities related to domestic cases and Personal Protection Orders. For example, we now screen all domestic case motions and conduct case reviews on motions regarding parenting time, custody, and change of domicile. And we review requests from citizens for Personal Protection Orders.

Q: In addition to your education and employment background, have you had any specialized training?

A: One thing the court has been very good about is providing staff training opportunities. I have been able to attend numerous seminars and training sessions, all of which have been helpful in my work.

Q: What part of your job do you find the most rewarding?

A: Giving young people the opportunity to change behaviors in diversion without having to have a formal delinquency record gives me the most satisfaction. Also, with the added responsibilities that have resulted from court unification, I have an opportunity to be responsive to a broader range of people who come to the court for assistance.

Q: What part of your job do you find the least rewarding?

A: Responding to, and working under, legislation requirements that are not well thought out. The political posturing that results in legislation being made without consulting the people that will be responsible for implementing it causes a great deal of frustration. An example of this is the victim rights legislation involving restitution. Although I'm in favor of restitution, you cannot treat juveniles like adults in that regard; but that's what the legislature has tried to do. Other questionable legislation, in my opinion, includes the registration of sex offenders, school safety legislation, changes in waiver statutes, and the development of "designation cases." While these efforts may be well intentioned, they contain some aspects that are detrimental to young people, and changes should be considered.

Q: What skills and abilities would you tell college students they should cultivate to work in your field?

A: Patience, resiliency, the ability to accept diversity, and flexibility are some of the main personality features anyone going into my field needs to cultivate. In addition, those who perform this job needs to be able to think on their feet and have excellent oral and written communication skills. An intake worker spends a great deal of time communicating with the public, police, victims, juveniles, family members, and the court. Finally, an intake officer must be a person who can think independently, but work under supervision.

Q: What particular strengths and abilities do you feel you bring to the position?

A: I'm patient, have the ability to see the "big picture," and I am flexible enough to respond to a changing work environment. I believe I have good communication skills and an ability to lead and motivate people without being coercive. I also understand the fundamentals of each of my employees' positions, which aids me in effective supervision.

Q: What are the major challenges that you face as intake supervisor?

A: Certainly my most challenging problem is dealing with serious delinquents in the system on the one hand, while continuing to advocate for prevention and diversion programs. I genuinely believe in the importance and the effectiveness of prevention and rehabilitation, but the serious delinquents take the public's focus off diversion. In addition, the development of domestic case intake processes has certainly been one of the biggest challenges we have faced.

Q: Who do you report to, and where are you in the court administration?

A: I report to the associate court administrator, who is in charge of intake and probation services. I'm considered a department head and am solidly in middle management.

Q: What unique programs does your department offer that you believe are particularly effective?

A: Our youth diversion program is unique in that it is run by the court, not a youth services bureau outside of the court. With good caseworkers and the backup of the court, it has proven to be very effective. Our inquiry process also works very well. We have four workers with nonlegal backgrounds, but who have been specially trained, and meet with youth and parents to discuss pending delinquency charges. There are a full range of alternatives available, including formal court action, and these workers do a nice job of screening out and diverting cases. Without them, the court would be overwhelmed with cases. Our prosecuting attorney's office has been very cooperative in allowing our staff this discretion also. They trust our judgment, and we certainly listen to them in dealing with serious delinquents. The relationship we have developed with the schools in Kalamazoo County has proven effective in trying to intervene early in some families, particularly our system for processing educational neglect petitions. Also, we have recently been working with a number of community agencies in the development and implementation of programs based on the philosophy of Balanced and Restorative Justice.

Q: What types of offenses are you seeing cross your desk on a regular basis?

A: Sadly, assaultive cases have overtaken retail fraud—shoplifting—as the number one offense that we see, and many of these assaults involve domestic violence where juveniles are referred for assaulting their parents. These cases concern everyone at the court and they require the development of interventions specifically for these cases.

Q: What is your salary as intake supervisor?

A: $62,000.

Q: What are the most serious issues that you see today in the area of juvenile justice?

A: Effective programming for the serious delinquent offender is very important, while at the same time not losing focus on the need for early intervention to prevent the circumstances that give rise to the serious delinquent. Another serious issue is dwindling resources for all systems that work with children and families, particularly Community Mental Health. We are seeing an increase in the number of severely emotionally disturbed juveniles involved with the court.

Q: Do you have much hands-on contact with juveniles in your position?

A: Not as much as I'd like, but I conduct inquiries on all school-related cases (truancy and educational neglect) as well as neighborhood nuisance offenses, mostly curfew violations, at a community-based diversion program operated out of the Boys and Girls Club. I also handle crisis walk-ins and coordinate our Teen Jury program.

Q: Is there anything else you would like to tell me about your role at the court?

A: My job is still vibrant and exciting, even after 25 years of involvement with the court. Every day brings new challenges. You never know what is at the other end of the phone call or waiting outside the door.

petition The request for formal court action that contains the written accusation against a juvenile and specifies the criminal or delinquent charges being made, and is properly verified as to age and venue by the complainant (the person bringing the charge).

For certain types of cases, the intake decision is routine. Indeed, in some cases formal processing is mandatory. For example, intake guidelines typically specify that certain types of serious offenses cannot be diverted and must be petitioned to court. A **petition** is a request for formal court action that contains a statement of charges that is presented at the adjudication or trial stage of the juvenile justice process. In other cases, the intake officer has considerable discretion in determining how a case should be handled. The intake officer may decide that dismissing the case or warning the juvenile is the best course of action. Indeed, information contained in the police report may make it clear that the complainant does not desire formal action, or the complaint may indicate that the matter was so trivial that an expenditure of court resources is unwarranted or inappropriate. Some courts operate special programs for certain types of offenders, such as youths who commit minor shoplifting or property damage offenses, and eligible youths are routinely referred to these programs by intake officers. In other instances, intake officers may refer youths, and in some cases youths and their families, to a variety of community agencies for assistance. Unfortunately, in many communities there are very few good diversion programs that court personnel can rely on.

The Adjudication Is the Juvenile Court Equivalent to the Trial

Adjudication is the trial, final plea, or fact-finding stage of the juvenile process, the stage at which the hearing official determines formal guilt or innocence. It is the juvenile court equivalent to the trial in adult court. Adjudication involves a "formal" hearing that is (usually) recorded or transcribed and at which due process protections are afforded the juvenile (i.e., the right to counsel, the right to jury trial [where available], the right to confrontation and cross-examination, the right to subpoena witnesses, and the right to testify or remain silent).

adjudication The formal court process or hearing at which the judge or judicial officer determines whether the juvenile comes within the court's jurisdiction due to a guilty plea or a finding of guilt at trial.

Although many complaints receive rather routine screening at the intake stage, others require considerably more of the intake officer's time. Because police complaints typically contain little information other than that pertaining directly to the alleged offense, intake officers often attempt to collect additional data that they feel will assist them in making appropriate screening decisions. Consequently, it is not uncommon for an intake officer to arrange a **case conference** with the juvenile and his or her parent(s). The purpose of the case conference is to allow the intake officer an opportunity to collect additional information that might aid him or her in making a screening decision. In other instances, the intake officer might contact a victim or talk with school or social agency personnel in an effort to collect information that they feel will help them in making a screening decision.

case conference A meeting between an intake officer, the juvenile, and his or her parents that is designed to collect additional information to aid the intake officer in screening a case.

Statutes or Court Rules Govern the Time Frames for Holding Adjudications

In most states, statutes or court rules set time limits for holding adjudications once a petition is filed. In situations where a youth is detained, an adjudication must be held within 10 to 180 days (depending on the state) from the date that the petition is filed. In situations where a youth is not detained, an adjudication must be held within 30 to 180 days.[2]

Although the probation officer or intake officer model is the traditional model employed in juvenile justice, in recent years there has been a move toward increasing the involvement of the prosecuting attorney in the intake process. In fact, in some jurisdictions (e.g., Colorado, Arizona, and Washington), intake screening is now the domain of the district or prosecuting attorney's office.[3] In other jurisdictions (e.g., Louisiana), the prosecutor has been given greater authority over intake decision making.[4]

The Purpose of Intake Is to Make Decisions About How to Handle Cases

Regardless of who performs the intake function—a probation officer who works for the juvenile court, a separate state agency, or a member of the district attorney's staff—the purpose of intake is to make decisions regarding how cases should be processed by the juvenile court.

punitive model of juvenile justice The increasing involvement of prosecuting attorneys in the intake function of the juvenile court to circumvent the traditional discretion of intake officers. The punitive model of juvenile justice employs the same criteria used in charging adults, and the focus is on punishment rather than rehabilitation.

The increasing involvement of prosecuting attorneys in the intake process reflects the increasing legalization and **punitiveness** of the juvenile justice process found in a number of jurisdictions. The increased use of prosecuting attorneys is seen as a way to respond forcefully to what are believed to be serious juvenile offenders and to circumvent the discretion employed by probation officers in the traditional intake-screening model. For example, the state of Washington has placed the responsibility for intake screening in the hands of the prosecutor for all felony and most misdemeanor cases.

In Michigan, as in a number of states, the prosecuting attorney is the only person who can petition the juvenile court in delinquency matters.[5] Furthermore, prosecutorial screening decisions are based on explicit criteria: offense seriousness, prior record, and the age of the youth. This removes traditional discretionary decision making from probation staff, although police agencies continue to "exercise their traditional discretion regarding whether to refer or adjust incidents involving juveniles."[6]

Discretion Is Involved in Intake Screening

Regardless of who makes intake decisions, discretion is still involved. Recent trends toward shifting intake decision making to the prosecuting attorney's office simply move the discretionary power from one office to another. Prosecuting attorneys still exercise considerable discretion in deciding which cases to petition, therefore the issue is not so much whether discretion is exercised but who exercises it and what the outcome of the decision making will be.

Although some degree of discretion is present in the juvenile court intake process, limits on discretion are imposed by state statutes and juvenile court rules of procedure, which often note the types of cases that can be handled informally and those that must be handled formally. For example, the California Rules of Court, Rule 1405 indicates that intake staff should consider 10 factors in deciding whether a matter can be handled at intake. These factors include an indication that the matter or condition that brought the youth to court is serious; whether or not the matter is a temporary problem; if another agency in the community is better suited to serve the needs of the child, the parent or the guardian; the attitude of the child, parent, or guardian; the age and maturity of the child; and the attitudes of affected persons.[7] These and similar criteria allow intake officers to employ considerable discretion in deciding how cases should be handled.

The Role of Discretion in the Intake Process

Those who make intake decisions determine who will penetrate further into the formal juvenile justice process. Although intake guidelines require intake officers to send certain cases forward for adjudication, intake decision makers often have considerable discretion in deciding how the majority of cases referred to the juvenile court should be handled. Options might include dismissing the case, having the offending youth and the parents in for a case conference to collect additional information to assist in making the intake decision, referring the youth to a diversion program (e.g., informal probation or counseling at a community agency), filing a petition (a request for formal court action that represents a legal version of the police complaint and indicates the specific charges used in the adjudication), and waiver or transfer of the case to the adult court.

Many Cases Are Diverted from Juvenile Courts by Intake

Of the estimated 1,673,000 delinquency cases referred to juvenile courts in 1999, approximately 57% had petitions filed. However, only 39% of those cases not petitioned were dismissed. Sizable percentages of cases not petitioned were dealt with in a variety of ways, such as placement on informal probation (34%) or some other type of disposition or sanction (27%).[8]

Intake Legal Requirements

Regardless of who is responsible for making intake decisions for the juvenile court, certain things must happen. First, a decision must be made, based on the documentation presented (usually a police report), that there are sufficient facts to justify an identifiable criminal charge. Consequently, the intake screener, prosecuting attorney, or whoever is making this decision needs to be familiar with the elements of criminal offenses and determine not only that a crime or act of delinquency occurred, but also what the exact charge should be.

Intake Often Involves Screening Cases for Legal Sufficiency

A common practice in instances where probation officers make intake-screening decisions is to have someone from the prosecuting attorney's office assess the complaint for legal sufficiency (i.e., if the elements for a chargeable offense are present).

Second, a petition or complaint must be prepared and filed with the court. One of the fundamental requirements is that the petition be verified—that someone signs it who vouches as to the accuracy of the charges. By signing the petition and vouching for the accuracy of the charges, the "petitioner" also agrees to be responsible for going forward with the case. Essentially, this step commits the prosecuting or state's attorney to represent the state's interest in court.

Third, the petition or complaint must be presented to the court for the court to take action. The petition can be filed and a detention hearing requested, at which time the juvenile and his or her family can be advised of the charges and of their legal right to a hearing, to an attorney, to confront adverse witnesses, to make a statement in court, to have bail set (where bail is used for juveniles), and to have a formal trial. Juveniles may also present evidence at detention hearings and contest their detention.

If no detention hearing is set, then the petition can be discussed with the juvenile at an informal (off the court record) case conference or preliminary inquiry. The case conference is a meeting between a court official and the juvenile and his or her parents or guardian (or other responsible adult). The purpose of the case conference is to obtain additional information that can assist the intake officer in making a decision about how a case should be processed. At the case conference, the juvenile and his or her parents or guardian may be advised of alternatives, such as diversion or trial, and of various rights that they have, such as a right to have an attorney represent them before the court, and they may ask questions about their case. In some instances, juveniles may have an attorney represent them at the case conference, although this is not common.

Several states use a pretrial hearing in which a **plea** is taken. At this hearing the youth typically admits to the charges, does not contest the charges, or denies the allegation(s) contained in the petition (the plea). The hearing typically occurs after the prosecutor and the attorney for the juvenile have met to discuss the case and have negotiated a plea, which may be the result of a plea bargain. Far more delinquency cases are settled by pleas than by trial, and the primary purpose of a pretrial hearing is to provide time for plea negotiations. If no plea can be agreed upon at the hearing, then the attorneys can exchange names of witnesses to be called at trial, set any pretrial motions for decision, and delineate for the court any legal or factual issues of importance.

plea The juvenile's formal response to criminal charges

Research on Intake Decision Making

A number of factors have been found to influence intake screening decisions, including the seriousness of the offense,[9] the youth's prior record,[10] and the youth's demeanor.[11] Youths who have committed serious offenses, those with prior records, and those who are uncooperative are more likely to be petitioned. In addition, variables such as age, socioeconomic status, race, and gender have been found to influence intake decision making in some jurisdictions, but the influence of these variables is not completely clear.[12] For instance, in one study of court decision making, both legal factors such as the seriousness of the offense and "nonlegal" factors such as gender, race, and socioeconomic status were found to influence intake decisions. The findings of this study indicated that the seriousness of the offense was related to the decision to file a petition, but it appeared to be a more important factor for lower-class African American males with a prior record than for other groups.[13] Another study that examined decision making at each step of the juvenile justice process found that minorities received harsher treatment at each decision-making point, including intake.[14]

Overall, the research on intake decision making suggests that, at least in some communities, lower-class minority males are more likely than other groups to receive more formal responses at the intake stage. In addition, there is considerable evidence that female status offenders are likely to be treated more harshly than their male counterparts, at least in some jurisdictions.[15] Although this research indicates that bias in the intake process is not a problem in every jurisdiction, it raises serious questions about the quality of justice available to juveniles in some communities.

Decisions Made at Intake Influence Decisions Made at Later Stages

Intake decision making is important because there is evidence that decisions made at the earliest stages of the juvenile justice process influence decisions made at later stages. Moreover, as noted, a substantial body of literature exists indicating that there are disparities in the treatment of juveniles in some jurisdictions, at least some of the time. This means that decisions to treat minority youths more severely at the intake stage of the court process will often result in more severe treatment at subsequent stages such as the adjudication.[16]

■ Additional Juvenile Justice Preadjudication Processes

In addition to intake screening, a variety of preadjudication processes can occur prior to the adjudication or trial. These include hearings such as waiver or transfer hearings (examined in detail in Chapter 9), detention hearings, preliminary hearings, arraignments, plea bargaining negotiations, and bail determinations. The remainder of this chapter describes the operation of these preadjudication processes, discusses legal issues that influence their operation, and examines relevant research on the preadjudication stage of the juvenile justice process.

Detention Intake and Detention Hearings

A special type of intake decision making occurs when police or other juvenile justice agents request the detention of a minor. As noted in Chapter 6, police officers, as well as other juvenile justice agents such as probation officers, at times request that youths be detained in juvenile detention units or, in some cases, adult jails. In these instances, an intake officer makes a decision about the appropriateness of the request for detention. It is important to note that it is the intake officer, acting as an officer of the juvenile court, who makes the decision to detain a youth, not the police officer.

The decision to detain a youth, however, is followed up in practice by a **detention hearing**. Each state's family or juvenile code mandates that a hearing be held when a minor is deprived of his or her liberty. These hearings are conducted by juvenile court judges, referees, or other hearing officers. At these hearings, the youth and his or her parents are advised of their rights, an attorney may be assigned to represent the youth, and several important determinations are typically made. These determinations include assessing the facts leading to the youth's custody to determine if probable cause exists to hold the youth. In addition, a decision regarding the placement of the youth is made. The court may decide to continue holding the youth in detention, place the youth in the custody of his or her parents

detention hearing A required hearing in front of a judicial officer after a juvenile has been detained or in order to get a juvenile detained. At this hearing, the juvenile has a right to be present, have his or her parents or significant adult present, be informed of the charges, have his or her legal rights explained, make a plea, have a trial date set, and request bond (where permissible) or guardian, or make some other placement decision. Finally, the court may authorize the filing of a petition and set dates for subsequent hearings to review the case or for an adjudication.

Trends in the Use of Detention

Although detention is used for a variety of reasons (e.g., holding youths who are awaiting trial, disposition, or transfer to another jurisdiction), there are clear indications that the use of detention for delinquency cases has grown in recent years. For example, data on national detention trends reveal an 11% increase in the use of detention for delinquency cases between 1990 and 1999. Although drug offense cases accounted for the greatest increase in detention cases during this period, youths charged with property offenses accounted for the greatest number of detention cases, followed by youths charged with person and public order offenses, and then drug offenses.[17]

A troubling reality that plagues juvenile justice in many communities is *disproportionate minority confinement* (DMC). DMC exists when "the proportion of juveniles detained or confined in secure detention facilities, secure correctional facilities, jails, and lockups who are members of minority groups . . . exceeds the proportion such groups represent in the general population."[18] In most states minorities are disproportionately represented at each stage of the juvenile justice process and they are disproportionately represented in detention and other secure placements.[19] In an effort to deal with this problem, federal legislation in the form of the 1988 amendments to the Juvenile Justice and Delinquency Prevention Act of 1974 was passed that requires states to address DMC in order to receive formula grants for juvenile justice. As a result, many states have begun to develop plans for reducing DMC, although it continues to be a significant problem around the country.[20]

In addition to minorities, males and older juveniles are also overrepresented in detention cases. Males make up the majority of cases involving detention. In 1996, for example, males accounted for 83% of all juvenile cases involving detention.[21] However, the number of females being detained has been increasing in recent years. Indeed, the number of females detained grew by 50% between 1990 and 1999 due in great part to a substantial increase in the number of females detained for person offenses.[22] Also, detention was used more frequently for older juveniles than for those who are younger. For instance, in 1999, youths 16 and older accounted for almost half of all detention cases.[23]

In contrast to the increase in the use of detention for criminal or delinquent offenses, detentions for status offenses have declined significantly in recent years. Between 1986 and 1990, detentions for status offenses declined by 35%.[24] As a result, only a small percentage of youths petitioned for status offenses were detained between 1990 and 1999. During the 1990s only 12% of petitioned runaway cases and 2% of truancy cases involved detention.[25] The decline in the detention of status offenders occurred for several reasons:

- The increasing volume of youths charged with serious offenses has "squeezed" available bed space in detention facilities. The result is fewer beds available for status offenders.

- New state and federal laws restrict the use of detention for runaways, truants, and incorrigible children.
- Licensing requirements for detention facilities in some states set age limits for the detention of juveniles. Because many status offenders are young, these licensing requirements result in a reduction in the number of status offenders detained.[26]

Juvenile Detention Regulations

All states mandate that a detention hearing be held within a time frame specified by state law, typically within 24 to 72 hours. In addition, most states specify a time frame within which the adjudication must be held when a youth is detained. As noted earlier, the time frame for adjudication ranges from 10 to 180 days.[27] The question remains as to how long a juvenile should be detained before being given his or her day in court. This question has been answered differently in different states. Although it is important to have a speedy trial, there should be enough time prior to trial to allow the attorney to prepare a meaningful defense. (A lack of time to adequately prepare is a problem that plagues many attorneys who practice in juvenile court.)[28]

Requiring a juvenile to stand trial within 10 to 14 days after being detained may sound enlightened, but if the juvenile is charged with a serious offense, 14 days may not be enough time for preparing and presenting an adequate defense at trial. Note that defense attorneys, particularly those who do public defender work, often have large caseloads.[29] In addition, a variety of tasks must be accomplished in order to construct an adequate defense. Witnesses need to be interviewed, police reports need to be read, and motions may need to be filed. Because attorneys who represent clients in juvenile court often lack adequate resources for performing these tasks,[30] forcing a trial on short notice encourages plea bargaining.

Clearly, the use of detention for youths charged with a delinquency offense has increased substantially in recent years. This increase in detention is no doubt helped by the ability of juvenile courts to use **preventive detention**. In an important U.S. Supreme Court case, *Schall v. Martin*, the court ruled that preventive detention was appropriate in juvenile cases in order to protect both the juvenile and society from offenses that the juvenile might commit prior to the adjudication. (For a discussion of this case, see Chapter 6.)

Preliminary Hearings and Arraignments

Another type of preadjudicatory hearing found in the juvenile court is the **preliminary hearing** or **arraignment**. Although the terms for these hearings and their purposes vary somewhat from state to state, they are primarily designed to take pleas from juveniles and consider the appropriate placement of juveniles prior to adjudication. In some jurisdictions, a preliminary hearing or arraignment is used to advise the youth of his or her rights and to determine if the youth plans to contest the charges contained in the complaint or referral. If the charges are being contested, then a date for the adjudication is often set. In instances where the charges are not contested, then a plea may be accepted at that time and a disposition entered by the court. In other jurisdictions, a preliminary hearing is typically used to consider the removal of the juvenile from his or her home. Preliminary

preventive detention Detention of an alleged offender prior to conviction so as to prevent the alleged offender from committing any crimes during the preadjudication period.

preliminary hearing See *detention hearing* and *arraignment*.

arraignment The initial "coming before" the judge or judicial officer at which time the juvenile is informed of the charges pending and of his or her constitutional rights; a plea is entered; a trial date is set, if necessary; and bail is determined, if needed.

hearings or arraignments are key events in the juvenile justice process. For example, the decision by the court to deprive a youth of his or her liberty is a potentially ominous determination. As noted earlier, processing decisions made at early stages of the juvenile justice process appear to affect decisions made at later stages.[31] Likewise, a decision by a juvenile to not contest a charge or to enter a guilty plea means that the juvenile court may assume formal jurisdiction over the youth. In these circumstances, the court determines that it will take jurisdiction over the youth, which subjects the juvenile to the range of dispositions open to the court, including removal from the home and institutional placement.

The Disposition Is the Juvenile Court Equivalent to the Sentencing in Criminal Court

The *disposition* is the juvenile court analog of the sentencing in criminal court. At the disposition, the court makes a determination about how it will formally respond to a juvenile who has been adjudicated and comes within the jurisdiction of the court. The most common disposition is probation.

According to a report prepared by the American Bar Association Juvenile Justice Center, the Juvenile Law Center, and the Youth Law Center,

> It is critical that counsel appear early in the life of a case. At first appearances in court, if judges ask about events surrounding alleged offenses, the circumstances of arrests, the roles of other youth involved, or clients' other contact with the juvenile justice system, and attorneys do not have answers, they lose the initial opportunity to present clients' cases in a favorable light. Judges are left to review the uncontradicted allegations in the charging petitions. Based on incomplete reviews, judges make early determinations regarding detention that may influence cases all the way until their dispositions.[32]

Given the gravity of the decisions made at preliminary hearings or arraignments, it is common for juveniles to be represented by attorneys at these hearings in many jurisdictions. However, the extent to which attorney representation is typical is not known. Indeed, there is evidence that attorney representation is more the exception than the rule in some courts and that the quality of representation that is provided to juveniles in some jurisdictions is inadequate.[33] Indeed, some of the research that has examined attorney effectiveness in juvenile court has found that youths who were represented by counsel were more likely to receive harsher punishments than youth without attorneys.[34] (The issue of attorney representation in the juvenile court process is discussed in more detail in Chapter 10.)

■ The Use of Bail in Juvenile Justice

Given the U.S. Supreme Court's endorsement of preventive detention in juvenile cases, it is not surprising that there is considerable variability in the use of bail in juvenile justice. The wide discretion that juvenile courts have in using preventive detention stands in marked contrast to most adult courts, which make bail available to offenders except in instances in which offenders are charged with capital crimes, they have jumped bail in the past, or they have committed another offense while on bail.[35] The use of bail in juvenile justice is less common than in the adult criminal justice system. Indeed, very little attention has been given to the use of bail in juvenile cases. Typically, juvenile codes ether allow for bail at the discretion of the court, deny youths the right to bail, or ignore it altogether.[36] Presently, some states, such as Georgia, Colorado, and Oklahoma, indicate that juveniles have the same right to bail as adults. Still others, such as Nebraska, Tennessee, Vermont, and Minnesota, make bail available to juveniles at the discretion of the judge. In other jurisdictions, such as Hawaii, Oregon, and Connecticut, juveniles have no right to bail.[37] Court rulings that deny juveniles bail are usually predicated on the assumption that because the actions of juvenile courts are intended to correct or help young offenders, because juvenile courts are able to release youths to their parents, and because detention hearings are common in juvenile justice, bail is unnecessary.[38]

Bail Is Sometimes Used in Juvenile Justice

"Bail or a bail bond is a surety in the form of money or property that may be posted by a bonding company or others, including defendants themselves, to obtain their temporary release from custody and to ensure their subsequent appearance at trial" (emphasis added).[39]

Even in states that allow bail for juvenile offenders, it may not be widely used.[40] Indeed, the use of bail in juvenile cases raises a number of questions. One question is, who should be able to post bond for a juvenile? Parents or relatives seem like logical choices, but what about a boyfriend or girlfriend? One possible rule is that the juvenile should only be released to a person who is legally responsible for his or her care, custody, and control. However, not all parents are in a position to post bail, and in other instances parents may be unwilling to post bail. Where does a juvenile get money for bail when the parents are poor or are angry and unwilling to post it? Would it be appropriate for bail bondsmen to provide bail money for juveniles?[41] If for some reason the juvenile absconds on bail, who should be responsible, the child or the parent? Because the purpose of bail has traditionally been to ensure the appearance of the alleged offender at subsequent hearings, if the parent posts bail, would this be sufficient encouragement for the juvenile to appear? These questions reflect the ambiguity that surrounds the issue of bond in the juvenile courts. Despite these questions, however, many courts have attempted to develop guidelines for setting bond in juvenile cases.

■ Plea Bargaining in Juvenile Justice

plea bargain
An agreement entered into between a prosecuting attorney and a juvenile, usually through the juvenile's attorney, according to which the juvenile will admit to a lesser charge in exchange for the reduction of the charge or will admit to some charges in exchange for the dismissal of other charges.

<u>Plea bargaining</u> is another preadjudicatory process that plays a role in juvenile justice. It consists of negotiations between the prosecution and the defense counsel, often with the consent of the court, whereby the defendant agrees to plead guilty in return for the prosecuting attorney's promise to reduce the charge or to drop some of the charges pending against the defendant.[42] The idea is that each party gives up something in order to achieve a reasonably certain outcome—an outcome that could not be guaranteed if the case went to adjudication. On the one hand, the plea agreement reached guarantees the prosecution a conviction, although the prosecution gives up the opportunity to have the defendant adjudicated on a greater number of charges or on more serious charges. On the other hand, the defendant gives up his or her right to contest the charges at trial as well as other constitutional rights. Further, the defendant gives up the opportunity to avoid juvenile court jurisdiction altogether by voluntarily submitting to the court's jurisdiction through his or her plea. Indeed, as a part of the plea agreement, the defendant admits to fewer offenses or to less serious charges.

Plea bargaining and the process through which juveniles make guilty pleas have received scant attention in the research literature. Although plea bargaining appears to be rare in some juvenile courts,[43] in other jurisdictions it is a common occurrence. Today, most states attempt to exert some control over plea bargaining. Twenty-five states have established formal rules that govern the plea negotiation process, 13 have passed legislation that spells out a judge's responsibilities when accepting a guilty plea, and 3 attempt to regulate plea negotiations through the appellate review process. Only 10 states to date have failed to establish guidelines or legislation that regulates plea bargaining.[44]

Although plea bargaining is recognized as a part of the juvenile justice process in most states, in practice the extent and quality of regulation varies considerably across jurisdictions. As Joseph Sanborn found in a study of the plea negotiation process across the country, some jurisdictions do not require judges to remind defendants that they are giving up constitutional rights, make them aware of possible dispositional outcomes, determine if the pleas were made voluntarily, and verify that the pleas had a factual basis. Indeed, only two states, Delaware and California, require judges to discuss each of these issues with juvenile defendants.[45] Moreover, Sanborn's research revealed that urban courts are more likely than suburban and rural courts to employ formal procedures governing the plea bargaining process. As Sanborn notes, his research raises important questions about the continuing informality of many juvenile courts at this crucial stage of the juvenile justice process. Given the increasing punitiveness of at least some juvenile courts, it is essential that defendants clearly understand the potential consequences of the decisions that they make or that are made on their behalf.[46]

Plea Bargaining Is Common in Many Juvenile Courts

Although plea bargaining is a common practice in many juvenile courts, most juvenile cases involve open admission to the charges. Although the percentage of cases across the nation that include plea negotiations is not known, Joseph Sanborn's research on juvenile courts in Philadelphia found that about 20% of the cases processed resulted in a negotiated plea. The majority of the negotiations focused on reducing the sentence, usually probation in lieu of an out-of-home placement, or, in cases where there was an out-of-home placement, on securing a less restrictive placement.[47]

■ Legal Issues

One legal issue worthy of attention is the advisability of having a nonjudicial officer make the detention decision. Indeed, the decision to detain a juvenile is the most important decision that the juvenile justice system can make. According to the Michigan Supreme Court in *Reist v. Bay County Circuit Judge* (396 Mich 326, 241 NW2d 55 [1976]):

> *The interest of a parent and child in their mutual support and society are of basic importance in our society and their relationship occupies a basic position in this society's hierarchy of values.* Clearly any legal adjustment of their mutual rights and obligations affects a fundamental human relationship. *The rights at stake are protected and encompassed within the meaning of 'liberty' as used in the due process clause (emphasis added).*

due process The constiutionally guaranteed right to be treated with fundamental fairness by the law.

The central issue in the removal decision is whether to remove a child from the family home. Without question, this is a "legal adjustment" of the family's mutual rights and obligations. However, should this fundamentally important decision—one that can trigger a variety of **due process** rights for children and their families—be left to someone who has a legal background or instead to someone who has social work training? If the detention decision is made by a supervisor of the detention facility without legal advice or without the benefit of clear legal guidelines or court policies, is this not allowing the "gatekeeper" to determine who should be detained? Moreover, doesn't this raise the issue of conflict of interest that was considered in *Gault* (see Chapter 5)?

Another legal issue that should be addressed at the preadjudicative stage of the juvenile justice process concerns the setting of bail. Most states have few or no guidelines for setting bail in juvenile cases, and the lack of guidelines raises serious questions about the fairness or equity of the bail determination process.

A third legal issue concerns the validity of plea bargaining and sentence bargaining in juvenile courts. There are many arguments made by proponents and opponents of plea bargaining and **sentence bargaining**, but most fail to address the fundamental reason behind the use of these "tools." The law is not just or fair in itself and the operations of the "system" are not just or fair; only people can do justice. Importantly, plea and sentence bargaining introduces flexibility in the system and allows people to see that justice is done. This flexibility is necessary because justice can only be achieved on an individual and case-by-case basis by looking at the offender as a human being, by looking at the victim as a human being, and by then reaching a humane resolution. The other reasons for bargaining—that the system could not try all the cases and that prosecutors routinely overcharge—are good reasons for bargaining, but none of these is the fundamental basis for plea bargaining.

sentence bargaining Negotiations between a juvenile, usually through his or her attorney, and the court over the punishment or other consequences that the juvenile will receive in return for a guilty plea.

■ Chapter Summary

This chapter examined the pretrial or preadjudicative process in juvenile court. This process includes a variety of hearings, conferences, and decision-making events whose purpose is to determine the best methods of handling juvenile court referrals. Although the hearings, conferences, and negotiations that take place at the preadjudication stage are less glamorous and less formal than the adjudication and the disposition, decisions made at this stage can have a profound effect on the later court hearings. For example, during the intake screening process, a variety of critical decisions are made about cases referred to juvenile courts. These decisions determine which charges against the defendant will be authorized and whether the case will go forward to the adjudication and face the prospects of a formal juvenile court response, be transferred to an adult criminal court for trial, or be dismissed.

Another type of intake decision regularly made in juvenile courts concerns the detention of juvenile suspects pending further court action. Initial decisions to detain juveniles are typically made by intake officers, but state laws require that such decisions be reviewed in a detention hearing within a short time after the juvenile is detained (typically 24 to 72 hours). Similarly, preliminary hearings and arraignments, which are used in some jurisdictions, consider some of the same issues addressed in detention hearings. These hearings are primarily designed to determine if the youth plans to contest the charges contained in the complaint or referral, to take pleas from the youth, and to consider the appropriate placement of the youth prior to the adjudication. If the charges are being contested, then a date for the adjudication is often set. In instances where the charges are not contested, then a plea may be accepted at that time and a disposition entered by the court. These hearings play an important role in the processing of juvenile cases, because a variety of important decisions are made that can influence subsequent case processing. For example, it is at these hearings that juveniles may have an attorney appointed to represent them, where decisions about the appropriateness of filing a petition are made, and where decisions about bail (where it is used in juvenile cases) are rendered. Indeed, not only do decisions made at this stage of the juvenile court process determine restrictions on the youth's freedom, but they can increase the likelihood of more severe dispositions in subsequent hearings.

Because the preadjudicative stage of the juvenile justice process has such significant consequences, the juveniles' rights must be protected during this stage. However, there is considerable evidence that some courts provide insufficient protections. For example, in some jurisdictions many juveniles are not represented by attorneys, and in other jurisdictions attorney representation is inadequate during the preadjudicative phase of juvenile court processing. The lack of adequate attorney representation, of course, raises serious concerns about the quality of justice that juveniles receive.

KEY CONCEPTS

adjudication The formal court process or hearing at which the judge or judicial officer determines whether the juvenile comes within the court's jurisdiction due to a guilty plea or a finding of guilt at trial.

arraignment The initial "coming before" the judge or judicial officer at which time the juvenile is informed of the charges pending and of his or her constitutional rights; a plea is entered; a trial date is set, if necessary; and bail is determined, if needed.

arrest The deprivation of a person's freedom by legal authority; the apprehension of the juvenile by the police; the physical seizing of a person by the authorities for the purpose of making that person answer for a criminal charge.

case conference A meeting between an intake officer, the juvenile, and his or her parents that is designed to collect additional information to aid the intake officer in screening a case.

detention hearing A required hearing in front of a judicial officer after a juvenile has been detained or in order to get a juvenile detained. At this hearing, the juvenile has a right to be present, have his or her parents or significant adult present, be informed of the charges, have his or her legal rights explained, make a plea, have a trial date set, and request bond (where permissible).

due process The constitutionally guaranteed right to be treated with fundamental fairness by the law.

intake The process of introducing youths who come to the police and court's attention into the juvenile justice system and determining how far they go into the system.

intake screening The process of determining whether youths should be diverted from the juvenile system or be referred for formal or informal services. Screening criteria usually include seriousness of the offense(s), family history, prior delinquency record or court contacts, legal mandates regarding diversion, attitude, age, school performance, parental cooperation, employment history, and restitution issues. Unfortunately, screening criteria may also include, despite their inappropriateness, race, gender, and socioeconomic status.

legal sufficiency The requirement that an alleged criminal offense and the alleged perpetrator meet certain criteria before the alleged perpetrator can be charged with a criminal act. A prosecuting attorney, district attorney, or state's attorney assesses the criminal charges and the characteristics of the alleged perpetrator to ensure the criteria are met.

petition The request for formal court action that contains the written accusation against a juvenile and specifies the criminal or delinquent charges being made, and is properly verified as to age and venue by the complainant (the person bringing the charge).

plea The juvenile's formal response to criminal charges.

plea bargain An agreement entered into between a prosecuting attorney and a juvenile, usually through the juvenile's attorney, according to which the juvenile will admit to a lesser charge in exchange for the reduction of the charge or will admit to some charges in exchange for the dismissal of other charges.

preliminary hearing See **detention hearing** and **arraignment**.

preventive detention Detention of an alleged offender prior to conviction so as to prevent the alleged offender from committing any crimes during the preadjudication period.

punitive model of juvenile justice The increasing involvement of prosecuting attorneys in the intake function of the juvenile court to circumvent the traditional discretion of intake officers. The punitive model of juvenile justice employs the same criteria used in charging adults, and the focus is on punishment rather than rehabilitation.

sentence bargaining Negotiations between a juvenile, usually through his or her attorney, and the court over the punishment or other consequences that the juvenile will receive in return for a guilty plea.

REVIEW QUESTIONS

1. What purpose does an intake unit serve in a juvenile court?

_____

_____

_____

_____

_____

2. What are the typical factors used to determine whether a juvenile is referred for formal court action?

_____

_____

_____

_____

_____

3. What factors may influence the intake decision? Which of these factors are appropriate to use?

_____

_____

_____

_____

_____

4. How are case conferences used in the intake process?

_____

_____

_____

_____

_____

5. Why are some jurisdictions using prosecuting attorneys to make intake decisions for the juvenile court?

_____

_____

_____

_____

_____

6. What is a detention (or preliminary) hearing, and how does it fit into the intake process?

_____

_____

_____

_____

_____

7. What is an arraignment, and how does it fit into the intake process?

_____

_____

_____

_____

_____

8. Does plea bargaining exert a positive or negative influence on the intake process?

_____

_____

_____

_____

_____

9. What legal rights must be observed by the court before a juvenile can be detained?

_____

_____

_____

_____

_____

10. What is bail, and how does it work?

_____

_____

_____

_____

_____

11. What impact has *Schall v. Martin* had on the preadjudication process?

_____

_____

_____

_____

_____

12. What is a sentence bargain?

_____

_____

_____

_____

_____

ADDITIONAL READINGS

B. C. Feld. 1999. *Bad Kids: Race and the Transformation of the Juvenile Court*. New York: Oxford University Press.

C. E. Pope and W. H. Feyerherm 1990. "Minority Status and Juvenile Justice Processing: An Assessment of the Research Literature," *Criminal Justice Abstracts* 22: 327–335.

Puritz, Patricia et al. 1995. *A Call for Justice: Assessment of Access to Counsel and Quality of Representation in Delinquency Proceedings*. Washington, DC: American Bar Association.

Sanborn, Joseph. 1992. "Pleading Guilty in Juvenile Court: Minimal Ado About Something Very Important to Young Defendants," *Justice Quarterly* 9: 126–150.

NOTES

1. B. Lubow and D. Barron, "Resources for Juvenile Detention Reform." *OJJDP Fact Sheet* (Washington, DC: Office of Juvenile Justice and Delinquency Prevention, 2000).
2. H. N. Snyder and M. Sickmund, *Juvenile Offenders and Victims: A National Report* (Washington, DC: Office of Juvenile Justice and Delinquency Prevention, 1995).
3. B. C. Feld, *Cases and Materials on Juvenile Justice Administration* (St. Paul, MN: West Group, 2000); H. T. Rubin, "The Emerging Prosecutor Dominance of the Juvenile Court Intake Process," *Crime and Delinquency* 26 (1980): 299–318.
4. National Conference of State Legislatures, *1993 State Legislature Summary* (Washington, DC: National Conference of State Legislatures, 1993).
5. See Mich. Comp. Laws 712A.2(a)(1).
6. A. L. Schneider and D. D. Schram, "The Washington State Juvenile Justice Reform: A Review of Findings," *Criminal Justice Policy Review* 2 (1986): 211–235.
7. Feld, *Cases and Materials on Juvenile Justice Administration*.
8. C. Puzzanchera et al., *Juvenile Court Statistics 1999* (Pittsburgh, PA: National Center for Juvenile Justice, 2003).

9. D. Bell and K. Lang, "The Intake Dispositions of Juvenile Offenders," *Journal of Research in Crime and Delinquency* 22 (1985): 309–328; L. E. Cohen and J. R. Kluegel, "Determinants of Juvenile Court Dispositions: Ascriptive and Achieved Factors in Two Metropolitan Courts," *American Sociological Review* 43 (1978): 162–176; C. R. Fenwick, "Juvenile Court Intake Decision Making: The Importance of Family Affiliation," *Journal of Criminal Justice* 10 (1982): 443–453; K. I. Minor, D. J. Hartmann, and S. Terry, "Predictors of Juvenile Court Actions and Recidivism," *Crime and Delinquency* 18 (1997): 295-318; B. R. McCarthy and B. L. Smith, "The Conceptualization of Discrimination in the Juvenile Justice Process: The Impact of Administrative Factors and Screening Decisions on Juvenile Court Dispositions," *Criminology* 24 (1986): 41–64.

10. Cohen and Kluegel, "Determinants of Juvenile Court Dispositions"; Fenwick, "Juvenile Court Intake Decision Making"; McCarthy and Smith, "The Conceptualization of Discrimination in the Juvenile Justice Process."

11. Bell and Lang, "The Intake Dispositions of Juvenile Offenders"; Fenwick, "Juvenile Court Intake Decision Making."

12. Bell and Lang, "The Intake Dispositions of Juvenile Offenders"; Cohen and Kluegel, "Determinants of Juvenile Court Dispositions"; McCarthy and Smith, "The Conceptualization of Discrimination in the Juvenile Justice Process"; T. P. Thornberry, "Sentencing Disparities in the Juvenile Justice System," *Journal of Criminal Law and Criminology* 70 (1979): 164–171.

13. C. W. Thomas and C. M. Sieverdes, "Juvenile Court Intake: An Analysis of Discretionary Decision-Making," *Criminology* 12 (1975): 413–432; see also Bell and Lang, "The Intake Dispositions of Juvenile Offenders"; McCarthy and Smith, "The Conceptualization of Discrimination in the Juvenile Justice Process"; Thornberry, "Sentencing Disparities in the Juvenile Justice System."

14. J. Fagan et al., "Blind Justice? The Impact of Race on the Juvenile Justice Process," *Crime and Delinquency* 33 (1987): 224–258.

15. For a review of this literature, see M. Chesney-Lind and R. G. Shelden, *Girls, Delinquency and Juvenile Justice*, 3rd ed. (Belmont, CA: Thompson/Wadsworth, 2004), 198–203.

16. C. E. Pope and W. H. Feyerherm, "Minority Status and Juvenile Justice Processing: An Assessment of the Research Literature," *Criminal Justice Abstracts* 22 (1990): 327–335.

17. Puzzanchera et al., *Juvenile Court Statistics 1999*.

18. Juvenile Justice and Delinquency Prevention Act of 1974, 1988 Amendments, cited in P. Devine et al., "Disproportionate Minority Confinement: Lessons Learned from Five States," *Juvenile Justice Bulletin* (Washington, DC: Office of Juvenile Justice and Delinquency Prevention, 1998): 2.

19. Office of Juvenile Justice and Delinquency Prevention, "Minorities in the Juvenile Justice System," *1999 National Report Series, Juvenile Justice Bulletin* (Washington, DC: Author, 1999).

20. C. E. Pope et al., "Disproportionate Minority Confinement: A Review of the Research Literature from 1989 through 2001," *Juvenile Justice Bulletin* (Washington, DC: Office of Juvenile Justice and Delinquency Prevention, 2002).

21. H. N. Snyder and M. Sickmund, *Juvenile Offenders and Victims: 1999 National Report* (Washington, DC: Office of Juvenile Justice and Delinquency Prevention, 1999).

22. P. Harms, "Detention in Delinquency Cases, 1990–1999," *OJJDP Fact Sheet* (Wasnington, DC: Office of Juvenile Justice and Delinquency Prevention, 2003).

23. Puzzanchera et al., *Juvenile Court Statistics 1999*.

24. J. Austin et al., *Juveniles Taken into Custody: Fiscal Year 1993*, Statistics Report (Washington, DC: Office of Juvenile Justice and Delinquency Prevention, 1995).

25. Puzzanchera et al., *Juvenile Court Statistics 1999*.

26. See Mich. Comp. Laws 712A.15(3); Mich. Court Rules 5.935(D); West's California Welfare and Institutions Code, sec. 632; Illinois S. H.A., chap. 37, para. 703-1; Juvenile Justice and Delinquency Prevention Act of 1974; 1977 IJA-ABA Joint Commission on Juvenile Justice Standards.

27. Snyder and Sickmund, *Juvenile Offenders and Victims: A National Report*.

28. P. Puritz et al., *A Call for Justice: Assessment of Access to Counsel and Quality of Representation in Delinquency Proceedings* (Washington, DC: American Bar Association, 1995).

29. Puritz et al., *A Call for Justice*.

30. Puritz et al., *A Call for Justice*.

31. W. C. Bailey and R. D. Peterson, "Legal versus Extra-Legal Determinants of Juvenile Court Dispositions," *Juvenile and Family Court Journal* 32 (1981): 41–59; S. H. Clarke and G. G. Koch, "Juvenile Court: Therapy or Crime Control, and Do Lawyers Make a Difference?" *Law and Society Review* 14 (1980): 263–308.

32. Puritz et al., *A Call For Justice*, 9.

33. B. C. Feld, "In *re Gault* Revisited: A Cross-State Comparison of the Right to Counsel in Juvenile Court," *Crime and Delinquency* 34 (1988): 393–424.

34. S. H. Clarke and G. G. Koch, "Juvenile Court: Therapy of Crime Control, and Do Lawyers Make a Difference?"; Feld, "*In re Gault* Revisited,"; G. W. Burruss and K. Kempf-Leonard, "The Questionable Advantage of Defense Counsel in Juvenile Court," *Justice Quarterly* 19 (2002): 37–68.

35. J. J. Senna and L. J. Siegel, *Juvenile Law: Cases and Comments*, 2nd ed. (St. Paul, MN: West, 1992).

36. Feld, *Cases and Materials on Juvenile Justice Administration*.

37. S. M. Cox et al., *Juvenile Justice: A Guide to Theory and Practice*, 5th ed. (New York: McGraw-Hill, 2003).

38. H. T. Rubin, *Juvenile Justice: Policy, Practice and Law*, 2nd ed. (New York: Random House, 1985).

39. D. J. Champion, *The Juvenile Justice System: Delinquency, Processing and the Law* (New York: Macmillan, 1992), 120.

40. Champion, *The Juvenile Justice System*.

41. C. E. Simonsen, *Juvenile Justice in America*, 3rd ed. (New York: Macmillan, 1991).

42. Michigan Judicial Institute, *Handbook of Legal Terms* (Lansing, MI: Michigan Judicial Institute, n.d.).

43. Senna and Siegel, *Juvenile Law*.

44. J. Sanborn, "Pleading Guilty in Juvenile Court: Minimal Ado About Something Very Important to Young Defendants," *Justice Quarterly* 9 (1992): 126–150.

45. Sanborn, "Pleading Guilty in Juvenile Court."

46. Sanborn, "Pleading Guilty in Juvenile Court."

47. Sanborn, "Pleading Guilty in Juvenile Court."

48. H. C. Black, *Black's Law Dictionary*, 4th ed. (St. Paul, MN: West, 1968), 140.

Chapter Resources

The Transfer of Juveniles to Criminal Court

Chapter Objectives

After studying this chapter, you should be able to:

- Describe what is meant by waiver or transfer of juvenile offenders from juvenile court to adult criminal court
- Describe the different methods by which youths are transferred or waived to the adult justice system
- Describe the general criteria that must be satisfied in order to transfer a juvenile to the adult justice system
- Present a detailed discussion of the reasons why transfer is used from the perspective of the community, the political and law-making bodies, and the juvenile and adult courts
- Describe recent trends in the use of transfer
- Present a detailed discussion of the legal and due process issues involved in transfer
- Describe how transferred juveniles are handled by the adult criminal courts
- Present arguments on the pros and cons of waiver

Chapter Outline

- Introduction
- The Purpose of Transfer to Criminal Court
- Mechanisms for the Transfer of Juveniles to Criminal Court
- Trends in the Use of Transfer
- Due Process and Transfer Decisions
- Research on the Use and Effectiveness of Transfer

■ Introduction

With the establishment of the first formally organized juvenile court in Illinois in 1899 and the spread of juvenile courts to other states, jurisdiction over juveniles charged with law violations shifted from the criminal courts to the juvenile courts. However, many states recognized that in certain circumstances some juveniles, such as those who committed serious offenses, might be more appropriately dealt with by criminal (or adult) courts. Consequently, most early juvenile codes contained provisions for transferring some types of juvenile cases to criminal courts, although the criteria used to make such decisions were often vague and subjective.[1]

Today, all states and the District of Columbia make it possible for some juveniles to be tried in criminal courts.[2] The process by which juveniles are bound over to criminal courts for trial is called __waiver__, *certification, remand, bindover,* or *transfer.* The use of transfer has increased dramatically in recent years as more policy makers have adopted a "get tough" approach in dealing with juvenile crime. However, the decision to try juveniles in criminal courts frequently appears to be a political and emotional decision that critics claim receives little support from research on the effectiveness of transfer.

Regardless of the reasons for the transfer of a juvenile case to a criminal court, the decision to go forward with the transfer can have profound consequences for youths who are subject to transfer, for their family and friends, and the community. Historically, juvenile courts have attempted to shield youths from the negative consequences associated with being prosecuted in open criminal courts and being exposed to more mature and hardened offenders in the adult criminal justice process. Furthermore, juvenile courts have sought to offer youths "protection against loss of civil rights, against disqualification for public employment, and against personal status degradation and restriction of legitimate opportunities that often follow criminal conviction (p.180)".[3] Once a youth is transferred to a criminal court, however, these protections are lost, which can have important long-term consequences for the juvenile and his or her significant others. Moreover, if the transfer decision fails to influence the juvenile in ways that reduce the likelihood of recidivism, then community safety is also jeopardized.

The option of transferring juveniles to criminal courts raises important questions about how best to respond to adolescent offenders, particularly those who

waiver The process by which a juvenile is moved from juvenile court jurisdiction to adult criminal court jurisdiction for an offense committed while still a juvenile. Also known as bindover, certification, remand, and transfer.

have committed serious offenses. In particular, to what extent does the transfer of juveniles to criminal courts benefit the youths themselves or the communities most will return to at some point? This question invariably arises in discussions of transfer, and it needs to be carefully addressed if effective policies for dealing with serious juvenile crime are to be developed.

This chapter examines the use and consequences of transfer. Specifically, it examines the purpose of transfer, different mechanisms by which juveniles may be transferred to criminal courts for trial, the legal requirements for the transfer of juveniles to criminal courts, trends in the use of transfer, and the effectiveness of criminal court handling of juveniles. It concludes by examining a number of problems associated with the transfer of juveniles to criminal courts.

■ The Purpose of Transfer to Criminal Court

Transfer occurs when jurisdiction over a juvenile case is turned over to a criminal court. The waiver or transfer of jurisdiction from juvenile court to criminal court is predicated on the assumption that some juveniles are not appropriate for processing in juvenile court and can be more effectively dealt with by criminal courts. In essence, it is believed that although juveniles, as a class, are treatable and therefore appropriate for juvenile court handling, particular juveniles, by virtue of their actions, do not merit the protected status given to youths in the juvenile court. Instead, protection of the community from such youths demands that they be identified and transferred to the adult criminal justice system.[4] Once a juvenile is transferred to a criminal court for trial, the juvenile legally becomes an adult and is subject to the same types of correctional responses as any other adult, including the death penalty in those jurisdictions with capital punishment statutes.[5]

Until 2005, Juveniles Could Receive the Death Penalty

The execution of people less than 18 years of age is not unprecedented in the United States. Since colonial times, there have been at least 366 executions of juveniles in America, and 22 were carried out between 1973 and 2003.[6] Despite this history, the constitutionality of capital punishment was in doubt until the late 1980s, when the United States Supreme Court heard two landmark cases, *Thompson v. Oklahoma* (1988) and *Stanford v. Kentucky* (1989). The effect of these two cases was to ban the execution of youths under the age of 16 but permit the execution of those who commit capital offenses at age 16 or older.[7] However, the Supreme Court in *Roper v. Simmons* (2005) has recently declared unconstitutional executions of people under 18 years of age.[8]

In order to determine which juveniles are appropriate for criminal court jurisdiction, states have established various criteria. Typically, the juvenile has to meet certain age and offense criteria. For example, in Michigan, under its traditional waiver statute, a juvenile has to be at least 14 years of age (the maximum age of original juvenile court jurisdiction in Michigan is 17) and must be charged with a felony offense before he or she can be waived to criminal court. In addition, a juvenile court hearing must be held to determine if probable cause exists and if transfer of the case to adult court is appropriate.

Mechanisms for the Transfer of Juveniles to Criminal Court

Traditionally, the decision to transfer a case to criminal court was made in a juvenile court hearing. Today, however, there are several mechanisms (i.e., **judicial waiver**, **prosecutorial waiver**, and **legislative waiver** or statutory exclusion) by which juvenile cases are transferred to adult courts. Indeed, some states have more than one mechanism by which the transfer or waiver of a juvenile case can be achieved; for example, in a number of states, such as Michigan, Virginia, Arkansas, Colorado, Wyoming, and the District of Columbia, both judicial and prosecutorial waiver mechanisms are in place. In other states (Arizona, Florida, Georgia, Louisiana, Montana, Oklahoma, and Vermont) all three mechanisms exist.

In most states (Connecticut, Massachusetts, Nebraska, New Mexico, and New York are exceptions) juveniles can be transferred to an adult criminal court by means of a separate waiver or transfer hearing in the juvenile court to determine the appropriateness of the waiver.[9] This is the traditional way in which juvenile cases were transferred to criminal courts, and can be called *judicial waiver*. However, the exact process by which judicial waiver occurs and the way in which the waiver hearing in juvenile court is conducted varies from state to state. Nevertheless, these waiver hearings typically encompass four elements: (1) a determination of probable cause that the accused juvenile committed the crime(s) charged, (2) a consideration of the potential threat the youth presents to the community, (3) an evaluation of the extent to which the offender is amenable to treatment by existing juvenile justice programs, and (4) a consideration of the types of programs in the adult system that might better meet the juvenile's needs. In determining the potential threat a youth poses to the community, the court considers the seriousness of the offense and the juvenile's offense history. In evaluating the extent to which a youth is amenable to treatment, the results of previous juvenile court dispositions and information about existing juvenile justice and adult corrections programs may be looked at.

A potential problem with basing waiver decisions on a youth's amenability to treatment in juvenile justice programs or adult programs is that such determinations can be highly subjective. Even if a youth has recidivated after participating in previous juvenile justice programs, how can one determine that other juvenile justice programs employing different interventions would not be successful? Furthermore, how can one determine the effectiveness of adult programs for juveniles prior to their placement? Such determinations cannot be made based on

judicial waiver A type of waiver decided by a juvenile court at a hearing. It is the traditional type of waiver.

prosecutorial waiver A type of waiver that is within the power of the local prosecutor or state attorney to choose and implement.

legislative waiver A type of waiver that results when certain threshold criteria specified by state laws are met.

empirical evidence. Instead, they are predicated on predictions—some of which will invariably be wrong—that judges, attorneys, probation officers, and others make about individual cases.

The initiation of a waiver hearing in juvenile court usually occurs when a prosecuting attorney, believing that the transfer of a case to criminal court is the appropriate course of action, requests such a hearing.[10] The basis of the request may be the violence of the crimes committed, but prosecuting attorneys also seek transfer for nonviolent or even minor acts of delinquency. In a few states, juveniles or their parents may request a waiver hearing.[11] One reason a juvenile may ask for a waiver is to ensure due process protections, which are not always available in juvenile courts.

Typically, Juvenile Courts Are Less Formal than Adult Courts

As noted in Chapter 5, juveniles who are processed in juvenile courts still do not enjoy all of the due process protections afforded to adults in criminal trials. Indeed, many juvenile courts still operate in an informal manner that would not be acceptable in criminal courts.

Since the 1970s, a number of states have developed additional mechanisms for transferring juvenile cases to criminal courts. One of these is referred to as *legislative waiver*. Some states have passed laws that indicate that youths who meet certain age and/or offense criteria will automatically be tried as adults, thus bypassing the juvenile justice process altogether. For example, in Georgia, Illinois, Oklahoma, and New York, juveniles who are at least 13 and are charged with murder are automatically tried in criminal courts. In North Carolina and Mississippi, juveniles who are least 13 and are charged with certain felonies are automatically tried in criminal courts. In other states, juveniles who have reached a certain age and have prior felony adjudications can be excluded from juvenile court jurisdiction.[12]

There is another way in which legislatures have excluded a large number of juvenile cases from juvenile court jurisdiction. When state laws establish an upper age limit for juvenile court jurisdiction of 15 or 16 years of age, they in effect transfer cases to criminal court jurisdiction. Indeed, data collected by the National Center for Juvenile Justice indicate that as many as 218,000 youth under the age of 18 were tried in criminal court in 1996 because they were defined as adults under state laws. Unfortunately, there are no national data on the number of juveniles who are transferred to criminal courts by means of these different legislative waiver mechanisms.[13]

A third method of waiver that has become possible in some jurisdictions is **prosecutorial waiver**. As of 1997, 14 states and the District of Columbia had passed legislation that makes **concurrent jurisdiction** in some cases possible.

concurrent jurisdiction
Both the juvenile and adult court share original jurisdiction. Consequently, in states with concurrent jurisdiction, prosecutors are given discretion to file cases that meet certain age and offense criteria in juvenile or criminal court.

prosecutorial waiver
A type of waiver that is within the power of the local prosecutor or state attorney to choose and implement.

When concurrent jurisdiction exists, the prosecutor is given discretion to file certain cases directly in the adult court when a youth meets certain age and offense criteria.[14] For example, in Florida, the prosecutor has discretion to file a felony case in either the juvenile or adult court when the alleged juvenile offender has reached 16 years of age. In these instances, original jurisdiction is shared by both the juvenile and criminal courts. In Michigan, prosecutors can automatically waive juveniles at the age of 14 if they commit specified felonies.[15] For these particular crimes, the prosecuting attorney can choose to file a petition in juvenile court or file a complaint and warrant in adult court. Examples of specified juvenile violations include

- arson of a dwelling
- assault with intent to commit murder
- assault with intent to maim
- armed assault with intent to rob
- attempted murder
- first-degree murder
- second-degree murder
- kidnapping
- first-degree criminal sexual conduct
- armed robbery
- car jacking
- bank or safe robbery
- delivery or manufacture of cocaine or a narcotic drug (650 grams or more)
- possession of cocaine or a narcotic drug (650 grams or more)
- first-degree home invasion with a dangerous weapon
- assault with intent to do great bodily harm with a dangerous weapon
- escape from certain juvenile facilities[16]

These newer waiver statutes are supported by legislators who think the juvenile courts are "soft" on youth crime and are intended to circumvent the perceived leniency of the juvenile courts. The existence of different perspectives on juvenile crime is not surprising when one considers that different government entities have different roles and responsibilities and to some extent constituencies they must satisfy. For example, the executive branch is concerned about social control; the legislature is concerned about being perceived by the community as tough on crime; while courts have to balance community protection, fairness, rehabilitation, and due process. As a result, some see automatic waiver as a tough practical response to serious juvenile crime. Others, however, view it as a politically motivated response that wins votes and upsets the delicate balance of powers between the executive, legislative, and judicial branches of government by usurping a role traditionally played by the courts.

Box 9-1

Interview: The Honorable Carolyn Williams, Presiding Judge, Family Division, Kalamazoo County Circuit Court

Q: How long have you been on the bench, and what was your career prior to taking the bench?

A: I have served as a juvenile court and family court judge for 18 years. Prior to my taking the bench, I was in private law practice with an emphasis on personal injury defense. This was my second experience in private practice. Early on in my career, I worked for two years doing criminal defense and plaintiff's personal injury representation. In between the private practice experiences, I worked for the Department of Social Services as an administrative law judge for 10 years and took time to be a mother to my two sons, who are now grown.

Q: What is your educational background?

A: I have a BA in political science with an emphasis on international relations from George Washington University in 1964; a Juris Doctoris degree from George Washington University in 1968 that I earned with honors. I am also a graduate of the National Judicial College, where I have also served as a faculty member.

Q: What are the parts of your job that you find particularly rewarding?

A: I particularly enjoy those cases in which the parent and/or child have made positive changes that will enable them to be successful. To have a part in reaching people and helping them to change or, when necessary, motivating them to change is very rewarding.

Q: What are the major challenges that you see the juvenile justice system facing today?

A: I see two major challenges: out of control children, and the increasing public hostility towards all children due to this small group that is out of control. Society must recognize the causes for these children being out of control—changing family structure, substance abuse, and domestic violence. Society needs to accept its share of responsibility for this and work toward dealing with the causes. Instead, it seems that the only response by many politicians and the media is open hostility. Society seems to want to focus on punitive sanctions rather than rehabilitative solutions. In an enlightened society, I would hope that we would direct our resources toward early intervention, not toward detention.

Q: As a concept or ideal, do you believe in rehabilitation?

A: Yes, absolutely. It is what sets the juvenile/family court apart from the adult criminal system. If there is no rehabilitation, there is no hope. Interestingly over the past 8 years, I have noted an increase in the use of principles of "therapeutic jurisprudence" in both the adult and juvenile justice systems. Michigan now funds specialized courts which focus on helping adults, juveniles, and families deal with drug and alcohol dependence. Other states have created courts which focus on assuring compliance with mental health problems of those people involved with the criminal justice system. Communities across the country are also taking responsibility for helping to assure community safety and offender accountability using principles of balanced and restorative justice. Drug courts, therapeutic jurisprudence, [and] restorative justice are all ideas and principles that the juvenile courts have traditionally engaged in even though they may not have been identified using those words or phrases. It is truly gratifying to have been around long enough to see that those principles are now being recognized as important to the adult criminal justice system as they have always been in dealing with juveniles. As I said before, and it is as applicable to adults as it is with juveniles, any system that simply locks people away and offers them no hope, no hope of changing their behavior, or re-integrating into society, cannot in my opinion be justified or cost effective. Without rehabilitation there is no hope!

Q: Focusing on the issue of transfer or waiver, do you believe that waiver of 13 and 14 year olds, in general, is a good idea?

A: No, regardless of whatever awful act the child committed, to face a lengthy incarceration in the adult system at that age is "throwing them away" far too soon. Children at that age are impulsive. They are not adults; they don't think like adults; and, for the most part, they don't act like adults. Society needs to focus on giving children chances, not taking them away.

Q: What is your opinion of the concept of waiver or transfer?

A: Waiver, plain and simple, is giving up on kids!

Q: What do you think about the "criminal track" alternative sentencing that exists in Michigan?

A: It, as with many tools, has benefits and drawbacks. I'm sure that for some youth, the potential of facing an adult sentence can be an incentive to work harder in the juvenile system. On the other hand, the fact that it, unlike waiver, has no lower age limit allows for very young children to face adult criminal punishment. This gives validity to the popular cry for punishment and retribution and focuses the public away from the more important focus of rehabilitation.

Q: What kinds of programming or services do you see serious delinquent juveniles needing?

A: This is not my area of expertise; but, in my experience, there are some common denominators with children. At all and any level of treatment, education must be a priority, with dollars spent on technological resources so that the children are receiving a marketable education. Second of all, in view of all the dysfunctional families we deal with, alternative families, foster care homes, and especially group homes are needed. Not only can adults in these homes model healthy relationships, but in group homes, many children feel more comfortable in not having to identify the foster parents directly in competition with their own.

Q: How do you think society should deal with serious and violent delinquents?

A: Early intervention is key. We must focus resources on parenting education, giving young people reasons to defer becoming parents, and on preventing those factors in homes that contribute to violent kids and serious delinquents—poverty, substance abuse, and domestic violence. At the other end of the spectrum, we need secure and therapeutic placements for those delinquents that must be separated from society.

Q: Do you think that separate facilities should be built to house these serious delinquents?

A: I have no objection to this concept if the goal is rehabilitation with the focus on education, substance abuse prevention, resocialization, life skills, etc. I would strongly object to any placement where a child was simply warehoused!

Q: What advice would you give to the politicians that are continually calling for stiffer penalties for juvenile offenders?

A: First, I would challenge them to promote community awareness of the underlying causes of delinquency rather than settling for the incarceration alternative. They need to get away from the political rhetoric and the "sound bite" and work at being proactive in their communities. Money needs to be spent on at-risk children at the front end, in their formative years, so that we can avoid the unsocialized, out-of-control juveniles that cause such public outcry. I would recommend a close study of the European child-care systems as models for early intervention and socialization.

Q: During your career, you have received a number of honors and awards for your work with delinquents and their families. Could you tell me about these?

A: I have received several awards, including Omega Psi Phi (Upsilon Pi Chapter) William H. Hastie Award for Significant Contributions in Law (1997); Service to Children Award of the Guidance Clinic (1993); the Kalamazoo Rotary Red Rose Citation (1997); the Glass Ceiling Award of Greater Kalamazoo Network (1997); [and] the YWCA Woman of Achievement Award (1998).

There are no comprehensive national data on the number of juvenile cases that are subject to prosecutorial waiver each year. Nevertheless, the number of such cases appears to outstrip the number that are judicially waived. Florida alone reports that, on average, nearly 5,000 juvenile cases are transferred to criminal courts via prosecutorial transfer annually.[17] Moreover, there are indications that in states that have both judicial and prosecutorial waiver mechanisms, prosecutorial transfer is accounting for much of the growth in waiver. For example, in 1981, Florida, which has both judicial and prosecutorial transfer mechanisms, filed two cases directly in criminal courts for every case that was judicially waived. However, by 1992, for every case that was judicially waived in Florida, more than six cases were transferred to criminal courts through prosecutorial waiver.[18]

■ Trends in the Use of Transfer

Since the mid-1970s, state legislatures have taken steps to expedite the transfer of juveniles to criminal courts.[19] Typically, states have made it easier to transfer youths to criminal court by doing the following: (1) enacting statutory exclusions, (2) lowering the minimum age for waiving youths, (3) expanding the range of offenses that can result in transfer, or (4) making judicial waiver *presumptive*[20] (i.e., where transfer for certain cases is assumed unless there are clear grounds for handling the case differently; the burden of proof for keeping the case in the juvenile court falls on the juvenile). By 1999, 29 states had passed legislative transfer (statutory exclusion) statutes that exclude some offenses from juvenile court jurisdiction; other states have expanded the number of offenses excluded from juvenile court jurisdiction. In 15 states, prosecutors have become more prominent players in transfer decisions because of the enactment of prosecutorial transfer (concurrent jurisdiction) laws that allow them to decide whether to file certain cases in the juvenile court or the criminal court.[21] As a result, there are three ways that juveniles can be waived to criminal court in some states.[22] Finally, some states have simplified the judicial waiver process so that it is much easier to transfer some cases to adult court for trial.

Historically, transfer to criminal court was a judicial decision, and the burden of demonstrating that a youth was not amenable to treatment in the juvenile justice system and that the adult system held more promise was on the prosecution. In some states, statutes (or court rule in Arizona) specify that certain cases are appropriate for waiver even when it is a judicial decision, and the burden is now on the defense to demonstrate that a juvenile program is more suitable for the youth and that the adult system is inappropriate.

As a result of the increased emphasis placed on "getting tough" with juvenile crime in some jurisdictions and the enactment of laws making it easier to transfer juvenile cases to adult criminal courts, there has been a substantial increase in the number of juveniles tried in criminal courts since the 1970s.[23] Indeed, between 1985 and 1994, the number of cases judicially waived to adult courts increased by 71%, from 7,200 to 12,300 cases,[24] although it declined to 7,500 by 1999.[25] However, although the number of judicially transferred cases has been declining recently, far more juveniles were tried in adult courts as a result of legislative and prosecutorial waiver statutes. Indeed, a study of waivers in 18 jurisdictions found that 85% of the decisions to try youths as adults were not made by judges, but resulted from statutory exclusion or prosecutorial discretion.[26]

As previously noted, it is estimated that as many as 218,000 youths under 18 years of age were tried in criminal courts in 1996 alone because they were defined as adults under state laws. Unfortunately, the total number of youths under 18 who are being tried in criminal courts as a result of various waiver mechanisms is not known.[27] Not only has there been a substantial change in the number of juvenile cases transferred to criminal courts, but the types of cases transferred to criminal courts, at least via judicial transfer, have also changed over time. In 1989 drug offense cases were the most likely to be judicially transferred, but by 1993 the largest group of cases transferred involved offenses against people, a trend that continued through at least 1999.[28]

■ Due Process and Transfer Decisions

As discussed in Chapter 5, the first juvenile court case that drew U.S. Supreme Court attention, *Kent v. United States*, addressed the issue of transfer. In this case, Morris Kent, a juvenile, was transferred to adult criminal court without a hearing or any formal notice. In its opinion, the Court set forth minimum due process standards for a valid waiver:

- Any juvenile facing waiver is entitled to an attorney.
- The juvenile is entitled to a meaningful hearing, even if informal and not on the record.
- The juvenile is entitled to access to any reports, records, and so on, used by the court in deciding waiver.
- The juvenile is entitled to know the reasons for the waiver decision.

The first requirement speaks for itself. A meaningful hearing is one in which the juvenile can present evidence, challenge evidence against him- or herself, and do this all before an impartial decision maker. This requirement is consistent with the Court's allowing the juvenile to challenge reports or other information that purport to be the basis for waiver. However, the Supreme Court did not require the hearings to be "on the record," and thus transcripts that indicate in detail what went on in the waiver process may not be available. Nevertheless, the Court did require a "statement of reasons" for waiver, and this could form the factual basis for an appeal.

Due to the serious nature of waiver cases, it is not surprising that the Supreme Court has heard other cases that involve this issue. One of these was *Breed v. Jones* (1975).[29] In this case, a 17-year-old, accused of armed robbery, was adjudicated to be a delinquent at an adjudication hearing in juvenile court. Two weeks later, but before deciding on any sentence or disposition, the juvenile court determined at a "fitness" hearing that the juvenile was not amenable to juvenile court "care, treatment, and training," and ordered the juvenile to be prosecuted as an adult. At the subsequent adult trial, the juvenile was convicted of first-degree robbery.

In its decision, the Supreme Court held that the juvenile had been "put in jeopardy" at the original juvenile court adjudicatory hearing and that, once found within the jurisdiction of the juvenile court, he could not then be tried as an adult for the same offense. To try him as an adult would be to violate his right to be free from **double jeopardy**. *Breed* had several ramifications. First, the Supreme Court recognized the seriousness of a juvenile court adjudication and gave it the same status afforded criminal trials for the purposes of double jeopardy protection. Second, and most important, the Court required the state to choose waiver "up front" in the process, prior to any juvenile court adjudication, denying the state multiple opportunities to try juveniles. Third, the decision put a strong emphasis on the transfer or waiver hearing in the juvenile court process because its outcome determines the sanctions possible for the juvenile.

double jeopardy
As applicable to transfer cases, a principle that requires the state authority to choose at the outset whether a juvenile is to be tried as a juvenile or as an adult. The choice must be made at the preliminary hearing or arraignment stage of the proceedings.

■ Research on the Use and Effectiveness of Transfer

Despite the importance of transfer and its long history of use in juvenile justice, only recently has there been much research devoted to the way transfer is used in different jurisdictions around the country and the effectiveness of transfer decisions.

The Use of Transfer

Although there is much about transfer that is not known, there is clear evidence that the use of transfer varies considerably across jurisdictions. Some jurisdictions are much more likely to transfer juveniles to adult courts than other jurisdictions. A study by Barry Feld found that juvenile court judges exercised considerable discretion in making transfer decisions and did not administer transfer statutes in an evenhanded manner. According to Feld, "Within a single jurisdiction, waiver statutes are inconsistently interpreted and applied from county to county and from court to court."[30] Another study, done by Tammy Poulos and Stan Orchowsky, examined transfers in Virginia and found that juvenile offenders in metropolitan areas were less likely to be transferred than those whose cases were processed in nonmetropolitan juvenile courts. Although the authors were not able to provide a definitive explanation for their findings, they noted the following:

> Juvenile court judges serving metropolitan jurisdictions may be less likely to send young offenders to the criminal courts for a number of reasons. Because they see so many serious offenders, their threshold for defining an offense as serious enough to warrant transfer may be higher than that of their rural counterparts. On the other hand, metropolitan judges may have at their disposal more dispositional options at the juvenile court level than their rural counterparts and thus rely less heavily on the last resort of transfer. (p. 14)[31]

Although one might expect that the level of serious crime in a jurisdiction has a direct influence on the number of juvenile cases waived to criminal courts, research indicates that the use of transfer is determined by a variety of factors. Likely candidates for transfer include juveniles who commit serious offenses such as murder, manslaughter, and rape; juveniles who sell drugs; older juveniles who use firearms in the commission of crimes and have prior records; and juveniles with lengthy juvenile court histories.[32] A study that examined waiver in Maricopa County, Arizona, found that previously being waived was the strongest predictor of transfer.[33]

There is evidence that the increased use of transfer in some jurisdictions is a reflection of the growing popularity of a more punitive philosophy toward juvenile offenders in general and the desire of political decision makers to appear tough on juvenile crime, rather than a direct response to higher levels of serious juvenile offending. After examining the case histories of 214 juveniles transferred to criminal court and interviewing key juvenile justice decision makers, M. A. Bortner concluded that organizational and political factors accounted for high rates of transfer. According to Bortner,

> Although remand represents a course of action ostensibly directed toward protection of the public, it is a policy integrally related to the juvenile system's interest in organizational maintenance. . . . Remand protects institutional authority by removing cases from juvenile court responsibility, thus reducing potential organizational troubles and public criticism that might damage institutional legitimacy. . . . In an era of fiscal uncertainty when the juvenile justice system is confronted with the necessity of reasserting its worth, maintaining its uniqueness, and redefining its mission, remanded juveniles provide

*a symbolic avenue for the reaffirmation of the juvenile system's commit-
ment to public safety and rehabilitation for most juveniles. In evidencing
a willingness to relinquish its jurisdiction over a small percentage
of its clientele, and by portraying these juveniles as the most intractable
and the greatest threat to public safety, the juvenile justice system
creates an effective public gesture of retribution and punishment in
the name of responsiveness to public concerns. . . .*

*The present analysis suggests that political and organizational factors,
rather than concern for public safety, account for the increasing
rate of remand. In evidencing a willingness to relinquish jurisdiction over a small percentage of its clientele, and by portraying these
juveniles as the most intractable and the greatest threat to public
safety, the juvenile justice system not only creates an effective symbolic
gesture regarding protection of the public but it also advances its terri-
torial interest in maintaining jurisdiction over the vast majority
of juveniles and deflecting more encompassing criticisms of the
entire system. (pp. 66–69)*[34]

 Efforts to "Get Tough" with Juveniles Continue

As noted in Chapter 2, there is no evidence that juvenile crime has become
significantly more serious over time. Indeed, arrests of youths under 18 for
Index Violent Offenses has been declining since 1994. However, efforts to
develop more punitive responses to juvenile crime, including the development of additional mechanisms for waiving juveniles to criminal courts, began
in the mid-1970s and continue today.

Although transfer is believed by many to be reserved for only the most serious juvenile offenders, in many jurisdictions a large percentage of the cases waived
involve nonviolent offenses. Moreover, waiver cases brought against some juveniles are not very strong.[35] As noted earlier, property offense cases accounted for
the greatest percentage of cases judicially waived in 1999. For example, data
collected by the National Center for Juvenile Justice indicate that in 1999 the
most serious offense committed by 34% of offenders whose cases were waived
was a person offense. In comparison, the most serious offense committed by
40% of offenders whose cases were transferred was a property offense. The offense
profiles of cases transferred to criminal court for selected years from 1985 through
1999 can be seen in **Table 9-1**.

Table 9-1	Offense Profile of Deliquency Cases Transferred to Criminal Court		
Offense	**1985**	**1994**	**1999**
Person	33%	44%	34%
Property	54%	37%	40%
Drugs	5%	11%	16%
Public Order	9%	8%	11%
Transferred Cases	7,200	12,300	7,500

Source: Adapted from *Juvenile Court Statistics,* (Washington, DC: Office of Juvenile Justice and Delinquency Prevention, 1994, p. 13); C. M. Puzzanchera, *Delinquency Cases Waived to Criminal Court, 1990–1999*, OJJDP Fact Sheet (Washington, DC: Office of Juvenile Justice and Delinquency Prevention, 2003).

Much less is known about the types of cases that are being waived under prosecutorial and legislative transfer laws than about judicially transferred cases. What is known is that the prosecutorial and legislative transfer statutes enacted in many states have resulted in more youth being subjected to criminal court processing.[36] In addition, there is research indicating that, at least in some jurisdictions, a large percentage of those transferred by prosecutorial transfer are nonviolent offenders. In a study of felony cases from a nationally representative sample of 300 counties, some of which were transferred by prosecutorial and legislative waiver statutes, it was discovered that only 53% of the cases transferred to and convicted in adult court involved violent offenses. In this study 24% of criminal court convictions involved property offenses, 13% were drug offenses, and 10% involved other offenses.[37] In another study of prosecutorial waiver in two Florida counties, Donna Bishop and her colleagues found that, despite the fact that prosecutors indicated that the youths who were transferred were dangerous offenders, only 29% had committed a violent felony.[38] In fact, the majority (55%) had committed property felonies. Moreover, Bishop and her colleagues found that the tendency had been for prosecutors to transfer greater proportions of nonviolent offenders, particularly youths charged with felony drug offenses, and misdemeanants over time. After examining the characteristics of juveniles transferred in the two counties they studied, Bishop and her colleagues concluded that "few of the juveniles transferred to criminal court via prosecutorial waiver would seem to be the kinds of dangerous offenders for whom transfer is most easily justified."[39]

MYTH VS.

REALITY **Youths Are Sometimes Transferred to Criminal Court**

Myth—Transfer to criminal court is reserved for only the most serious juvenile offenders.

Reality—Older youths, youths who commit serious offenses, and those with previous juvenile court histories are more likely to be waived to adult courts.[40] There is also evidence that minority youths are disproportionately represented in transfers to adult court.[41] However, it is important to note that transfer is frequently used for youths who have committed property offenses and drug offenses, and it is sometimes used for juveniles who have committed public order offenses. In fact, transfer may have more to do with political and/or economic reasons than the usual reason given—protection of the community. Furthermore, juvenile courts, frustrated with juvenile offenders who resist rehabilitation, may opt for the waiver of these youths even though they have not committed serious offenses.

The Effectiveness of Transfer

One common rationale for the transfer of juveniles to criminal courts is that it serves as a tougher response to serious juvenile offenders and is more likely to protect community safety than juvenile court processing. As noted earlier, however, the juveniles transferred are not always serious or violent offenders. This section examines the extent to which transfer is effective in protecting community safety.

Early research on transfer indicated that when juvenile offenders were transferred to criminal court, they were often treated leniently. One suggested explanation for the lenience was that the juveniles who appeared in adult courts were often young and were seen as first-time offenders.[42] In a 1978 national survey of the transfer of juveniles to criminal courts, Donna Hamparian and her associates found that only 46% of cases waived by juvenile court judges and 39% of cases transferred to criminal court by prosecutorial waiver resulted in a criminal court sanction that involved incarceration.[43] Similarly, Bortner's study, which looked at 214 juveniles waived in a western state, found that 63% of juveniles whose cases were waived received probation. Indeed, only 32% of the juveniles whose cases were waived received jail or prison sentences, 1% were given fines, and 4% had their cases dismissed.[44]

Although many juveniles transferred to adult courts have been treated leniently, others are treated more harshly than similar offenders in juvenile court. In some jurisdictions, incarceration, not leniency, appears to be the norm. For example, a study conducted by Rudman, Hartstone, Fagan, and Moore looked at 138 youths who were charged with violent offenses and were considered for transfer between 1981 and 1984 in Boston; Newark, NJ; and Phoenix. They found that 94% of these youths were convicted of violent crimes in criminal court and that 90% of the convicted youths were incarcerated. Moreover, they found that, among youths who were sentenced to a period of incarceration, those convicted and sentenced in adult courts received substantially longer sentences than those convicted and sentenced in juvenile courts.[45] Likewise, a study by Jeffrey Fagan that compared 15- and 16-year-old felony offenders in New York (where they are excluded from juvenile court jurisdiction) to similar youths in New Jersey (where they are under juvenile court jurisdiction) found that youths in New York were more likely to be incarcerated.[46]

Studies that have examined transfer indicate that there is considerable variability in how youths are treated when they are convicted in adult courts. The exact reasons for this are not completely clear, but it may be due in large part to the types of offenses juveniles are charged with. In the studies by Hamparian and colleagues and by Bortner, many of the youths transferred to criminal court were property offenders. In contrast, the youths transferred to criminal court in the study by Rudman and colleagues were convicted of violent offenses. This suggests that if juveniles are transferred to adult courts for minor property offenses, they are likely to be treated leniently, but if they are transferred to adult courts for serious offenses against people, they are likely to receive severe sanctions. Indeed, in the research study noted earlier that examined juveniles convicted of felonies in criminal courts, juveniles who were convicted of murder were actually sentenced to *longer* prison terms than older people convicted of the same offense.[47]

MYTH VS. REALITY — Sometimes Juveniles Are Treated *More* Harshly than Adults

Myth—Juveniles are able to use their youth as a *mitigating factor* to convince courts that they should not be punished for serious offenses, which results in the lenient treatment of serious juvenile offenders.

Reality—Like some adults, juveniles are sometimes treated leniently for serious offenses. However, it should also be noted that youths' age is sometimes used as an *aggravating factor* that paints them as very dangerous offenders and results in punishments that are more serious than those sometimes given adults. This occurs when juries or hearing officers are convinced by prosecutors that if a youth has committed a serious offense as a juvenile he or she will likely be even more dangerous in the future.[48]

Contrary to what some might expect, juveniles transferred to adult courts are not always treated more severely than youths whose cases are adjudicated in juvenile courts. Indeed, in some instances juveniles receive more severe dispositions in juvenile courts. Jeffrey Fagan found that in 1981 and 1982, youths processed in juvenile courts in New Jersey were half as likely to receive sentences mandating incarceration as youths of a similar age convicted in adult courts in New York. For example, 18% of the youths convicted of robbery in New Jersey juvenile courts were incarcerated, compared with 46% of youths convicted of robbery in criminal courts in New York. In 1986 and 1987, however, Fagan found juveniles convicted of robbery in New Jersey were more likely to be incarcerated than those handled in New York. During those 2 years, the percentage of juveniles convicted of robbery who were incarcerated in New Jersey increased to 57%, compared with 27% for New York.[49] In another study comparing youths processed in juvenile courts and those tried in criminal courts, the researchers found that, on average, criminal courts sanctioned offenders more severely for the same offense. However, the researchers also discovered that the differences were due, in part, to differences in the prior records of the offenders. What they discovered was that youths who had prior records and whose cases were adjudicated in juvenile courts were treated far more severely than youths with prior records who were tried in adult courts.[50]

MYTH VS. REALITY — Youths May or May Not Be Treated Harshly by Adult Courts

Myth—Youths who are transferred to criminal courts typically receive severe sanctions, such as long periods of incarceration.

Reality—Many youths transferred to criminal courts are treated severely and receive longer periods of incarceration than similar youths convicted in juvenile courts.[51] However, many others are treated leniently and even receive probation.[52] The differences in treatment may be due to differences in the types of offenses. Courts may be likely to treat juveniles who have committed violent offenses more severely than juveniles who have committed minor offenses.

It is also worth noting that the research evidence on transfer does not demonstrate that public safety is enhanced by waiver, nor does it clearly show that alternative treatments are usually exhausted before youths are transferred to criminal court.[53] Even when youths are subjected to incarceration in criminal courts, there is little evidence that the punishment has a deterrent effect. For example, studies of legislative waiver in New York and Idaho failed to uncover a deterrent effect associated with the waiver statutes of these two states.[54] Also, Jeffrey Fagan's study of youths charged with felony burglary or robbery in both juvenile and criminal courts found that, among youths charged with robbery, those processed in juvenile courts were significantly less likely to be rearrested and reincarcerated than those tried in adult courts. Furthermore, among recidivists, the length of time before they were rearrested was significantly longer for youths processed in juvenile courts than for those tried in criminal courts.[55] In a 1-year follow-up that compared case outcomes for close to 3,000 juveniles transferred to criminal court in Florida with youths retained in juvenile court, Donna Bishop and her colleagues discovered that transferred youths had a higher rearrest rate, were more likely to be rearrested for more serious offenses, and were rearrested within a shorter time frame than youths who were processed in juvenile court.[56]

Although juveniles who had not been transferred had equaled transferred youths with respect to the proportion that had been arrested nearly 6 years after the initial study, transferred youths who were rearrested were reapprehended in a shorter time period and were rearrested more times, on average, than youths retained in juvenile court.[57] Unfortunately, only a small amount of research has compared the effects of juvenile court sanctions and the effects of criminal court sanctions. Nevertheless, the research that does exist fails to demonstrate that criminal court sanctions have more of an impact on recidivism than sanctions levied by juvenile courts.[58]

■ Other Developments in the Use of Transfer

Designation

designation An action by a prosecutor or juvenile court that makes available adult court sentencing alternatives even though the juvenile is being tried under the jurisdiction of the juvenile court.

In addition to the passage of prosecutorial and legislative transfer laws, states have implemented other laws and procedures related to the transfer of juveniles to adult court. Michigan has developed an alternative to prosecutorial waiver called **designation** that allows a juvenile to be tried in juvenile court in the same manner as an adult is tried in criminal court, including having a probable cause, preliminary type hearing, judgment by a jury of 12, and a formalized sentencing hearing. A juvenile can be designated in two ways. The prosecutor can designate a youth who has committed a "specified juvenile violation" or the court itself can designate a youth for any other delinquent offense. If the prosecutor designates, then the juvenile is on the criminal track. If the court designates, then the juvenile is entitled to a designation hearing to contest the referral to the criminal track. At a designation hearing, the court is required to balance the best interests of the juvenile and the public by considering a variety of factors such as the seriousness of the offense, the culpability of the juvenile committing the offense in light of aggravating or mitigating factors, the juvenile's prior delinquency record, the juvenile's history in previous correctional programs, the adequacy of the punishments and/or programs available in the juvenile justice system, and the different disposition options available to the court. These are essentially the same factors that the juvenile court would consider in determining the appropriateness of judicial waiver.

After designation, the juvenile has essentially an adult criminal trial in juvenile court, with all of the accompanying due process rights. Upon conviction, the juvenile court has to hold a sentencing hearing to review the available alternatives. These include the following:

- sentencing the juvenile as a juvenile
- sentencing the juvenile as an adult
- sentencing the juvenile on a delayed basis (if the disposition is not successful, the court can resentence the juvenile as an adult)
- sentencing the juvenile to a boot camp for a period of 90–180 days, with 120–180 days in aftercare

In order to impose a delayed adult sentence, the court must hold a hearing in which the rehabilitation of the juvenile is balanced against the juvenile's risk to public safety. The court is required to consider the following at the sentence review hearing:

- the extent of the juvenile's conformity to the directives in the original disposition, including participation in education, counseling, work programs, and so on
- the juvenile's willingness to accept responsibility for his or her delinquent behavior
- the juvenile's behavior in his or her current placement
- the juvenile's prior record, character, and physical and mental maturity
- the juvenile's potential for violent conduct as shown by prior behavior
- recommendations from the caretaker institution
- other information submitted by the juvenile and/or the prosecutor

Certainly, designation gives the juvenile court another option in dealing with serious juvenile offenders. One of the concerns about the Michigan statute, however, is that there is no lower age limit for designation. In other words, a juvenile of any age can be designated. Although designation became effective January 1, 1997, it does not appear to be widely used. Moreover, to date there has been no evaluation of its effectiveness.

Reverse Waiver

At least 23 states have laws that make it possible to transfer juvenile cases from adult court to juvenile court. This makes it possible for the criminal court to transfer cases to juvenile court that have been initially waived via prosecutorial transfer, legislative waiver, or, in some cases, judicial transfer. Generally, reverse waiver occurs after a hearing in criminal court where the criminal court judge considers the appropriateness of handling the case in criminal court or in the juvenile court. At this hearing, the adult court employs the same "best interest" standards used in juvenile courts to determine the appropriateness of transfer to the criminal court.[59]

Once an Adult, Always an Adult

At least 31 states also have "once an adult, always an adult" requirements built into state law. Most state statutes that contain these requirements mandate that once a youth is transferred to criminal court for trial, all subsequent offenses against that individual will be heard in criminal court. In some states, however, only a specified range of subsequent offenses must be heard in adult court once a youth has been waived. Also, most state laws require that the youth must be convicted of the initial offense in criminal court before "once an adult, always an adult" requirements take effect, although in some states (e.g., California and Delaware) conviction on the initial charge is not always necessary.[60]

Blended Sentencing/Determinate Sentencing

A number of states also have **blended sentencing**/determinate sentencing provisions that allow either the juvenile court or the adult court to render a disposition that involves placement in a juvenile or adult correctional institution or both. These laws have often resulted from concerns over perceived increases in serious juvenile crime and a belief that young juvenile offenders who commit serious offenses need additional supervision beyond that normally possible by the juvenile court.[61] Under blended sentencing or determinate sentencing options,

blended sentencing
Allows either the juvenile court or the criminal court to impose a disposition that involves placement in a juvenile correctional institution or an adult institution, or to serve periods in both juvenile and adult programs.

depending on the state, either the juvenile or criminal court can sentence youths to longer periods in juvenile facilities than was traditionally possible under previous law, can sentence a youth to placement in an adult facility, or can sentence a youth to a juvenile facility for a period of time followed by placement in an adult facility. For example, under Texas's determinate sentencing law, youths can be sentenced for up to 40 years' incarceration by the juvenile court. They would serve their sentence in juvenile facilities operated by the Texas Youth Commission until they reach their eighteenth birthday, at which time they would be transferred to the Texas Department of Corrections to serve the remainder of their sentence.[62]

Standards Governing Transfer Decisions

Although there is undoubtedly a trend toward making it easier to transfer juveniles to adult courts, the transfer of many nonviolent juvenile offenders to criminal court appears to be encouraged by a lack of clear standards for transfer decisions in some jurisdictions. In their study of prosecutorial waiver in Florida, Bishop and her colleagues found that most juvenile division chief prosecutors in Florida had not established formal policies to guide transfer decisions. Instead, they relied on informal guidelines to direct the waiver decisions made by attorneys under their supervision. Moreover, many of the prosecutors indicated that they thought the standards used by other division chiefs to limit attorney discretion were too low.[63]

The available evidence suggests that the increased use of transfer to criminal courts is the product of a number of factors. To some extent, the increased use is likely due to periodic fluctuations in the incidence and seriousness of juvenile crime and the difficulty that overburdened and underfunded juvenile justice agencies have in dealing with the juvenile crime problem. It is also likely due, in part, to the perception that the juvenile courts do not impose severe enough sanctions on some juvenile offenders and that they often lack reliable data information systems capable of tracking offenders involved in serious and repetitive criminality.[64] In addition, factors such as offense seriousness, age, offense history (including a history of previous waiver), a lack of clear criteria for transfer, and a desire among policy makers to appear tough on juvenile crime have also contributed to the increased use of transfer as a response to juvenile crime.[65]

MYTH VS. REALITY

Waiver Laws Appear to Do Little to Deter Juvenile Crime

Myth—Threatening youths with waiver and sending more youths to adult courts will deter youths from delinquent behavior.

Reality—Studies that have examined the potential deterrent effects of state transfer laws have failed to find that they deter juvenile crime.[66]

■ Correctional Programming for Juveniles Convicted in Criminal Court

The increase in the number of adolescents being subjected to incarceration in adult correctional facilities has, in some instances, led to the development of different strategies for responding to the juvenile offender population. At least in some jurisdictions, the influx of juvenile offenders has placed increasing pressure on those who are responsible for the development of adult institutional correctional policy to implement programs that better respond to the needs of youthful prisoners, who are typically more vulnerable than inmates traditionally handled by these facilities. Around the country, four basic approaches to responding to adolescent offenders sentenced to periods of incarceration in the adult system presently exist: (1) straight adult incarceration, (2) <u>graduated incarceration</u>, (3) <u>segregated incarceration</u>, and (4) designation of certain offenders as <u>youthful offenders</u>.[67]

Straight Adult Incarceration

Most states make it possible to place juveniles who are sentenced as adults in adult correctional facilities, either with younger adult offenders or in the general correctional population, providing the juvenile has reached a certain age (e.g., in North Dakota and California, a youth must be at least 16 to be placed in an adult correctional facility). Only six states (Arizona, Hawaii, Kentucky, Montana, Tennessee, and West Virginia) prohibit the placement of juveniles in adult correctional facilities or require juveniles to be segregated from adults.[68] On June 30, 2003, counts of state prison and jail inmates indicated that there were 3,006 prisoners housed in state prisons and 6,869 inmates in adult jails who were under 18 years of age.[69]

Graduated Incarceration/Blended Sentencing/Determinate Sentencing

The development of new waiver mechanisms and the increased use of transfer has taxed the ability of the adult system to provide an appropriate environment for transferred youths. One response to this problem has been the development of graduated incarceration programs designed to mitigate the harshness of adult punishments. In states that have this option, juveniles who are sentenced to periods of incarceration as adults begin their incarceration in a juvenile facility or in a separate facility operated by the adult department of corrections until they reach a specified age, usually 18. Once they turn 18, they may be transferred to a traditional adult facility to serve out the remainder of their sentences or they may be released. States utilizing this option include Delaware, Georgia, Maryland, Missouri, North Dakota, Ohio, Oregon, Tennessee, Texas, Utah, Washington, and West Virginia.[70]

Segregated Incarceration

Some states, such as Florida and South Carolina, provide separate programs within their adult correctional systems for younger adult inmates. In these states, the department of corrections operates separate facilities for adolescent adult offend-

graduated incarceration
An incarceration strategy according to which a convicted juvenile begins his or her sentence in a juvenile facility and then moves to an adult facility upon reaching the appropriate age.

segregated incarceration
The practice of providing a separate correctional facility within the adult justice system designated for juvenile offenders.

youthful offender status
A special legal status existing in some jurisdictions that allows young offenders to receive special consideration as regards sentencing, privacy, and rehabilitation, even as first-time adult offenders.

ers within a certain age range (18 to 21 years of age or 18 to 25 years of age). When juveniles in these states are convicted and sentenced as adults, they may become eligible for placement in these facilities because of their age.[71]

Designation of Certain Juveniles as Youthful Offenders

Another strategy used in some states is to designate certain juveniles as "youthful offenders" (other terms may also be used). Such a designation often provides special legal protections to a juvenile so designated, such as the sealing of the court record if the juvenile successfully completes his or her sentence. In addition, special programming for these youths may be provided in the juvenile or adult correctional systems. States that presently have some version of the youthful offender designation are Colorado, Kentucky, New Mexico, Wisconsin,[72] and Michigan.

In Some States, Youths May Be Designated as Youthful Offenders

One example of a youthful offender statute is Michigan's Holmes Youthful Trainee Act. This act, originally adopted in 1967, recognizes that youthful offenders, even if tried as adults, should not necessarily be treated as hardened criminals.[73] According to this statute, "youthful trainee status" is available only once to a youth charged with a felony offense (excluding offenses that result in life imprisonment or offenses that constitute a major drug offense). Originally, this act applied to youths between 17 and 20 years of age. However, amendments to the original act moved the upper age limit to 21 and the lower age limit to 16, making youthful trainee status available to youths who are transferred to adult court. Upon a guilty plea, the adult criminal court does not enter a conviction but assigns the youth to youthful trainee status. This status, which is revocable at the court's discretion, usually involves supervision and probation for up to 3 years. The focus, even if the juvenile is sent to jail as one of the conditions, is on work and education. If the youth completes the program, the criminal proceedings are dismissed, no criminal conviction is entered, and all of the proceedings are closed to public inspection, in recognition of the stigma that attaches to a person accused of serious crimes, no matter what the eventual legal outcome.

■ Legal Issues

From a purely philosophical perspective, it can be asked whether it is proper to transfer juveniles to the adult criminal justice system. Moreover, a growing body of scientific evidence indicates that juveniles are developmentally different than adults, and thus, should be treated differently.[74] If a fundamental premise of juvenile justice is that juveniles should be treated differently than adults, why should any exceptions be made and waiver allowed? A counterargument can be made that waivers are necessary because the juvenile justice process has failed to do an adequate job of rehabilitating the types of youths that are the subjects of waiver. For those children who are waived at the beginning of their juvenile court history, it can be argued that the system, be it in the form of a child welfare agency or a juvenile court, failed to recognize them as problem children at an early enough age for meaningful intervention to take place.

Is the increasing trend toward legislative, prosecutorial, or presumptive waiver a statement by the legislative and executive branches that the judicial branch either is not competent to make these decisions or is consistently making poor decisions? Under either theory, the advent of nonjudicial transfer is a clear attack on the judicial branch. Any time one branch of our tripartite government invades the province of another branch, it should be cause for grave concern. Clearly, many legislators and prosecutors believe, despite the lack of supporting statistics, that the juvenile courts are too easy on serious delinquents.

The crux of the matter seems to be politics. Being hard on serious delinquents is a popular stand, and one that is hard to oppose. Because the courts cannot easily or ethically move into the political arena to protect their authority and discretion, they are vulnerable to attack. Elected officials' first responsibility is to get reelected. Politically popular stands—even those without a factual basis—garner votes. Yet it is important to recognize that political grandstanding does not always help youths or protect the community.

■ Chapter Summary

Prior to the development of a separate court for juveniles at the end of the 19th century, youths accused of illegal behavior were subject to the jurisdiction of adult criminal courts. With the establishment of juvenile courts, however, jurisdiction over the criminal behavior of youths shifted to the newly organized courts, which were intended in part to shield youths from the harmful effects of criminal court processing and commitment to adult correctional institutions. Although jurisdiction over criminal and status offense behavior shifted to juvenile courts once they were established, state juvenile codes still made it possible to try some juveniles as adults. The transfer of a juvenile to a criminal court, which was typically accomplished by holding a juvenile court hearing, was generally thought to be appropriate if the juvenile had committed a serious offense.

Today, each state and the District of Columbia has a mechanism by which some juveniles can be sent to adult courts for trial (called bindover, waiver, transfer, remand, or certification, depending on the state). In fact, the ways in which juveniles can be transferred to adult courts have expanded in recent years. Traditionally, transfer or waiver was accomplished by holding a juvenile court hearing to consider the appropriateness of the action. Some states, in an effort to "get tough" on juvenile crime, have implemented prosecutorial waiver mechanisms that allow prosecuting attorneys the discretion to waive juveniles who meet certain age and offense criteria. Other states have established legislative waiver mechanisms that automatically transfer some juveniles to adult courts for trial if they meet certain age and offense criteria. In addition, some states have developed presumptive waiver laws as well as blended and determinate sentencing laws that make it easier to transfer certain youth to adult courts, maintain jurisdiction over juveniles for longer periods of time, and sentence them to adult correctional facilities.

The "get tough" movement within juvenile justice and the resulting efforts to develop additional transfer mechanisms have produced a substantial increase in the number of juveniles being tried as adults. Today, many serious juvenile offenders are tried in adult courts. However, a lack of clear criteria to guide prosecutorial decisions about transfer in some jurisdictions has contributed to the transfer of many juveniles who have engaged in nonserious crimes, such as property crimes.

As the number of youths tried in adult courts increases, so will the number of youths placed in adult correctional programs. Unfortunately, there is a lack of convincing evidence that adult correctional programs are equipped to meet the needs of juvenile prisoners. Indeed, there is little evidence that criminal court responses to juvenile offenders serve their best interests or increase the likelihood that they will be rehabilitated or deterred from subsequent offending. Consequently, claims about the importance of transferring youths to adult courts as a way of protecting the community, at least in the long run, do not currently rest on a sound scientific basis. Yet, despite the paucity of supporting evidence, further increases in the use of waiver are likely to occur given present "get tough" trends.

KEY CONCEPTS

blended sentencing Allows either the juvenile court or the criminal court to impose a disposition that involves placement in a juvenile correctional institution or an adult institution, or to serve periods in both juvenile and adult programs.

concurrent jurisdiction Both the juvenile and adult court share original jurisdiction. Consequently, in states with concurrent jurisdiction, prosecutors are given discretion to file cases that meet certain age and offense criteria in juvenile or criminal court.

designation An action by a prosecutor or juvenile court that makes available adult court sentencing alternatives even though the juvenile is being tried under the jurisdiction of the juvenile court.

double jeopardy As applicable to transfer cases, a principle that requires the state authority to choose at the outset whether a juvenile is to be tried as a juvenile or as an adult. The choice must be made at the preliminary hearing or arraignment stage of the proceedings.

graduated incarceration An incarceration strategy according to which a convicted juvenile begins his or her sentence in a juvenile facility and then moves to an adult facility upon reaching the appropriate age.

judicial waiver A type of waiver decided by a juvenile court at a hearing. It is the traditional type of waiver.

legislative waiver A type of waiver that results when certain threshold criteria specified by state laws are met.

prosecutorial waiver A type of waiver that is within the power of the local prosecutor or state attorney to choose and implement.

segregated incarceration The practice of providing a separate correctional facility within the adult justice system designated for juvenile offenders.

waiver The process by which a juvenile is moved from juvenile court jurisdiction to adult criminal court jurisdiction for an offense committed while still a juvenile. Also known as bindover, certification, remand, and transfer.

youthful offender status A special legal status existing in some jurisdictions that allows young offenders to receive special consideration as regards sentencing, privacy, and rehabilitation, even as first-time adult offenders.

REVIEW QUESTIONS

1. What is meant by the terms *waiver*, *transfer*, *certification*, *remand*, and *bindover*?

_____

_____

_____

_____

_____

2. What is judicial waiver, and how does it compare with legislative waiver and prosecutorial waiver?

_____

_____

_____

_____

_____

3. What are the advantages and disadvantages of prosecutorial waiver?

_____

_____

_____

_____

_____

4. How has the double jeopardy provision of the U.S. Constitution been applied to juvenile court transfer proceedings?

_____

_____

_____

_____

_____

5. What are the minimally required due process rights of a juvenile at a judicial waiver hearing?

_____

_____

_____

_____

_____

6. What is designation, and how does it differ from waiver?

_____

_____

_____

_____

_____

7. What is youthful offender status, and how does it differ from waiver?

_____

_____

_____

_____

_____

8. What are graduated incarceration and segregated incarceration?

_____

_____

_____

_____

_____

9. What evidence exists to indicate that waiver benefits juveniles or the community?

_____

_____

_____

_____

10. What are the thresholds that juveniles must cross in order to be eligible for waiver?

_____

_____

_____

_____

11. What are the "political" issues that influence legislative policies regarding transfer of juveniles to adult courts?

_____

_____

_____

_____

12. In what ways has the use of transfer changed over time?

_____

_____

_____

_____

13. How does the juvenile justice system benefit from transfer?

_____

_____

_____

_____

_____

14. What does concurrent jurisdiction mean?

_____

_____

_____

_____

_____

ADDITIONAL READINGS

Coordinating Council on Juvenile Justice and Delinquency Prevention. 1996. *Combating Violence and Delinquency: The National Juvenile Justice Action Plan*. Washington, DC: Office of Juvenile Justice and Delinquency Prevention.

Snyder, H. N. and M. Sickmund. 1999. *Juvenile Offenders and Victims: 1999 National Report*. Washington, DC: Office of Juvenile Justice and Delinquency Prevention, pp. 102–109, 170–183.

Torbet, P. et al. 1996. *State Responses to Serious and Violent Juvenile Crime*. Research Report. Washington, DC: Office of Juvenile Justice and Delinquency Prevention.

Wizner, S. 1995. "Discretionary Waiver of Juvenile Court Jurisdiction: An Invitation to Procedural Arbitrariness." In *The New Juvenile Justice*, ed. M. L. Forst. Chicago: Nelson-Hall Publishers.

CASES CITED

Thompson v. Oklahoma, 487 U.S. 815 (1988).
Stanford v. Kentucky, 492 U.S. 361 (1989).
Breed v. Jones, 421 US 519 (1975).

NOTES

1. M. L. Forst, ed., *The New Juvenile Justice* (Chicago: Nelson-Hall, 1995).
2. H. N. Snyder and M. Sickmund, *Juvenile Offenders and Victims: 1999 National Report* (Washington, DC: Office of Juvenile Justice and Delinquency Prevention, 1999).
3. D. M. Bishop et al., "Prosecutorial Waiver: Case Study of a Questionable Reform," *Crime and Delinquency* 35 (1989): 179–201.

4. M. C. Gardner, Jr., "Due Process and Waiver of Juvenile Court Jurisdiction," *Washington and Lee Law Review* 30 (1973): 591–613; J. Gasper and D. Katkin, "A Rationale for the Juvenile Court's Power to Waive Jurisdiction," *Pepperdine Law Review* 7 (1980): 937–951.

5. M. A. Bortner, *Delinquency and Justice: An Age of Crisis* (New York: McGraw-Hill, 1988).

6. V. L. Streib, *The Juvenile Death Penalty Today: Death Sentences and Executions for Juvenile Crimes*, January 1, 1973–June 30, 2003. Retrieved July 20, 2004, from http://www.law.onu.edu/faculty/streib/juvdeath.htm.

7. K. C. Hass, "Too Young to Die? The U.S. Supreme Court and The Juvenile Death Penalty." In *Juvenile Justice: Policies, Programs, and Services*, ed. A. R. Roberts (Chicago: Nelson-Hall, 1998).

8. Legal Information Institute. Supreme Court Collection, *Roper v. Simmons*. Retrieved March 5, 2005, from http://supct.law.cornell.edu/supct/html/03-633.ZS.html.

9. Snyder and Sickmund, *Juvenile Offenders and Victims: 1999 National Report*.

10. H. N. Snyder and M. Sickmund, *Juvenile Offenders and Victims: A National Report* (Washington, DC: Office of Juvenile Justice and Delinquency Prevention, 1995).

11. Snyder and Sickmund, *Juvenile Offenders and Victims: A National Report*.

12. Snyder and Sickmund, *Juvenile Offenders and Victims: 1999 National Report*; Snyder and Sickmund, *Juvenile Offenders and Victims: A National Report*.

13. Snyder and Sickmund, *Juvenile Offenders and Victims: 1999 National Report*.

14. Snyder and Sickmund, *Juvenile Offenders and Victims: 1999 National Report*.

15. Snyder and Sickmund, *Juvenile Offenders and Victims: 1999 National Report*; Mich. Comp. Laws 712A.2(a)(1).

16. Mich. Comp. Laws 769.1(1).

17. Snyder and Sickmund, *Juvenile Offenders and Victims: 1999 National Report*.

18. Snyder and Sickmund, *Juvenile Offenders and Victims: A National Report*.

19. Coordinating Council on Juvenile Justice and Delinquency Prevention, *Combating Violence and Delinquency: The National Juvenile Justice Action Plan* (Washington, DC: Office of Juvenile Justice and Delinquency Prevention, 1996); B. Krisberg and J. F. Austin, *Reinventing Juvenile Justice* (Newbury Park, CA: Sage, 1993).

20. H. N. Snyder, M. Sickmund, and E. Poe-Yamagata, *Juvenile Transfers to Criminal Court in the 1990's: Lessons Learned from Four Studies* (Washington, DC: Office of Juvenile Justice and Delinquency Prevention, 2000).

21. C. M. Puzzanchera, "Delinquency Cases Waived to Criminal Court, 1990–1999." *OJJDP Fact Sheet* (Washington, DC: Office of Juvenile Justice and Delinquency Prevention, 2003).

22. Snyder and Sickmund, *Juvenile Offenders and Victims: 1999 National Report*.

23. B. Krisberg et al., "The Watershed of Juvenile Justice Reform," *Crime and Delinquency* 32 (1986): 5–38.

24. J. A. Butts, "Delinquency Cases Waived to Criminal Court, 1985–1994," *OJJDP Fact Sheet* (Washington, DC: Office of Juvenile Justice and Delinquency Prevention, 1997); Snyder and Sickmund, *Juvenile Offenders and Victims: A National Report*.

25. Puzzanchera, "Delinquency Cases Waived to Criminal Court, 1990–1999."

26. J. Juszkiewicz, *Youth Crime/Adult Time: Is Justice Served?* Retrieved August 31, 2001, from http://www.buildingblocksforyouth.org/ycat/ycat.html.

27. Snyder and Sickmund, *Juvenile Offenders and Victims*: 1999 National Report.

28. J. A. Butts et al., *Juvenile Court Statistics, 1993* (Washington, DC: Office of Juvenile Justice and Delinquency Prevention, 1996); Snyder and Sickmund, *Juvenile Offenders and Victims: 1999 National Report*.

29. *Breed v. Jones*, 421 US 519, 95 SCt 1779, 44 L Ed2d 346 (1975).

30. B. C. Feld, "The Juvenile Court Meets the Principle of the Offense: Legislative Changes in Juvenile Waiver Statutes," *Journal of Criminal Law and Criminology* 78 (1987): 471–533.

31. T. M. Poulos and S. Orchowsky, "Serious Juvenile Offenders: Predicting the Probability of Transfer to Criminal Court," *Crime and Delinquency* 40 (1994): 3–17.

32. J. Fagan et al., "Racial Determinants of the Judicial Transfer Decision," *Crime and Delinquency* 33 (1987): 259–286; E. Nimick et al., *Juvenile Court Waiver: A Study of Cases Transferred to Criminal Court* (Pittsburgh, PA: National Center for Juvenile Justice, 1986); Poulos and Orchowsky, "Serious Juvenile Offenders."

33. L. Lee, "Factors Determining Waiver in a Juvenile Court," *Journal of Criminal Justice* 22 (1994): 329–339.

34. M. A. Bortner, "Traditional Rhetoric, Organizational Realities: Remand of Juveniles to Adult Court," *Crime and Delinquency* 32 (1986): 53–73.

35. Juszkiewicz, *Youth Crime/Adult Time: Is Justice Served?*

36. Bishop et al., "Prosecutorial Waiver."

37. J. Brown and P. Langan, *State Court Sentencing of Convicted Felons*, 1994 (Washington, DC: Bureau of Justice Statistics, 1998); Snyder and Sickmund, *Juvenile Offenders and Victims: 1999 National Report.*

38. Bishop et al., "Prosecutorial Waiver."

39. Bishop et al., "Prosecutorial Waiver," 193.

40. Snyder and Sickmund, *Juvenile Offenders and Victims: 1999 National Report.*

41. Juszkiewicz, *Youth Crime/Adult Time: Is Justice Served?*

42. Snyder and Sickmund, *Juvenile Offenders and Victims: A National Report.*

43. D. Hamparian et al., "Youth in Adult Courts: Between Two Worlds," *Major Issues in Juvenile Justice Information and Training* (Washington, DC: Office of Juvenile Justice and Delinquency Prevention, 1982).

44. Bortner, "Traditional Rhetoric, Organizational Realities."

45. C. Rudman et al., "Violent Youth in Adult Court: Process and Punishment," *Crime and Delinquency* 32 (1986): 75–96.

46. J. Fagan, *The Comparative Impacts of Juvenile and Criminal Court Sanctions on Adolescent Felony Offenders* (Washington, DC: National Institute of Justice, 1991).

47. Brown and Langan, *State Court Sentencing of Convicted Felons*, 1994; Snyder and Sickmund, *Juvenile Offenders and Victims: 1999 National Report.*

48. For an examination of this issue see A. Dobbs, "The Use of Youth as an Aggravating Factor in Death Penalty Cases Involving Minors," *Juvenile Justice Update* 10 (2004): 1–2, 13–16.

49. J. Fagan, *The Comparative Impacts of Juvenile and Criminal Court Sanctions on Adolescent Felony Offenders.*

50. P. A. Greenwood, *Youth Crime and Juvenile Justice in California: A Report to the Legislature* (Santa Monica, CA: Rand Corporation, 1983).

51. Rudman et al., "Violent Youth in Adult Court."

52. Bortner, "Traditional Rhetoric, Organizational Realities."

53. See also L. K. Gillespie and M. D. Norman, "Does Certification Mean Prison: Some Preliminary Findings from Utah," *Juvenile and Family Court Journal* 35 (1984): 23–35.

54. E.L. Jensen and L.K. Metsger, "A Test of the Deterrent Effect of Legislative Waiver on Violent Juvenile Crime," *Crime and Delinquency* 40 (1994): 96–104; S. Singer and D. McDowall, "Criminalizing Delinquency: The Deterrent Effects of the New York Juvenile Offender Law," *Law and Society Review* 22 (1988): 521–535.

55. Fagan, *The Comparative Impacts of Juvenile and Criminal Court Sanctions on Adolescent Felony Offenders.*

56. D. M. Bishop et al., "The Transfer of Juveniles to Criminal Court: Does It Make a Difference?," *Crime and Delinquency* 42 (1996): 171–191.

57. L. Winner et al., "The Transfer of Juveniles to Criminal Court: Reexamining Recidivism Over the Long-Term," *Crime and Delinquency* 43 (1997): 548–563.

58. Snyder and Sickmund, *Juvenile Offenders and Victims: 1999 National Report.*

59. P. Griffin et al., *Trying Juveniles as Adults in Criminal Court: An Analysis of State Transfer Provisions* (Washington, DC: Office of Juvenile Justice and Delinquency Prevention, 1998).

60. Griffin et al., *Trying Juveniles as Adults in Criminal Court.*

61. M. R. Podkopacz and B. C. Feld, "The Back-Door to Prison: Waiver Reform, 'Blended Sentencing,' and the Law of Unintended Consequences," *The Journal of Criminal Law and Criminology* 91 (2001): 997–1071.

62. D. P. Mears and S. H. Field, "Theorizing Sanctioning in a Criminalized Juvenile Court," *Criminology* 38 (2000): 983–1019.

63. Bishop et al., "Prosecutorial Waiver."

64. Coordinating Council on Juvenile Justice and Delinquency Prevention, *Combating Violence and Delinquency.*

65. Bishop et al., "Prosecutorial Waiver"; Bortner, "Traditional Rhetoric, Organizational Realities"; Fagan et al., "Racial Determinants of the Judicial Transfer Decision"; Lee, "Factors Determining Waiver in a Juvenile Court"; Nimick et al., *Juvenile Court Waiver*; Poulos and Orchowsky, "Serious Juvenile Offenders."

66. Jensen and Metsger, "A Test of the Deterrent Effect of Legislative Waiver on Violent Juvenile Crime"; Singer and McDowall, "Criminalizing Delinquency."

Chapter Resources

Chapter Resources

67. P. Torbet, *State Responses to Serious and Violent Juvenile Crime* (Washington, DC: Office of Juvenile Justice and Delinquency Prevention, 1996).

68. Lis, Inc., "Offenders Under Age 18 in State Adult Correctional Systems: A National Picture," in *Special Issues in Corrections* (Longmont, CO: Department of Justice, NIC Information Center, 1995).

69. P. M. Harrison and J. C. Karberg, *Prison and Jail Inmates at Midyear 2003*, Bureau of Justice Statistics Bulletin (Washington, DC: U.S. Department of Justice, 2004).

70. Torbet, *State Responses to Serious and Violent Juvenile Crime*.

71. Torbet, *State Responses to Serious and Violent Juvenile Crime*

72. Torbet, *State Responses to Serious and Violent Juvenile Crime*.

73. Mich. Comp. Laws 762.11; Mich. Stat. Ann. 28.853(11).

74. See E. S. Scott and T. Grisso, "The Evolution of Adolescence: A Developmental Perspective on Juvenile Justice Reform," *Journal of Criminal Law and Criminology* 88 (1998): 137–189.

The Contemporary Juvenile Court

10

Chapter Objectives

After studying this chapter, you should be able to

- Delineate the ways in which the juvenile court is the center and focus of the juvenile justice process
- Describe how the juvenile court, through its leadership and employees, acts as an advocate for children and families
- Describe the structure of the modern juvenile court and where the key personnel fit in that structure
- Describe the primary functions of key juvenile court personnel
- Describe the purpose and importance of the adjudication and disposition hearings conducted in the contemporary juvenile court
- Describe the due process protections that juveniles presently have when they appear in juvenile court
- Explain the role of attorneys in the adjudicative and dispositional hearings
- Describe how the juvenile court's place in the local court hierarchy can affect the resources made available to the juvenile court
- Describe the roles that legal and extra legal factors play in juvenile court dispositions

Chapter Outline

- Introduction
- The Operation of the Juvenile Court
- The Structure of the Juvenile Court
- Juvenile Court Personnel: The Key Players
- The Adjudication and Disposition Hearings

- Chapter Summary
- Key Concepts
- Review Questions
- Additional Readings
- Notes

juvenile court The court responsible for holding hearings and making decisions regarding the disposition of juveniles who have entered the juvenile justice process. This court plays many roles, including children's advocate, program leader, fundraiser, consensus builder, parent, and protector of youths and the community.

■ Introduction

The <u>juvenile court</u> is the heart of the juvenile justice process. It is in the juvenile court that previous decisions made by law enforcement agencies, probation officers, child welfare workers, school personnel, and prosecuting attorneys are supported or altered. It is there that additional decisions—ones that can have a lasting effect on children, their families, and the community at large—are made. Indeed, the juvenile court is the most powerful institution within juvenile justice. Not only does the juvenile court determine the outcomes of individual cases (outcomes that can result in the removal of children from their homes and, in some cases, the termination of parental rights), but also through its legal authority it can determine how other institutions and agencies respond to children. For example, by interpreting laws, issuing various orders, and developing policies, juvenile courts can determine which categories of youths are subjected to and which are diverted from formal court processing.

In addition, the juvenile court typically plays a central role in the child welfare system. Because the juvenile court is often seen as the institution having the greatest responsibility for responding to delinquent youths, it is often looked to for leadership in efforts to understand juvenile crime and develop more effective responses to delinquency. In many instances, the burden of acting as an advocate within the political arena on behalf of families and children falls on the shoulders of the juvenile court. Children have little political power, especially children who are poor, and frequently judges and other court personnel testify in front of legislative committees on issues affecting children and families. Local and state bar organizations often have family law and juvenile law committees made up of judges or referees, who are in positions to influence juvenile law practice. Furthermore, many charitable organizations, such as the United Way, seek input from juvenile court personnel about children and family concerns.

Many juvenile courts also find themselves in the role of consensus builder, fostering agreement among the various community agencies, such as social service and mental health agencies, that work with children and families. Much of the difficulty in dealing with child and family problems involves the scarcity of funding for meeting the needs of children and families. In some communities juvenile courts are more consistently and better funded and staffed than local social service agencies. As a result, these agencies regularly look to the courts for assistance in their efforts to serve clients.

Juvenile Courts Perform Multiple Roles

Today, juvenile courts and their personnel perform multiple advocacy roles for children and families. Courts and their personnel act as legal experts, political advocates, community consensus builders, and consultants on children's issues.

Like other social institutions, the juvenile court is made up of many individuals who perform a variety of functions. Ultimately, the individuals who occupy various roles within the juvenile court strive (at least theoretically) to protect public safety, serve the **best interests** of children and families, and ensure the smooth and efficient operation of the juvenile justice process. Yet, like other institutions, it is sometimes unable to meet its goals, and occasionally some of its goals are displaced or overridden by other goals. This chapter examines the structure, organization, and operation of the contemporary juvenile court. It also examines critical decision-making events in the juvenile court, the legal context within which juvenile court decisions are made, and the important players who influence juvenile court practice.

best interests of the child A catch phrase that serves as a reminder that the primary focus of a juvenile court should be on the rehabilitation of the children who come before it.

■ The Operation of the Juvenile Court

Case Trends and Types of Cases Processed in Juvenile Court

Most delinquency cases are referred to courts by law enforcement agencies, and these agencies have been sending increasing numbers of cases to juvenile courts. For example, in 1999 juvenile courts processed almost 1.7 million delinquency cases. This represents a 27% increase in the number of delinquency cases processed since 1990. Moreover, the number of drug law violation cases increased 169%, public order offense cases increased 74%, and person offense cases increased 55%, although the number of property offense cases declined 9% (see **Figure 10-1**).[1] As the above data indicate, the increase in the number of cases being formally processed by juvenile courts appears to be partially due to the increasing numbers of youths who are being referred to courts for drug offenses, public order offenses, and person offenses. On a more positive note, however, the number of juvenile court cases involving Index Violent Offenses and Index Property Offenses declined 37% and 27%, respectively, between 1995 and 1999.[2] This trend mirrors the trends in juvenile arrests for these offenses noted in Chapter 2.

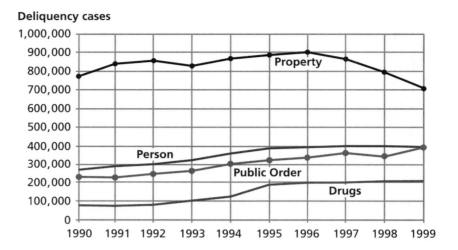

Figure 10-1 Caseloads increased between 1990 and 1999 for person, property, drug law violations, and public order offenses.

Adapted from C. Puzzanchera et al. *Juvenile Court Statistics, 1999* (Pittsburgh, PA: National Center for Juvenile Justice, 2003).

The increase in the number of person, drug offense, and public order cases being handled by juvenile courts is an important trend. Although juvenile courts are, in general, more likely to formally process serious offenses than nonserious offenses, the increase in court processing of drug and public order offenses suggests a willingness by courts to formally process certain types of nonviolent cases. In fact, in 1999, the percentage of drug offense cases handled formally by juvenile courts (61%) slightly exceeded the percentage of person offense cases (60%).[3]

Another fact worthy of note is that case referrals to court increase with the age of the juvenile. For example, the *case referral rate* (the number of cases referred to court for every 1,000 youths 10 years or older) was 5.3 per 1,000 youths for 10-year-olds compared to 113.7 for 17-year-olds. Moreover, as National Center for Juvenile Justice data indicate, a larger proportion of youths over 15 years of age are referred for drug offenses[4]—offenses that have been receiving additional scrutiny from courts and juvenile justice policy makers in recent years. However, juvenile courts are more likely to handle cases involving youths who are 16 as opposed to 17 years old because in some states 17-year-olds are considered adults.[5]

Although juvenile courts are handling increasing numbers of cases, including cases involving crimes against persons, drug offenses, and public order offenses, in other respects the cases currently being processed by juvenile courts are similar to those handled by juvenile courts in the past. In 1999, property offenses accounted for 42.2% of the delinquency cases handled by juvenile courts, person offense cases accounted for 23.2%, public order offenses accounted for 23.2%, and drug offenses accounted for 11.4%.[6] Moreover, juvenile courts continue to handle a variety of status offense cases such as truancy, running away from home, and incorrigibility.

The Philosophy behind Case Processing in the Juvenile Court

The early juvenile courts were characterized by informality,[7] and they paid little attention to due process protections—protections recognized as the cornerstone of adult criminal court operation because they help prevent governmental abuse of powers. In the juvenile courts, however, an emphasis on due process protections was thought to be unnecessary and possibly counterproductive. It was thought to be unnecessary because the courts had been created to serve the best interests of children. It was thought to be potentially counterproductive because it could hinder the efforts of the courts to respond quickly to the needs of children.

Although contemporary juvenile court operation is still marked by considerable informality in many jurisdictions, U.S. Supreme Court rulings such as *Kent v. United States*, *In re Gault*, *In re Winship*, and *Breed v. Jones*, as well as state court rulings, have led to more formal procedures in juvenile courts. For example, in many jurisdictions, it is now common practice for juveniles to be represented by counsel in juvenile court hearings, for records to be made of juvenile proceedings, for courts to carefully detail the rights of youths and parents when they appear before the court, and for courts to follow the same procedural rules used in adult criminal courts. Nevertheless, it is also true that the extent to which courts employ more formalized procedures and protect due process rights varies considerably across jurisdictions. Indeed, in some states many youths still appear before juvenile courts without representation by counsel.[8] Furthermore, the quality of representation that many youths receive is inadequate.[9]

Attorneys Do Not Always Provide Adequate Representation to Juvenile Clients

Unfortunately, youths and their parents often get what they pay for. Most attorneys that appear in juvenile courts are not retained by the parents of the juvenile to represent their child or themselves. Most are court appointed, and the pay is minimal. New attorneys, recently admitted to the bar, typically try to get on court-appointed lists as a way to get court experience while gaining some income. This may explain the inadequate representation that juveniles sometimes receive in juvenile courts.[10] Of course, effective and competent representation is found in the juvenile courts. Many veteran attorneys practice in juvenile courts as appointed counsel, not so much for the money but because they want to have a positive impact on young lives. Many attorneys take the position that if they can help juvenile offenders straighten out their lives, they may not have to provide legal representation to these same individuals once they become adults.

These realities highlight the sharp differences in juvenile court operation found in the United States. These differences include not just variations in the quality of representation, but also differences in the structure of the courts, as explained in the following section.

■ The Structure of the Juvenile Court

family court A unified trial court where all cases involving families are heard, including divorce, adoption, custody, guardianship, paternity, neglect and abuse, and delinquency cases. Family court advocates claim that when all family matters are handled by one court, there will be less chance for overlapping services, redundant reports, and fragmented intervention.

Each state, as well as the District of Columbia, has at least one court with authority over minors who engage in illegal behavior. Generally, courts that handle delinquency cases are either part of the highest court of general trial jurisdiction or part of a lower trial court where less serious criminal cases or limited-claim civil matters are heard. In some states, however, delinquency cases are heard in separately organized **family courts** by judges who specialize in juvenile and domestic relations matters.

Although a court that hears delinquency cases is generally referred to as "the juvenile court," in most states this term is not used as the official name of the court. In fact, the structure of courts with juvenile jurisdiction varies considerably from state to state, and can even vary within a state. A court with jurisdiction over juveniles may be part of a district, superior, circuit, county, or family court. It typically has a separate division that handles juvenile cases involving criminal or status offense behaviors, and in addition it may handle abuse and neglect cases as well as adoption, termination of parental rights, and emancipation of minors.[11] For example, the juvenile court in North Carolina is part of the district court, a lower general trial court within the state. One or more district court judges, depending on the size of the jurisdiction, volunteer to specialize in juvenile cases, or if there are no volunteers, one or more judges may be appointed by the chief district court judge to hear juvenile cases. In California and Alaska, the juvenile court is a part of the superior court, which is also a court of general trial jurisdiction, whereas in Rhode Island, the juvenile court is part of a separately organized statewide family court system. In Michigan, the juvenile court was a division of the probate court until 1998, at which time it became part of the family division of circuit court, the general trial court in that state. Colorado has a hybrid model: Denver has a separately organized municipal juvenile court, but in the remainder of the state the juvenile court is part of the district court, which is a court of general trial jurisdiction.

As noted earlier, in some states juvenile courts are part of lower trial courts. National standards recommend that states establish family courts as a division of the general jurisdiction trial court. This would allow these courts to hear a wide range of juvenile and family-related issues and theoretically would allow for better coordination of cases when families are involved with multiple courts. However, there has been only limited movement in this direction to date. Some states (e.g., Rhode Island, New York, Delaware, and South Carolina) have placed jurisdiction for delinquency matters in a family court.[12] One state that has recently taken some steps in this direction is Massachusetts. In Massachusetts, there is a statewide juvenile court under the Administrative Office of the Trial Court that has 9 divisions around the state (eventually there will be 11) and has its own chief justice and 40 associate justices. The Massachusetts juvenile court has general jurisdiction over delinquency, children in need of services (CHINS), care and protection **petitions**, adult contributing to the delinquency of a minor cases, adoption, guardianship, termination of parental rights proceedings, and youthful offender cases.[13] Also, a number of other states (e.g., Alaska, California, Colorado, Florida, Illinois, and Wisconsin) have moved jurisdiction over delinquency cases to the highest court of general trial jurisdiction. Altogether, a variety of juvenile court models exist—indeed sometimes more than one type of court structure can be found within the same state (e.g., Colorado and Massachusetts).

According to H. Ted Rubin, who has studied courts extensively, "The structure of any court is significant because it affects the status of the court, in part the quality of the judges of the court, and frequently the budget and the adequacy of the staff of the court."[14] Within each state, the *state supreme court* is the court of highest status. At the trial level, however, the court of highest status is the **general trial court**, whatever it happens to be called (e.g., circuit court, superior court, district court, court of common pleas). The general trial court hears felony cases and civil claims that have no maximum dollar limit.[15]

When a juvenile court is part of the general trial court, it occupies a prestigious position within the state court structure. As a result, it is in a better position to attract a larger share of state resources than lower courts, which command less prestige. It also has higher paid personnel, including judges and others who staff the court; better facilities; and more support services than it would if it were part of a lower trial court. In some cases, the additional resources allow more effective and efficient court administration.[16] Conversely, when a juvenile court is part of a lower trial court of limited jurisdiction, it may have difficulty attracting sufficient resources to support its mission.

When the early juvenile codes were first enacted, few felt there was a need for a full-time court to handle delinquency matters. Consequently, juvenile judges only spent part of their time hearing juvenile cases. These judges were generalist judges who heard all types of legal matters. Indeed, in many instances judges who had the least seniority or had the lowest status among the sitting judges were assigned to hear juvenile cases, regardless of their level of interest in juvenile law. The chance of a juvenile case being heard by a judge with little interest in juvenile law was especially high in small one-judge jurisdictions.[17]

Despite the trend toward placing juvenile courts within more prestigious general trial courts, some juvenile courts are still part of lower trial courts, perhaps because of a belief held by some that lower trial courts may be able to relate more effectively to parents and youths within the community.[18] The more likely reason, however, is that state legislatures, as well as the legal profession itself, have traditionally seen juvenile courts as less important than the courts that deal with adult crime. In addition, as mentioned earlier, the juvenile courts' clients have never been politically powerful, and the courts' prestige may have suffered as a result. Furthermore, the early juvenile courts were viewed as playing a parental role and had the responsibility not only of dealing with juvenile crime, but also of protecting children. Thus, many attorneys practicing in juvenile courts viewed themselves more as **guardians ad litem** (persons appointed by courts to represent children and serve their best interests) than as ardent defenders of their clients' legal rights.

petition A pleading to initiate a matter in juvenile court. A petition sets forth the alleged grounds for the court to take jurisdiction of a case and requests court intervention.

general trial court A court of high status within a state's legal system that hears felony criminal cases and civil cases with unlimited dollar amounts.

guardian ad litem A person, usually an attorney, who is appointed by the court to represent the best interests of a child involved in legal proceedings and with social service agencies.

As a result of the *Kent* and *Gault* decisions in the 1960s and the introduction of due process into the juvenile courts, attorneys, judges, litigants, and legislators are likely to view these courts as "real" or "normal." Even so, more intense scrutiny of the juvenile courts by the U.S. Supreme Court and the public has its drawbacks. For example, many legislators and the federal government are questioning the effectiveness of juvenile courts in dealing with serious juvenile offenders. In response, many juvenile courts are attempting to implement a more **balanced approach** to juvenile crime, one that protects the rights of juveniles and families but lets the courts address the needs of other community groups such as victims and those who feel that juveniles should be held accountable for their actions. This balance will be discussed later in the section on adjudication hearings.

Regardless of whether a juvenile court is part of an upper or lower trial court, generalist judges can still be found on the bench. This raises the question of whether juvenile matters are being given adequate attention by these judges. Indeed, if a judge is required to hear juvenile matters but is not particularly interested in the types of cases that come before the juvenile court, these cases will likely get less attention than they deserve. Even if a generalist judge has an interest in juvenile matters, he or she may have difficulty developing expertise in juvenile law, not to mention adolescent and family issues, because of the variety of cases the judge must hear.[19]

Concerned that generalist judges are devoting insufficient attention to juvenile matters, leading juvenile court advocates have called for a separate juvenile or family court with specialist judges. These advocates maintain that specialization is required if judges are to understand the complex legal, family, and adolescent issues involved in juvenile cases. They also argue that specialist juvenile court or family court judges, unlike generalists, will be able to more effectively oversee juvenile cases that sometimes take considerable time and resources.[20]

An innovation beginning to receive attention is the idea of "one judge, one family." The thinking behind this idea is that keeping one judge with a family allows that judge to get to know the entire dynamics of that family and to thus be able to take a "holistic" approach in handling the case. Those who support the "one judge, one family" concept argue that, even though a judge may not be a family court expert, he or she could become an expert on a particular family and its problems. Of course, each judge would be responsible for multiple families, but even then the support staff, referees, and caseworkers would have the ability to know the families individually. Instead of trying cases, judges would preside over families. Although this is an intriguing concept, there are some clear drawbacks. One drawback is the possibility that a judge could lose his or her objectivity and become unable to treat the family fairly.

Another trend has been toward the establishment of a **unified trial court**. In this approach, a single trial court capable of dealing with all matters requiring legal intervention would be established in each community. Within this single trial court would be more specialized courts dealing with family matters, delinquency, traffic, civil litigation, and adult criminal behavior. Advocates of the unified trial court maintain that tying separately organized courts together would make court administration more efficient and effective, would eliminate overlapping jurisdictions, and would allow for better utilization of court personnel and

balanced approach to juvenile justice
The belief that juvenile justice should give equal attention to public safety, offender accountability, the needs of victims, and the correction and treatment of juvenile offenders.

unified trial court Comprehensive courts encompassing several specialist courts, such as courts devoted to delinquency and family matters, as well as other types of courts. Also see family courts.

other resources.[21] Some argue that a unified court encompassing several specialist courts, such as courts devoted to delinquency and family matters, would be better positioned to deal with delinquency cases than a general trial court. Although there is little evidence of a trend toward the development of separately organized juvenile courts, there has been some movement toward more specialized courts dealing with juvenile and family matters within a unified trial court.

Still another type of specialized juvenile court that has been established in some jurisdictions is the **juvenile drug court**. The first juvenile drug court was implemented in 1989 in Miami, Florida. Since that time juvenile drug courts have been started in a number of states. Drug courts are special courts that handle cases "involving substance-abusing offenders through comprehensive supervision, drug testing, treatment services and immediate sanctions and incentives."[22] What is important to note about drug courts is that not only do they hear cases and deal with legal issues around substance possession and use, but they also play a critical role in the coordination of treatment for youths with substance abuse problems. Indeed, these courts follow the therapeutic jurisprudence model that is designed to: (1) provide immediate intervention, (2) use nonadversarial adjudication, (3) employ active judicial intervention, (4) utilize treatment programs, and (5) have clear rules and goals.[23] Although these courts hold youths accountable for behaviors, the judge, attorneys, probation staff, and treatment providers also work as a team with the youth and family to encourage behavioral changes.

juvenile drug court Specialized juvenile court that handles cases involving substance-abusing offenders through comprehensive supervision, drug testing, treatment services, and immediate sanctions and incentives.

■ Juvenile Court Personnel: The Key Players

The juvenile court, of course, is more than just a structure. It is made up of a number of individuals who work within the organizational framework of the court. These individuals perform a variety of roles—roles that regularly require them to make important decisions about the lives of youths and families. Collectively, these roles and the decisions made determine the quality and quantity of justice dispensed by the court. The following sections describe some of the key decision makers who make up the juvenile court.

The Juvenile Court Judge

The most important decision maker in the juvenile court is the judge. The judge has ultimate responsibility for the operation of the court and for the legal direction the court takes. The judge exerts influence through (1) **administration** of the court and (2) **judicial leadership**.

As the ultimate leader of the court organization, the juvenile or family court judge may be responsible for hiring and firing of court personnel, court policies, work rules, and the level and priority of court staffing. Usually, a court administrator is hired to handle the day-to-day operation of the court, hire and fire staff, do the budget work, and perform other administrative and policy tasks. The amount of authority delegated to the court administrator and other middle management personnel in the court, however, often depends on the philosophy of the judge. If the judge takes a "hands-on" approach, he or she will take an active interest in the day-to-day operation of the court. The judge may want

judicial administration The daily and long-range management of a court. The judge, as head of the court, is ultimately responsible for judicial administration but can choose to delegate many of the particular duties to others, such as a court administrator.

judicial leadership Guidance provided by a judge in the areas of programming, personnel, and budget administration. Also seen in his or her role as children's advocate and consensus builder within the community.

to review all personnel, policy, and budget decisions before they are implemented. Furthermore, court staffing may well reflect the judge's view as to what the court's emphasis should be. For example, if the judge believes strongly in probation and the development and operation of programs for youths, then extra resources might go to hiring probation officers and developing and staffing various treatment programs, assuming that the court budget makes these activities feasible. In another court, the emphasis might be on community protection, and the judge might want money spent on a detention center and out-of-home placements. The main disadvantage of a hands-on approach is that the judge's involvement in administrative matters takes time away from his or her courtroom duties. It can also lead to frustration on the part of the court administrator and other staff, who may resent a judge's constant incursions into their professional domains.

If the judge views his or her role as setting broad administrative guidelines and policies and letting the hired managers run the court, a different atmosphere is present. A hands-off approach allows court managers to do what they do best—run the court. The advantages are obvious. People in the court are able to perform the roles their professional training prepared them for. The disadvantages include the possibility that the judge will become isolated from the daily life of the court and lose touch with the employees. Another concern is that administrators, because they are hired and not elected, may not be attuned to community priorities and concerns to the same extent as judges. Thus, the ideal is for the judge to set broad guidelines for the administrator and the other managers but to continue to actively review the progress and outcomes of court programs.

In the judicial arena, the judge is paramount. He or she is the role model for all of the quasi-judicial officers such as referees, masters, and magistrates employed in the court. The judge decides how the courtroom will be run, what cases have priority, what outcomes are preferred, and the extent of due process for juveniles and families. For example, if the judge decides that status offenders are not going to be dealt with by the court, by that decision the judge sets a priority for the court that affects not only the types of cases heard, but also the types of staff the court will need.

Judicial leadership is not simply relegated to the court, but is found in the community as well. Judicial leadership carries into the community, as judges commonly serve on numerous community boards and committees. How the citizens of the community perceive the judge and react to that perception will have much to do with their support for the court as a whole and for individual court programs. The judge must maintain favorable relations with business and community leaders and elected officials. Many times the court's budget is controlled by the county board or council, and how they feel about the judge may be translated into dollars and cents. Importantly, most juvenile court judges are elected officials. Once elected, they are always "running" for re-election when in the public eye, and their ability to deal with the political pressure they experience is crucial to their success and the success of the court. For instance, they are often under considerable public pressure to take a tough stand on juvenile crime, which can sometimes conflict with their desire to help youths.

Box 10-1

Interview: The Honorable Gerald W. Hardcastle, 8th District Judicial Circuit Court, District Judge Assigned to Family Division

Judge Hardcastle spent 18 years practicing family law; is a former prosecuting attorney in Clark County, Nevada; and has served on the family court bench for 12 years in a combined family court system that hears all neglect, delinquency, adoption, divorce, custody, and guardianship cases.

Q: Judge, how did you come to serve on the family court bench?

A: The six elected Family Division district court judges chose among themselves who will serve in each area of the family court's responsibility. My assignment for 8 years has been to juvenile court.

Q: What types of cases do you hear?

A: All child abuse and neglect cases. In Nevada, we have referees, sometimes called masters, who help judges with the caseload.

Q: How heavy is your caseload?

A: Last year, myself and three referees heard 64,000 cases! Las Vegas and Clark County contain 65% of the population of the state of Nevada, and our caseloads have grown tremendously in the recent past.

Q: What is your educational background?

A: I graduated from Weber State University as an undergrad and attended law school at the University of Utah.

Q: Do you support the family court concept as the best way to make the system better for families and children?

A: Yes and no. I certainly believe that matters involving families need to be dealt with in one place and by one court for the sake of convenience for the public and efficiency, but I believe that delinquency cases and child abuse and neglect cases need to be separated from the other family court areas. Special judicial expertise is required in handling these cases, and our present system of 3-year rotation does not allow any one judge an opportunity to acquire and cultivate that expertise.

Q: Could you expand on your concerns about the need for specific judicial expertise in delinquency and neglect cases?

A: Certainly. Juvenile court judges have a number of roles to fulfill. They are looked at by the community to be leaders in advocating for the needs of children. It takes some time to understand these needs and the political and economic factors influencing them. Any court needs consistent policies, and if these change every 3 years with a new judge—which they tend to do—the court processes are continually disrupted. For example, if the outgoing judge advocates for foster placement of delinquents and the new judge advocates for institutionalization, the whole course of services and funding is impacted.

Q: In Clark County does the juvenile court have responsibility for delivering services to children and families?

A: No, and this is frustrating to a certain extent. Judges, as elected officials, have priorities and are expected by the public to possibly have more control over children than we do. I'm lucky, however, to have an excellent family services director in Kirby Burgess, and he and I work closely in planning for services.

Q: One of the "burning" issues in juvenile justice today is the issue of waiver or transfer of juveniles to adult courts. What are your opinions on this issue?

A: I am glad you asked. One of the most important challenges facing juvenile courts today is the need to effectively deal with the chronic serious delinquent offender. Juvenile courts need to accept this challenge and find ways to reach these kids and protect the community. Unfortunately, legislatures across the country are taking any opportunity for the juvenile courts to meet this challenge away from them with automatic transfer legislation. This is done even though there is ample evidence that the adult criminal justice system has no better solution to the problem. It doesn't work with adults, why would anyone think it will work with juveniles?

Q: Judge, could you share your thoughts on how you would change the system?

A: Of course. I believe the most effective intervention is on the local or neighborhood level. The juvenile court should use neighborhood centers for intake and mediation resolution. Only if this did not work would the formal court process be used. Along with this idea, the only way to treat families is over the long term, with a consistent continuum of services. In this area, the court needs to work closely with the Department of Family and Youth Services.

Q: What have been your most gratifying experiences while sitting as a juvenile court judge?

A: I believe that any success with families comes about when dedicated workers try any number of things to help kids and families and when the court has the flexibility to be willing to try a number of things. We have good workers in the court and the department who really care about what they do. They have a commitment to children from the director on down. My goal is to help children. My focus is to be sure that they're treated fairly. The judge must be the person who establishes the perception of fairness.

In large urban counties with several juvenile court judges, the judges typically elect one of their number to be chief judge, usually on the basis of seniority,[24] although the chief judge in some jurisdictions is chosen by the state's highest court. In a statewide juvenile court system like that found in Massachusetts, there is a chief justice who is appointed by the governor and acts as the overall administrator for the juvenile court divisions around the state. Regardless of how the chief judge is chosen, and the range of their administrative authority (e.g., county or statewide system), the chief judge has the ultimate responsibility for court administration and judicial leadership, although daily court operations may be delegated to a court administrator. In large courts, the burden of this responsibility can be extremely taxing, and therefore good administration is even more important in these courts, whether administration is directed by the judge, an administrator, or a judge and administrator team.

Most Juvenile Court Judges Are Elected

Juvenile court judges are selected in a variety of ways. In most states, they are elected in partisan or nonpartisan elections. In other states, they are appointed by the governor from a list of candidates chosen by a screening board. In Connecticut, Virginia, and South Carolina, the legislature appoints juvenile court judges. Also, about a dozen states have adopted the Missouri plan. Under this plan, an elected official, usually the governor, appoints a candidate from a list compiled by a commission. Once appointed to the bench, however, incumbent judges must run on their records in nonpartisan and uncontested elections.[25] Although there are many dedicated and competent judges who sit on the juvenile court bench, present methods of selecting juvenile court judges do have some drawbacks. Because successful political campaigns require substantial amounts of money, a heavily bankrolled politician can prevail over a more qualified candidate. Moreover, the appointment of candidates to the bench is often a highly political affair, and work for the party in office may be treated as more important than judicial qualifications.[26]

Juvenile court judges wield tremendous power over juveniles and families who come before the court, and they exercise wide discretion in how they respond to cases. This is most clearly seen when judges remove children from their homes. Indeed, the ability to take juveniles from their homes and detain them or impose some other out-of-home placement represents a conspicuous example of the power vested in judges. Yet juvenile court judges exercise tremendous power in a variety of other ways—they issue orders that require youths to seek treatment, obey their parents, avoid unsavory persons or places, attend school, cooperate with probation officers, adhere to curfews, and engage in other actions judges feel are appropriate. In addition, juvenile court judges may also order parents to engage in (or refrain from) certain actions. For example, a judge may order parents to attend counseling, transport their child to court-ordered counseling, clean up their home, ensure that their child attends school, cooperate with probation officers, and pay some or all of the costs of the services provided by the court. Furthermore, a juvenile court judge may hold the parents or guardian of a child in contempt of court and have them jailed for not complying with court orders.

Clearly, juvenile court judges have considerable power. Not surprisingly, some juvenile court judges abuse their power and act like tyrants or dictators when on the bench. Many others, however, are dedicated and able jurists. These judges are careful to protect the rights of juveniles and families before the court, and they strive to balance the best interests of children and families against the need to protect public safety.

Juvenile Court Referees and Other Quasi-Judicial Hearing Officers

Among the critical players in many contemporary juvenile courts are individuals appointed by judges as **quasi-judicial hearing officers**. These individuals, who are usually attorneys, are referred to by such titles as "referee," "commissioner," "master," "administrative law judge," and "magistrate." No matter what they are called, their primary role is to hear cases. Indeed, in some jurisdictions, the great majority of cases heard in juvenile courts are presided over by quasi-judicial hearing officers who are not judges.[27]

Although a variety of factors often enter into the selection of referees and similar quasi-judicial hearing officers, three stand out: (1) knowledge of and expertise in juvenile law, (2) judicial demeanor and interpersonal skills, and (3) ability to assist the court in handling its caseload. As noted earlier, the vast majority of judges are elected. Unfortunately, however, judicial elections are often treated as less important than those for other offices. Furthermore, voters frequently have a hard time distinguishing one candidate from another because ethical considerations prevent judges from campaigning "against" their opponents. As a result, judges are often elected because of name recognition or political connections rather than expertise in the appropriate area of law.

Referees or similar quasi-judicial hearing officers who are appointed by judges, on the other hand, are frequently chosen for their expertise. In many instances they constitute the true repository of knowledge regarding the applicable case and statutory law in the area of juvenile justice, and in some courts they have far more experience on the juvenile court bench than the sitting judges. In addition to their legal knowledge, their judicial demeanor (i.e., how they conduct themselves on the bench) and their interpersonal skills can be highly valued by juvenile court judges.

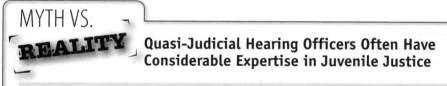

MYTH VS. REALITY

Quasi-Judicial Hearing Officers Often Have Considerable Expertise in Juvenile Justice

Myth—Juvenile court referees (or masters, commissioners, and magistrates, as they are called in some states) lack the expertise of juvenile court judges.
Reality—Quasi-judicial hearing officers may be hired by a juvenile court judge precisely because of their knowledge of juvenile and family law. In addition, some quasi-judicial hearing officers have far more experience on the juvenile court bench than some juvenile court judges.

It has long been recognized in many states that judges alone cannot handle the large volume of cases that are referred to the juvenile court. There are also a number of minor judicial tasks that, from a practical and an economic point of view, could better be performed by someone else. Consequently, as the volume of work in juvenile courts has increased, the use of quasi-judicial hearing officers has expanded to keep pace. In many courts, these people preside over most of the same types of hearings as judges, and their recommendations are treated as having the same force and effect as the judges' orders. Nevertheless, as Rubin notes, most states place some limits on the authority of referees and other quasi-judicial hearing officers, such as preventing them from conducting waiver hearings or presiding over jury trials where they are available.[28]

Quasi-Judicial Hearing Officers Perform Various Roles in Juvenile Justice

Referees and other quasi-judicial hearing officers are often **on call** or on weekend duty because of state laws that require a judicial review to be conducted within 24 to 48 hours following the placement of a juvenile in detention. Their role as on-call judicial officers gives referees substantial authority to determine which juveniles should be kept in detention and for what types of offenses. They also conduct preliminary hearings or arraignments, where they make important decisions about bond and the need for further court action. This screening function is crucial for the efficient processing of cases.

on call Availability for emergency service. Some jurisdictions require there always be a judicial officer available to make emergency decisions about the placement of alleged juvenile offenders.

Administratively, referees and other quasi-judicial hearing officers are often looked to as the people to go to for legal advice in the court. Increasingly, they are required to be attorneys, and in many courts they are more accessible to the line staff than are the judges. In some courts, quasi-judicial hearing officers act as legal advisors to the court administrators and thus have a significant influence on court policy. They are also popular in many jurisdictions because they are uniformly less expensive than judges. Their salaries are usually significantly less, and they may require less support staff. For example, a judge may need a personal secretary and a court recorder. Referees frequently end up doing their own recording and using the court's clerical pool for the typing of orders, reports, and other legal documents. As caseloads and docket pressures have increased in many jurisdictions, the creation of quasi-judicial hearing officer positions has become a popular way of dealing with these pressures, and they are seen as economically efficient alternatives to the creation of new judgeships.

The use of referees and other quasi-judicial hearing officers in juvenile courts does have some potential problems. In jurisdictions where quasi-judicial hearing officers hold a majority of the hearings, they are arguably "judicial substitutes rather than judicial supplements,"[29] despite claims that they are intended to assist judges, not replace them.[30] Moreover, heavy reliance on referees may send a message that juvenile court matters are not sufficiently important to merit more judges, thus diminishing the stature of the juvenile court. Still another problem is that judges, in situations where they are required to review the findings and orders of referees, may do little more than rubber-stamp them. Finally, because referees are hired by judges and serve at their pleasure, a newly hired referee would not have the same degree of power and independence that a judge in a newly created judgeship would have.[31]

Box 10-2

Interview: R. Scott Ryder, Juvenile Division Director/Referee, 45th Circuit Court, Family Division, St. Joseph County, Michigan

Age: 55
Education: BA History, Wittenberg University; JD, Indiana University
Salary: $60,500.00

Q: How long have you been the Juvenile Division Director in St. Joseph County?

A: Effective May 21, 2004, I retired from my previous job as a family court referee with the 9th Circuit Court Family Division, a position that I had for more than 25 years. I then began my employment as the court director/referee for the Juvenile Court in St. Joseph County beginning May 24, 2004. It was a brief retirement!

Q: What are your new responsibilities and how do you like them?

A: Although I still do the occasional hearing—I have presided over more than 20,000 hearings in my career—primarily I work on financial issues, including preparing and overseeing the budget; personnel issues, hiring, firing, and contract negotiating; court scheduling and docket control; legal research and policy preparation; acting as a liaison between the court and other public and private agencies dealing with juveniles; and finally just managing the day-to-day activities of the court. I am enjoying the challenge and the opportunity to see the operation of the court from a different perspective.

Q: You worked for over 25 years as chief referee in a juvenile court and you still hear cases. What do you like about the referee position?

A: I like the challenge of making decisions that can have a positive impact on children and families. I also like the challenge of articulating community standards and the challenges that arise in the course of hearings, and I enjoy very much the people I work with.

Q: What were the most difficult challenges you faced as chief referee?

A: Being in a supervisory position is always a challenge, especially when you are responsible for supervising independent judicial officers, because there is not always a simple "right" decision. My approach has always been to ensure that the right procedures and policies are followed. Of course, any time one works in government there are a variety of political issues you must face. Unfortunately, political decisions that influence juvenile justice are often made without input from people who work in the field. Another challenge is working with families. There is a lot of poor parenting, and it can be difficult trying to get parents to be more responsible and positively involved in their children's lives.

Q: What is a typical presenting problem in the cases you see in court?

A: An absence of responsible parenting. Many of the parents I see don't properly supervise their children, impart appropriate values to them, or support their children in positive activities. Too many parents are so caught up in their own lives that they fail to play a positive role in the lives of their children.

Q: What qualities does it take to be a successful hearing officer?

A: You have to know family and child law, how the court works (what it can and cannot do), and how it fits into a continuum of services for youths and families. You have to know the fundamentals of appropriate parenting. You must be able to be confrontational when necessary, but you also need to be patient and understanding. I also think it is important to have a deep compassion for those involved in the court and a genuine desire to help them make their situations better.

Q: What do you feel is unique to the referee position in the court?

A: Referees can bring some unique qualities to the court that, unfortunately, are not always recognized. Because referees are not elected, they can be more independent. Also, referees can bring tremendous experience to the court that can take judges years to develop.

Q: What are the new challenges that you face in your new position?

A: St. Joseph County is a small rural county in southwestern Michigan. The entire court budget is slightly under $1 million including employees' salaries and fringe benefits. The court always has to be diligent in overseeing how the money is being spent because economic resources influence our ability to help children and families. Consequently, the court has to be creative in utilizing community resources and do a better than "good job" at determining which youths threaten community safety and should be removed from the community. It also means the court has to be very proactive, not only in developing local programs, but also in networking with local public and private agencies to coordinate services for youths.

Q: What qualities does it take to be successful in your position?

A: You have to know family and child law, how the court works, legally what it can and cannot do. You must understand the local services culture, which agency does what, who you need to contact so that resources can be directed to those who need them most. Most of all you must understand how to manage, motivate, and lead people. You must understand the strengths and weaknesses of your employees, how to best motivate them, and what incentives or discipline works with each. You must be a good listener, open always to both sides of the story, and, finally, you must be decisive. Nothing paralyzes an organization more than leaders who do not lead. I am a "middle" manager, so I have to know my position in the county hierarchy and understand that my real bosses are the citizens and taxpayers of the county. Political understanding, "moxey," can never be overlooked in importance. Finally, you have to understand the judge you work for, how he or she works, their philosophy, and how they intend to implement it.

The Juvenile Court Administrator

The court administrator is the manager of the court. He or she has primary responsibility for (1) personnel, (2) the budget, and (3) programming. The court administrator hires and fires employees, interviews new employees, oversees employee performance evaluations, negotiates with any collective bargaining units, and acts as a liaison with other government agencies on employment-related matters. For example, many juvenile courts are subject to the financial controls of state or local governments (county or city). Usually these governments have personnel offices and directors who work with local employees in animal control, police, fire, mental health, and parks and recreation departments. In some jurisdictions, court employees are also local government employees and are subject to the same policies as other such employees. These situations require the court administrator to act as a liaison and coordinator between the court and the local government. If the court is a separate government unit, then the court administrator is primarily responsible for developing and implementing personnel policies, conducting and overseeing employee evaluations, and engaging in other employment-related functions. His or her actions, however, may be subject to the final approval of the judge.

In addition to their employment-related responsibilities, juvenile court administrators also have budget responsibilities. Most juvenile courts have large budgets, and it is the administrator's job to create a responsible budget and then make sure the court lives within that budget. Budgetary items needing consideration would typically include (1) capital expenses for buildings and grounds, (2) employee salaries and fringe benefits, (3) placement costs for out-of-home care, (4) juvenile detention center costs, (5) court-appointed attorney fees, (6) witness fees, (7) security costs, (8) equipment costs and amortization, (9) training costs, and (10) mileage and travel costs. It is evident from this list that a great deal of economic tradeoff and balancing must be done to meet the needs of children, families, and the community without forgetting the employees of the court who provide services to the court's clientele.

Finally, the court administrator must exercise leadership in developing and implementing programs to serve children and families and to accomplish the mission of the court. To do this effectively, the administrator must be familiar with the needs and resources of the community, the preferences of the judges, and the economic constraints of the court budget. Consequently, the competent court administrator must be knowledgeable in a variety of areas, including (1) community corrections, (2) diversion alternatives, (3) detention resources and secure placement options, (4) group and foster care programs, (5) substance abuse treatment, (6) mental health options, (7) domestic violence programs and shelters, and (8) institutional placement options. In many jurisdictions, the court actually operates some of these programs and must fit them with other programs to provide a **continuum of care** for children. The creation of a continuum of care requires not only staffing and managing court-operated programs, but also connecting with other state and local resources and agencies that operate programs for children and families. The administrator must be able to develop needed programs, see that they are properly staffed and managed, evaluate their effectiveness, and know when resources outside of the court must be used to aid in the treatment of juveniles and to protect the community.

continuum of care
A range of comprehensive and connected services for families and children.

The Prosecuting Attorney

The prosecuting district, state, or commonwealth attorney is the chief law enforcement officer of the county, district, or local government. All police work goes to the prosecutor for review, and in most cases only the prosecutor can issue charges of delinquency against juveniles (see **Figure 10-2**). Indeed, the prosecutor plays a key role in determining which cases will go to court, what the specific charges will be, which cases will be considered for waiver to adult courts, and what the disposition of each case will be.[32] Given the amount of discretion vested in the office of the prosecutor, the prosecutor can virtually control the juvenile court's delinquency docket. Moreover, by deciding what kinds of cases are to be charged, the prosecutor helps determine community standards of acceptable conduct. The decision, for example, to enforce a community curfew makes a statement to youths and their parents about when minors should be at home in the evening. By enforcing a standard for behavior, the prosecutor articulates the community standard, sets forth consequences for violation of that standard, and influences the behavior of some youths and their parents.

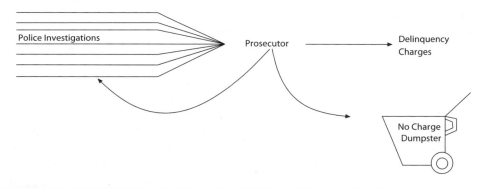

Figure 10.2 The role of the prosecuting attorney.

Prosecutors, in many states, also have legislatively mandated obligations to victims. In a very real sense, when prosecuting a crime, a prosecutor becomes the victim's attorney. Only recently have victims been recognized as more than just necessary witnesses. As a result, the prosecutor may have an obligation to seek restitution on behalf of a victim or the victim's estate and to take special measures to protect the victim both in the court proceedings and out of court.

Prosecutors also play a crucial role in ensuring that the court operates as efficiently as possible. Any court has a finite amount of courtroom time available. An attempt by a prosecutor to charge each juvenile and adjudicate each case would more than use up available courtroom time and would completely exhaust judges, referees, and other court personnel. By careful use of plea negotiating, the prosecutor can save the state's resources for the most important cases and keep the juvenile court docket from getting clogged up.

**Prosecutors Should Balance Juveniles'
Interests with Those of the Community**

According to standards developed by the National District Attorney's Association, prosecutors are obligated to protect the safety and welfare of the community, including crime victims. However, they should also consider the interests and needs of juveniles to the extent possible without compromising their obligations to the community.[33]

The Defense Attorney

It is the defense attorney's job to represent the wishes of his or her client (e.g., an accused juvenile), ensure that the client's rights are not violated, and present the client's case in the most favorable light, regardless of the client's involvement in illegal activity.[34] To provide an effective defense, the juvenile defense attorney must establish a good relationship with the client in order to understand the client's needs and wishes.[35] Theoretically, the attorney must follow the wishes of the client absolutely and vigorously defend the client,[36] but many attorneys believe they also should act as an advisor. As mentioned earlier, a defense attorney typically wants to get each client needed help, but he or she must ensure that the court has legal grounds for exercising authority over the client.[37]

What usually happens is that the defense attorney insists on the juvenile's due process rights at the initial detention hearing and continues to insist on them through the adjudicative or trial stage. Once a conviction by trial or plea has been obtained, however, the defense attorney may become more of a guardian *ad litem*, focusing on the juvenile's best interests. A guardian *ad litem* is a person appointed by the court to represent the best interests of a child involved in legal proceedings and with social service agencies. When the attorney assumes this role, then he or she gives more consideration to the needs of the client as a child than to the expressed wishes of the client as a legal defendant.

Defense attorneys can also influence the court's docket by deciding when they will take a case to trial and when they will attempt to negotiate a plea. For example, in 1997, Michigan enacted a sexual offender registration act that requires all people who are convicted of a sexual offense to be placed on a list available to local law enforcement agencies. Before the registration law, many attorneys would plea bargain sexual offense cases out of concern for their clients' future and possible need for treatment. Now most of these cases are going to trial because of attorneys' desire to keep their clients from suffering the stigmatization that goes along with appearing on the sexual offender list. This is a good example of the unintended consequences associated with new legislation. The legislature wanted communities to know who convicted sexual offenders were, for obvious public safety reasons, but the unintended results of the law appear to include an increase in sexual offense trials and a corresponding increase in court expenses.

To understand the role of defense attorneys in juvenile court, it is necessary to distinguish the ways attorneys become involved in the court. Usually, attorneys who appear in delinquency cases are (1) client retained, (2) court appointed, (3) legal aid or public defender attorneys, or (4) members of a consortium of attorneys.

Client-Retained Attorneys

Not much needs to be said about privately retained attorneys, although often they know little about how the juvenile court operates. In fact, it may be to a client's advantage to have a court-appointed attorney, because an attorney who practices regularly in juvenile court knows the court process and personnel. On the other hand, court-appointed attorneys may have relationships with the court or prosecution that prevent them from acting as zealous advocates for their clients.[38]

Court-Appointed Attorneys

Court-appointed attorneys usually must apply to the chief judge or court administrator to be placed on a list of attorneys available for representing clients in the juvenile court. Minimum levels of training and expertise are often required for qualification, and continuing education is frequently mandated. Court-appointed attorneys are usually called by the court on a rotating basis, and parents of children represented are often required to meet with the court's finance officer to determine the family's ability to reimburse the court for attorney fees.

Legal Aid or Public Defender Attorneys

These attorneys often appear in juvenile courts because of a contract between the court and their agency to provide legal services. The extent of the services provided is determined by the contract.

Attorney Consortiums

These attorneys belong to a group of attorneys who have joined together in order to offer their services to the court for a set amount of money. Although a contract between the court and the consortium dictates the terms of the representation, these private attorneys have a specific interest in juvenile court practice.

Probation and Other Casework Staff

In addition to hearing officers and attorneys, a variety of caseworkers play important roles in the juvenile court process, including court hearings. One of the most important is the probation officer. Probation officers perform six basic roles in juvenile courts: (1) screening cases in intake, (2) conducting presentence investigations, (3) supervising and monitoring youths' adherence to their rules of probation, (4) providing assistance to youths placed on probation, (5) providing ongoing assessments of clients' needs, and (6) completing a variety of job-related administrative tasks. In other words, probation officers are the people who may visit youths' homes, locate kids who are not where they are supposed to be, confront parents when they fail to assist their children, check with school and other agency personnel to monitor youths' behavior and progress, and when necessary, request that a youth's probation be revoked. They may also conduct investigations, work with attorneys in developing dispositional plans, serve summons, and make arrests.

Probation officers or other caseworkers may be employed in specialized programs operated by or on behalf of courts. For example, juvenile courts or other government agencies may operate a variety of programs such as diversion programs and foster care. Furthermore, they may also operate their own shelter care units, group homes, and detention centers. Probation officers or other caseworkers may be used to staff these programs or be given responsibility for providing casework services to youths in these programs.

Probation and other caseworkers who work for or with juvenile courts clearly perform a number of vital tasks. Indeed, those who occupy these positions can influence which youths penetrate further into the juvenile justice process. They can influence youths' involvement in a variety of juvenile justice programs, from diversion to probation to institutional placement. They can also influence how long youths are involved with the juvenile court. In short, probation officers and other caseworkers act as the eyes and ears of the court. These individuals gather and analyze different types of information and make important recommendations to the court that almost always carry weight with judges and other juvenile justice decision makers.

■ The Adjudication and Disposition Hearings

All of the key players in the juvenile court—judges and other quasi-judicial hearing officers, prosecuting and defense attorneys, probation officers, and other caseworkers—make decisions that can have a profound effect on youths and families. Although decisions about youths and families are made at many points in the juvenile justice process, they typically are made either in preparation for critical events or during critical events. Previous chapters examined several of these critical events, including the arrest, the juvenile court intake process, and the detention hearing. This section examines two more critical events in juvenile justice: the adjudication hearing and the disposition hearing.

The Adjudication Hearing

adjudication The juvenile court equivalent of an adult court trial. It is at the adjudication that the juvenile court attempts to ensure due process, find the facts, uphold the law, and make decisions concerning jurisdiction over the juvenile.

The **adjudication** hearing is a critical event in the juvenile justice process that serves as the juvenile court equivalent to a criminal trial.[39] It is the adjudication hearing that determines whether the juvenile comes within the jurisdiction and under the authority of the court. If the juvenile is adjudicated, the court has the power to issue orders affecting the juvenile and his or her parents (or custodian or guardian). Without an adjudication, the court has no legal authority to intervene in the life of the juvenile and his or her family.

The adjudication hearing can take one of two forms. It can be a plea-taking hearing, in which the juvenile admits to a delinquency offense, or it can be a trial. A plea-taking hearing is by far the most common type of adjudication hearing.[40] Ideally, at a plea taking, the purpose and potential outcomes of the plea and the rights that the juvenile and parents have are explained to them. If the plea is the result of a plea negotiation (or plea bargain) between the prosecuting attorney and the juvenile, which is common in juvenile justice, the terms of the negotiation must be placed on the court record, including what the juvenile gets from the plea and what he or she gives up.[41] Usually the juvenile admits to a less serious delinquency charge or a lesser number of charges, the tradeoff being that the juvenile avoids the chance of a more severe sentence and the prosecuting attorney is assured of a conviction.

Plea Bargains Are Common Events in Juvenile Justice

A <u>**plea bargain**</u> is a deal between the defense attorney (who represents the defendant) and the prosecuting attorney (who represents the state), according to which the defendant agrees to plead guilty to a lesser charge or to fewer charges than originally at issue. Like other deals, both sides get something and give up something. What the prosecutor gets is a conviction and the court's jurisdiction and power over the juvenile. What the prosecutor gives up is the opportunity at a trial to convict the juvenile of all of the charges or of the more serious charge. What the juvenile gets is an adjudication on a less serious charge or on fewer charges. What the juvenile gives up is the opportunity to have his or her day in court and be found not guilty.

plea bargaining
Negotiation of an agreement between the prosecuting and defending attorneys, often with the consent of the court, to have the accused plead guilty to a reduced charge or to fewer charges than originally brought. The prosecutor benefits by gaining a certain conviction and the defendant benefits by avoiding the possibility of a more severe sentence.

The juvenile's due process rights at the adjudicative phase are virtually identical to those of adults at trial. Consequently, in order to take a valid plea, a complete advice of rights should be given and an understanding waiver or "giving up" of those rights should be placed on the record. Unlike in an adult proceeding, however, the juvenile's parents should also be consulted about whether they agree with the waiver of rights, because it is a very important event in the life of a child and the court will likely acquire authority over the parents as well once it acquires jurisdiction over their child.

If the juvenile chooses to contest the delinquency charges, the only resolution of the matter is to have a contested adjudication hearing or trial. In many jurisdictions, contested adjudications proceed much like trials in criminal courts, although there can be considerable variation in the court rules and procedures that govern juvenile court adjudications. In some jurisdictions, many, if not most, of the procedural rules used in criminal trials are used in juvenile proceedings. In other states, there are no clear statutory requirements regarding rules of evidence in the areas of "<u>**opinion testimony**</u>, <u>**hearsay evidence**</u>, rules of impeachment, and so forth."[42] As a result, hearsay evidence from social investigations is sometimes used in juvenile court adjudications. This practice would not be acceptable in criminal courts, because information about how children are doing in school, their peer associations, and their family circumstances are not relevant to their guilt or innocence.[43] In still other states, civil law rules of evidence are used in delinquency matters.[44] There are also differences between jurisdictions in the extent to which adult process such as <u>**pretrial discovery**</u> is afforded to juveniles. In Michigan, for example, juveniles are entitled to limited pretrial discovery;[45] they have the right to use the defense of alibi, insanity, diminished capacity, or mental illness;[46] and motion practice is governed by virtually the same rules as in adult criminal proceedings.[47]

In regards to due process protections, there can also be differences between juvenile adjudications and criminal trials. For example, in many states juveniles

opinion testimony
Testimony based on one's opinions as opposed to objective facts.

hearsay evidence
Second-hand evidence, such as a witness's testimony that he or she heard someone say something.

discovery Part of the pretrial process where the defense and the prosecution request information from the other side in an effort to discover pertinent facts regarding a case.

have most but not all of the due process protections available to adults, such as the right to be proven guilty "beyond a reasonable doubt,"[48] to have the assistance of counsel, to receive written and timely notice, to cross-examine witnesses, and to remain silent.[49] However, in an important U.S. Supreme Court case, *McKeiver v. Pennsylvania*, the Court refused to guarantee a jury trial to juveniles,[50] although a number of states have granted juveniles this right. Even in these states, jury trials rarely occur, however, and they typically result in the same outcomes as noncontested cases.[51]

MYTH VS. REALITY — **Juveniles Have Some But Not All of the Same Rights as Adults**

Myth—Juveniles have the same due process protections at trial as adults.
Reality—Juveniles do not have a right to a jury trial in all states, and even in states where this right is available, it is rarely exercised. Moreover, juvenile hearings are often characterized by practices that would not be acceptable in criminal court proceedings.

Given the "get tough" approach toward many juvenile offenders and the increasingly harsh punishment alternatives available to juvenile courts, it is important that juvenile defendants be afforded meaningful due process protections at adjudication. (A complete list of the due process rights afforded juveniles in one Michigan Court is found in **Exhibit 10-1**). Undoubtedly, the most important of the protections available to juveniles at the adjudication is the right to be represented by counsel. Representation by counsel is critical because attorneys are trained in the law and have an obligation to ensure that the rights of their clients are protected. Despite the obvious importance of this due process right, many juveniles appear in juvenile court without attorney representation. For example, in a study based on juvenile court data in six states, Barry Feld found that, in cases where a petition had already been filed, only about half of the juveniles were represented by counsel.[52] Similar findings have been presented in other studies of attorney representation in juvenile courts.[53] These findings raise the question of whether due process protections are in fact a reality in many juvenile courts, despite U.S. Supreme Court rulings mandating these protections.

Judge/Referee _____ APA _____ Date _____

❑ Minor waives Attorney   ❑ Parents/Guardian agree with attorney waiver

Minor's Attorney _____   ❑ Parents/Guardian present

❑ Proceedings on the Record:

Plea of   ❑ Guilty   ❑ Original Charge   ❑ Lesser _____
      ❑ Nolo Contendere
      1. ❑ Court has stated why plea is appropriate.
      2. ❑ Evidence presented to support finding of guilt.

ALL PLEAS—MINOR PERSONALLY ADVISED

❑ Nature of charge
❑ Disposition Court could impose
❑ Minor on probation
❑ Plea admits violation of probation

❑ Parents/Guardian contest verbally to
   minor waiving these rights. (Be sure
   both parents state so on record.)

MINOR ADVISED THAT BY PLEA HE GIVES UP

❑ Jury trial
❑ Trial by just judge without jury
❑ Presumption of innocence
❑ Proof beyond reasonable doubt
❑ Compulsory process
❑ Confront witness
❑ Remain silent
❑ No adverse inferences from silence
❑ Right to testify

DISPOSITION AGREEMENT

❑ Terms on record
❑ Petitioner agrees   ❑ Minor agrees   ❑ Minor's attorney agrees
        ❑ Minor's parents/guardian agree
❑ Court states if prior agreement in plea or disposition

MINOR ASKED

❑ Promised anything beyond stated disposition
❑ Threatened
❑ His choice to plead   ❑ Parents agree

COURT SATISFIED

❑ Plea freely, voluntarily made
❑ Crime committed (Question minor for elements)
❑ Minor involved or took part in
❑ Plea accepted

Disposition Date _____

Source: Courtesy of Probate Court for the County of Kalamazoo, Michigan, Juvenile Division.

Exhibit 10.1 Probate Court for the County of Kalamazoo, Michigan, Juvenile Division: Delinquency Guilty Plea Checklist

Hearing officers in juvenile courts must be sufficiently cognizant and protective of the due process rights of the juvenile if the court is to have credibility with those it is intended to serve. If, after a conviction at trial, the juvenile or his or her family believe that the trial was unfair or that their rights were slighted by the court, cooperation at the disposition will be affected. Even though many juveniles will continue to proclaim their innocence after trial, it is crucial that adjudication be conducted in a clear and fair manner in order to facilitate subsequent court decisions made at the disposition hearing.

disposition The sentencing phase of the juvenile court process. During this phase, the court tries to establish individualized plans for juveniles that balance rehabilitation and community safety.

The Disposition Hearing

The **disposition** hearing is "the primary feature that distinguishes the juvenile system from the adult criminal court."[54] Unlike the sentencing in criminal court, where a sanction is applied to the offender, the juvenile court disposition is intended to assist the youth and protect the community. It is at the disposition hearing that formal plans designed to meet the various needs of the youth, the family, and the community are initiated. It is also at this hearing that the judge or other hearing officer attempts to balance the "best interests" of the youth and the need for community safety. Judges and other quasi-judicial hearing officers often have great latitude and discretion in making dispositional decisions.

In formulating a disposition, the court usually seeks a great deal of input and information, which is gathered and interpreted for the hearing officer by a caseworker (e.g., a probation officer) in the form of a predisposition report or social history. In preparing the predisposition report, the caseworker pulls together information from various sources in order to present a detailed social history of the youth and the family. The sources may include (1) school reports, (2) victim impact reports and restitution reports, (3) psychological evaluations, (4) substance abuse assessments, (5) financial statements and tax returns, (6) letters from the friends and family of the juvenile, (7) criminal histories of the juvenile and other family members, (8) child abuse and neglect history, and (9) the caseworker's own observations and conclusions. The purpose of this report, which basically outlines "problems" perceived by the caseworker and recommended responses to those problems, is to ensure the juvenile receives "individualized" justice.[55]

Juvenile court hearing officers often rely heavily on the information and recommendations in these reports in determining how best to respond to juvenile offenders. The amount and type of information contained in these reports is determined by state laws, court procedures,[56] and local custom. In some jurisdictions, predispositional reports are very lengthy and reflect a detailed investigation into the background and present circumstances of the juvenile and the family. In other jurisdictions, only limited information about the juvenile's background is included.[57]

In most instances, hearing officers follow the recommendations contained in the predisposition report.[58] However, this may be because probation officer recommendations are typically well within the hearing officer's beliefs about what is appropriate in particular types of cases. Indeed, as discussed in the next section, there is considerable evidence that both legal factors such as the seriousness of the offense, the youth's prior record, and earlier decisions to detain a youth and nonlegal factors such as race, ethnicity, and gender play a significant role in court dispositions.

The Predisposition Report Is an Important Document in the Juvenile Court

Although predisposition reports aid judges and quasi-judicial hearing officers in deciding on appropriate dispositions, they contain a considerable amount of opinion and hearsay evidence. Indeed, much of the information contained in a typical predispositional report would not be legally admissible during the adjudicatory hearing. Consequently, the hearing officer must not see any of this information prior to the trial or plea. Furthermore, the caseworker should not begin his or her investigation and preparation of the report until after jurisdiction is obtained at the adjudication. Initiating a sentencing investigation, which clearly implies a presumption of guilt, before any guilt has been formally determined undermines the presumption of innocence and gives the juvenile reason to doubt the court's objectivity. Unfortunately, predisposition investigations are sometimes started prior to adjudication, and hearing officers sometimes view the information gathered prior to adjudication.

Given the importance of the disposition hearing, it is critical that the juvenile receive competent representation by counsel. Moreover, it is important that the juvenile and his or her counsel have access to copies of all documents that are considered by the court in formulating the disposition—a practice supported by the U.S. Supreme Court in *Kent v. United States*[59] (see Chapter 5). Also, all documents considered by the court should be marked as exhibits and entered as evidence, on the record, at the dispositional hearing. Despite the obvious importance of attorney representation, in some jurisdictions many if not most juveniles are not represented by counsel at the disposition hearing.[60] Moreover, site visits by researchers from the American Bar Association Juvenile Justice Center, the Juvenile Law Center, and the Youth Law Center revealed that when attorneys did represent youths at dispositions, they often felt ill-prepared because of high caseloads and the lack of resources and support staff.[61]

Having a Record of Juvenile Court Proceedings Is an Important Due Process Safeguard

The term **on the record**, when used to describe hearing procedures, means that an audio or audiovisual recording was made of the hearing. Having this record of the proceedings is a fundamental due process safeguard for the juvenile, because any appeal of the adjudication or disposition is based on this record. Having such a record helps prevent judges and quasi-judicial hearing officers from acting capriciously or inappropriately.

on the record Recorded in some fashion. Making audio or audiovisual recordings of critical hearings, including adjudication and disposition hearings, can facilitate the appeal of a verdict and in general provides due process protection.

least restrictive alternative The placement that restricts a juvenile offender's freedom least while at the same time ensuring the safety of the juvenile and the community.

Although the predisposition report plays an important role in formulating the disposition, the hearing officer is not bound by the recommendations of the caseworker. Furthermore, state statutes, local resources, and funding place some limits on the dispositional authority of juvenile courts. Nevertheless, hearing officers often have rather broad discretion in determining the disposition,[62] although courts are usually required to use the **least restrictive available alternative** that meets the child's needs and ensures public safety. Dispositional alternatives available to most courts include

- probation in the juvenile's own home (the most common disposition employed by courts)[63]
- placement in a relative's home on probation, or placement in a foster home on probation
- probation with restitution to the victims or probation with community service
- commitment to the state for placement in a state facility
- detention for a specified time period, then release on probation
- placement in a private institutional setting funded by the court and/or the state or a state agency
- placement on intensive probation or house arrest
- placement in a boot camp program
- in-state or out-of-state placement in a private correctional facility
- some combination of the above alternatives

Research on Factors Influencing Dispositional Decision Making

Research on juvenile court dispositions suggests that a variety of factors may play a role in dispositional decision making, including *legal factors* (e.g., prior record, severity of offense, and prior juvenile justice processing decisions) and *nonlegal factors* (e.g., race, gender, social class, family structure, and age). The research results consistently show that prior record and severity of offense are strongly related to severity of disposition.[64] In addition, there is considerable evidence that previous juvenile justice decisions such as the decision to detain a youth influences juvenile court dispositions.[65]

Research on the effect of nonlegal variables on dispositions has produced mixed findings. Although the effects of age on dispositions has rarely been examined, an early study by Robert Terry found that older youths were more likely to receive severe dispositions.[66] As regards the effects of race, social class, and gender, some studies indicate that minority youths, lower-class youths, and females are more likely to receive harsher dispositions.[67] However, other studies have failed to find that these variables influence dispositional decisions.[68] At least one study found that males are more likely to be treated more severely than females for criminal offenses but that females receive harsher dispositions for status offenses.[69]

Overall, the research on disposition decision making reveals mixed results. Although some studies fail to find evidence of bias in the dispositional phase of the juvenile court process, other studies suggest that, in some jurisdictions, nonlegal variables such as gender, race, and social class influence dispositions, raising serious questions about bias in dispositional decision making.

■ Chapter Summary

As noted at the beginning of this chapter, the juvenile court is the heart of the juvenile justice process. It is a place where previous decisions made by personnel in other agencies, including law enforcement officers and prosecuting attorneys, are supported or altered and where a variety of decisions are made that can have an indelible effect on children, their families, crime victims, and the community at large. Furthermore, not only does the juvenile court determine the outcome of individual cases, but through its legal authority it can determine how other institutions and agencies respond to children.

Although the juvenile court occupies a central position in juvenile justice, its position as a legal institution is far from secure. Indeed, there are various trends underway that will impact the juvenile court in the near future. For example, the number of cases being processed by juvenile courts continues to grow, and this growth will tax the personnel, financial, and other resources of the juvenile court for some time to come.

Another trend is toward reformation of juvenile court organizational structures. Presently, a variety of structures are found around the country. Unfortunately, some do not allow the juvenile court the status or the resources that it deserves. Moreover, some jurisdictions still employ generalist judges who hear a variety of cases and whose commitment to juvenile and family matters is suspect. Although some jurisdictions have moved toward family courts or unified court structures that may lend more prestige to juvenile and family law matters, the extent to which these changes will continue is unknown.

Of course, juvenile courts, like other juvenile justice agencies, are made up of people who perform a variety of roles. Key personnel include the judges, a variety of quasi-judicial hearing officers (referees, masters, or commissioners), court administrators, prosecuting and defense attorneys, and probation officers and other caseworkers. Clearly, the most important figure in the juvenile court (and juvenile justice in general) is the juvenile court judge. Indeed, the juvenile court judge makes decisions that not only influence the lives of youths and their parents, but also affect the operation and mission of the juvenile court as well as other juvenile justice agencies. Of course, the juvenile court judge and other hearing officers do not make these decisions without assistance. They are surrounded by a supporting cast of professionals who exercise considerable discretion and who themselves make a number of important decisions, including decisions that determine which youths move to the adjudicatory and dispositional stages of the juvenile justice process.

The adjudication and disposition hearings determine which youths come within the formal jurisdiction of the court. Moreover, it is in these hearings that attorneys, probation officers, and others provide information that assists the court in determining guilt and innocence and in making decisions about how to respond to the adjudicated delinquent. Yet, despite the significance of these hearings, there is considerable evidence that youths are not always treated fairly: In some jurisdictions, youths are not afforded representation by counsel; when youths are represented by counsel, that representation is sometimes inadequate; adjudicatory and dispositional decisions are based on opinion and hearsay evidence; and gender, race, and social class bias sometimes impacts judicial decision making.

Clearly, serious problems confront juvenile courts in the United States, and in some jurisdictions these problems undoubtedly affect the quality of justice that juveniles receive. However, it should not be forgotten that most juvenile court personnel, from judges to caseworkers, are highly motivated and skilled individuals who are committed to serving the best interests of children and protecting community safety—and who are frequently able to achieve both of these sometimes conflicting goals.

KEY CONCEPTS

adjudication The juvenile court equivalent of an adult court trial. It is at the adjudication that the juvenile court attempts to ensure due process, find the facts, uphold the law, and make decisions concerning jurisdiction over the juvenile.

balanced approach to juvenile justice The belief that juvenile justice should give equal attention to public safety, offender accountability, the needs of victims, and the correction and treatment of juvenile offenders.

best interests of the child A catch phrase that serves as a reminder that the primary focus of a juvenile court should be on the rehabilitation of the children who come before it.

continuum of care A range of comprehensive and connected services for families and children.

discovery Part of the pretrial process where the defense and the prosecution request information from the other side in an effort to discover pertinent facts regarding a case.

disposition The sentencing phase of the juvenile court process. During this phase, the court tries to establish individualized plans for juveniles that balance rehabilitation and community safety.

family court A unified trial court where all cases involving families are heard, including divorce, adoption, custody, guardianship, paternity, neglect and abuse, and delinquency cases. Family court advocates claim that when all family matters are handled by one court, there will be less chance for overlapping services, redundant reports, and fragmented intervention.

guardian ad litem A person, usually an attorney, who is appointed by the court to represent the best interests of a child involved in legal proceedings and with social service agencies.

general trial court A court of high status within a state's legal system that hears felony criminal cases and civil cases with unlimited dollar amounts.

hearsay evidence Second-hand evidence, such as a witness's testimony that he or she heard someone say something.

judicial administration The daily and long-range management of a court. The judge, as head of the court, is ultimately responsible for judicial administration but can choose to delegate many of the particular duties to others, such as a court administrator.

judicial leadership Guidance provided by a judge in the areas of programming, personnel, and budget administration. Also seen in his or her role as children's advocate and consensus builder within the community.

juvenile court The court responsible for holding hearings and making decisions regarding the disposition of juveniles who have entered the juvenile justice process. This court plays many roles, including children's advocate, program leader, fundraiser, consensus builder, parent, and protector of youths and the community.

juvenile drug court Specialized juvenile court that handles cases involving substance-abusing offenders through comprehensive supervision, drug testing, treatment services, and immediate sanctions and incentives.

least restrictive alternative The placement that restricts a juvenile offender's freedom least while at the same time ensuring the safety of the juvenile and the community.

on call Availability for emergency service. Some jurisdictions require there always be a judicial officer available to make emergency decisions about the placement of alleged juvenile offenders.

on the record Recorded in some fashion. Making audio or audiovisual recordings of critical hearings, including adjudication and disposition hearings, can facilitate the appeal of a verdict and in general provides due process protection.

opinion testimony Testimony based on one's opinions as opposed to objective facts.

petition A pleading to initiate a matter in juvenile court. A petition sets forth the alleged grounds for the court to take jurisdiction of a case and requests court intervention.

plea bargaining Negotiation of an agreement between the prosecuting and defending attorneys, often with the consent of the court, to have the accused plead guilty to a reduced charge or to fewer charges than originally brought. The prosecutor benefits by gaining a certain conviction and the defendant benefits by avoiding the possibility of a more severe sentence.

pre-trial discovery See *discovery*.

quasi-judicial hearing officer Hearing officer empowered by the court to hear cases. Hearing officers are often attorneys and hear a variety of cases, although state law typically places some limits on their authority.

unified trial court Comprehensive courts encompassing several specialist courts, such as courts devoted to delinquency and family matters, as well as other types of courts. Also see *family courts*.

REVIEW QUESTIONS

1. What is the basis for claiming the juvenile court is the center of the juvenile justice process?

_____

_____

_____

_____

_____

2. What responsibility does the juvenile court judge have as an advocate for children?

_____

_____

_____

_____

_____

3. How may being part of a lower trial court influence juvenile court operations? How may being part of an upper-level trial court influence juvenile court operations?

_____

_____

_____

_____

_____

4. Who are the key personnel in the juvenile court, and what functions do they perform?

_____

_____

_____

_____

_____

5. What is an adjudication hearing?

_____

_____

_____

_____

_____

6. What due process protections exist at the adjudication hearing?

_____

_____

_____

_____

_____

7. What is the disposition hearing, and what type of information is usually presented to the court at this hearing?

_____

_____

_____

_____

_____

8. What is a family court, and how does it differ from a juvenile court?

_____

_____

_____

_____

_____

9. What is the role of the prosecuting attorney in the juvenile court?

_____

_____

_____

_____

_____

10. What is the role of the defense attorney in the juvenile court?

_____

_____

_____

_____

_____

11. What does it mean to say that a judge is a "hands-on administrator"? Is this a good thing to be?

_____

_____

_____

_____

_____

12. What function does the court administrator have in the juvenile court?

_____

_____

_____

_____

_____

13. How many different roles or functions does a caseworker (such as a probation officer) have in the juvenile court?

_____

_____

_____

_____

_____

14. What are the multiple factors that impact the disposition decision? How do courts weight them?

_____

_____

_____

_____

_____

ADDITIONAL READINGS

Bernard, T. J. 1992. *The Cycle of Juvenile Justice.* New York: Oxford University Press.

Feld, B. C. 1993. *Justice for Children: The Right to Counsel and the Juvenile Courts.* Boston: Northeastern University Press.

Puritz, P. et al. 1995. *A Call for Justice: An Assessment of Access to Counsel and Quality of Representation in Delinquency Proceedings.* Washington, DC: American Bar Association Juvenile Justice Center.

Rubin, H. T. 1985. *Behind the Black Robes: Juvenile Court Judges and the Juvenile Court.* Beverly Hills, CA: Sage.

Senna, J. and L. J. Siegel. 1992. *Juvenile Law: Cases and Comments.* St. Paul, MN: West Publishing Co.

NOTES

1. C. Puzzanchera et al., *Juvenile Court Statistics, 1999* (Pittsburgh, PA: National Center for Juvenile Justice, 2003).
2. Puzzanchera et al., *Juvenile Court Statistics, 1999.*
3. Puzzanchera et al., *Juvenile Court Statistics, 1999.*
4. Puzzanchera et al., *Juvenile Court Statistics, 1999.*
5. Puzzanchera et al., *Juvenile Court Statistics, 1999.*
6. Puzzanchera et al., *Juvenile Court Statistics, 1999.*
7. T. J. Bernard, *The Cycle of Juvenile Justice* (New York: Oxford University Press, 1992); I. M. Schwartz, *(In)Justice for Juveniles: Rethinking the Best Interests of the Child* (Lexington, MA: Lexington Books, 1989).
8. B. C. Feld, "In re Gault Revisited: Cross-State Comparison of the Right to Counsel in Juvenile Court," *Crime and Delinquency* 34 (1988): 393–424; B. C. Feld, *Justice for Children: The Right to Counsel and the Juvenile Courts* (Boston: Northeastern University Press, 1993); J. B. Jones, "Access to Council," *Juvenile Justice Bulletin* (Washington, DC: Office of Juvenile Justice and Delinquency Prevention, 2004).
9. Feld, "In re Gault Revisited"; B. C. Feld, "The Right to Counsel in Juvenile Court: An Empirical Study of When Lawyers Appeal and the Difference They Make," *Journal of Criminal Law and Criminology* 79 (1989): 1187–1346; P. Puritz et al., *A Call for Justice: An Assessment of Access to Counsel and Quality of Representation in Delinquency Proceedings* (Washington, DC: American Bar Association Juvenile Justice Center, 1995).
10. Puritz et al., *A Call for Justice.*
11. H. N. Snyder and M. Sickmund, *Juvenile Offenders and Victims: A National Report* (Washington, DC: Office of Juvenile Justice and Delinquency Prevention, 1995).
12. H. T. Rubin, "The Juvenile Court Landscape," in *Juvenile Justice: Policies, Programs, and Services,* 2nd ed., ed. A. R. Roberts (Chicago: Nelson-Hall, 1998).
13. Administrative Office of the Trial Court, Juvenile Court Department. Retrieved August 20, 2004, from http://www.mass.gov/courts/courtsandjudges/courts/juvenilecourt/#Jurisdiction.

14. Rubin, *Juvenile Justice*, 350.
15. Rubin, *Juvenile Justice*.
16. L. J. Siegel and J. J. Senna, *Juvenile Delinquency: Theory, Practice, and Law*, 6th ed. (St Paul, MN: West, 1997).
17. Rubin, *Juvenile Justice*.
18. Siegel and Senna, *Juvenile Delinquency*.
19. Rubin, "The Juvenile Court Landscape," 2nd ed.
20. Rubin, "The Juvenile Court Landscape," 2nd ed.
21. Rubin, "The Juvenile Court Landscape," 2nd ed.
22. National Association of Drug Court Professionals, "Facts on Drug Courts." Retrieved August 23, 2004, from http://www.nadcp.org/whatis/.
23. P. F. Hora et al., "Therapeutic Jurisprudence and the Drug Treatment Court," *Notre Dame Law Review* 74 (1999): 439–537.
24. Rubin, *Juvenile Justice*.
25. Siegel and Senna, *Juvenile Delinquency*.
26. For an insightful view of juvenile court judges, see H. T. Rubin, *Behind the Black Robes: Juvenile Court Judges and the Juvenile Court* (Beverly Hills, CA: Sage, 1985).
27. Rubin, "The Juvenile Court Landscape," 2nd ed.
28. Rubin, "The Juvenile Court Landscape," 2nd ed.
29. H. T. Rubin, "The Juvenile Court Landscape," in *Juvenile Justice: Policies, Programs, and Services*, ed. A. R. Roberts (Chicago: Dorsey Press, 1989), 114.
30. R. W. Drowns and K. M. Hess, *Juvenile Justice*, 2nd ed. (St. Paul, MN: West, 1995).
31. Rubin, "The Juvenile Court Landscape."
32. S. M. Cox et al., *Juvenile Justice: Guide to Theory and Practice*, 5th ed. (New York: McGraw-Hill, 2003).
33. National District Attorney's Association, *National Prosecution Standards*, 2nd ed. (Alexandria, VA: Author, 1991).
34. F. L. Bailey, *The Defense Never Rests* (New York: Signet, 1971).
35. Puritz et al., *A Call for Justice*.
36. R. H. Mnookin and D. K. Weisberg, *Child, Family and State: Problems and Materials on Children and the Law* (Boston: Little, Brown & Co., 1989).
37. Puritz et al., *A Call for Justice*.
38. Puritz et al., *A Call for Justice*.
39. Puritz et al., *A Call for Justice*.
40. Bernard, *The Cycle of Juvenile Justice*; S. J. Fox, *The Law of Juvenile Court in a Nutshell*, 2nd ed. (St. Paul, MN: West, 1977).
41. Puritz et al., *A Call for Justice*.
42. J. Senna and L. J. Siegel, *Juvenile Law: Cases and Comments* (St. Paul, MN: West, 1992), 130.
43. B. Krisberg and J. F. Austin, *Reinventing Juvenile Justice* (Newbury Park, CA: Sage, 1993).
44. Senna and Siegel, *Juvenile Law*.
45. Mich. Court Rules 5.922 (A).
46. Mich. Court Rules 5.922 (B).
47. Mich. Court Rules 5.922 (C).
48. In re Winship, 397 US 385 (1970).
49. In re Gault, 387 US 1, 87 S.Ct. 1428.
50. McKeiver v. Pennsylvania, 403 US 528 (1971).
51. A. R. Mahoney, "Jury Trial for Juveniles: Right or Ritual?" *Justice Quarterly* 2 (1985): 553–565.
52. B. Feld, "In re Gault Revisited."
53. Puritz et al., *A Call for Justice*; Also see J. B. Jones, "Access to Counsel," *Juvenile Justice Bulletin* (Washington, DC: Office of Juvenile Justice and Delinquency Prevention).
54. Puritz et al., *A Call for Justice*, 36.
55. Mnookin and Weisberg, *Child, Family and State*.
56. Siegel and Senna, *Juvenile Delinquency*.
57. Siegel and Senna, *Juvenile Delinquency*.
58. M. D. Jacobs, *Screwing the System and Making It Work: Juvenile Justice in the No-Fault Society* (Chicago: University of Chicago Press, 1990); Siegel and Senna, *Juvenile Delinquency: Theory, Practice, and Law*, 8th ed. (Belmont, CA: Wadsworth/Thompson, 2003).

Chapter Resources

59. Kent v. United States, 383 US 541 (1966).

60. Puritz et al., *A Call for Justice*; J. D. Walter and S. A. Ostrander, "An Observational Study of a Juvenile Court," *Juvenile and Family Court Journal* 33 (1982): 53–69.

61. Puritz et al., *A Call for Justice*.

62. Senna and Siegel, *Juvenile Law*.

63. Puzzanchera et al., *Juvenile Court Statistics*, 1999.

64. T. Carter, "Juvenile Court Dispositions: Comparison of Status and Nonstatus Offenders," *Criminology* 17 (1979): 341–359; S. H. Clarke and G. G. Koch, "Juvenile Court: Theory or Crime Control and Do Lawyers Make a Difference?" *Law and Society Review* 14 (1980): 263–308; L. Cohen and J. R. Kluegel, "Determinants of Juvenile Court Dispositions: Ascriptive and Achieved Factors in Two Metropolitan Courts," *American Sociological Review* 43 (1978): 162–176; D. Dannefer and R. K. Schutt, "Race and Juvenile Justice Processing in Court and Police Agencies," *American Journal of Sociology* 87 (1982): 1113–1132; B. R. McCarthy and B. L. Smith, "The Conceptualization of Discrimination in the Juvenile Justice Process: The Impact of Administrative Factors and Screening Decisions on Juvenile Court Dispositions," *Criminology* 24 (1986): 41–64; W. G. Staples, "Law and Social Control in Juvenile Justice Dispositions," *Journal of Research in Crime and Delinquency* 24 (1987): 7–22; C. W. Thomas and R. Cage, "The Effect of Social Characteristics on Juvenile Court Dispositions," *Sociological Quarterly* 18 (1977): 237–252.

65. W. C. Bailey and R. D. Peterson, "Legal versus Extra-Legal Determinants of Juvenile Court Dispositions," *Juvenile and Family Court Journal* 32: 41–59; Clarke and Koch, "Juvenile Court: Theory or Crime Control and Do Lawyers Make a Difference?"

66. R.M. Terry, "Discrimination in the Handling of Juvenile Offenders by Social Control Agencies," *Journal of Research in Crime and Delinquency* 4 (1967): 218–230.

67. D. M. Bishop and C. E. Frazier, "Race Effects in Juvenile Justice Decision-Making: Findings of a Statewide Analysis," *Journal of Criminal Law and Criminology* 86 (1996):392–414; Carter, "Juvenile Court Dispositions: Comparison of Status and Nonstatus Offenders,"; M. Chesney-Lind and R. G. Shelden, Girls, *Delinquency and Juvenile Justice*, 3rd ed. (Belmont, CA: Wadsworth/Thompson, 2004); F. Scarpitti and R. M. Stephenson, "Juvenile Court Dispositions: Factors in the Decision-Making Process," *Crime and Delinquency* 17 (1971): 142–151; Cohen and Kluegel, "Determinants of Juvenile Court Dispositions"; B. C. Feld, *Bad Kids: Race and the Transformation of the Juvenile Court* (New York: Oxford University Press, 1999); R. M. Terry, "Discrimination in the Handling of Juvenile Offenders by Social Control Agencies"; T. P. Thornberry, "Race, Socioeconomic Status, and Sentencing in the Juvenile Justice System," *Journal of Criminal Law and Criminology* 64 (1973): 90–98.

68. Carter, "Juvenile Court Dispositions: Comparison of Status and Nonstatus Offenders,"; Clarke and Koch, "Juvenile Court"; A. Horwitz and M. Wasserman, "Formal Rationality, Substantive Justice, and Discrimination," *Law and Human Behavior* 4 (1980):103–115.

69. F. Scarpitti and R. M. Stephenson, "Juvenile Court Dispositions."

Community-Based Correctional Programs for Juvenile Offenders

11

- Linking Institutional and Community-Based Corrections: Aftercare Programs
- Legal Issues
- Chapter Summary
- Key Concepts
- Review Questions
- Additional Readings
- Notes

■ Introduction

nonformal processing Includes various ways of processing cases that do not involve an adjudication and the initiation of a formal juvenile court record for the youth.

formal processing Case processing that results in an adjudication and initiation of a formal juvenile court record.

community-based corrections Noninstitutional correctional programs and interventions.

When youths are adjudicated, a number of dispositional options may be exercised by juvenile courts, although the options available in many jurisdictions are fairly narrow. Regardless of the number of options available to courts, however, they fall into three categories. One category of options is **nonformal processing** which includes dismissal of the case, used in a small percentage of cases; referral to a community diversion program (note that some states limit the types of crimes that can be referred to diversion programs); and placement on nonformal probation. The two categories of **formal processing** options are placement in a community-based correctional program and placement in an institutional correctional program. This chapter examines the operation and effectiveness of placement in a community-based correctional program.

The term **community-based corrections** refers to diverse types of supervision, treatment, reintegration, control, and supportive programs for offenders.[1] These programs are based on the belief that the most effective way to encourage law-abiding behavior on the part of youths is to help them assume legitimate roles within the community.[2] Examples of community-based programs are diversion, pretrial release, probation, foster care, group home placement, and parole. Some of the programs are designed to provide services to youths in their own homes, and these are referred to as *at-home community-based programs*. Other community-based programs provide services to youths who are removed from their homes, at least for short periods of time; these are referred to as *out-of-home community-based programs*.

Like other correctional interventions, community-based programs are intended to accomplish a variety of objectives, although the objectives actually pursued by different programs vary to some extent. These objectives include controlling and sanctioning youths, allowing youths to maintain existing ties with the community, helping them restore ties and develop new and positive ones with the community (reintegration), helping the individual avoid the negative consequences of institutional placement, providing a more cost-effective response to those who violate the law, and reducing the likelihood of recidivism.[3]

■ At-Home Community-Based Programs for Juvenile Offenders

Probation

<u>Probation</u> is the most frequently used correctional response to youths who are adjudicated in juvenile courts. For example, in 1999, 4 out of every 10 delinquency cases and over half (59%) of all juvenile cases were given probation.[4] Probation is the conditional release of an adjudicated youth into the community under the supervision of the court. The conditions under which a youth is released constitute the rules of probation. Typical rules of probation require the probationer to obey all laws; follow home rules; be home each day by a certain time; meet with the probation officer when requested; and, if the youth is of school age, attend school each day and obey the rules of the school. However, a court may make other rules of probation if it thinks they are necessary to assist the youth and protect the community. The court, for example, might demand that the youth avoid certain people or places, attend counseling, and make restitution to the victim. The role of the probation officer is to monitor the youth and/or assist the youth in his or her efforts to adhere to the rules of probation.

> **probation** The conditional release of an adjudicated delinquent into the community under the supervision of the court.

This description of probation, however, does not reflect the disparities that exist in the practice of probation. In fact, considerable variability in probation practice can occur in the same state and even in the same probation department. Some juvenile courts have more than one type of probation program for youths under their supervision. These different programs—sometimes called standard, moderate, and intensive probation—are intended to reflect different levels of contact between probation officers and their clients. Other courts or probation departments do not make a formal distinction between types of probation programs. Whether probation is called standard, moderate, or even intensive probation in some cases, the level of contact between a probation officer and a client can range from regular to infrequent. There are also jurisdictions that, because of large caseloads, are able to provide only "file drawer probation," which is characterized by a lack of meaningful and regular contact between probation officers and probationers. Moreover, there is considerable variability in the training, skills, motivation, and workload of individual probation officers. Some probation officers are well trained, highly skilled, motivated, and work with small caseloads. Others have received relatively little training, lack motivation,and have large caseloads, which of course prevents them from having regular contact with each of their clients.[5] In order to determine the number of probation officers available for working with (or for) juvenile courts, some states use population formulas. In other states, the determination of the number of available probation officers is made locally and may be based on the level of community resources and various political factors.[6]

The Level of Probation Supervision Varies Across Jurisdictions and Clients

Many probation officers develop their own levels of probation supervision for their caseload or their own system for prioritizing their cases, based on their assessment of need. For example, at any one time, some cases on a probation officer's caseload will be in crisis (e.g., because of a family fight, the suspension of a youth from school, or because the youth lost his or her job) and will require considerable expenditure of time and energy on the part of the officer. In contrast, other cases will be relatively trouble free and close to discharge from probation. As a result, the probation officer will have considerably more contact with some probationers than with others.

Not only do probation officers differ in terms of their ability and willingness to supervise or work with clients, but communities also vary in the number and quality of services available for use by probation officers and probationers. Some communities have a range of high-quality programs appropriate for probationers, such as alternative education programs, educational support services, job training and placement programs, substance abuse treatment programs, counseling, and structured recreational programs. Unfortunately, many other communities lack such programs or have only a few high-quality programs that are accessible to court-involved youths. The availability of high-quality programs in a community is important to probation officers, because they often serve as *service brokers* who attempt to link clients with appropriate community services and thereby reduce the likelihood of further law-violating behavior. Juvenile courts and other governmental units responsible for funding or administering proba-

Probation Can Mean Regular Supervision and Provision of Services

Myth—Most juvenile probation officers have large caseloads that prevent them from effectively monitoring their clients, thus endangering public safety.
Reality—Although some juvenile probation officers have caseloads of more than 200 clients, the average juvenile probation caseload is about 41 active cases (11 more than the optimal caseload of 30 suggested by probation officers themselves).[7] Moreover, most youths placed on probation are nonviolent offenders. Only 23% of cases placed on probation in 1999 involved person offenses.[8]

tion departments also vary in their support of probation services. Some probation departments receive considerably more resource support than others. Desired resources and supports include probation and support staff, staff training, pay and benefits, and effective supervision and administration of the department.

Although the actual practice of probation varies, probation officers usually perform six important functions in the juvenile justice process: (1) they do intake screening; (2) they conduct presentence investigations; (3) they supervise offenders and monitor the extent to which youths adhere to their probation orders; (4) they provide assistance to youths placed on probation; (5) they provide ongoing assessments of clients' needs; and (6) they complete a variety of job-related administrative tasks. In some jurisdictions, all of these functions are the responsibility of the same individual, whereas in other jurisdictions, probation officers specialize in one or more.

Probation Officers Perform a Number of Functions

It is estimated that there are roughly 18,000 juvenile probation officers in the United States. Approximately 85% of these individuals are line staff involved in providing basic intake, investigation, and supervision services; the remaining 15% are involved in the administration of probation offices or the supervision of probation staff.[9]

As noted in Chapter 8, *intake screening* involves making decisions about the most appropriate ways of processing cases referred to the juvenile court. It is at the intake stage that decisions to dismiss, petition, or handle a case in some nonformal way are made. Moreover, those who perform intake screening employ considerable discretion in making intake decisions. In addition to this, the **presentence investigation** plays a critical role in juvenile court dispositions because hearing officers rely on these investigations to decide upon the most appropriate way of handling formal cases.

The supervision and assistance roles of probation officers involve monitoring probationers to ensure that they are complying with the rules of probation and providing various types of help to them. The supervision role requires checking to ensure that probationers are at school, work, home, or other places at designated times or monitoring youths assigned to electronic or officer-monitored house arrest programs. It may also require collecting urine samples from youths to enable screening for substance use. The assistance role may encompass service brokerage and the provision of direct services to clients. When acting as a service broker, the probation officer attempts to link youths and possibly families with community agencies. The officer might also provide direct services such as basic counseling, job search assistance, basic family counseling, and

presentence investigation An investigation that is typically completed by a probation officer prior to the disposition and that is intended to assist the hearing officer in deciding the best response to an adjudicated youth.

crisis intervention services as well as act as an advocate for youths in dealings with teachers, school administrators, employers, and social service personnel. In cases where probation officers have an opportunity to have regular contacts with clients, they are in a position to better understand problems that may have contributed to the youths' involvement with the court. They can also conduct investigations into youth and family needs. Such investigations may uncover problems that were passed over or have arisen since the presentence investigation. By becoming aware of current problems, the probation officer will be in a better position to recommend or develop, possibly with the court's assistance, strategies for helping the youth successfully complete probation and avoid subsequent delinquent activity.

In addition to the these roles, probation officers engage in a variety of job-related administrative tasks, such as filing orders and paperwork related to the processing of cases and preparing reports of casework activity.

There Is Considerable Variability in Probation Practice

Although probation involves doing intake screening, conducting presentence investigations, supervising cases, providing assistance to clients, conducting ongoing evaluations of clients' needs, and completing routine job-related administrative tasks, the amount of time that probation officers spend on the different tasks varies considerably. In some jurisdictions, probation officers spend a majority of their time dealing directly with clients or making client-related contacts designed to assist youths on probation. In other jurisdictions, much of their time is spent preparing court reports and engaging in other administrative tasks.

Probation officers may be employed in specialized programs operated by or on behalf of courts. For example, juvenile courts or other government agencies may have a variety of programs, such as diversion and foster care programs, and also operate their own shelter-care units, group homes, and detention centers. Probation officers may be used to staff these programs or may be responsible for providing basic casework services to youths in these programs.

In performing their various duties, probation officers may experience conflict between their supervision and assistance roles (sometimes referred to as their "law enforcement" and "social work" roles).[10] As social control agents, probation officers have a responsibility to protect the community and to make sure that their clients follow the rules of probation. This is their *law enforcement role*. On the other hand, they may be expected to work closely with other agencies and with parents in assisting their clients. Moreover, they are expected to work closely with their clients in order to help them make appropriate choices and avoid further law-violating behavior. This is their *social work role*. Their law enforcement and social work roles are not always compatible,[11] and individual officers may deal with the role conflict by emphasizing one role over another or by attempting to balance the roles,[12] often emphasizing different roles with different probationers. However, large caseloads cause many probation officers, regardless of their own beliefs about their roles, to operate in a crisis intervention mode. When clients' problems come to their attention, they attempt to react to these crises as best they can, but they may have very little time left over to spend in providing proactive assistance to clients.

The Organization of Juvenile Probation

The organization of juvenile probation varies from state to state. In 23 states and the District of Columbia, juvenile probation is administered by the local juvenile court or by the state administrative office of courts. In 14 states, probation is administered by a combination of agencies. In these states, juvenile probation is typically administered by the juvenile court in urban counties and by a state executive agency in smaller counties. Although juvenile probation is most often organized under the judiciary, recent trends in state court organization have resulted in moving probation services from under the direction of local juvenile courts to a state court judicial agency.[13]

Research on Standard Juvenile Probation Effectiveness

Research on the effectiveness of juvenile probation is complicated by the wide variability in probation practices across and within jurisdictions, by the lack of research on probation in general, and by the failure of studies of probation to clearly document the types of interventions used by probation officers and the quality of services provided to youths. The lack of research on probation effectiveness is somewhat surprising considering that juvenile probation has existed since the mid-1800s. Nevertheless, some studies indicate that probation is effective at both reducing and prolonging recidivism[14] and represents a cost-effective alternative to incarceration,[15] particularly for moderately delinquent youths.[16] Other studies, however, suggest that probation does not appear to be particularly effective at reducing recidivism, particularly for those involved in serious delinquency,[17] thus questioning its ability to protect the public from chronic or serious offenders. Other research indicates that community-based programs have often failed to provide needed services to clients and have failed to reduce correctional costs.[18]

 Probation Is Often as Effective as Institutional Placement

Myth—Youths who are placed on probation usually engage in additional crimes while on probation or when they are discharged.
Reality—Research indicates that youths on probation in many communities avoid offending while on probation as well as subsequent to court discharge, thus indicating that probation is at least as effective at reducing recidivism as institutional placement.[19]

house arrest See *home confinement.*

intensive supervision program A program intended to ensure that there is regular contact between probationers and probation officers. Although the contact is regular, its frequency varies considerably across programs.

intermediate sanction A diverse range of responses (e.g., intensive supervision, day treatment, house arrest) to offenders that are more restrictive than standard probation but less restrictive than institutional placement.

Recent Trends in Juvenile Probation

A recent trend in probation involves the development of intensive probation (supervision) programs, which in some jurisdictions can also involve **house arrest** (confinement). Intensive supervision programs are intended to ensure that there is regular contact between probationers and probation officers. They also serve as an **intermediate sanction** for youths that is more restrictive than standard probation but less restrictive than incarceration. However, as in the case of standard probation programs, the frequency of contact between probation officers and probationers varies considerably across intensive supervision programs.[20]

Although there is wide variability in what intensive supervision means in different jurisdictions, there is some indication that programs that provide frequent supervision of offenders as well as services are as effective as incarceration at reducing recidivism.[21] Moreover, there is evidence suggesting that intensive programs are more cost-effective than incarceration, providing they actually divert a sizable number of youths from institutions.[22] However, other research indicates that intensive supervision programs may *not* be as cost-effective as other interventions that keep youths in the community, such as standard probation and programs using

MYTH VS. REALITY

Probation Is Effective with Some Youths

Myth—Probation is not an effective response to more serious juvenile offenders, cannot be used for youths who otherwise would be placed in institutions, and endangers community safety.

Reality—Although standard probation programs do not appear to reduce recidivism among chronic or serious offenders, probation can be successfully used as an alternative to institutionalization for nonserious offenders who account for a sizeable proportion of those youths placed in institutional settings.[25] In addition, research that has looked at some intensive supervision programs found they were just as effective in reducing recidivism as institutional placement. In one study, William Barton and Jeffrey Butts examined a population of Michigan youths who were randomly assigned to participate in an intensive supervision program or placed in state institutions. Those assigned to the intensive supervision program were part of small caseloads, were closely monitored, and were given a variety of support services, including individual counseling, social skills training, parent counseling, and involvement in youth groups and recreational activities. After being in the program for a 12-month period, the youths were placed on standard probation. In their evaluation, Barton and Butts found that, controlling for time in the community, the intensive supervision program was as effective at reducing recidivism as institutionalization. Moreover, they found that the intensive supervision program was considerably less expensive.[26]

cognitive behavioral treatment.[23] Moreover, failure rates in these programs can be high, particularly among younger property offenders.[24] This is not surprising because youths placed on intensive supervision programs are often more serious offenders, and enhanced monitoring of these youths increases the probability that violations of program rules or further law violations will be uncovered.

Intensive supervision sometimes employs **home confinement** or *house arrest*, which began to be used with juvenile offenders in the 1970s[27] and has grown in popularity since that time. Today, home confinement or house arrest programs use two mechanisms to monitor youths: frequent probation officer contacts and **electronic monitoring** (which requires the offender to wear a tamper-resistant electronic device that automatically notifies the probation department if the juvenile leaves his or her home or another designated location). Although home confinement programs have grown in popularity around the country, there has been limited research on their effectiveness. Nevertheless, the research that has been done is somewhat encouraging. For example, in one national evaluation of seven home confinement programs that included officer monitoring, the researchers found that the programs were able to prevent both recidivism and violations of program rules. Moreover, when participants did run into problems, it was usually for violating program rules, not for committing new criminal offenses.[28]

Electronic monitoring of juvenile offenders began in the 1980s and has become more prevalent since that time.[29] However, because electronic monitoring technology is new, diverse, and being rapidly developed, relatively little research has focused on its ability to reduce recidivism and the types of clients that are likely to fare well on electronic monitoring programs. Nevertheless, there is some evidence that recidivism is no higher in electronic monitoring programs than in other community-based programs, costs can be lower, and they may help reduce institutional crowding. However, evidence suggests that serious felony offenders—those with substance abuse problems, repeat offenders, and those with long sentences—are least likely to be successful in these programs.[30]

Research has documented some of the reasons why some jurisdictions find electronic monitoring attractive: It is a way to ease the problem of detention overcrowding; it allows youths to participate in counseling, educational, and vocational programs without endangering public safety; it allows youths to live in a natural environment, albeit with supervision; and it allows court workers to better assess the ability of youths to live in the community under standard probation after they leave the program.[31]

home confinement
The state of being confined to home. A home confinement program requires the participants to spend certain hours at home. Supervision in these programs is done by probation officers who make random checks on clients or monitor compliance by electronic monitoring. Also called *home arrest* or *house arrest*.

electronic monitoring
The ability to determine an individual's movement by having him or her wear a tamper-resistant electronic device that allows communication between the device and a computer housed at the monitoring agency.

Box 11-1

Interview: James McIntyre, Juvenile Court Counselor, North Carolina Department of Juvenile Justice and Delinquency Prevention

Education: BS, Criminal Justice, Western Carolina University

Q: How long have you been a juvenile court counselor?

A: 15 years

Q: How did you get interested in juvenile justice?

A: Since I was about seven or eight, I knew I wanted to work in the justice field. At first I wanted to work in law enforcement, but while I was in college I worked as a volunteer with child-serving groups and really liked it. Also, after I graduated I worked at a children's home. My job title was "resident counselor." I was part of an interdisciplinary team that provided crisis intervention, basic counseling, and supervision to emotionally disturbed kids who were placed out of their homes. In many instances these children were the victims of emotional, physical, and sexual abuse and neglect. That position provided me with an excellent experiential base for my present job.

Q: Does your department have specialized units?

A: Units no, there are only three of us in this county. But we do have specialized tasks and responsibilities.

Q: What is your role?

A: I am responsible for delinquency intake as well as maintaining a moderately sized supervision caseload of 15–18. I also serve on numerous boards and committees such as the Juvenile Crime Prevention Council and Communities-In-Schools. One of my co-workers is responsible for undisciplined intake while also maintaining a slightly larger supervision caseload of 22–27. My other co-worker is responsible for a supervision caseload of 30–35. Like me, my two co-workers serve on several boards.

Q: Tell me about your caseload and how you spend your time?

A: Most of my time is focused on delinquency intake. I receive a report from law enforcement, which I then review and enter into our department's database. I then create a petition. Soon thereafter I contact the victim and then meet with the juvenile and his or her family. That interview concerns the allegations facing the juvenile as well as how things are at home, at school, and in the community. I then make a decision as to whether the juvenile should be approved for court or if a lesser community-based intervention can be utilized to address the situation. If the juvenile is adjudicated delinquent I then prepare a report on the child and family, which includes an evaluation of risk and need factors. I also recommend to the court what measures should be taken to protect the community as well as how best to assist the juvenile and family in correcting and moving away from the problems that contributed to the juvenile's current situation. From that point the case is supervised by myself or one of my co-workers until the child is no longer under the jurisdiction of the court (usually one year).

Q: From your perspective, why do children get involved with the court?

A: There is no easy answer to that. For most children, part of the equation is certainly the breakdown of the family unit, which many times is multi-generational in nature. This breakdown often manifests itself through a lack of appropriate parenting skills. Let's face it, if these parents were not exposed to appropriate skills such as consistent discipline and structure by their own parents, how can they accomplish that for their own children? Many of the children I see come from single parent homes where the fathers are absent. This can have a negative impact on areas such a self-esteem and is often a root cause of many behavioral problems such as poor anger management and substance abuse. A significant number of these families face serious economic problems as well.

Q: How dangerous is your work?

A: I don't perceive much of a risk/threat from the children or their families. However, many of these families live in dangerous neighborhoods, so it's important that we keep our eyes and ears open when we visit the area.

Q: What do you like about your work?

A: I like working with children and families. I know that sounds ambiguous; I like the challenge and I don't take the responsibility lightly. I really like trying to serve the child's best interests, along with doing what is in the best interest of the community. I also enjoy collaborating with other agencies. We all work together in order to maximize our opportunities to successfully serve the community and its children and families. I also like the flexibility in my schedule, and setting my daily agenda—except for court hearings, you don't have a choice about that.

Q: What do you not like?

A: Being told that there are no "funding streams" available to finance services for a juvenile and their family. Money should be the last consideration, but is often the first and highest hurdle we must overcome.

Other At-Home Community-Based Interventions for Juveniles

Aside from probation, juvenile courts in some jurisdictions may employ a variety of other types of community-based interventions, such as restitution or community service programs, wilderness probation programs, and day treatment programs. In practice, probation is often combined with one of these other community-based interventions. For example, it is common for courts to place a youth on probation and order him or her to make restitution to the victim.

Restitution

Restitution programs require offenders to compensate victims for damage to property or for physical injuries.[32] The primary goal of restitution programs is to hold youths accountable for their actions. In addition, restitution programs may

restitution program
A program that requires offenders to compensate victims for damages to property or for physical injuries.

also seek to provide services or treatment to juveniles and/or provide services or reparation to victims.[33] There are several types of restitution. *Monetary restitution* involves making a cash payment to the victim for harm done. *Victim service restitution* requires that the offender provide some service to the victim. The last type, *community service restitution*, requires the offender to provide some assistance to a community organization.[34] Each of these types has its own rationale. Monetary restitution is intended to help the juvenile offender realize the economic costs incurred by the victim as a result of the offender's behavior. In-kind or in-person victim restitution is intended, among other things, to help humanize the victim. Community service restitution is intended to help the juvenile offender repay the community for the harm it suffered as a result of the youth's actions.

Like research on the effectiveness of other juvenile justice programs, research on the effectiveness of juvenile restitution programs shows mixed results. Data collected as part of a national evaluation of juvenile restitution programs found extremely high rates of compliance with restitution orders. This study found that 95% of youths participating in various federally sponsored restitution programs successfully carried out their restitution orders.[35] However, compliance rates appear to be considerably higher in well-managed and -structured programs than in informal and less structured programs, where offenders are simply ordered to make restitution but where there is little follow-up. One evaluation of structured and unstructured programs conducted in Wisconsin found that only 45% of those youths involved in unstructured programs completed their restitution, compared with 91% of the youths in formal programs.[36]

Other research indicates that restitution programs can also collect substantial amounts of monetary restitution for victims and can produce substantial amounts of community service and victim service hours. One federal assessment of 85 restitution programs found that the programs had collected $2,593,581 in monetary restitution and had gotten participants to perform 355,408 community service hours and 6,052 victim service hours over a 2-year period.[37]

Some studies show that restitution programs can have an impact on recidivism when they are used either prior to or after adjudication. For example, a study of Utah restitution programs found that, when used at both the preadjudicatory stage and as part of formal probation, youths involved in restitution relapsed less than those who received other nonformal responses or youths who were placed on probation alone.[38] A study of four restitution programs located in Boise, Idaho; Washington, D.C.; Clayton County, Georgia; and Oklahoma City, Oklahoma, found that the programs in Washington and Clayton County produced significantly lower rates of recidivism than traditional probation programs in those two jurisdictions. In the two remaining programs, the differences were not significant.[39] Still another study of six experimental restitution programs found that juveniles who were randomly assigned to a restitution program were less likely to relapse than youths placed on probation or given short terms of incarceration.[40]

Despite the growing popularity of juvenile restitution programs,[41] as well as the ability of some programs to effectively reduce recidivism, there are problems, or potential problems, with these programs. Some programs (those that are informal or poorly managed) have low compliance rates,[42] some programs experience high recidivism rates,[43] and some hearing officers make unrealistic orders that cannot be complied with by juveniles. Moreover, concern over who should be responsible for restitution is not uncommon. Is restitution the youth's responsibility or the responsibility of the parents or guardian? The existence of such problems and concerns indicates that not all restitution programs are holding youths responsible for their actions or resulting in satisfactory outcomes from the point of view of the youths involved, their parents, or the victims. Indeed, such problems and concerns can lead to negative perceptions of juvenile justice.

Two additional problems are that restitution programs can contribute to **net-widening**,[44] and restitution requirements can be subject to discretionary abuse.[45] There is always the potential that restitution programs will subject youths to juvenile court control who would otherwise not be subjected to such control. Moreover, once youths fall under the control of a court, they are at risk for additional coercive treatment. Although controlling some youths is appropriate and can even help them and protect the community, in other cases court control is clearly counterproductive: It may result in harm to youths and increased danger to the community. Finally, youths who commit similar offenses may receive substantially different restitution orders, although some jurisdictions have established restitution guidelines in an attempt to remedy this problem.[46]

Wilderness Probation (Outdoor Adventure) Programs

Wilderness probation, or outdoor adventure programs for juvenile offenders, are based, in part, on ideas derived from Outward Bound. A basic assumption underlying these programs is that learning is best accomplished by acting within an environment where there are consequences for one's actions.[47] Consequently, these programs involve youths in a physically and psychologically challenging outdoor experience intended to help participants develop confidence in themselves, learn to accept responsibility for themselves and others, and develop trust in program staff.[48] The participants engage in such activities as camping, backpacking, rock climbing, canoeing, and sailing, and, in many programs, they complete a solo wilderness experience that lasts one or more nights.

Evaluations of wilderness probation programs have indicated that some programs may produce positive effects, such as an increase in self-esteem and a decrease in criminal activity, both during and after the program.[49] In a relatively recent meta-analysis of wilderness programs, Sandra Wilson and Mark Lipsey found that youths in wilderness programs engaged in less post-program delinquency than comparison youths. Moreover, they found that short-term programs lasting up to 6 weeks appeared to be the most effective, along with those with distinct therapeutic components.[50] However, research also indicates that the positive effects associated with these programs may diminish over time[51] or may be no greater than the effects produced by probation programs in which regular and meaningful contact occurs between probation officers and probationers.[52] Moreover, as Wilson and Lipsey note, the evaluations done on these programs have primarily focused on white males already in the system, so the effects of these programs for other populations of youths is unknown.[53]

Day Treatment Programs

Day treatment programs for juvenile offenders are operated in a number of jurisdictions around the United States. These programs, which usually target serious offenders who would otherwise be candidates for institutionalization, provide

net-widening This occurs when youths, who would otherwise not be subjected to formal social control, are subjected to such control because they are placed in a new program. This increases the likelihood that these youths will be subjected to increasingly punitive control.

wilderness probation program A program designed to reform the behavior and attitudes of juvenile offenders by involving them in wilderness activities, including solo camping. The philosophy behind such programs is partly based on the philosophy of Outward Bound.

day treatment program A program that provides treatment or services to youths during the day and allows them to return home at night.

treatment or services to youths during the day and allow the youths to return home at night. Because they are viewed as alternatives to incarceration, they are thought to be cost-effective, and because they provide highly structured programs for youths during the day, they are felt to protect community safety as well. The range of services or treatments may include academic remediation, individual and group counseling, job skills training, job placement services, and social skills training.

An example of a day treatment program is Project New Pride, which was developed in Denver, Colorado, and has been replicated in a number of other cities around the country. This program is intended to provide intensive services to youths who have committed serious offenses. Program services include diagnostic assessment; remedial education; special education for youths with learning disabilities; cultural, physical, and health education; job preparation; job placement; volunteer support; intensive supervision (during the initial 6 months of the program); and follow-up with clients after the initial intensive supervision phase of the intervention.[54]

The few studies done of day treatment programs such as New Pride have not produced clear evidence of their effectiveness. Although there is some evidence that such programs may be as effective or more effective than institutional placement, there is also evidence that some programs may be no more effective at reducing recidivism than standard probation.[55] A key component in many day treatment programs is involvement of the child's parents, guardian, or custodian. This involvement may range from something as simple as aiding in overnight monitoring to undergoing intensive family counseling. Parental involvement is key to the success of these programs, because any lack of involvement or indifference shown by the parents gives a clear message to the child that cooperation is not important. At present, there is a need to conduct sound research on these programs in order to evaluate their effectiveness.

MYTH VS. REALITY

Community-Based Interventions Can Be Effective With Some Serious Juvenile Offenders

Myth—Community-based interventions are not appropriate for serious juvenile offenders.

Reality—Some community-based interventions have been proven to be effective with some serious and chronic offenders. For example, multisystemic therapy (MST) has been found to be a highly effective treatment for many youth engaged in serious and chronic delinquency. MST is a theory- and research-based comprehensive treatment approach that provides intensive family-focused services and is designed to address individual, family, peer, school, and neighborhood factors that encourage serious antisocial behavior. Moreover, MST is designed to foster collaborative relationships with families and other social institutions and to mobilize these institutions in an effort to change behaviors that foster delinquency. Importantly, controlled studies have found that MST is a cost-effective treatment approach that is associated with decreased criminal activity and lower levels of incarceration among program participants.[56]

■ Out-of-Home (Residential) Community-Based Placements for Juvenile Offenders

Although many community-based programs are designed to provide services to youths in their own homes, other programs provide services to youths who are removed from their homes and placed in some type of residential setting. Typically, these <u>community-based placement programs</u> are intended to provide services to youths who are placed out of their homes for a short period of time—from several days to several months. A few programs are longer term and may last for many months or even years.

Foster Homes

<u>Foster homes</u> have an extensive history in the United States. For example, in colonial villages children over the age of 7 who were not appropriately supervised by their parents or who engaged in "the sin of idleness" were placed with other parents.[57] Also, as we noted in Chapter 4, <u>placing out</u> was used as a response to many destitute and problem children throughout much of the 1800s and early 1900s. Indeed, the major response of child welfare organizations, whether private or public, has been to remove children from parents when those parents have been deemed unfit, and foster care has often been the result.[58] Today, foster care is still a popular means for dealing with children whose parents fail to care for them or abuse them, children who come from chaotic families, and children who exhibit behavioral problems. In 2001, some 805,000 children spent time in foster care.[59]

community-based placement An out-of-home placement for a youth offender, often lasting from one day to several months, although occasionally of longer duration.

foster home A family-like setting to which youths are temporarily assigned. In foster care programs, foster parents are licensed to provide care for one or more youths (usually no more than three) and are paid a daily rate (per diem) for the cost of care.

placing out An effort to respond to the growing numbers of problem children in the cities during the 1800s. It consisted of the placement of children on farms and ranches in the Midwest and West.

Box 11-2

Interview: Connie Loviska, Foster Parent

Q: How many years have you been a foster parent?

A: I have been a foster parent for 34 years. My husband and I started with the Kalamazoo County Juvenile Court in 1969.

Q: Who are the members of your family?

A: We have five sons, now all grown, and we adopted a girl who was a foster child. My husband and I were married for 43 years when he died suddenly in December 2002, but I continue to be a foster mother.

Q: What impact has being a foster parent had on your children?

A: A very positive impact. My children are very compassionate people who are very supportive of individuals who are disadvantaged or who are having day-to-day problems. They want to help everyone get a fair chance in life; but, at the same time, they are not tolerant of people who play the "victim" all the time. They all have a well-developed social conscience and sense of community service.

Q: What led you to become a foster parent?

A: Originally, we had relatives that needed help, and we would take care of their children off and on. One of my son's friends was having some difficulty, and the juvenile court got involved. We contacted the court about this child and ended up being licensed by the court as foster parents.

Q: What do you see as the main day-to-day challenges of being a foster parent?

A: The most difficult thing is making the foster children realize that they must obey the rules of the home. In many instances, they come from dysfunctional families that have no boundaries or inconsistent rules. We have to show them that normal families have rules, that normal people all have rules to follow. Closely related to this, of course, is the challenge of holding the foster child accountable for rule violations. Often they don't understand about accountability or responsibility. Last, it is often difficult to cooperate with the foster child's natural family and not come across as judgmental of them. As soon as you become judgmental of the child's family—even if the judgment is accurate—you have erected a barrier and given the child ammunition to use against you.

Q: What are the rewards of being a foster parent?

A: Children that succeed! By success I mean becoming a normal functioning adult, not dependent on welfare, not involved in the criminal justice system. It's the satisfaction of knowing that you helped break the cycle of abuse and dysfunction that brought the child into your home in the first place.

Q: What types of children have you fostered, and do you specialize?

A: We have neglected, emotionally fragile, and delinquent teenagers in our home. We specialize in delinquent teens.

Q: What do you do to positively influence your foster children?

A: We teach them about families and proper family relationships, interaction, and behavior by including them in our family. We work together, we play together, we travel together. Through these family activities, we teach.

Q: Have you ever been victimized by a foster child?

A: Yes, but rarely over the long haul. I have been physically attacked, threatened, and have had property damaged. It's the property damage that's really upsetting. I can understand a flash of anger from a child I'm confronting who might physically lash out at me, but "coolly" calculated property damage hurts all the kids in the home, not just me.

Q: How do you see foster care in the overall continuum of care for juveniles?

A: Of course I believe that foster care is very important for kids in the system. The problem is that it's not used soon enough. In recent years, the courts are leaving kids in dysfunctional homes too long, which makes my job as a foster parent much more difficult. The kids' attitudes become entrenched; and the longer they go being irresponsible, the harder it is for me to teach them to be responsible!

Q: What special training or expertise do you have?

A: No training can replace 34 years of experience. However, I regularly attend conferences to learn new techniques in dealing with difficult children. I go to eight or nine trainings each year through our licensing agency and the state foster care association. I also keep up with the available literature and with TV shows that help.

Q: How much money do you earn from being a foster parent?

A: Foster parenting is not a money-making proposition. The per diem can never reimburse you for all of the little things that are involved in raising children in a family. All of those expenses, from varsity jackets to music, cannot be predetermined or reimbursed. I don't do it for the money, nor did my husband and I do it for the money. We valued what would be called the "nonmonetary" rewards that make foster parenting so worthwhile.

Q: What advice would you give to a person who might want to consider becoming a foster parent?

A: Be certain that you have the support of your entire family and friends. Foster parenting is a lot of work, and it's easy to become isolated. Make it a family decision. When everyone is part of the decision, they aren't so ready to draw lines between "us" and "them," which always leads to conflict.

Q: In your opinion, what are the major challenges facing foster parents in the future?

A: Allegations of improper conduct. It's great that kids today have been educated about abusive behavior and dangerous situations, but this does not mean that every allegation is true or that their opinion about what is abusive or appropriate is accurate. Every foster parent lives in fear of a foster child making false allegations in an attempt to manipulate a situation. Be sure to know why you are a foster parent. As time passes, re-evaluate. You may want to change age groups or types of problems in the children you have, or you may need to take a break. It's okay to need time for yourself.

Q: Is there anything else you would like to say?

A: I intend to stay with it as long as I can. I still love the children and the challenges they bring. The lasting rewards are all of the wonderful relationships that still continue.

Unlike institutional or some other types of community-based programs, foster care is intended to resemble, as much as possible, life in a family-like setting. Foster parents are licensed to provide care for one or more youths (usually one to three) and are paid a daily rate (*per diem*) for the cost of care. Foster placements are often used by courts when a youth's home life has been particularly chaotic or harmful. In these cases, foster homes are used to temporarily separate the conflicting parties—the youth and the parents or guardian—in an effort to resolve those problems that resulted in the youth's removal. However, foster homes are also used instead of more restrictive institutional placements, such as detention or assignment to a state training school, for some nonviolent delinquent offenders. Programs that use foster homes in these two ways can be classified as *halfway-in programs.* Programs that use foster homes as transitional placements for youths who are returning from an institutional placement can be classified as *halfway-out programs.*

Although foster homes are widely used by juvenile courts in some jurisdictions, there are few sound evaluations of the effectiveness of foster home placements for juvenile offenders. Clearly, many foster homes provide a supportive and caring environment for youths and help them lead law-abiding lives. However, the limited research that exists indicates that foster care is not generally effective and may even be counterproductive as a strategy for reducing delinquency.[60] These findings parallel other efforts to examine foster outcomes for non-delinquent populations that demonstrate that many foster care children fare poorly on indicators of well-being such as neurological and cognitive development, and they tend to exhibit behavioral and emotional problems compared to other youths.[61] Although some foster home placements undoubtedly help youths, the short length of time youths often spend in foster homes, the use of multiple foster care placements for youths, the poor or inadequate treatment given some youths in foster care,[62] and the problems that children may bring with them may account for less than favorable results in too many other cases.

Foster Care Is Less Expensive Than Institutional Placements

It is not unusual to pay $100 or more per day to house a youth in an institutional setting, whereas foster parents are often paid $15 per day for the care of a child in nonspecialized foster care and up to $50 for specialized foster care.

The effectiveness of foster care depends in large part on the abilities of the individual foster parents, the characteristics of the youths, and the ability of the foster parents to establish meaningful individual relationships with their foster children. The operation of foster care programs, unfortunately, can act as a barrier to effective or humane care. According to child advocate Ben Wolf, an attorney and the director of the American Civil Liberties Union Roger Baldwin Foundation Children's Initiative Project, foster parents are often mistreated by child welfare agencies. For example, they are often told they will be reimbursed for expenses but remain unpaid, they are often denied opportunities for respite or vacations, they are often denied the training and support needed to work with troubled youths, they may be given little or no information about a child's health or other needs, and they may receive very few backup services.[63]

Although there is good evidence that typical foster care placements may not adequately serve children's needs, there is evidence that *treatment foster care* (TFC) can be an effective response to some youths, including serious and chronic juvenile offenders. One such program was developed by the Oregon Social Learning Center. The program provides specialized training for foster parents and they receive support from a program case manager. The program is designed to reinforce youths' positive behaviors, provide close supervision, assist youths in following rules and providing appropriate consequences for rule violations, support the development of youths' academic skills and work habits, and improve family communication skills. Importantly, evaluations of TFC indicate that it can be more effective than traditional foster or group care programs as well as many other community-based programs at reducing delinquency, and it is more cost effective.[64] Thus, TFC may represent a viable community-based program for some (including serious and chronic) juvenile offenders.

Group Homes and Halfway Houses

Similar to foster homes, group homes and halfway houses are open, nonsecure community-based placement settings[65] that may be used as halfway-in programs (i.e., as a last alternative before institutional placement) or halfway-out programs

group home An open, nonsecure community-based setting, larger and often less "familylike" than a foster home. Group homes provide a place for youths to reside temporarily without having to accept "substitute parents," which can be beneficial in cases where a youth is in the process of developing emotional independence from parental figures.

(i.e., for youths leaving institutional settings). Consequently, group homes and halfway houses are another type of intermediate sanction that is found in many communities. However, group homes and halfway houses are somewhat larger and are often more impersonal and less family-like than foster homes.[66] Typically, these facilities are operated by staff who work shifts, as opposed to "parents" who live in the foster home. In fact, the purpose of many group homes and halfway houses is to avoid requiring youths to accept "substitute parents," partly because many youths who need placement are in the process of developing emotional independence from parental figures.[67] Nonetheless, staff in these facilities serve as important role models, and the facilities themselves are intended to be less impersonal than more closed institutions and are often designed to individualize treatment as much as possible. They are also less expensive than more restrictive institutional placements, and because they are often located in urban or suburban neighborhoods, residents have an opportunity to take advantage of a variety of community services. For example, youths who reside in group homes and halfway houses may go to school in the community or work in the community.[68] They may also participate in recreational and cultural activities in the community and use various types of treatment and educational services. Group home and halfway house treatment programs typically arrange for individual and group counseling to be provided by group home staff or outside counselors. In addition, residents may receive academic instruction and participate in social and vocational skills-building courses within the facility or in the community.

Although group homes are used extensively in juvenile justice, there is relatively little sound research that has examined the effectiveness of these programs. One well-known study of a group home program, the Silverlake Experiment, did find that youths placed in this program for approximately 6 months were neither more nor less likely to engage in subsequent delinquency than similar youths who were randomly assigned to a more restrictive institutional setting.[69]

There is no doubt, however, that group home programs often face a number of obstacles that act as barriers to the effective treatment of youths. These barriers include poor pay and benefits for staff, a lack of staff training on how to work with troubled youths, a paucity of support services for staff, and local resistance to allowing group homes into the community. Moreover, short-term group home placements do not allow sufficient time to affect youths in a positive way, and large group homes are not able to offer the types of individualized treatment that many youths need.

Another problem that sometimes plagues group homes, halfway houses, and other types of residential placement settings is the victimization of residents. Although the extent of this problem is not clear, there have been instances where youths residing in such settings have been victimized by other residents or even staff. Robert Mutchnick and Margaret Fawcett, in a study of nine group homes in southwestern Pennsylvania, found that some of the homes were more likely to have high levels of resident victimization than others. Group homes with high levels of resident victimization were characterized by extremely open or closed environments, a lack of program structure, staff who displayed little concern for security, and residents who expressed attitudes of mistrust and negativism. In contrast, homes with low levels of resident victimization were characterized by open but structured environments, close staff supervision, consistent enforcement of policies and rules, staff who demonstrated concern for residents, and residents who exhibited relatively positive attitudes.[70]

Shelter-Care Placements

Shelter-care settings are like group homes, although they often hold youths for shorter periods of time. A *shelter-care program* is a type of nonsecure program that typically provides a temporary residence for youths prior to returning them home or determining the suitability of some other placement. Often shelter care is used for youths who have run away from home, who have been abused by their parents, who have engaged in minor types of delinquency, or who need temporary emergency placement. Shelter care typically involves short-term crisis intervention services, basic counseling, and basic educational programming.

Family Group Homes

Family group homes are similar to both foster homes and group homes—they are run by foster parents but house more youths than a regular foster home. Although family group homes are designed to provide youths with parent substitutes and a more family-like environment than might be found in a traditional group home, and although they have sometimes been advocated as a way of reducing community opposition to larger institutional facilities, little is known about the effectiveness of family group home programs. Moreover, these programs can suffer from the same problems that confront traditional foster care and group home programs.

■ The Effectiveness of Community-Based Corrections for Juvenile Offenders

As indicated earlier, the research that has been done on community-based correctional programs has produced mixed results. Studies of some community-based programs indicate they are as effective as more restrictive institutional placements at reducing recidivism. There is also evidence that community-based programs are far less costly than institutional programs, in some cases only half as costly.[71] It should be noted, however, that evaluations of other programs indicate that they have little effect on subsequent offending. Such findings are not surprising, however, given the diversity in both the types of programs and the quality of programs that are in operation around the country. Despite the mixed research results on the effectiveness of community-based programs, the efforts of some states to deinstitutionalize their juvenile offender populations and move toward a more community-based system have met with considerable success.[72]

A good example of a state that has moved toward a more community-based correctional system is Massachusetts. Under the leadership of a new commissioner of youth corrections, Dr. Jerome Miller, Massachusetts began closing its state training schools in the 1970s. At first, Miller attempted to implement reforms in the Massachusetts institutions where he observed frequent abuse of residents. When these reforms were resisted, he began closing the training schools and developed or used community-based programs for youths who were released from the institutions. Despite the abuse and inhumane conditions within Massachusetts' training schools, Miller continued to face strong opposition to his efforts to deinstitutionalize the juvenile offender population. Moreover, many politicians and others allied with the existing system of institutions argued that deinstitutionalization would lead to an increase in juvenile crime as "hardened" young offenders were released into the community. However, despite the protestations of those opposed to deinstitutionalization, a juvenile crime wave did not sweep the state.[73] In fact, evaluations of deinstitutionalization in Massachusetts conducted by Harvard's Center for Criminal Justice and by the National Council on Crime and Delinquency found that the community-based programs were effective alternatives to institutions and did not threaten public safety.[74] In communities that had a variety of quality programs for offenders, recidivism decreased rather than increased, contrary to what skeptics had argued would happen.[75] The Massachusetts experiment initiated by Miller demonstrated that a network of high-quality community-based programs can serve as a viable alternative to a system of juvenile correctional institutions.

MYTH VS. REALITY — Many Youths in Institutions Could Be Better Served in Community-Based Programs

Myth—Most youths housed in juvenile correctional institutions pose an immediate threat to public safety.

Reality—As Jerome Miller notes in his book *Last One Over the Wall: The Massachusetts Experiment in Closing Reform Schools*, many youths in juvenile correctional facilities could be more effectively served in community-based programs without endangering community safety. Miller points out that in Massachusetts only about 25% of the youths committed to state correctional facilities had committed offenses against persons. Further, many of these supposedly violent offenses did not actually involve physical violence or the threat of physical violence. This shows that many of the youths who are placed in correctional facilities, including those confined for "violent offenses," do not pose a grave threat to public safety.[76]

Linking Institutional and Community-Based Corrections: Aftercare Programs

Aftercare involves the provision of services intended to help youths successfully make the transition from juvenile institutions to life back in the community. The services are the same as those provided by the community-based programs described earlier. Indeed, aftercare may encompass foster care, shelter care, or group home placement; home placement; or efforts to help youths live on their own (independent living). Parole, which consists of the conditional release of a youth from a correctional facility, is one form of aftercare. Youths on parole are required to

aftercare The provision of services intended to assist youths in successfully making the transition from juvenile institutions to life in the community.

parole The conditional release of a youth from a correctional facility (one form of aftercare).

adhere to certain conditions, such as avoidance of illegal behaviors, attendance at school, and attendance at meetings with their parole officers upon request. If youths violate these conditions, they may be returned to an institution. Parole services are typically provided by a parole officer who is a state employee. However, a variety of aftercare services are provided to youths, and these services may be provided by court probation staff or even by staff hired by an institution to provide follow-up services once a youth returns to the community.

Unfortunately, the quality of many aftercare programs is questionable, and, in addition, many youths fail to receive any services after institutional release. Parole supervision may actually involve very little contact between parole officers and parolees. Moreover, large caseloads carried by aftercare workers may preclude the provision of meaningful services. In other instances, the supervision that occurs may be primarily focused on surveillance and control and may provide little in the way of treatment or service provision. Indeed, research on parole has not found it to be particularly effective. In one study of juvenile parolees in California, it was discovered that there were no differences in the likelihood of a subsequent arrest among youths randomly assigned to parole supervision or to discharge without supervision. However, those who were given parole supervision were more likely to be arrested for a serious offense.[77] In another California study, youths who were placed on intensive parole supervision after a short institutional stay were as likely to have their parole suspended for illegal activity as youths placed in California Youth Authority facilities.[78] (Keep in mind that efforts to deinstitutionalize correctional populations have not led to increases of recidivism when a range of quality community-based programs are available.)[79]

More recently, the Office of Juvenile Justice and Delinquency Prevention (OJJDP) has supported an effort to develop an effective approach to aftercare called the *intensive aftercare program* (IAP). Developed by David Altschuler and Troy Armstrong, the IAP is intended to be a theory-driven comprehensive approach to juvenile aftercare that overcomes the institutional boundaries that frequently prevent institutions and community-based programs from developing and implementing sound collaborative plans for youths' return to the community. A key component of the IAP model is the concept of *overarching case management* that consists of five elements:

- Risk assessment designed to identify and focus services on high-risk youths
- Individualized case planning that includes both institutional and community-based staff as well as family and community members and addresses youths' needs within their families, peers, schools, and other socializing institutions
- A mix of both intensive surveillance and services
- The use of a graduated system of incentives for positive behavior and sanctions for violations of program rules
- The development of links with community resources that can support program success[80]

Presently the IAP model is being implemented in demonstration sites; early evaluation results indicate that the IAP model has received considerable support and has resulted in improved relationships between institutional and aftercare staff. However, outcome results are not yet available. Nevertheless, there is some evidence that these programs as well as similar programs may represent a more viable form of aftercare than has traditionally been provided to youths who are released from institutional settings.[81]

■ Legal Issues

Community service restitution programs carry with them a possibility of civil liability. If a juvenile is injured doing community service, the supervising court, the probation officer, or the community agency could be held legally responsible for any supervisory negligence. They could also be held legally responsible if the juvenile, during the course of doing the community service, injures someone or damages property in a negligent or reckless manner.

The monitoring of delinquent youths in a community-based corrections program raises the issue of the legal authority of probation and parole officers, foster parents, and group home personnel to apprehend, take into custody, and generally control youths under their supervision. In many states, probation officers are considered to be government officials with the power to arrest, apprehend, detain, or place a child.[82] Probation officers are usually "sworn" employees of the court or government agency. Foster parents, unless they are sworn probation officers, which would be unusual, are not empowered as probation officers and would have to rely on a liaison probation officer or law enforcement officer to help them control youths placed in their homes. Group homes can be privately run or can run under court auspices. If private, then group home personnel would not be considered government officials. If the group home was run by the court, the employees could be considered probation officers.

Juvenile courts by their very nature deal primarily with juveniles. A juvenile court has a legitimate expectation that, for each juvenile it deals with, there will be a responsible adult or adults, either parents, guardians, or custodians, in the home. Just as a juvenile on probation is subject to orders of the court, in many jurisdictions adults who are responsible for the care of a juvenile can also be subject to court orders. As mentioned earlier, in many jurisdictions, adults can be ordered to make monetary restitution for their children's acts, and this restitution would necessarily be court ordered. The payment of restitution by minors raises the issue of the means or method a juvenile court can use to enforce its orders over an adult. The most commonly used means is to cite the adult for contempt of court for failure to obey a court order. Two kinds of contempt are generally recognized, civil and criminal:

1. Criminal contempt usually results from an act done in the presence of the court during a hearing or while the court is in session, such as swearing at the judge. This kind of contempt does not require a hearing but may be summarily determined by the judge, and sanctions, including jail, may be levied.

2. Civil contempt is usually associated with an individual's failure to obey a court order that has been legally served on the individual. In a case of restitution, the responsible parent would be given an opportunity in a court hearing to "show cause" as to why he or she is not in contempt

civil contempt Willful failure of an individual to obey a court order that has been legally served on the individual. Consequences may include monetary fines, court costs, and jail for limited time periods.

and could usually "purge" him- or herself of contempt by obeying the order—even after a hearing that resulted in the parent being found in contempt of court. Usual consequences for contempt include monetary fines, court costs, and jail for limited time periods. One of the harsh realities of the adult criminal justice system is that, due to chronic jail overcrowding, persons held for contempt are often the first to be released if there is a shortage of beds.

■ Chapter Summary

This chapter examined the operation and effectiveness of community-based correctional programs for juvenile offenders. These types of correctional programs consist of (1) at-home programs, such as probation, restitution, community service, day treatment, house arrest, and aftercare or parole programs, which are typically designed for youths who reside in their own homes, and (2) out-of-home programs, such as foster care, group homes, shelter care, and family group home programs, which provide services to youths who are removed from their own homes and placed in residential settings.

Like other correctional programs, community-based programs attempt to accomplish a number of objectives. Although individual programs vary in the extent to which they emphasize these objectives, they typically include sanctioning and controlling youths, helping them maintain existing ties within the community, helping them restore ties that have been severed as a result of their delinquent behavior, assisting their efforts to develop new and positive ties to others in the community, providing a correctional response that avoids the negative consequences of institutional placement, providing a more cost-effective response to juvenile crime, and reducing the likelihood of recidivism.

Since the development of the juvenile court in the late 1800s, the most popular correctional response to juvenile offenders has been probation. Despite its long history, however, little is known about the effectiveness of probation. Moreover, what is known suggests that, although standard probation alone is reasonably effective with nonserious juvenile offenders, it is not very effective with chronic or serious offenders. As a result, many jurisdictions have developed intensive supervision programs that are intended to increase levels of offender supervision and service. In addition, a number of jurisdictions have developed a variety of other interventions (e.g., restitution and community service, day treatment, and house arrest) that are often used in combination with probation. Although research on these interventions has produced mixed results, there is evidence that some programs that employ these interventions help youths avoid subsequent offending behavior. Moreover, there is sound evidence that some community-based interventions, even when used with chronic and serious juvenile offenders, can be effective in reducing recidivism and avoiding the negative aspects of incarceration.

Similar to studies of at-home community-based programs, those that have examined residential community-based programs have also produced mixed results. Nevertheless, at least some studies indicate that residential community-based programs are as effective as or more effective than institutionalization at reducing recidivism. In addition, research indicates that residential community-based corrections are considerably less expensive.

Because of the paucity of research, it is difficult to make definitive statements about the effectiveness of the great majority of community-based correctional programs. There is good evidence that some programs not only provide high-quality services and successfully address a variety of client needs, but also do an effective job of protecting public safety. There is, unfortunately, good evidence that other programs fail to do any of these things. Nevertheless, community-based programs clearly play a critical role in the juvenile correctional system, and they will continue to play an even more important role in the future.

KEY CONCEPTS

aftercare The provision of services intended to assist youths in successfully making the transition from juvenile institutions to life in the community.

civil contempt Willful failure of an individual to obey a court order that has been legally served on the individual. Consequences may include monetary fines, court costs, and jail for limited time periods.

community-based corrections Noninstitutional correctional programs and interventions.

community-based placement An out-of-home placement for a youth offender, often lasting from one day to several months, although occasionally of longer duration.

day treatment program A program that provides treatment or services to youths during the day and allows them to return home at night.

electronic monitoring The ability to determine an individual's movement by having him or her wear a tamper-resistant electronic device that allows communication between the device and a computer housed at the monitoring agency.

formal processing Case processing that results in an adjudication and initiation of a formal juvenile court record.

foster home A family-like setting to which youths are temporarily assigned. In foster care programs, foster parents are licensed to provide care for one or more youths (usually no more than three) and are paid a daily rate (per diem) for the cost of care.

group home An open, nonsecure community-based setting, larger and often less "familylike" than a foster home. Group homes provide a place for youths to reside temporarily without having to accept "substitute parents," which can be beneficial in cases where a youth is in the process of developing emotional independence from parental figures.

home confinement The state of being confined to home. A home confinement program requires the participants to spend certain hours at home. Supervision in these programs is done by probation officers who make random checks on clients or monitor compliance by electronic monitoring. Also called *home arrest* or *house arrest*.

house arrest See **home confinement**.

intensive supervision program A program intended to ensure that there is regular contact between probationers and probation officers. Although the contact is regular, its frequency varies considerably across programs.

intermediate sanction A diverse range of responses (e.g., intensive supervision, day treatment, house arrest) to offenders that are more restrictive than standard probation but less restrictive than institutional placement.

net-widening This occurs when youths, who would otherwise not be subjected to formal social control, are subjected to such control because they are placed in a new program. This increases the likelihood that these youths will be subjected to increasingly punitive control.

nonformal processing Includes various ways of processing cases that do not involve an adjudication and the initiation of a formal juvenile court record for the youth.

parole The conditional release of a youth from a correctional facility (one form of aftercare).

probation The conditional release of an adjudicated delinquent into the community under the supervision of the court.

placing out An effort to respond to the growing numbers of problem children in the cities during the 1800s. It consisted of the placement of children on farms and ranches in the Midwest and West.

presentence investigation An investigation that is typically completed by a probation officer prior to the disposition and that is intended to assist the hearing officer in deciding the best response to an adjudicated youth.

restitution program A program that requires offenders to compensate victims for damages to property or for physical injuries.

wilderness probation program A program designed to reform the behavior and attitudes of juvenile offenders by involving them in wilderness activities, including solo camping. The philosophy behind such programs is partly based on the philosophy of Outward Bound.

REVIEW QUESTIONS

1. What is the definition of community-based juvenile corrections?

_____

_____

_____

_____

_____

2. What types of at-home and out-of-home (residential) community-based corrections programs are discussed in the text? Describe each one.

_____

_____

_____

_____

_____

3. What are the objectives of community-based corrections programs?

_____

_____

_____

_____

_____

4. What is probation?

_____

_____

_____

_____

_____

5. What are some typical rules of probation?

_____

_____

_____

_____

_____

6. What factors influence the operation of probation between and within jurisdictions?

_____

_____

_____

_____

_____

7. What are the basic roles performed by juvenile court probation officers? What do these roles entail?

_____

_____

_____

_____

_____

8. According to the research, how effective is standard probation?

_____

_____

_____

_____

_____

9. What are some recent trends in juvenile probation?

_____

_____

_____

_____

_____

10. How effective are community-based correctional programs at reducing recidivism?

_____

_____

_____

_____

_____

Chapter Resources

11. hat types of legal authority are possessed by juvenile probation officers?

_____

_____

_____

_____

_____

12. What legal issues are raised by juvenile court-ordered restitution?

_____

_____

_____

_____

_____

ADDITIONAL READINGS

Empey, L. T. and M. L. Erickson. 1972. *The Provo Experiment: Evaluating Community Control of Delinquency*. Lexington, MA: Lexington Books.

Empey, L. T., and S. G. Lubeck. 1971. *The Silverlake Experiment: Testing Delinquency Theory and Community Intervention*. Chicago: Aldine.

Gendreau, P. 1994. "Intensive Rehabilitation Supervision: The Next Generation in Community Corrections?" *Federal Probation* 58: 72–78.

Gies, S. V. 2003. "After Care Services." *Juvenile Justice Bulletin*. Washington, DC: Office of Juvenile Justice and Delinquency Prevention.

Jacobs, M. D. 1990. *Screwing the System and Making It Work: Juvenile Justice in the No-Fault Society*. Chicago: University of Chicago Press.

Jones, M. A. and B. Krisberg. 1994. *Images and Reality: Juvenile Crime, Youth Violence, and Public Policy*. San Francisco: National Council on Crime and Delinquency.

Miller, J. G. 1991. *Last One over the Wall: The Massachusetts Experiment in Closing Reform Schools*. Columbus, OH: Ohio State University Press.

Schneider, A. L. 1986. "Restitution and Recidivism Rates of Juvenile Offenders: Results from Four Experimental Studies." *Criminology* 24: 533–552.

Sherman, L. W. et al. 1997. *Preventing Crime: What Works, What Doesn't, What's Promising*. Report to the U.S. Congress. Washington, DC: National Institute of Justice.

NOTES

1. E. J. Latessa and H. E. Allen, *Corrections in the Community* (Cincinnati, OH: Anderson, 1997).
2. L. E. Ohlin et al., *Juvenile Correctional Reform in Massachusetts* (Washington, DC: U.S. Government Printing Office, 1977).
3. B. R. McCarthy and B. J. McCarthy, Jr., *Community-Based Corrections*, 2nd ed. (Pacific Grove, CA: Brooks/Cole, 1991).

4. C. M. Puzzanchera, "Juvenile Delinquency Probation Caseload, 1990–1999." *OJJDP Fact Sheet* (Washington, DC: Office of Juvenile Justice and Delinquency Prevention, 2003).

5. See M. D. Jacobs, *Screwing the System and Making It Work: Juvenile Justice in the No-Fault Society* (Chicago: University of Chicago Press, 1990), 125–129; H. T. Rubin, "The Juvenile Court Landscape," in *Juvenile Justice: Policies, Programs, and Services*, ed. A. R. Roberts (Chicago: Nelson-Hall, 1998).

6. See Mich. Comp. Laws Ann. 712A.9; Mich. Stat. Ann. 27.3178(598.9).

7. P. M. Torbet, "Juvenile Probation: The Workhorse of the Juvenile Justice System," *Juvenile Justice Bulletin* (Washington, DC: Office of Juvenile Justice and Delinquency Prevention, March 1996).

8. Puzzanchera, "Juvenile Delinquency Probation Caseload, 1990–1999."

9. Torbet, "Juvenile Probation."

10. Jacobs, *Screwing the System and Making It Work*; W. A. Reese et al., "Juvenile Justice as People Modulating: A Case Study of Progressive Delinquent Dispositions," *Journal of Research in Crime and Delinquency* 26 (1989): 329–357.

11. L.E. Ohlin et al., "Major Dilemmas of the Social Worker in Probation and Parole," *National Probation and Parole Officer Journal* 2 (1956): 211–225; J. T. Whitehead and S. P. Lab, *Juvenile Justice: An Introduction* (Cincinnati, OH: Anderson, 1990).

12. Whitehead and Lab, *Juvenile Justice*.

13. Torbet, "Juvenile Probation."

14. L. T. Empey and M. L. Erickson, *The Provo Experiment: Evaluating Community Control of Delinquency* (Lexington, MA: Lexington Books, 1972); C. A. Murray and L. A. Cox, *Beyond Probation: Juvenile Corrections and the Chronic Delinquent* (Beverly Hills, CA: Sage, 1979); National Institute of Mental Health, *Community-Based Corrections* (Washington, DC: U.S. Government Printing Office, 1973); A. Reiss, "Delinquency as the Failure of Personal and Social Control," *American Sociological Review* 16 (1951): 196–207; F. R. Scarpitti and R. M. Stephenson, "A Study of Probation Effectiveness," *Journal of Criminal Law, Criminology, and Police Science* 59 (1968): 361–369; J. D. Wooldredge, "Differentiating the Effects of Juvenile Court Sentences on Eliminating Recidivism," *Journal of Research on Crime and Delinquency* 25 (1988): 264–300.

15. Empey and Erickson, *The Provo Experiment*.

16. Scarpitti and Stephenson, "A Study of Probation Effectiveness."

17. S. P. Lab and J. T. Whitehead, "An Analysis of Juvenile Correctional Treatment," *Crime and Delinquency* 34 (1988): 60--83; D. Lipton et al., *The Effectiveness of Correctional Treatment: A Survey of Treatment Evaluation Studies* (New York: Praeger, 1975); J. Robison and G. Smith, "The Effectiveness of Correctional Programs," *Crime and Delinquency* 17 (1971): 67–80.

18. D. F. Greenberg, "Problems in Community Corrections," *Issues in Criminology* 10 (1975): 1–33; J. H. Hylton, "Rhetoric and Reality: A Critical Appraisal of Community-Correctional Programs," *Crime and Delinquency* 28 (1982): 341–373.

19. Empey and Erickson, *The Provo Experiment*; Scarpitti and Stephenson, "A Study of Probation Effectiveness"; Wooldredge, "Differentiating the Effects of Juvenile Court Sentences on Eliminating Recidivism."

20. T. L. Armstrong, "National Survey of Juvenile Intensive Supervision," parts 1 and 2, *Criminal Justice Abstracts* 20 (1988): 342–348, 497–523.

21. W. H. Barton and J. A. Butts, "Viable Options: Intensive Supervision Programs for Juvenile Delinquents," *Crime and Delinquency* 36 (1990): 238–256; R. G. Wiebush, "Juvenile Intensive Supervision: The Impact on Felony Offenders Diverted from Institutional Placement," *Crime and Delinquency* 39 (1993): 68–89; E. Walker, "The Community Intensive Treatment for Youth Program: A Specialized Community-Based Program for High-Risk Youth in Alabama," *Law and Psychology Review* 13 (1989): 175–199.

22. Barton and Butts, "Viable Options: Intensive Supervision Programs for Juvenile Delinquents."

23. A. A. Robertson et al., "A Short-Run Cost-Benefit Analysis of Community-Based Interventions for Juvenile Offenders," *Crime and Delinquency* 47 (2001): 265–284.

24. J. Ryan, "Who Gets Revoked? A Comparison of Intensive Supervision Success and Failures in Vermont," *Crime and Delinquency* 43 (1997): 104–118.

25. H. N. Snyder and M. Sickmund, *Juvenile Offenders and Victims: A National Report* (Washington, DC: Office of Juvenile Justice and Delinquency Prevention, 1995).

26. Barton and Butts, "Viable Options: Intensive Supervision Programs for Juvenile Delinquents."

27. R. A. Ball, et al., *House Arrest and Correctional Policy: Doing Time at Home* (Newbury Park, CA: Sage, 1988).

28. T. M. Young and D. M. Pappenfort, *Secure Detention of Juveniles and Alternatives to Its Use: National Evaluation Report, Phase I* (Washington, DC: U.S. Government Printing Office, 1977).

29. D. Ford and A. K. Schmidt, "Electronically Monitored Home Confinement," *Research in Brief* (Washington, DC: National Institute of Justice, 1985); M. Renzema and D. T. Skelton, "Use of Electronic Monitoring in the United States: 1989 Update," *Research in Brief* (Washington, DC: National Institute of Justice, 1990).

30. S. Roy, "Five Years of Electronic Monitoring of Adults and Juveniles in Lake County, Indiana: A Comparative Study of Factors Related to Failure," *Journal of Crime and Justice* 20 (1997): 141–160.

31. J. B. Vaughn, "A Survey of Juvenile Electronic Monitoring and Home Confinement Programs," *Juvenile and Family Court Journal* 40 (1989): 1–36.

32. A. M. Durham, III, *Crisis and Reform: Current Issues in American Punishment* (Boston: Little, Brown & Co., 1994).

33. A. L. Schneider and J. S. Warner, *National Trends in Juvenile Restitution Programming* (Washington, DC: Office of Juvenile Justice and Delinquency Prevention, 1989).

34. L. J. Siegel et al., *Juvenile Delinquency: Theory, Practice and Law*, 8th ed. (Belmont, CA: Wadsworth/Thompson, 2003).

35. P. Schneider et al., *Two Year Report on the National Evaluation of the Juvenile Restitution Initiative: An Overview of Program Performance* (Eugene, OR: Institute of Policy Analysis, 1982).

36. A. Schneider and P. Schneider, "A Comparison of Programmatic and 'Ad Hoc' Restitution in Juvenile Courts," *Justice Quarterly* 1 (1984): 529–548.

37. P. Schneider, "Research on Restitution: A Guide to Rational Decision Making," in *Guide to Juvenile Restitution*, ed. A. L. Schneider (Washington, DC: Department of Justice, 1985).

38. J. A. Butts and H. N. Snyder, "Restitution and Juvenile Recidivism," *Research Update* (Washington, DC: Office of Juvenile Justice and Delinquency Prevention, 1992).

39. A. L. Schneider, "Restitution and Recidivism Rates of Juvenile Offenders: Results from Four Experimental Studies," *Criminology* 24 (1986): 533–552.

40. L. Ervin and A. Schneider, "Explaining the Effects of Restitution on Offenders: Results from a National Experiment in Juvenile Courts," in *Criminal Justice, Restitution, and Reconciliation*, ed. B. Galaway and J. Hudson (Monsey, NY: Criminal Justice Press, 1990).

41. Schneider and Warner, *National Trends in Juvenile Restitution Programming*.

42. Schneider and Schneider, "Comparison of Programmatic and 'Ad Hoc' Restitution in Juvenile Courts."

43. Schneider, "Research on Restitution."

44. B. Krisberg and J. Austin, "The Unmet Promise of Alternatives to Incarceration," *Crime and Delinquency* 28 (1982): 374–409.

45. Siegel et al., *Juvenile Delinquency*.

46. Siegel et al., *Juvenile Delinquency*.

47. G. L. Golins, *Utilizing Adventure Education to Rehabilitate Juvenile Delinquents* (Las Cruces, NM: Educational Resources Information Center, Clearinghouse on Rural Education and Small Schools, 1980).

48. Golins, *Utilizing Adventure Education to Rehabilitate Juvenile Delinquents*; K. I. Minor and P. Elrod, "The Effects of a Probation Intervention on Juvenile Offenders' Self-Concepts, Loci of Control, and Perceptions of Juvenile Justice," *Youth and Society* 25 (1994): 490–511.

49. R. Callahan, "Wilderness Probation: A Decade Later," *Juvenile and Family Court Journal* 36 (1985): 31–35; F. J. Kelly and D. J. Baer, "Physical Challenge as a Treatment for Delinquency," *Crime and Delinquency* 12 (1971): 437–445; J. Winterdyk and R. O. Roesch, "A Wilderness Experiential Program as an Alternative for Probationers: An Evaluation," *Canadian Journal of Criminology* 24 (1982): 39–49.

50. S. J. Wilson and M. W. Lipsey, "Wilderness Challenge Programs for Delinquent Youth: A Meta-Analysis of Outcome Evaluations," *Evaluation and Program Planning* 23 (2000): 1–12.

51. J. Winterdyk and C. Griffiths, "Wilderness Experience Programs: Reforming Delinquents or Beating Around the Bush?" *Juvenile and Family Court Journal* 35 (1984): 35–44; Winterdyk and Roesch, "A Wilderness Experiential Program as an Alternative for Probationers."

52. H. P. Elrod and K. I. Minor, "Second Wave Evaluation of a Multi-Faceted Intervention for Juvenile Court Probationers," *International Journal of Offender Therapy and Comparative Criminology* 36 (1992): 247–262.

53. Wilson and Lipsey, "Wilderness Challenge Programs for Delinquent Youth: A Meta-Analysis of Outcome Evaluations."

54. Office of Juvenile Justice and Delinquency Prevention, *Project New Pride: Replication* (Washington, DC: Office of Juvenile Justice and Delinquency Prevention, 1979).

55. See T. Palmer, *The Re-Emergence of Correctional Intervention* (Newbury Park, CA: Sage, 1992).

56. S. W. Henggeler et al., "Multisystemic Therapy," in *Blueprints for Violence Prevention*, ed. D. S. Elliott (Denver, CO: C&M Press, 1998).

57. R. Golden, ed., *Disposable Children: America's Welfare System* (Belmont, CA: Wadsworth, 1997).

58. Golden, *Disposable Children*.

59. S. Vandivere et al., "Children in Foster Homes: How Are They Faring?" *Research Brief* (Washington, DC: Child Trends, 2003).

60. H. P. Elrod and M. P. Brown, "Individual, Court and Offense-related Variables as Predictors of Subsequent Offense Activity among Juvenile Court Probationers: A Preliminary Analysis" (paper presented at the American Society of Criminology Annual Meetings, San Francisco, 1991); J. McCord et al., "The Effects of Foster-Home Placement in the Prevention of Adult Antisocial Behavior," in *Prevention of Delinquency*, ed. J. R. Stratton and R. M. Terry (New York: Macmillan, 1968).

61. Vandivere et al., "Children in Foster Homes: How Are They Faring?"

62. H. T. Rubin, *Juvenile Justice: Policy, Practice and Law*, 2nd ed. (New York: Random House, 1985).

63. B. Wolf, "Advocate's Narrative," in *Disposable Children: America's Welfare System*, ed. R. Golden (Belmont, CA: Wadsworth, 1997).

64. P. Chamberlain, "Comparative Evaluation of Specialized Foster Care for Seriously Delinquent Youths: A First Step," *Community Alternatives: International Journal of Family Care* 2 (1990): 21–36; P. Chamberlain and J. B. Reid, "Comparison of Two Community Alternatives to Incarceration for Chronic Juvenile Offenders," *Journal of Consulting and Clinical Psychology* 66 (1998): 624–633; for a review see P. Chamberlain, "Treatment Foster Care," *Juvenile Justice Bulletin* (Washington, DC: Office of Juvenile Justice and Delinquency Prevention, 1998).

65. J. O. Finckenauer, *Juvenile Delinquency and Corrections: The Gap between Theory and Practice* (Orlando, FL: Academic Press, 1984).

66. Finckenauer, *Juvenile Delinquency and Corrections*.

67. C. E. Simonsen, *Juvenile Justice in America*, 3rd ed. (New York: Macmillan, 1991).

68. Simonsen, *Juvenile Justice in America*.

69. L. T. Empey and S. G. Lubeck, *The Silverlake Experiment: Testing Delinquency Theory and Community Intervention* (Chicago: Aldine, 1971).

70. R. J. Mutchnick and M. Fawcett, "Group Home Environments and Victimization of Resident Juveniles," *International Journal of Offender Therapy and Comparative Criminology* 35 (1991): 126–142.

71. P. W. Greenwood et al., "Diverting Children from a Life of Crime: Measuring Costs and Benefits," rev. ed. (Santa Monica, CA: Rand, 1996); A. Rutherford and O. Benger, *NEP Phase I Summary Report: Community-Based Alternatives to Juvenile Incarceration* (Washington, DC: U.S. Government Printing Office, 1976), 27–31; H. Yoshikawa, "Prevention as Cumulative Protection: Effects of Early Family Support and Education on Chronic Delinquency and Its Risks," *Psychological Bulletin* 115 (1994): 1–27; H. Yoshikawa, "Long-Term Effects of Early Childhood Programs on Social Outcomes and Delinquency." *The Future of Children* 5 (1995): 51–75.

72. National Council on Crime and Delinquency, Juvenile Justice: Tough Enough (San Francisco: National Council on Crime and Delinquency, n.d.). Also see R. J. Lundman, *Prevention and Control of Juvenile Delinquency* (New York: Oxford University Press, 2001).

73. J. G. Miller, *Last One over the Wall: The Massachusetts Experiment in Closing Reform Schools* (Columbus, OH: Ohio State University Press, 1991).

74. L. Ohlin et al., *Report on Massachusetts Department of Youth Services, 1975* (Cambridge, MA: Center for Criminal Justice, Harvard University Law School, 1975), cited in Miller, *Last One over the Wall*, 220–222; B. Krisberg et al., *Working Juvenile Corrections: Evaluating the Massachusetts Department of Youth Services* (San Francisco: National Council on Crime and Delinquency, 1989), cited in Miller, *Last One over the Wall*, 223–225.

75. L. Ohlin et al., *Report on Massachusetts Department of Youth Services*, 1975, cited in Miller, *Last One over the Wall*, 220–222.

76. Miller, *Last One over the Wall*, 193–198.

77. P. Jackson, *The Paradox of Control: Parole Supervision of Youthful Offenders* (New York: Praeger, 1983).

78. P. Lerman, *Community Treatment and Social Control: A Critical Analysis of Juvenile Correctional Policy* (Chicago: The University of Chicago Press, 1975).

79. Miller, *Last One over the Wall*.

80. D. M. Altschuler and T. L. Armstrong, *Intensive Aftercare for High-Risk Juveniles: A Community Care Model. Summary.* (Washington, DC: Office of Juvenile Justice and Delinquency Prevention, 1994); S. V. Gies, "Aftercare Services," *Juvenile Justice Bulletin* (Washington, DC: Office of Juvenile Justice and Delinquency Prevention, 2003).

81. Gies, "Aftercare Services."

82. Mich. Court Rule 5.903(11).

Institutional Correctional Programs for Juvenile Offenders

12

Chapter Outline

■ Introduction

Institutional or residential placement is the most restrictive type of placement used by juvenile courts. Indeed, a distinguishing characteristic of institutional placement is that it restricts youths' access to the community to a greater degree than does assignment to a community-based program. Institutional and residential placement settings include a variety of residential facilities, such as juvenile detention facilities, jails, diagnostic and reception centers, forestry camps, ranches, boot camps, and state training schools, among others. Although many of these facilities house offenders who have committed a variety of offenses, others house specialized offender populations, such as sex offenders and youths with mental health and substance abuse problems.

Although institutions for juvenile offenders have existed since the early 1800s and have played an important role in juvenile corrections, they have become increasingly popular as tools for responding to juvenile crime in recent years. The growing popularity of institutional placement appears to be in part a result of increased violence among some juvenile offenders, but it also appears to reflect the belief by many policy makers that the best approach to juvenile crime is to "get tough" with juvenile offenders. Presently, institutional placement is used for many youths charged with violent juvenile offenses. However, today, as in the past, this type of placement is used not only for youths charged with and adjudicated for serious offenses, but also for those charged with and adjudicated for minor offenses. Indeed, in many instances it is used for youths who could be better and more cheaply served by community-based programs.

Despite the lengthy history of juvenile institutions and their increasing popularity, there is little evidence that the incarceration of youths is an effective correctional response to juvenile crime. As we noted in Chapter 4, early institutions for juveniles faced a number of problems, including overcrowding, mixing minor or status offenders with serious offenders, poor programming, inhumane treatment of residents, inadequate or nonexistent aftercare, and high recidivism rates. Although the state of institutional juvenile corrections has generally improved since the 1800s, none of these problems has been completely eliminated. As a result, critics charge that institutions have, in many cases, failed to achieve their stated goals of rehabilitating youths or protecting the public.

This chapter examines contemporary institutional placements for juvenile offenders. It begins with a general overview of juvenile correctional institutions, then examines types of juvenile correctional institutions found in the United States, with special attention paid to the two most common: detention centers and state training schools. Throughout, the chapter focuses on the purpose and operation of these institutions as well as their problems. The chapter concludes with an examination of trends in juvenile incarceration and the effectiveness of juvenile correctional institutions.

■ Overview of Institutional Corrections for Juveniles

As noted above, juvenile correctional institutions represent the most restrictive option available to juvenile courts. These institutions, however, vary in the extent to which they focus on custody and control. For example, some employ a variety of security hardware and structures, such as perimeter fencing or walls, and surveillance and detection devices, such as motion detectors, sound monitors, and security cameras.[1] These juvenile institutions are classified as secure facilities, and their purpose is to closely monitor the movements of residents within the facility and restrict their access to the community. Most public and private detention centers, reception and diagnostic centers, and state training schools are **secure facilities**.[2]

Other juvenile institutions rely much less on security devices. These facilities have no perimeter fencing, and some do not lock entrances or exits at night.[3] These institutions are classified as **open institutions**,[4] and they rely more heavily on staff than on security hardware, structures, and devices. Most private facilities as well as most public shelters, ranches, forestry camps, and farms are considered open institutions.[5]

Juvenile correctional institutions also differ in a number of ways besides the degree of reliance on security hardware, structures, and devices. For example, some institutions are *privately operated* and some are *publicly operated*. In fact, about two-thirds of juvenile correctional facilities are privately operated, although the majority of youths are held in public institutions. When the 1999 Census of Juveniles in Residential Placement (CJRP) was taken, approximately 71% of the youths in juvenile facilities were residing in public facilities and about 29% were residing in private facilities.[6]

secure facility A facility that closely constrains and monitors the residents' activities and physically restricts their access to the community.

open institution A facility that is not locked down, has no walls or fencing, and relies on staff to control inmates.

The CJRP Provides Valuable Data on Youths in Institutional Placements

The Census of Juveniles in Residential Placement (CJRP) is conducted every 2 years and contains detailed individual data on youths residing in juvenile correctional facilities on the fourth Wednesday in October in the year the census is conducted.

private facility A facility that is privately owned and operated. Private facilities can generally choose whom to accept, allowing them to focus on specific types of juveniles, such as sexual offenders or less serious offenders.

public facility A government-run facility. Public facilities have less ability than private facilities to screen out difficult and violent offenders and usually house more of both.

As one might expect, there are significant differences between **private facilities** and **public facilities**. For example, private facilities tend to be smaller than those that are publically operated.[7] Another difference is the types of youths that are placed in public and private institutions. Although some private facilities specialize in treating violent or seriously delinquent offenders, as a general rule public facilities contain a much higher percentage of youths who have committed delinquent offenses than do private facilities. Conversely, private facilities typically house a much higher percentage of youths who are in custody for committing status offenses and for other nondelinquent reasons.[8] The differences in the types of youths admitted to public and private facilities is likely due to several factors, including the ability of private facilities to choose who they will serve, the existence of specialized programming (e.g., substance abuse or mental health treatment) in many private facilities, as well as economic considerations that require private facilities to accept youths who have insurance or the ability to pay for placement.

Private Facilities Often House Different Populations than Public Facilities

In 1999, status offenders accounted for 10% of the residents of private facilities. Moreover, private facilities held the majority of youth who were in placement as part of a diversionary agreement and who had not been adjudicated in the juvenile court.[9] Indeed, a substantial percentage of youths admitted to private facilities are "voluntarily" admitted by their parents or by school officials.[10]

Another difference between public and private facilities is that public facilities contain greater proportions of nonwhite youths, males, and older youths. Minorities made up nearly two-thirds of the residents in public facilities but only about half of the residents in private facilities in 1999. Moreover, males account for 88 percent of youths in public facilities, and 84% of those in private placements.[11]

MYTH VS. REALITY — Public vs. Private Facilities

Myth—Private facilities provide more humane care for juveniles than do public facilities.

Reality—There are many well-run private facilities that provide excellent care for delinquent and problem youths. However, just because a juvenile corrections facility is privately operated does not automatically make it better than many publicly operated facilities. Indeed, some privately operated facilities have treated their charges as badly as the worst public facilities. Research by David Shichor and Clemens Bartollas that looked at public and private facilities used by a large probation department in southern California found that private facilities did not always provide promised services to clients, that many private facility staff members did not have appropriate qualifications, and that private facilities did not always separate violent and older youths from younger and less serious offenders.[12] Keep in mind, however, that there are many well-run publicly operated programs around the United States.

Juvenile correctional institutions vary in other ways as well. For example, the majority of both private and public institutions are small, housing 40 residents or less, although some state-operated institutions have a legal capacity of 800 or more residents.[13] Also, some institutions are coed institutions, whereas others are single-gender institutions.

Facilities also differ in the average length of time that residents stay. Typically, youths committed to public facilities by juvenile courts stay longer than those placed in private facilities.[14] In addition, some private and public facilities are designed for short-term placement, but others are intended for long-term stays. Detention centers, jails, and diagnostic and reception centers tend to be short-term facilities, and ranches, forestry camps, farms, and training schools are generally considered long-term facilities. However, there is considerable variation in the average length of time youths spend in institutions classified as long term.[15]

Minority Youths and Males Tend to Stay Longer in Institutional Placements

On average, among youths committed by juvenile courts after an adjudication, males and minority youths have longer stays. On average, committed males spend 3 weeks longer in institutional settings than females, and minorities spend 2 weeks longer.[16]

Another way that juvenile institutions differ is in the types of programming and the quality of institutional life they provide. Almost all juvenile correctional institutions offer basic educational and counseling programs for their residents. Moreover, over half of all institutions provide family counseling and employment counseling, hold peer group meetings, or use behavioral contracts or a point system. However, there is wide variability in the extent to which programming that goes beyond basic education and counseling is offered.[17]

As noted in Chapter 4, children have often been subjected to abuse and inhumane treatment in juvenile correctional institutions. Certainly the overall quality of institutional life in juvenile correctional facilities has vastly improved since youths were first placed in the houses of refuge and early reform and training schools. Many juvenile institutions are administered by competent, caring, and professional administrators who oversee skilled and caring staff involved in the delivery of a variety of quality services to residents. Yet, in other instances, many of the problems that have historically plagued juvenile correctional institutions are still evident. Some of the problems facing juvenile correctional institutions are considered later in this chapter.

■ Types of Juvenile Correctional Institutions

Juvenile detention centers, jails, diagnostic and reception centers, forestry camps, ranches, boot camps, and state training schools are among the main types of institutions used by the juvenile courts. These types of facilities differ from each other, of course, but there are often considerable differences between facilities of the same type. This section provides a general overview of juvenile institutions employed in juvenile justice. For purposes of organization, it divides juvenile institutions into two types: <u>preadjudicatory institutions</u>, which are primarily intended to house youths awaiting adjudication, and <u>postadjudicatory institutions</u>, which are more often used for youths after adjudication.

preadjudicatory institution A temporary institution for juveniles who are awaiting trial or awaiting sentencing.

postadjudicatory institution An institution to which a juvenile offender has been sentenced.

Preadjudicatory Institutional Placements

Detention Centers

Detention centers (sometimes called juvenile halls or juvenile homes) are common in the United States. Indeed, many large counties and even many small ones operate their own detention facilities for juvenile offenders. In addition, many states operate regional detention centers that may be used to house youths taken into custody around the state. Detention centers, whether local or regional, house youths in physically restrictive settings "pending juvenile court action, or following adjudication pending disposition, placement, or transfer" to some other setting.[18] Although they are designed primarily to house youths awaiting adjudication, disposition, or placement in another correctional setting, such as a state training school, they are also used for postdispositional placement in some jurisdictions.[19]

Detention centers constitute the most popular type of correctional institution employed in juvenile justice and account for the great majority of admissions to juvenile correctional facilities each year. Typically, youths spend only a short time in detention: Their average stay is around 2 weeks.[20] However, the number of youths admitted to detention facilities has been increasing in recent years. For example, the number of delinquency cases that resulted in detention increased 11% between 1990 and 1999.[21]

MYTH VS. REALITY — Youths May Be Detained for Minor Offenses

Myth—Only youths who have committed the most serious offenses are placed in detention centers.

Reality—Many youths who are detained have not committed serious crimes. In fact, most detentions are for property offenses. In 1999, property offenses accounted for 34% of all detentions, person offenses accounted for 27%, public order offenses 26%, and drug offenses 13%.[22]

Typically, youths are detained in one of two ways. First, they can be detained after a legal arrest by a police officer for an offense that is serious enough to warrant detention. Incarceration in a detention center is seldom automatic, however. During court hours, an arrested juvenile is usually brought to the court with the necessary paperwork, the juvenile is charged by the prosecutor or state attorney, and a detention hearing is held. After hours, an "on call" person designated by the judge typically makes the detention decision.

The second way juveniles are detained is by the issuance of a court order, sometimes called a warrant, complaint, pick up order, or apprehension and detention order (A & D). This court order allows for the arrest (apprehension) and incarceration (detention) of a juvenile, with a hearing to follow within a specified period of time, usually 24 hours. The legal grounds for these orders vary, but common grounds include these:

- The juvenile engaged in serious delinquent behavior that, if engaged in by an adult, would justify that adult's arrest and incarceration.
- The juvenile failed to appear in court after having been properly summoned by the court for a hearing.
- The juvenile ran away from a setting where he or she was ordered to be placed by the court (this may also be considered a probation violation).

Children found to be in situations that threaten their safety (e.g., neglected or abused children) can also be subject to an A & D order and be picked up by the police and placed in shelter care or foster care. These youths should not be placed in a detention center. Nevertheless this does happen in some instances.

Status offenders (see Chapter 13) should not be detained for running away unless they are under the court's jurisdiction and court ordered into a placement. Indeed, the Juvenile Justice and Delinquency Prevention Act of 1974, as amended, prohibits the detention of status offenders, although federal regulations have interpreted the act to permit detention for a period not to exceed 24 hours when status or nonoffenders first come into contact with the police. However, some courts have skirted the act by holding hearings and ordering youth to stay at home or attend school. If youths violate these orders then they are detained for violation of a legitimate court order, not the offense of running away or failing to attend school.

The key point to remember is that taking a child into custody may or may not be a judicial decision, but keeping a juvenile in secure detention must be a judicial decision.

The Juvenile Court, or Those Authorized By the Court, Make Detention Decisions

It should be noted that police officers and others must request that juveniles be detained. This is because the decision to detain a youth is typically made by the juvenile court or those authorized by the juvenile court to make detention decisions. In practice, however, such requests are rarely denied.

The detention of juveniles is supported on a number of legal grounds, including the following:

- The juvenile, if released, may commit other offenses, injuring other persons or property.
- The juvenile, if released, may be harmed or injured by others.
- The juvenile, if released, will fail to appear at future court hearings.

- Unless the juvenile is detained, an uncooperative parent may remove him or her from the state, making the juvenile unavailable for trial.
- There is no responsible adult willing or able to provide supervision or care for the juvenile.[23]

In addition to the statutorily authorized grounds for detention, there are a number of reasons that detention is favored as a juvenile justice response in some localities. According to H. Ted Rubin, who has extensive experience as a juvenile justice practitioner and researcher, these include the following:

- Detention is used as a form of punishment.
- Detention is used to fill up detention beds and demonstrate a need for detention.
- Detention represents a convenient response to juvenile offenders that can be exercised by juvenile justice agents.
- Detention placates law enforcement officials who demand tough responses to juvenile offenders.
- Detention is used as an emergency mental health placement because such placements are often in short supply.[24]

The Operation of Detention   Once a juvenile is brought to a detention facility, a detention intake or admittance procedure is followed. Although each procedure is unique, the procedures used at different facilities have some features in common. One of the first things done by the staff, if this has not already been done by the arresting officer, is to attempt to make contact with the minor's parents or guardian. At the same time, a variety of other steps are taken to ensure the safety of the juvenile and detention personnel and to acclimate the juvenile to the facility:

- The youth is searched. A strip search may be used to make sure the youth is not in possession of any contraband, drugs, or weapons.
- The youth's personal possessions are collected and stored until the youth is released.
- The youth is screened to determine if he or she is under the influence of drugs or alcohol, if any medical intervention is required, and if the youth has a medical condition that may affect other residents.
- Basic information about the youth, such as gender, age, address, complete name, date of birth, parents' names, school, and the reasons for detention, is collected.
- The rules that residents are expected to follow while in the facility are explained. The youth may be given a handbook that explains facility rules and may be required to take a test to demonstrate understanding of the rules prior to placement with the other residents.

Detention Admission Procedures Have Sometimes Been Abusive

Although the admission steps listed above seem reasonable given the need to protect the safety of residents and staff, there have been many instances of abuse associated with them, including dehumanizing searches intended to threaten and intimidate residents, such as pelvic examinations and vaginal searches of females, isolation of youths, and physical abuse.[25] Unfortunately, the extent to which such practices still occur is unknown.

detention program The total set of organized activities in which youths are involved while in a detention facility.

token economy A type of behavior modification program in which juveniles, in return for appropriate behaviors, earn points that can be redeemed for privileges, snack foods, and so on.

In a detention facility, the **detention program** "is the sum of all activities in which the child engages while in the detention home."[26] Two important components of a detention program are (1) the rules that youths are expected to follow while in detention and (2) the interventions, such as the educational curriculum and the behavior management system, used by detention personnel to control resident behavior.

Many detention facilities try to manage resident behavior through use of a **token economy**. In a token economy, youths earn tokens (or points) for exhibiting appropriate behavior. For example, youths may earn points for using appropriate language and gestures, helping clean the facility, completing homework assignments, and following other program rules. As they accumulate points, they are able to purchase or earn privileges, such as more recreational time, more telephone time, and a later bedtime. Supporters of token economy programs argue that they help reinforce positive behavior among youthful detainees. These programs are based on operant conditioning principles that assume that behaviors that are reinforced are more likely to be exhibited in the future.

When token economies are implemented by people who clearly understand their purpose and operation, they can help shape youths' behavior in the desired way. When these programs are not implemented properly, however, they can increase the potential for conflict among residents and also between residents and staff. A major problem in many facilities with token economy programs is inconsistency in the operation of the programs. Such inconsistency can cause residents to perceive some groups or individuals as being given differential or preferential treatment, which in turn can increase the level of hostility among residents and between residents and staff.

Another concern is that token economy programs can disguise punitive programming with a veil of scientific language. For example, many programs contain a large number of rules that can result in residents' losing points but relatively few ways for residents to earn points. As a result, the overall focus of these programs becomes punishment, not reward, and such punitive programs are inappropriate, according to critics, because many youths in detention have not been adjudicated.[27]

A variety of other interventions are found within detention facilities. For example, many detention centers, to be in compliance with state compulsory education laws, operate basic educational programs staffed by certified teachers who work for local school systems. These programs typically provide basic and individualized instruction for residents, although some facilities offer more specialized programs including vocational training or college-level instruction.[28] Unfortunately, the motivation and training of teachers in these programs and the resources devoted to them vary considerably.

It is also common for detention facilities to provide some basic health care services, such as health appraisals, instruction on how to access various health services, and routine screening for health problems.[29] Most detention centers have mental health professionals available to youths who require such services,[30] and some type of basic counseling is offered as well. However, information on the extent to which counseling is a part of detention center programming and the quality of existing counseling programs is not available.

Many detention centers also operate some type of recreational program for residents. Although almost all centers meet American Correctional Association standards, which require youths to be allowed at least one hour of recreation that involves large muscle activity (exercise) per day and at least one hour of leisure time per day, some operate more structured recreation programs (though the number of these is not known).

Problems Confronting Juvenile Detention Today juvenile detention faces a number of problems. One major problem is the absence of strict standards governing the use of juvenile detention and a lack of agreement over the purpose of detention in many jurisdictions.[31] Indeed, wide discretion to detain juveniles is the rule in many communities.[32] As a result, almost any youth charged with a criminal offense in some jurisdictions can be detained.[33]

Although the intent of the juvenile court is to provide for the care, protection, and rehabilitation of youths, abuse is a potential problem in any system intended to exercise restrictive control over youths. One type of abuse occurs when youths are detained without sufficient evidence that detention is necessary. Abuse also occurs when detention becomes a form of punishment. This is particularly troublesome prior to adjudication, because youths are presumed to be innocent prior to fact finding by the court. However, punishment after the adjudication can also be abusive when the express intention of the court is to "teach the youth a lesson," and even in cases where the intent is to treat or rehabilitate a youth. Importantly, efforts to treat or rehabilitate youths often involve some element of punishment. Furthermore, when juveniles are detained in order to ensure subsequent court appearances or to protect victims from harm, those making decisions to detain youths are making predictions about their future behavior. However, accurate predictions about juveniles' future behavior are difficult to make and are often based on criteria that have little relationship to their likelihood of appearing in court or engaging in subsequent delinquent acts.[34]

This is not to say, however, that using detention as a short-term solution cannot have some benefits. Most county or local detention facilities do not have the programming to keep juveniles for an extended period of time, but short-term detention that supports the authority of the court, probation, and programs such as house arrest can have a beneficial impact on the lives of some youths.

The effectiveness of detention, unfortunately, is continually being undermined by the chronic overcrowding of local detention facilities. Many facilities have licensed limits on their capacities, and exceeding those limits can result in loss of licensing, to say nothing about increased legal liability, increased stress on the staff and the residents, and the increased financial costs associated with detention. Indeed, crowding influences all aspects of institutional life. It is related to increased levels of anger and hostility between residents, and between residents and staff, that contribute to disciplinary infractions, escapes, and violence. It also negatively impacts residents' perceptions of safety and leads to a variety of resident responses such as joining gangs for protection or acting "crazy" so that they will be placed in isolation where they feel less threatened. Crowding also leads to a variety of staff responses such as greater focus on control and confinement of youths in locked rooms for longer periods of time, and it makes adequate classification and placement of youths within the facility more difficult. In short, crowding leads to a variety of responses on the part of residents and staff that exacerbate many of the problems that exist in institutional settings.[35]

Detention Overcrowding Is a Serious Problem

The number of overcrowded detention centers in the United States increased 632% between 1985 and 1995.[36] As a result, many youths today are held in facilities where the size of the resident population exceeds the design capacity.

Unlike the adult justice system, which makes bail available to most offenders, juvenile justice relies heavily on preventive detention. Most states allow judges to deny bail to adult suspects only if they have committed capital crimes (e.g., murder), if they have jumped bail in the past, or if they have committed another offense while on bail.[37] However, in the 1984 U.S. Supreme Court case *Schall v. Martin*, a majority of the justices ruled that the preventive detention of a juvenile is proper if there is reason to believe that the juvenile would commit additional crimes while awaiting adjudication, although the Court also ruled that juveniles have a right to a detention hearing and a statement indicating the reasons why they are being detained. In handing down its decision, the majority justified its decision by noting that such detention protects both the juvenile and society from the "hazards" related to pretrial crime. Furthermore, the Court argued that preventive detention was appropriate because juveniles, unlike adults, are always in some form of custody[38] (see the discussion of *Schall v. Martin* in Chapter 6). The Court also pointed out that many states allowed for preventive detention of juveniles, which indicates that the concept did not offend any fundamental due process right. So long as juvenile courts afford juveniles timely hearings to challenge detention decisions, the constitutional right to be free from "excessive bail," found in the Eighth Amendment, takes a back seat to society's interest in the supervision and control of children.

Unfortunately, one of the results of *Schall v. Martin* has been to serve as a legal support for the overuse of detention. The excessive use of detention occurs in many jurisdictions, despite the national organizations and commissions, such as the National Advisory Commission on Criminal Justice Standards and Goals, the National Advisory Committee for Juvenile Justice and Delinquency Prevention, the Institute of Judicial Administration, and the American Bar Association, that recommend that detention be used only as a last resort for youths who have allegedly committed serious delinquent acts.[39] Indeed, in many jurisdictions the great majority of youths who are detained pending a hearing are released after their hearing, often on probation. Many, then, question why it is necessary to detain these youths, when later it seems evident they present no danger to the community.

Although a number of grounds appear to be used to justify the detention of juveniles, research on detention decisions suggests that a variety of factors appear to influence them, although the factors vary across jurisdictions. Early studies suggested that the availability of detention facilities and a lack of alternative placements in a community influenced the level of detention use.[40] Still other research revealed that the willingness of parents or guardians to accept youths back into their homes after being taken into custody by the police influenced detention decisions.[41] Some research has found that race is related to detention decisions, particularly when minority youths live in disadvantaged neighborhoods.[42] According to an analysis by researchers at the National Center for Juvenile Justice, of delinquency cases processed in juvenile courts in 1999, 18% of cases where the offender was a white youth resulted in detention at some point between court referral and case disposition. Among black and other minority youths, the percentage was 25% and 23%, respectively. The percentage differences were particularly large in drug offense cases. Detention was used in 17% of drug cases involving white youths but in 38% of the cases involving blacks and in 21% of the cases involving youths of other races.[43]

Moreover, there is evidence that females in general and minority females in particular represent a growing population of youths in detention.[44] Historically, girls have been much more likely to be detained for status offenses, such as running away and incorrigibility, than males.[45] However, for person, property, drug, and public order offenses, males are more likely to be detained than females, and there is some indication that gender differences in detention for status offenses have declined in recent years.[46] Unfortunately, many girls who end up in detention have been victims of physical, sexual, or emotional abuse and fail to receive needed treatment while detained.[47]

Another problem uncovered by research is that youths who are detained are more likely to be treated more severely at the next stage of the juvenile justice process than similar youths who are not detained.[48] The explanation for the difference might be that youths who are detained are less able to assist in the preparation of their defense than youths released pending a hearing. Also, hearing officers at subsequent hearings may feel that youths who have been detained represent more of a threat to community safety, making the officers more likely to impose severe sanctions on these youths than those who are not detained. Judges who are elected (and hearing officers appointed by and responsible to those judges) are not oblivious to public sentiments regarding juvenile crime. Although judges and referees may protect juveniles' due process rights at trial, detention often proves to be an easy and attractive remedy that earns them community support. Being tough on delinquent juveniles is always a popular political stand.

Not only are there problems associated with decisions to detain youths, but there are also problems associated with the daily operation of many detention facilities. One serious problem is child neglect and abuse, which has occurred despite the fact that part of the mission of detention centers is to provide youths with a safe environment. As noted in Chapter 4, the history of juvenile institutions has been characterized by punishment and inhumane treatment touted as rehabilitation. Many youths in detention have been physically and sexually assaulted by other residents and staff, many have been subjected to intimidation and other forms of psychological abuse, and many have been exposed to a variety of degrading and dehumanizing conditions. Unfortunately, abuse and inhumane treatment have not been eradicated from all juvenile detention facilities.

Today, there is evidence that isolation, mechanical restraints (e.g., handcuffs, straitjackets, anklets, etc.), and physical restraints (i.e., tackling and bodyholds) are still overused in some facilities. For example, according to a large-scale study of confinement conditions in juvenile correctional facilities published in 1994, there is some evidence that isolation and "time out" is overused in many facilities. This same study found that almost three-quarters (72%) of detention facilities reported the use of mechanical restraints, usually handcuffs and/or anklets. Other problems found in some facilities include high numbers of juvenile injuries (which typically result from attacks by other residents), occasional escapes and escape attempts, and occasional self-destructive behaviors among youths.[49]

Escapes and Escape Attempts Are a Problem in Detention Facilities

According to a national study of juvenile correctional facilities, detention centers reported 108 successful escapes and 169 unsuccessful escape attempts during a typical month in 1991.[50]

Despite such problems there are well-run detention facilities managed and staffed by well-trained and caring employees who work diligently to ensure that children under their supervision live in a safe, supportive, and secure environment. Such facilities share a number of program characteristics, according to David Roush, a former detention director and now the director of research and professional development at the National Juvenile Detention Association at Michigan State University. They tend to have an adequate number of qualified staff, a concern with safety, a range of positive activities, strong leadership, a focus on staff education about detention, and a strong evaluation component.[51]

Jails

In some jurisdictions, juveniles may also be placed in adult jails. Like detention centers, jails represent a type of preadjudicatory institutional placement. The jailing of juveniles often occurs in rural jurisdictions, where there are few juvenile detention facilities. On the day that the annual survey of jails was taken in 2003, 6,869 youths under the age of 18 were detained in adult jails.[52] This statistic represents a recent decline in the numbers of youths placed in jails. Indeed, at least 7,000 youths under age 18 resided in adult jails each year that the survey was administered between 1995 and 2003.[53]

The federal government has taken an active role in attempting to have juveniles removed from adult jails since 1980, as a result of an amendment to the 1974 Juvenile Justice and Delinquency Prevention Act. In testifying before a U.S. House of Representatives subcommittee on the reauthorization of the act in 1980, the deputy attorney general of the United States, Charles B. Renfrew, noted the following:

> *The jailing of children is harmful to them in several ways. The most widely-known harm is that of physical and sexual abuse by adults in the same facility. Even short term, pretrial or relocation detention exposes juveniles to assault, exploitation and injury.*
>
> *Sometimes, in an attempt to protect a child, local officials will isolate a child from contact with others. Because juveniles are highly vulnerable to emotional pressure, isolation of the type provided in adult facilities can have a long-term negative impact on an individual child's mental health.*
>
> *Having been built for adults who have committed criminal acts, jails do not provide an environment suitable for the care and maintenance of delinquent juveniles or status offenders. In addition, being treated like a prisoner reinforces a child's negative self-image. Even after release, a juvenile may be labeled as a criminal in his community as a result of his jailing, a stigma which can continue for a long period.*
>
> *The impact of jail on children is reflected in another grim statistic—the suicide rate for juveniles incarcerated in adult jails during 1978 was approximately seven times the rate among children in secure juvenile detention facilities.*[54]

Under the Juvenile Justice and Delinquency Prevention Act (JJDPA), states that received federal funds under the act were required to separate juveniles from adults. Furthermore, as a result of a 1980 amendment to the act, they were required to remove juveniles from adult jails within five years or lose federal funds.[55] Additional legislation, however, indicated that 75% compliance by 1985 was acceptable.[56]

Juveniles May Be Placed in Adult Jails Under Certain Conditions

The JJDPA does not completely ban the jailing of juveniles. Subsequent court rulings and legislation permit juveniles to be incarcerated in adult facilities under certain circumstances—for example, where a juvenile is being tried as an adult or has been convicted of a felony. Also, a juvenile can be held in an adult jail or lockup for up to 6 hours while other arrangements are being made, providing there is "sight and sound separation" of adults and juveniles. Also, in rural areas a juvenile may be held in an adult jail or lockup for up to 24 hours if the following conditions are met:
- The minor is awaiting an initial court appearance.
- An alternative placement is not available.
- There is sight and sound separation of juveniles and adults.[57]

The Juvenile Justice and Delinquency Prevention Act has encouraged the removal of children from adult jails, at least in some states. For example, in 1991 the state of Indiana abolished the jailing of juveniles in that state. In writing about the jail removal efforts in Indiana, Barry Krisberg and James Austin, two respected juvenile justice researchers, noted the following:

> *For years, the state of Indiana placed hundreds of children in dangerous situations in adult jails. Yet a bipartisan coalition led by the Department of Corrections convinced state lawmakers that better alternatives existed. The jail removal movement in Indiana demonstrated again that the goals of child protection and public safety were not antithetical. Moreover, many Indiana communities have learned that sensible alternatives to jailing, such as temporary shelter care or home detention, can be created without requiring large new expenditures of public funds. Also, elected officials in Indiana and many other states learned that they could support humane policies for children without suffering the alleged political liability of appearing to be soft on crime. Successes in the jail removal area illustrate the power of legislative action to discourage harmful practices. These stories also suggest that the long standing inferior treatment of children can be quickly remedied.*[58]

Other states, such as Idaho, Ohio, and North Dakota, have also made significant headway in this area.[59]

Despite pressure from the federal government on the states, however, jail removal has not been completely successful. Although progress has been made in some jurisdictions, juveniles are still being incarcerated in adult jails. In fact, there is some pressure from "get tough on crime" advocates to place more juveniles in adult facilities.[60] As a result of such pressure, a failure of states to develop juvenile detention facilities, and an unwillingness to develop appropriate alternatives to the jailing of juveniles in some jurisdictions, a number of juveniles continue to be housed in jails each year. Moreover, get tough measures that have resulted in more youths being tried and sentenced as adults means that a number of persons under 18 are being jailed as adults. For example, at mid-year 2003, 5,484 persons under 18 were being held as adults in jails.[61] Although the optimistic assessment of Barry Krisberg and James Austin quoted earlier regarding the removal of juveniles from adult jails was clearly justified in the early 1990s and still holds true in some states, the jailing of juveniles continues to occur in many jurisdictions.

Problems Associated with Placing Juveniles in Adult Jails Incarcerating juveniles in adult jails creates a variety of problems for those juveniles who are incarcerated as well as those who are responsible for their safety. Indeed, many of today's jails face a number of serious problems: a deteriorated or poorly designed physical plant, overcrowding, a lack of services and constructive activities for inmates, gang conflicts, and violence. It is worth noting that many jails have difficulty meeting the needs of adult inmates, let alone the needs of juveniles.

One obvious problem facing juveniles in jails is the risk of physical and sexual assault. Although juveniles are supposed to be separated from adult prisoners, separation does not always occur. Moreover, there are often inmate groups in jails, such as trustees, who have access to juveniles even when they are separated from the regular adult population in the jail.[62] Furthermore, many jails do not do a good job of isolating juveniles who are at risk of victimization by older and more aggressive juveniles. Consider the case of Eric, a 13-year-old who was jailed for stealing a dirt bike. He was incarcerated "next to a 17-year-old who was being held for assault and who had a history of violent and sexual offenses." In 1986, a class action suit was filed by private attorneys in Portland, Maine, with help from the Youth Law Center in San Francisco, alleging that staff in the jail left the cell doors open, allowing the 17-year-old to assault Eric and force him to have oral intercourse.[63] Juvenile girls may be particularly vulnerable to sexual assaults by jail personnel. As Meda Chesney Lind notes, they may be "doubly disadvantaged" because they are females and juveniles.[64]

Physical and sexual assault are but two of numerous problems associated with placing juveniles in adult facilities. Even when juveniles do not come into direct contact with adults, they may be subjected to psychological abuse by adult inmates. In a 1967 report, the President's Commission on Law Enforcement and Administration of Justice noted that jail is characterized by "enforced idleness, no supervision, and rejection. It is a demoralizing experience for a youngster at a time when his belief in himself is shattered or distorted. Repeated jailing of youth has no salutary effect on the more sophisticated youngster; on the contrary, it reinforces his delinquency status with his peers and his self-identification as a criminal. Enforced idleness in a jail gives the sophisticated juvenile ample time and reason for striking back at society."[65]

The idleness and rejection experienced by youths in jail, coupled with inadequate supervision due to limited staff who are often poorly trained to deal with juveniles, often leave youths in a fragile psychological condition. Ironically, steps taken to protect youths, such as keeping them segregated from adults, often exacerbate the effects of jailing, because separation often amounts to solitary confinement. Under such conditions, depression is not uncommon, and the probability of youths engaging in self-destructive behaviors, such as suicide and self-mutilation, is increased. As noted earlier, suicide or attempted suicide is a well-known problem associated with placing juveniles in adult jails. Indeed, research clearly demonstrates that the rate of suicide among children held in adult jails is much greater than that among youths in juvenile detention centers or among youths in the general population of the United States.[66]

Jailing Juveniles Increases Their Risk for Self-Destructive Behavior

The suicide rate for juveniles in adult jails is at least 4.6 times higher than the rate for juveniles in the general population and 7.7 times higher than the rate for inmates of juvenile detention facilities.[67] There is some evidence that it may even be higher.[68]

Although juveniles of any age or gender can experience problems in jail, females may be particularly vulnerable. This is because female juveniles often have a history of sexual and physical abuse, which makes them susceptible to depression and self-destructive behaviors.[69] This type of history is often related to their running away from home and other status offenses—behaviors that may result in their incarceration in jails. There is considerable evidence that girls are often jailed for status offenses such as running away or because they are neglected or dependent youth.[70] In addition, they are often younger than male juveniles who are jailed. Typically, jails are even less adequately prepared to meet the needs of female juveniles than male juveniles, and they often end up becoming victimized a second time—by the criminal justice system supposedly there for their protection.

Not only does the jailing of children create serious problems for children, but it also creates problems for jail administrators and staff. First, the requirement to separate juveniles from adults causes serious difficulty for jail administrators, who already have the burden of segregating a variety of offender groups. For example, jails need to separate males and females and isolate inmates who have medical problems and inmates who are disruptive. They may also separate younger adults from older adults and separate offenders awaiting trial from those who are serving sentences. Another problem for administrators is that jails are often crowded and frequently lack adequate personnel and support services, such as mental health workers, to assist those in need of such services. A third problem is that the jailing of juveniles creates additional liability for jails and county or city governments. In short, when juveniles are admitted to jails, they simply increase the headaches experienced by jail staff, who have a diverse population to work with already.

The truth of the matter is that jails are often not equipped to deal with the needs of many adult offenders, much less children. Most have neither the physical facilities nor an adequate number of correctional and other staff to adequately care for juvenile offenders. Furthermore, correctional staff are rarely trained to deal with the special needs and problems presented by young adult or juvenile inmates, nor do jails typically have ancillary resources at their disposal to help them meet the needs of these inmates. The problems associated with jailing juveniles raise serious questions about the appropriateness of this practice. Despite these problems, however, the practice continues in many jurisdictions.

Postadjudicatory Institutional Placement

State Training Schools

<u>State training schools</u>, which are public facilities, often serve as the last resort for youths designated as serious or repeat offenders. Yet not all youths who reside in such facilities are there because they are violent. Indeed, in 2001, only 24% of the juveniles committed to state correctional institutions were held because they had committed a violent crime against a person.[71]

Some training schools actually resemble adult prisons. These institutions have a *closed environment.* They are physically secure; they can have various combinations of physical security devices such as high walls and fences, razor wire, motion detectors, and locked cell blocks; and they are often self-sufficient (i.e., the laundry, infirmary, school, and maintenance facility are run by institutional staff). They may also have solitary confinement or isolation rooms for those who consistently misbehave, they may employ mechanical restraints, and they place considerable emphasis on physical security.

Other training schools, however, use a <u>cottage system</u>. In these institutions, youths are housed in cottages, which are usually small and have a more "home-like" atmosphere. Each cottage is run by cottage "parents," or staff who work on shifts. These institutions have a more *open environment.* Rather than relying on fences and walls and other hardware for security, they rely on staff to maintain security within the institution.

state training school
A restrictive facility designed to deal with repeat juvenile offenders.

cottage institution
Facility in which juvenile offenders live in a "home-like" setting. Such institutions are typically more open and smaller than other types of juvenile correctional institutions, are run by "cottage parents," and have high staff-to-juvenile ratios.

A Majority of Youths Placed in Correctional Institutions Are Placed in Secure Facilities

While juvenile institutions vary in their use of security hardware and structures (e.g., walls and perimeter fences), the use of physical security devices is common. In 1999, 66 percent of juveniles who were committed to juvenile correctional facilities were housed in locked facilities.[72]

Operation Programming in state training schools usually consists of a combination of (1) academic education, (2) vocational education, (3) recreation, (4) basic counseling, and (5) behavior management. In addition, a variety of specialized treatment and education programs exist in some facilities. For example, over half of all training schools offer family counseling, health and nutrition counseling, substance abuse education, and sex offender treatment. Religious counseling and programs are also prevalent in juvenile training schools and are more common than any other type of volunteer program found in juvenile correctional facilities. It should be noted that training schools are more likely to offer programs that do not require counseling staff,[73] because institutional budgets often preclude hiring highly paid professional staff.

Many of the youths in training schools have fewer years of education than youths of the same age in the general population,[74] and many others are well below their grade level in academic performance. Indeed, recent research indicates that only about 23% of juveniles in correctional facilities read at their nominal grade level.[75] Consequently, the educational programs at these institutions focus on basic skills, and most are designed to help students obtain a general equivalency diploma (GED). The majority of youths in training schools have access to special education classes, and almost all have access to literacy or remedial reading programs. These programs are important because as many as 40% of youths in residential facilities may have some type of learning disability.[76] Some facilities also offer college programs for residents.[77]

Many Training Schools Do Not Meet ACA Recommendations for Educational Programs

Only 55% of juveniles confined in training schools in 1991 were in schools that met all of the American Correctional Association (ACA) recommendations for educational programs in juvenile facilities.[78] According to the ACA recommendations, facilities should provide a comprehensive educational program for juveniles, use licensed or accredited teachers, and have a student-to-teacher ratio of no more than 1:15. In addition, the ACA suggested that programs should focus on meeting residents' academic, vocational, and personal needs.[79]

Today, most youths confined in state training schools have access to vocational education programs. Like the academic educational programs offered in these facilities, the vocational programs offered are also intended to give youths basic skills. The most common types of vocational courses offered are carpentry and various building trades, food services, and auto mechanics, although printing, cosmetology, secretarial training, retail trades, and forestry/agriculture are found in at least 10% of training schools.[80] One problem with these programs is *sex typing*—a problem that may be more difficult to avoid in single-sex institutions.[81] Another problem is that they do not necessarily prepare youths for employment once they leave the institution. Once youths leave institutions, they may have difficulty joining unions or they may still lack the necessary education needed for many jobs.[82] Although some institutions provide job training, few offer job placement programs that attempt to match youths with jobs offering a chance of advancement.

Most training schools operate recreational programs for residents. Indeed, courts have ruled that the failure of institutions to allow children regular exercise and recreation violates juveniles' constitutional rights.[83] The ACA, in its *Standards for Juvenile Training Schools*, recommends that juveniles be given 2 hours of recreation each school day and an additional hour on nonschool days.[84] Most training schools have an outdoor playing field, an outdoor basketball court, an outdoor volleyball court, indoor table games, Ping-Pong tables, and television sets. Many also have an indoor gym. Although program administrators typically view recreational programming as an important responsibility, it is worth noting that 19% of juvenile training schools failed to meet the ACA recommended requirements for recreation in 1991.[85]

Another program component consists of counseling or treatment. According to correctional administrators, approximately half or more of the residents in their facilities are experiencing at least one of the following problems: family difficulties, drug or alcohol abuse, peer problems, parental abuse, disruptive behavior, and depression.[86] Consequently, most institutions offer basic individual counseling. This type of counseling is provided by a range of program staff and is intended to help youths deal with adjustment problems within the institution and general problems they are experiencing in their lives. Some institutions use more structured therapeutic counseling, such as psychotherapy or reality therapy. This type of counseling is conducted by trained therapists and is also designed to help youths adjust to the institution as well as resolve a variety of personal problems.

A Number of Youths Placed in Correctional Institutions Suffer from Mental Disorders

The rates of mental disorders are higher among youths in correctional institutions than among youths in the general population.[87]

In addition to individual counseling, many institutions use some type of group counseling, such as guided group interaction (GGI), positive peer culture (PPC), or milieu therapy. **Group counseling** is less costly than individual counseling, because counselors work with more than one youth at a time. The goal of group counseling is to develop a group that helps the members identify personal issues and supports individual change. Group counseling can be a powerful intervention when it is led by a skilled therapist. Unfortunately, there is little information on the quality of the different types of counseling programs found in juvenile correctional facilities and the qualification and training of staff who provide counseling to residents.

group counseling A type of therapy designed to use group members to help individuals identify personal issues and support individual change.

In many institutions, treatment and behavior management programs include a behavior modification component. Behavior modification is based on the idea that behaviors are learned and that current and future behaviors can be shaped by a system of rewards and punishments. Like detention centers and other types of correctional programs, training schools commonly employ a *token economy program,* in which youths earn or lose points based on their behavior. Indeed, release may be based on the earning of a specified minimum number of points, indicating the youth has achieved the treatment goals set out by the institution.

Problems Associated with Juvenile Training Schools One serious problem facing a majority of training schools is overcrowding. In crowded facilities, youths have less privacy, the probability of interpersonal conflict is increased, and "double bunking" (sleeping more than one resident to a room) increases the chances of physical or sexual assault. Despite the drawbacks of overcrowding, according to the 2000 Juvenile Residential Facility Census, almost 4 in 10 juvenile correctional facilities reported that the number of residents in the institution exceeded the number of standard beds.[88] Consequently, many juveniles in training schools are housed in rooms that do not meet ACA standards for living space and are housed in units that exceed the ACA recommended standard for residents per unit (25 or less) or do not conform to local fire codes.[89]

Contributing to institutional overcrowding is the inappropriate placement of many nonserious offenders in state training schools. As noted earlier, only 24% of youths committed to state institutions in 2001 were incarcerated for violent crimes against persons. Almost 11% had committed other crimes against a person, approximately 30% were held for property offenses, 10% for public order offenses, and 9% for drug offenses. Approximately 12% were incarcerated for technical violations, and almost 5% were admitted for status offenses.[90] Certainly, juvenile offenders that represent a clear threat to others need to be detained. However, one must question the extensive use of expensive and often ineffective state institutional placement for nonviolent offenders.

Another problem facing training schools is escapes. "Nationwide, juvenile training schools reported 454 successful escapes and 478 unsuccessful escape attempts during a typical month in 1991. This would translate into nearly 5,500 escapes and more than 5,700 unsuccessful attempts per year."[91] Escapes and escape attempts can result in injury to staff and residents, can result in injury to escapees on the run, and can become a public relations nightmare for correctional administrators, who frequently face some community resistance to institutional programs.

Still another problem faced by facilities is too few staff and treatment resources. Because of inadequate funding, many facilities have lower-than-recommended security staff ratios.[92] Many also have fewer treatment staff, particularly professional treatment staff, and other programming staff than they would like. Scarce resources may mean that staff do not receive adequate training. Unfortunately, when budgets are austere, treatment programming and staff training typically suffer, and more emphasis is placed on security than treatment.

Overcrowding, inappropriate placement of nonviolent offenders in institutional settings, inadequate treatment staff and programming, and poor staff training combine to adversely affect the correctional environment in many training schools.

At best, mere warehousing of youths is the result. At worst, residents are exploited by other residents and, in some cases, by staff. In either case, such institutions fail to rehabilitate youths, protect them, or respond to them in ways that ensure community safety. These problems represent only a sample of the difficulties that presently confront publicly operated juvenile training schools. Additional problems for juvenile institutions in general, including training schools, are examined later in this chapter.

Other Institutions Used for Postadjudicatory Placement

Besides publicly operated juvenile training schools, there are a variety of other institutions and programs in which juveniles are placed following adjudication. These include diagnostic and reception centers; ranches, forestry camps, farms, and outdoor programs; boot camps; and private institutions. This section describes these types of facilities.

Diagnostic and Reception Centers The goal of <u>diagnostic and reception centers</u> is to assess youths who have been committed to state care by juvenile courts and determine the best treatment plans and placements for them. Once a diagnostic and reception center completes an assessment, which on average takes 5 weeks,[93] the youth is placed in a training school or another correctional facility to complete the period of commitment. Although these centers are used in only a few states, where they do exist they play an important role in determining the placements and the treatment goals for youths under state care. In states without diagnostic and reception centers, many of the functions performed by them are performed at each individual institution.

The focus of diagnostic and reception centers is on determining each youth's short-term treatment needs. Typically, this is accomplished by conducting a general assessment that may encompass psychiatric and psychological evaluations, educational assessment, a social history, and behavioral observations as well as a review of information provided to the center by the juvenile court and other social agencies. For example, a psychiatrist may conduct a psychiatric evaluation and then treat the youth if he or she is determined to need psychiatric treatment. A clinical psychologist may administer a variety of psychological tests, such as IQ, personality, and other diagnostic tests. A social worker may compile a social history that examines the youth's family and social background. Educational staff may screen the youth for learning disabilities and other learning and academic problems. Unit supervisors may assess the youth's institutional adjustment and peer relationships. All the information acquired is compiled and summarized in order to determine the treatment needs and the best institutional placement for the youth.[94]

Ranches, Forestry Camps, Farms, and Outdoor Programs These long-term facilities and programs are for persons who do not require secure custody.[95] Although the facilities and program settings are not secure and residents have more access to the community than in some other types of institutions, their physical location serves as a barrier between them and the community. Perhaps because of their location, they rely less on formal security measures, such as institutional head counts and classification systems that separate groups of offenders, than on training schools, detention centers, and diagnostic and reception centers.

In addition, they typically have lower security-staff-to-resident ratios than other institutional settings.[96]

Ranches, forestry camps, farms, and outdoor programs are not new. Indeed, the basic philosophy behind these facilities and programs dates back to the placing out movement, active in the 1800s (see Chapter 4).[97] At that time, many reformers believed that children could be reformed if they were removed from the corrupting influence of the city and reared in wholesome rural environments where they could learn the value of honest work. Today, these programs focus on teaching youths responsibility and self-sufficiency and building their self-esteem. Like earlier programs, they are premised on the idea that "by placing a child in a totally foreign environment, old attitudes may be altered, and new, acceptable behavior patterns may be created."[98]

Boot Camps   Although military boot camps have existed for some time, their first modern correctional equivalent was founded in 1973 by a former marine, George Cadwalader. This boot camp, called the Penikese Island School, was based on the belief that serious young offenders could be influenced in a positive way by exposing them to rigorous military-style training.[99] It was not until the 1980s, however, that boot camps began to become popular for dealing with young adult offenders, and it was not until the 1990s that their popularity began to spread to juvenile justice.[100]

A number of **juvenile boot camps** have been developed around the United States. They exhibit considerable variability but share one essential feature: military-style training and discipline.[101] Consequently, drill instructors and other staff with military backgrounds play a critical role in these institutions, although teachers and case managers are also employed.[102] In addition, substance abuse treatment specialists and other treatment providers are beginning to play a more prominent role in many of these institutions.

Typical elements of a juvenile boot camp include a platoon structure (with platoons made up of 10 to 13 youths who move through the program together), military-style uniforms for residents and drill instructors, military customs and jargon, a strenuous daily routine that begins at 5:30 or 6:00 a.m. and ends with lights out at 9 or 10 p.m., immediate responses to minor rule infractions (e.g., push-ups) and progressive sanctions (including removal from the program for serious rule violations), and a public graduation ceremony.[103]

The idea of placing juvenile offenders into boot camps has become popular with some individuals, particularly those with military backgrounds. And of course the discipline they impose and the relatively low institutional costs of these programs have an obvious appeal to many policy makers. They have encountered, however, a number of difficulties. Low salaries, high staff turnover and burnout, and inadequate aftercare have plagued many of these programs.[104] In addition, there is no evidence suggesting that these programs are any more effective than traditional institutional placements at reducing recidivism.[105]

Private Institutions   Privately operated juvenile correctional institutions, which have a long if not always distinguished history, continue to play a significant role in juvenile corrections. As noted in Chapter 4, the first correctional institutions designed especially for children in this country, the houses of refuge, were

juvenile boot camp
A camp for juvenile offenders that uses military-style training and discipline to control and modify their behavior.

privately operated. Moreover, a number of juvenile justice interventions, such as placing out and juvenile probation, were first developed as private sector endeavors.

In 1997, private facilities made up almost two-thirds of all juvenile corrections facilities in the United States but held only about 28% of all youths in custody.[106] However, there is some indication that private institutions have begun to play an even more important role in juvenile justice because of the belief (not always supported by fact) of some policy makers that privately operated ventures, whether in business or public life, are more efficient and effective than those operated by government. Indeed, in recent years, admissions to privately operated facilities have been increasing.[107]

A Variety of Institutions Exist for Controlling Youths

A substantial number of the admissions to private facilities are so-called **voluntary admissions**, although they result from referrals by parents or school officials, and the juveniles in question hardly have a choice. Such admissions represent the growing "medicalization of defiance."[110] The result is that behaviors like "running away from home, habitual truancy, and being in conflict with one's parents are now being classified as emotional and psychiatric disorders"[111] that are dealt with by a growing medical industry that charges insurance companies and parents as much as $200 to $600 a day for inpatient psychiatric and chemical dependency treatment.[112]

voluntary admission
The placement of a juvenile into an institution by his or her parents or guardian without benefit of a court order or a review of due process. The cost of such a placement is borne, not by the public, but by the parents or guardian.

The large number of private institutions designed to deal with problem children make up what long-time juvenile justice researcher and practitioner Ira Schwartz refers to as a "hidden system" of control.[108] This system of control comprises detention facilities, shelter homes, group homes, training schools, drug and alcohol programs, ranches, farms, forestry programs, outdoor experiential programs, and mental health facilities, among others. In 1999, the Census of Juveniles in Residential Placement revealed that there were over 31,000 youths residing in these privately operated facilities.[109]

The quality of care given children in these institutions is as varied as the types of institutions that exist. A number of them provide outstanding care to their clients—better care, indeed, than many publicly operated facilities, owing to their lower staff-resident ratios, their greater flexibility, and the greater numbers of qualified professional staff they can afford to hire. In addition, many privately operated facilities are more cost effective than many publicly operated institutions,[113] and evidence indicates some may be more successful at reducing recidivism than many traditional training schools.[114] On the other hand, there is considerable evidence that many of the admissions to privately operated chemical dependency and mental health facilities are medically unnecessary.[115] There is additional evidence that the quality of care provided children in many private facilities is woefully inadequate, despite the exorbitant prices charged. Some programs appear to be more interested in profit than quality of care or treatment.[116] Consider the following examples cited by Ira Schwartz in his book *(In)Justice for Juveniles: Rethinking the Best Interests of the Child*: " 'One private psychiatric hospital [in Los Angeles County] served its new adolescent patients only a cheese sandwich and milk for each meal until they earned enough points to eat other items.' Another private hospital in the same community 'conducted routine strip searches, where female adolescents had to squat, en masse, and were, on occasion, subjected to involuntary pelvic exams for contraband.' "[117] Such practices ended only after a threat of legal action by local advocacy groups.[118]

■ Trends in Juvenile Incarceration

Incarceration has become an increasingly popular response to delinquency. According to the annual Children in Custody Census, which provides an estimate of the number of children placed in juvenile facilities on a specific day during a particular year, the number of youths placed in public and private juvenile facilities increased from approximately 74,270 in 1975 to approximately 134,011 in 1999, an increase of over 80%.[119] Although this number decreased to 104,413 in 2001, this still represents an increase of 41% over 1975 levels.[120]

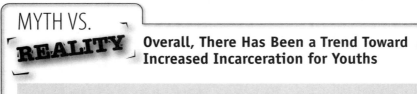

MYTH VS. REALITY

Overall, There Has Been a Trend Toward Increased Incarceration for Youths

Myth—The increase in juvenile incarceration is a product of the increasingly serious juvenile crime wave.

Reality—The overall juvenile crime rate has been relatively stable over time. Moreover, the violent juvenile crime rate did not show an increase until the late 1980s, well after the documented trend toward increased incarceration began, and it has been decreasing since 1994. It may well be that the increase in juvenile incarceration primarily reflects a change in many juvenile justice decision makers' beliefs about the best way to deal with delinquency.

Aside from the increase in the number of youths incarcerated, several other interesting trends in juvenile incarceration have been noted in recent years. One important trend has been the increasing privatization of juvenile corrections. Both the number of admissions and the rate of admissions to private juvenile facilities grew substantially between the late 1970s and the late 1980s. A similar trend was also evident in the 1990s. The number of youths committed to private facilities increased by 79% during the 1990s while commitments to public facilities increased by 41%.[121]

Another significant trend since the late 1970s has been an increasing number of minority youths placed in institutional settings. As a result, by 2001, 6 in 10 youths incarcerated in juvenile correctional facilities were minority youths. Indeed, minority youths made up 31% of the juvenile population but accounted for 60.5% of youths in custody. Moreover, *disproportionate minority confinement (DMC)* was evident in almost every state.[122]

DMC Is a Significant Problem

The 1992 amendments to the Juvenile Justice and Delinquency Prevention Act (JJDPA) require that states determine whether the proportion of juveniles in confinement exceeds the proportion of juveniles in the population. States are also required to determine whether there is disproportionate minority confinement (DMC) and, if so, to take steps to remedy the problem.[123]

■ The Effectiveness of Institutional Correctional Programs For Juveniles

Despite the long history of juvenile correctional institutions and their increasing popularity, little research on their effectiveness has been done. Moreover, what is known is not encouraging. Although some institutional programs for juveniles appear to be effective, the bulk of the evidence indicates that many juvenile institutions have little impact on recidivism. (Not surprising, perhaps, considering the quality of life characteristic of many juvenile institutions and the woeful lack of good treatment programs in many facilities.) One review of rearrest rates in states that rely heavily on juvenile institutions found that the percentage of youths rearrested ranged from 51% to more than 70%.[124] Another study, which followed up with almost 450 youths released from state training schools, discovered that the recidivism rate increased each year after discharge, reaching 54% in the fourth year.[125] An examination of the re-arrest rates of youths discharged from correctional facilities in New York state found that 51% were picked up by the police within a year of release.[126] Furthermore, a number of literature reviews have found that institutional placements and correctional treatments in general have little positive effect on recidivism.[127]

MYTH VS. REALITY — Some Institutional Placements Are Effective

Myth—Institutional programs for juvenile offenders are not effective.
Reality—Although some institutional programs are not effective, there is clear evidence that some programs are associated with substantial reductions in recidivism.

In contrast, other studies and literature reviews have presented a more favorable picture of the effectiveness of some institutional programs. For example, one large-scale review of the literature on residential programs found that such programs do produce a modest reduction in recidivism.[128] Moreover, a more recent review of institutional programs for serious juvenile offenders conducted by Mark Lipsey and David Wilson indicated that although some institutional programs have no appreciable effect on recidivism, others were associated with substantial reductions in subsequent offending.[129]

Overall, then, the research that has examined the effectiveness of institutional programs for delinquents has produced mixed results. There are quality institutional programs that treat youths humanely and that are operated by caring and professional staff. There is some evidence that small secure treatment facilities that target violent or chronic offenders and use **therapeutic**, **cognitive-behavioral**, and **skill-building interventions** are effective at reducing recidivism.[130] On the other hand, many institutional programs are characterized by a lack of genuine caring and concern for youths and poorly designed and implemented programs. Moreover, many institutions, particularly large state institutions, are little more than warehouses for youths and have little positive effect on the behavior of former residents. In fact, they may increase the likelihood that youths will engage in subsequent delinquency.

■ Challenges Facing Juvenile Correctional Institutions

A number of challenges presently face juvenile correctional institutions. One challenge is a lack of comprehensive information on these facilities. Although the Office of Juvenile Justice and Delinquency Prevention, with assistance from other organizations and individuals such as the National Center for Juvenile Justice, the National Council on Crime and Delinquency, and university researchers, collects and disseminates valuable information about juvenile institutions, there is still much that is not known about contemporary correctional institutions for youths. Indeed, relatively little is known about the institutional environments found in many facilities, the quality of programs and services provided to youths, and the effectiveness of those programs and services. Until the early 1990s, much of what was known came from studies of individual institutions or studies that looked at several institutions within one state. Unfortunately, regular, nationally representative studies of juvenile facilities were uncommon. What the existing research made clear, however, was that the quality of care in residential and institutional settings varied considerably. Indeed, this research, along with law suits filed against various institutions, showed that, at least in some facilities, the exploitation and victimization of residents, both male and female, occurred all too frequently.

To some extent, the lack of knowledge about juvenile corrections was alleviated in 1994 by the publication of *Conditions of Confinement: Juvenile Detention and Corrections Facilities*. Mandated by Congress as part of the amendments to the Juvenile Justice and Delinquency Prevention Act, this study represented an impor-

therapeutic interventions
Interventions intended to have some positive rehabilitative or habilitative effect on the individual.

cognitive-behavioral interventions
Interventions based on the premise that many offenders have not acquired the necessary cognitive (thinking) skills necessary for effective social interaction. Consequently, these approaches attempt to help youths identify problem thinking patterns, take responsibility for their thoughts and actions, and develop the skills necessary to engage in effective social interaction.

skill-building interventions
Interventions designed to help youths develop various social and interpersonal skills needed for effective social functioning. These skills may include anger management, conflict resolution, effective communication, dealing with authority, planning and decision making, and avoiding risk.

tant contemporary milestone in our understanding of juvenile institutions nationwide. Perhaps not surprisingly, it found that there are still a number of widespread problems confronting juvenile institutions, including the following:

- Crowding was a pervasive problem in juvenile facilities. As a result, a significant number of juveniles were housed in facilities where they had inadequate living space.
- A substantial number of juvenile facilities had lax security measures, which led to escapes and injuries.
- Suicidal behavior was a problem in juvenile institutions, and many institutions lacked adequate measures for preventing suicide.
- Many institutions failed to conduct health appraisals and health screenings in a timely manner. In addition, institutions often lacked good data on health care services given to residents.[131]

Other studies point to a variety of other difficulties confronting juvenile institutions. As noted earlier, one of these is the disproportionate confinement of minorities in institutional settings. For example, in 1999 the custody rate per 100,000 youths was 212 for whites, 1,004 for blacks, 485 for Hispanics, 632 for American Indians, and 182 for Asians.[132] In some facilities, minorities account for a large proportion of the institutional population.

Possibly not surprisingly, racial tension is another problem found in many settings. Racial tension is often exacerbated when institutions have sizable gang populations because gangs are often divided along racial lines. Racial and gang rivalries can feed on one another, creating the potential for institutional conflict. Importantly, a majority of administrators of state-operated secure facilities indicate that the youths being admitted to their facilities are more likely to be members of gangs than was true in the past.[133] The result is an increased probability of conflict within institutions.

Another challenge facing juvenile institutions is the development of an organizational climate and culture conducive to positive behavioral change. Too many institutions still foster an inmate code that operates to subvert institutional treatment goals. The **inmate code** consists of a set of norms and values that stand in opposition to the expressed norms and values of the institution. For example, in a study of institutionalized juveniles, Clemens Bartollas, Stuart Miller, and Simon Dinitz found an inmate code according to which exploiting others, maintaining distance from staff, not giving in to others, and not "ratting" on peers were proper ways to behave.[134] Other research has also confirmed the existence of inmate subcultures that influence youth behaviors within institutions.[135]

One characteristic of the organizational climate and inmate code of a number of institutions is support for the exploitation of others. Such exploitation ranges from the theft of goods from weaker residents by older or more physically imposing inmates to rape and other types of physical and sexual assault by residents or staff.[136] In their 1976 study of a male training school in Ohio, Bartollas, Miller, and Dinitz found that 90% of the 150 residents were being exploited by others or were exploiting another resident. Their research found a clear hierarchy of roles within the institution and a system of exploitation based on hierarchical position. This exploitation involved taking food, clothing, cigarettes, and sex from those in lower positions.[137] In a follow-up to this study, these researchers found a simi-

inmate code An unwritten set of norms and values that stand in opposition to the expressed norms and values of the institution.

lar institutional culture, although there was less sexual victimization (when residents engaged in sexual relations, these relations were more likely to be consensual).[138]

A clear challenge facing each juvenile institution is to develop a positive social organization characterized by a positive and supportive set of relationships that exist between those who work and live within the institution. As noted earlier, one type of adaptation can result in subcultures that promote violence and exploitation. However, other types of adaptations to institutional life are also possible. For example, institutionalized youths are deprived of normal heterosexual relationships, which can lead to a variety of adaptations. In some cases, a hierarchy of roles is created, and youths at the bottom of the hierarchy, who are generally less able to defend themselves, become the sexual objects of those higher up the hierarchy. In other cases, the development of sexual roles is based on the need to replace normal relationships that are denied youths who are institutionalized. The youths may develop kinship role systems, such as *make-believe families*, in which youths take on various family roles (e.g., mother, father, husband, wife, sister, and brother). Members of the make-believe family give each other mutual advice and assistance, look out for one another's interests, and participate in institutional activities together. In addition, some youths may develop homosexual alliances. In a study of three coed and four all-female training schools, Alice Propper found that 17% of institutionalized girls reported at least one homosexual experience, ranging from kissing to intimate sexual contact. Also, about half of the girls reported taking a make-believe family role. Propper found that make-believe families were as common in coed institutions as in all-female facilities.[139] Other research has produced similar findings.[140]

Exploitation and violence toward others, homosexual relationships, and make-believe families are examples of **secondary adjustments** that youths make to the organizational climate of the institution. These secondary adjustments are responses to the unique characteristics of the institutional environment and may not be seen by the juvenile as reasonable or appropriate in another context. For example, a youth may engage in a homosexual relationship with another youth within an institutional setting yet not view himself as a homosexual, nor consider the relationship conceivable in another social context.

The type of organizational culture that develops within an institution and the types of adaptations that residents make to that culture are not simply an outgrowth of placing "delinquents" in institutions. To a large degree, the **social organization** of an institution is a product of the management style and philosophy employed in the institution, the available resources, and the training and correctional philosophy of the staff.[141] For example, research indicates that inmates in custody-oriented institutions report higher levels of violence than those in treatment-oriented institutions.[142] Moreover, an institutional focus on custody has been found to encourage the development of an inmate code that runs counter to institutional goals.[143]

Of course, the social organization of an institution affects not only the youths who reside there, but also the correctional staff. Consequently, another challenge facing juvenile corrections is to develop a social organization that meets staff needs. Working in a juvenile facility can be a very exciting and rewarding

secondary adjustment A response that a juvenile makes to the unique characteristics of an institutional environment that may not be perceived as reasonable or appropriate in another context. Examples include violence, homosexual relationships, and make-believe families.

social organization Refers to the types and quality of relationships that exist between people who come together in a particular setting, such as those who live, work, or participate in a correctional institution or some other organizational or social setting.

experience, but it can also be a frustrating and stressful experience, particularly when staff are concerned about being assaulted by residents or feel that the residents lack needed services.[144]

Another problem that can contribute to staff stress is role conflict—conflict between the staff's custody and treatment responsibilities. Facility staff are often expected to enforce rules and to be confidants and helpers to youths under their care. However, these roles are not simultaneously compatible, and staff may emphasize one of these roles over the other, emphasize both of these roles, or emphasize neither. Unfortunately, role conflict can lead to confusion among both residents and staff about which roles are appropriate. In addition, other factors such as low pay, poor training, a lack of public understanding, and indifference from management can produce stress. The result can be staff burnout and high staff turnover, which are problems that confront many institutions. Consequently, it is important that staff be given the training, supervision, and support they need in carrying out their roles within the institution.

The role of institutional corrections, however, does not end when a youth is released from a corrections facility. Indeed, a major challenge facing contemporary corrections is the development of effective **aftercare** (often called parole) for released juveniles. As noted in the previous chapter, aftercare involves the provision of services intended to help youths successfully make the transition from juvenile institutions to life back in the community. Ideally, aftercare provides a planned transition from the institution to the community. Unfortunately, as Richard Snarr notes, "In many jurisdictions, aftercare (parole) is an after-thought. . . . Many times little or no thought and planning are given to the transition when a youth is released from an institution or residential setting and faced with living back in the community."[145] The failure to provide adequate aftercare for youths released from institutions is widely recognized as a major shortcoming in juvenile corrections, and the development of effective aftercare programs represents a significant challenge.[146]

Good aftercare programs do exist. For example, Kalamazoo County (Michigan) Juvenile Court has developed a day treatment program to provide aftercare for juveniles returning to the community from institutional placements. The program includes a strict phone and/or tether monitoring program, family counseling, and educational and vocational components. The program is partly based on the realization that, no matter to what extent the out-of-home placement changes the juvenile, if the home environment does not change, the chance of a successful return home will be diminished. Intensive family counseling helps all family members understand their roles in reintegrating the juvenile back into the family and into the community.

A final and serious challenge facing juvenile correctional institutions is the development of more programs that respond to children in humane ways and that protect the community. Ultimately, of course, this is the goal of most correctional programs—a goal that is, at present, unmet in far too many instances. Yet, there is some reason for optimism. There are model programs around the country that can serve as exemplars for those interested in programs that really do work. And, as discussed in Chapter 16, there are a variety of other reasons to believe that a more reasonable and effective response to juvenile offenders can be achieved.

aftercare Supervision given to children for a limited time after they are released from a correctional facility but are still under the control of the facility or the juvenile court.

■ Legal Issues

In some instances, the institutionalization of a child may be necessary in order to rehabilitate the child, deliver appropriate punishment, or protect the community. How institutional placement is accomplished and what happens to the juvenile while in the institution poses a number of important legal questions. Moreover, decisions to incarcerate youths and the conditions of their confinement influence the extent to which the juvenile justice process helps youths and protects the community.

Clearly, the decision to remove a child from his or her home is a serious one and affects not only the child, but also the family and others. Given the seriousness of this decision, it is important to consider a number of issues related to the incarceration of youths, including these:

- What rights do juveniles have at the time that treatment decisions are made?
- Do juveniles have a right to treatment?
- Do juveniles have the right not to be treated?
- What types of conditions should juveniles be subjected to, and what types of treatment should juveniles receive in institutions?
- How long should juveniles remain in institutions?
- Are adult facilities appropriate for the placement of juveniles?
- What are the liabilities that correctional institutions and institutional personnel are subject to?

What Rights Do Juveniles Have at the Time That Treatment Decisions Are Made?

The power to remove children from their homes is an ominous one and requires that children have some protection when removal is a possibility. The importance of such protection is clearly articulated in the *Gault* ruling, which stated that juveniles have a right to counsel when facing the prospect of institutionalization. But what if the parents are the ones initiating commitment on grounds other than delinquency? One case that examined this issue was *Heryford v. Parker*,[147] in which it was determined that before a child could be committed to a training school for the "feeble minded," he or she was entitled to representation by counsel. Moreover, this ruling held that representation was necessary even though the parents had initiated the commitment. Consequently, court rulings have indicated that juveniles have a right to counsel when facing institutionalization even when their parents have initiated commitment proceedings, but these rulings tell us nothing about the quality of assistance that children should receive.

Do Juveniles Have a Right to Treatment?

On the most fundamental level, the placement of a juvenile in an institutional setting against his or her will raises a question about the purpose of incarceration. In order to pass constitutional muster, there must be a legal justification for every deprivation of liberty, regardless of age. Historically, the placement of children in institutions by the juvenile courts rests on the *parens patriae* obligation of the state to provide care for children in need. If the state assumes the responsibility for the child's welfare and then fails to protect that welfare, the child's due process rights are violated.

Another persuasive argument for a juvenile's right to treatment is found in the **equal protection** safeguard. Juveniles are often "sentenced" to institutional placements for a much longer time than an adult convicted of a similar offense. The only way that this disparity can be justified is if the term of incarceration of the juvenile is for treatment, not punishment. Of course, the obligation "to treat" carries with it the presumption that treatment is available. Certainly, a juvenile should be able to make sure of the availability and suitability of treatment before being committed to that treatment. Consequently, juvenile courts may also be required to consider additional factors regarding placement, such as the location of the institution. Is it in state or out of state, and does the distance and the available transportation make it convenient or inconvenient for family members to be involved in treatment? If the placement is out of state or far away, are equivalent closer facilities available? Will the placement cause the child or the family undue hardship?[148]

equal protection
The constitutional right of all persons to have equal access to the government and courts and to be treated equally by the law and courts.

Do Juveniles Have the Right Not to Be Treated?

"Treatment" is a nebulous concept, and the line between treatment and punishment is often nonexistent. In reality, treatment usually involves some level of discomfort or punishment regardless of the intentions of treatment providers. Moreover, there are many instances where treatment cannot be reasonably justified because of the abusive conditions that youths experience in some institutional settings. Consequently, juveniles have legitimate concerns about the actual type of treatment they are asked to receive. If asked about their treatment needs, many youths might choose a different type of treatment. Still others might opt for confinement so that they could "do their time" and be done with it.

Certainly, juveniles have the "right" not to be placed into programs that degrade or dehumanize them or that are designed to embarrass or humiliate them. The American Bar Association has promulgated standards that would allow children to refuse services except for school attendance, services designed to prevent harm to their health, and services that are court ordered as conditions for noninstitutional placement (e.g., probation). The common denominator of these services is that they are "services" that society would reasonably expect to be accepted.[149] However, standards like the American Bar Association's invariably result in conflicts between those who feel that treatment providers and others have an obligation to provide for children's needs (i.e., that children have a right to treatment) and those who are concerned about the coercive nature of service provision.

What Types of Conditions Should Juveniles Be Subjected to and What Types of Treatment Should Juveniles Receive in Institutions?

Because children have the right to be treated, the type of treatment they receive is of utmost importance. The Eighth Amendment forbids treatment that is so foul, so inhumane as to violate basic societal concepts of decency. For example, long-term solitary confinement, inadequate hygiene facilities, dirty or unsanitary rooms or cells, and failure to allow for proper clothing have all been determined by appellate courts to be in violation of the Eighth Amendment. Unfortunately, there is ample evidence that many juvenile institutions have dealt with children in abusive and neglectful ways.

In addition, because the state has even a higher degree of responsibility for children than for adults, some maintain that juveniles could suffer cruel and unusual punishment if they are denied rehabilitative opportunities such as therapy, school, and job training programs.[150] Without these, any placement becomes punishment, which raises equal protection and due process issues as well.

How Long Should Juveniles Remain in Institutions?

Determining when a juvenile has been rehabilitated is not an exact science. Rehabilitation as an outcome is as individualized as the person who is being rehabilitated. Determining that a juvenile needs to be in placement, in many instances, is not very difficult. Determining when a youth is rehabilitated is often more difficult. To keep a child in an institution longer than necessary to protect the community and effect rehabilitation would be an unjustifiable deprivation of liberty[151] and a waste of public tax dollars. This issue takes on new importance in view of the increasing privatization of treatment institutions. If profit is the goal of a placement, there will be pressure on the institution to keep the child longer than necessary in order to achieve financial objectives. Most, if not all, juvenile court statutes and applicable court rules have provisions for periodic reviews to check on the welfare and progress of juveniles removed from home. These review requirements provide for "windows of opportunity" for return home and act as safeguards against "extra" or unnecessary time in placement.[152]

Are Adult Facilities Appropriate for the Placement of Juveniles?

As noted earlier in this chapter, the jailing of juveniles presents a variety of problems for correctional personnel as well as juveniles. In recognition of the potentially harmful effects of jailing juveniles, the Juvenile Justice and Delinquency Prevention Act was amended in 1980 to abolish the confining of juveniles in adult jails and lockups in states that received funds under this act.[153] The power of the purse can be an excellent persuader, but it has no impact on states that choose to opt out of participation. Moreover, as noted earlier in this chapter, there are circumstances under which juveniles can be placed in adult jails. As a result, juveniles continue to be placed in jails in some jurisdictions. Moreover, youths under 18 are also committed to state prisons each year. For example, there were 5,600 court commitments of persons under 18 years of age to state prisons in 1999, and 38% of these commitments were for nonviolent offenses.[154] This occurs despite the fact that good evidence exists that such commitments neither help youths nor reduce the likelihood that these youths will recidivate upon release. See Chapter 9 for a thorough discussion of the benefits of transfer to adult court.

What Are the Liabilities That Correctional Institutions and Institutional Personnel Are Subject To?

Should correctional personnel be liable for harm to juveniles in their care? Generally, states are protected from liability based on the doctrine of sovereign immunity. However, in a Kansas case, *C.J.W. by and through L.W. v. State*, the state was found liable under the state's tort claims act.[155] At issue in this case was whether the state had an obligation to ensure the safety of minors by making reasonable efforts to convey known information about an aggressive and sexually deviant juvenile inmate to protect other youths with whom this youth came into contact. The state was found liable for injuries suffered by a 12-year-old inmate who was attacked and sexually abused by an older youth on two separate occasions.[156]

■ Chapter Summary

A variety of types of institutions and programs are used for the placement of juvenile offenders, including detention centers, jails, training schools, ranches, farms, forestry camps, outdoor programs, chemical dependency units, and mental health facilities. Even institutions and programs of the same type can have substantial differences. For example, contemporary correctional facilities can vary significantly in their operation, in the types of programs they offer, in their social organization, and in their effectiveness. Moreover, some of these facilities are categorized as open institutions, whereas others are closed. Some institutions, such as detention centers, reception and diagnostic centers, and training schools, house a mixed population of offenders, whereas other facilities focus on specialized offender groups, such as youths who are chemically dependent or who have mental health problems. Some are publicly operated, whereas others are privately operated (and these arguably constitute a growing "hidden system" of juvenile control).

The use of institutional placement has grown significantly since the mid-1970s, despite the lack of scientific evidence that juvenile crime increased during much of this time. Indeed, its increased use appears to be driven by a punitive "get tough" approach to delinquents. In any case, it is a significant trend in juvenile justice. However, it has not affected all groups of children equally. As has always been the case, poor children, minorities, and girls are disproportionately represented in institutional populations. Particularly striking is the overrepresentation of minorities in publicly operated juvenile institutions.

There is little scientific evidence that indicates incarceration in general has a significant impact on the level of juvenile crime or a positive effect on juvenile offenders. A possible reason is that many juvenile correctional institutions are ill equipped to respond to the needs of juvenile offenders. They typically lack the monetary resources, facilities, treatment interventions, aftercare programs, and political support needed to effectively respond to children's needs. The worst institutions expose youths to inhumane treatment, victimizing those they are supposedly there to help and protect. These institutions protect neither youths nor the community.

Clearly, there are many challenges facing juvenile institutions, such as the provision of adequate living space, the development of better security measures, the discovery of better ways to prevent self-destructive behaviors, and the provision of better health screening. The inappropriate placement of juveniles in institutional environments, including jails and adult prisons; the overrepresentation of minorities in institutional settings; negative social organization; a lack of sound aftercare programs; and meager evaluation of existing programs are some of the problems that currently exist. In spite of these problems, however, there are institutional programs that appear to make a difference in the lives of youths and thus offer the promise of protecting the community as well. And there is no question that institutions capable of handling juvenile offenders are needed. What must be determined is which youths need to be placed in institutions and what types of institutions these should be.

KEY CONCEPTS

aftercare Supervision given to children for a limited time after they are released from a correctional facility but are still under the control of the facility or the juvenile court.

cognitive-behavioral interventions Interventions based on the premise that many offenders have not acquired the necessary cognitive (thinking) skills necessary for effective social interaction. Consequently, these approaches attempt to help youths identify problem thinking patterns, take responsibility for their thoughts and actions, and develop the skills necessary to engage in effective social interaction.

cottage institution Facility in which juvenile offenders live in a "homelike" setting. Such institutions are typically more open and smaller than other types of juvenile correctional institutions, are run by "cottage parents," and have high staff-to-juvenile ratios.

detention program The total set of organized activities in which youths are involved while in a detention facility.

diagnostic and reception center A facility that holds juveniles temporarily while they are evaluated to determine the best ultimate placement for them.

equal protection The constitutional right of all persons to have equal access to the government and courts and to be treated equally by the law and courts.

group counseling A type of therapy designed to use group members to help individuals identify person issues and support individual change.

inmate code An unwritten set of norms and values that stand in opposition to the expressed norms and values of the institution.

juvenile boot camp A camp for juvenile offenders that uses military-style training and discipline to control and modify their behavior.

open institution A facility that is not locked down, has no walls or fencing, and relies on staff to control inmates.

postadjudicatory institution An institution to which a juvenile offender has been sentenced.

preadjudicatory institution A temporary institution for juveniles who are awaiting trial or awaiting sentencing.

private facility A facility that is privately owned and operated. Private facilities can generally choose whom to accept, allowing them to focus on specific types of juveniles, such as sexual offenders or less serious offenders.

public facility A government-run facility. Public facilities have less ability than private facilities to screen out difficult and violent offenders and usually house more of both.

secondary adjustment A response that a juvenile makes to the unique characteristics of an institutional environment that may not be perceived as reasonable or appropriate in another context. Examples include violence, homosexual relationships, and make-believe families.

secure facility A facility that closely constrains and monitors the residents' activities and physically restricts their access to the community.

skill-building interventions Interventions designed to help youths develop various social and interpersonal skills needed for effective social functioning. These skills may include anger management, conflict resolution, effective communication, dealing with authority, planning and decision making, and avoiding risk.

social organization Refers to the types and quality of relationships that exist between people who come together in a particular setting, such as those who live, work, or participate in a correctional institution or some other organizational or social setting.

state training school A restrictive facility designed to deal with repeat juvenile offenders.

therapeutic interventions Interventions intended to have some positive rehabilitative or habilitative effect on the individual.

token economy A type of behavior modification program in which juveniles, in return for appropriate behaviors, earn points that can be redeemed for privileges, snack foods, and so on.

voluntary admission The placement of a juvenile into an institution by his or her parents or guardian without benefit of a court order or a review of due process. The cost of such a placement is borne, not by the public, but by the parents or guardian.

REVIEW QUESTIONS

1. What are the distinguishing characteristics of institutional or residential placements?

_____

_____

_____

_____

_____

2. What are the four most common problems of institutional placements?

_____

_____

_____

_____

_____

3. Describe the fundamental differences between secure facilities and open facilities.

_____

_____

_____

_____

_____

4. How do private institutional placements differ from public institutional placements?

_____

_____

_____

_____

5. What is the most common preadjudicatory institutional placement?

_____

_____

_____

_____

6. In what ways do juveniles commonly access detention?

_____

_____

_____

_____

7. What is a token economy?

_____

_____

_____

_____

_____

8. What were the major issues and decisions of the U.S. Supreme Court in the case of _Schall v. Martin_?

_____

_____

_____

_____

_____

9. What are the arguments against using adult jails to house juveniles?

_____

_____

_____

_____

_____

10. How are training schools commonly used?

_____

_____

_____

_____

_____

11. How are diagnostic and reception centers used?

_____

_____

_____

_____

_____

12. What are the distinguishing characteristics of juvenile boot camps?

_____

_____

_____

_____

_____

13. Describe the recent trends involving institutional and residential placements for juveniles.

_____

_____

_____

_____

_____

14. Is aftercare necessary?

_____

_____

_____

_____

_____

15. What are secondary adjustments?

_____

_____

_____

_____

_____

16. What legal rights do juveniles have while incarcerated or institutionalized?

_____

_____

_____

_____

_____

_____

ADDITIONAL READINGS

Bartollas, C. et al. 1976. "The Exploitation Matrix in a Juvenile Institution." *International Journal of Criminology and Penology* 4: 257–270.

Giallombardo, R. 1974. *The Social World of Imprisoned Girls: A Comparative Study of Institutions for Juvenile Delinquents*. New York: Wiley.

Lipsey, M. W. and David B. Wilson, 1998. "Effective Intervention for Serious Juvenile Offenders," in *Serious and Violent Juvenile Offenders, Risk Factors and Successful Interventions*, eds. R. Loeber and D. P. Farrington (Thousand Oaks, CA: Sage).

Miller, J. G. 1991. *Last One over the Wall: The Massachusetts Experiment in Closing Reform Schools*. Columbus, OH: Ohio State University Press.

Parent, D. G. et al. 1994. *Conditions of Confinement: Juvenile Detention and Corrections Facilities*. Washington, DC: Office of Juvenile Justice and Delinquency Prevention.

Polsky, H. W. 1963. *Cottage Six: The Social System of Delinquent Boys in Residential Treatment*. New York: Russell Sage Foundation.

Rothman, D. J. 1971. *The Discovery of the Asylum*. Boston: Little, Brown & Co.

Schwartz, I. M. 1989. *(In)Justice for Juveniles: Rethinking the Best Interests of the Child*. Lexington, MA: Lexington Books.

Wooden, K. 1976. *Weeping in the Playtime of Others*. New York: McGraw-Hill.

NOTES

1. P. Elrod et al., "Toward an Understanding of State Operated Juvenile Institutions: A Preliminary Descriptive Analysis of a Survey of Chief Correctional Administrators" (paper presented at the annual meeting of the American Society of Criminology, New Orleans, 1992).

2. T. P. Thornberry et al., *Office of Juvenile Justice and Delinquency Prevention Report on Children in Custody 1987: A Comparison of Public and Private Juvenile Custody Facilities* (Washington, DC: Office of Juvenile Justice and Delinquency Prevention, 1991).

3. Elrod et al., "Toward an Understanding of State Operated Juvenile Institutions."

4. Thornberry et al., *Office of Juvenile Justice and Delinquency Prevention Report on Children in Custody 1987.*

5. Thornberry et al., *Office of Juvenile Justice and Delinquency Prevention Report on Children in Custody 1987.*

6. M. Sickmund, "Juveniles in Corrections." *Juvenile Offenders and Victims National Report Series* (Washington, DC: Office of Juvenile Justice and Delinquency Prevention, 2004).

7. Sickmund, "Juveniles in Corrections."

8. Sickmund, "Juveniles in Corrections"; H. N. Snyder and M. Sickmund, *Juvenile Offenders and Victims: A National Report* (Washington, DC: Office of Juvenile Justice and Delinquency Prevention, 1995).

9. Sickmund, "Juveniles in Corrections."

10. Snyder and Sickmund, *Juvenile Offenders and Victims.*

11. Sickmund, "Juveniles in Corrections."

12. D. Shichor and C. Bartollas, "Private and Public Juvenile Placements: Is There a Difference?" *Crime and Delinquency* 36 (1990): 286–299.

13. Elrod et al., "Toward an Understanding of State Operated Juvenile Institutions."

14. H. N. Snyder and M. Sickmund, *Juvenile Offenders and Victims: 1999 National Report* (Washington, DC: Office of Juvenile Justice and Delinquency Prevention, 1999).

15. Thornberry et al., *Office of Juvenile Justice and Delinquency Prevention Report on Children in Custody 1987.*

16. Snyder and Sickmund, *Juvenile Offenders and Victims: 1999 National Report.*

17. Thornberry et al., *Office of Juvenile Justice and Delinquency Prevention Report on Children in Custody 1987.*

18. D. G. Parent et al., *Conditions of Confinement: Juvenile Detention and Corrections Facilities* (Washington, DC: Office of Juvenile Justice and Delinquency Prevention, 1994), 1.

19. Snyder and Sickmund, *Juvenile Offenders and Victims: A National Report.*

20. Parent et al., *Conditions of Confinement.*

21. P. Harms, "Detention in Delinquency Cases, 1990–1999." *OJJDP Fact Sheet* (Washington, DC: Office of Juvenile Justice and Delinquency Prevention, 2003).

22. C. Puzzanchera, *Juvenile Court Statistics 1999* (Pittsburgh, PA: National Center for Juvenile Justice, 2003).

23. H. T. Rubin, *Juvenile Justice: Policy, Practice and Law*, 2nd ed. (New York: Random House, 1985).

24. Rubin, *Juvenile Justice.*

25. M. Chesney-Lind and R. G. Shelden, *Girls, Delinquency and Juvenile Justice* (Belmont, CA: Thomson/Wadsworth, 2004); Parent et al., *Conditions of Confinement*; P. Puritz and M. A. Scali, *Beyond the Walls: Improving Conditions of Confinement for Youth in Custody* (Washington, DC: Office of Juvenile Justice and Delinquency Prevention, 1998).

26. K.A. Griffiths, "Program Is the Essence of Juvenile Detention," *Federal Probation* 28 (1964): 31–34.

27. Rubin, *Juvenile Justice.*

28. Parent et al., *Conditions of Confinement.*

29. Parent et al., *Conditions of Confinement.*

30. Parent et al., *Conditions of Confinement.*

31. S. Burrell et al., *Crowding in Juvenile Detention Centers: A Problem Solving Manual* (Richmond, KY and San Francisco, CA: National Juvenile Detention Association and the Youth Law Center, 1998); D. M. Pappenfort and T. M. Young, *Use of Secure Detention of Juveniles and Alternatives to Its Use, Reports of the National Juvenile Justice Assessment Centers* (Washington, DC: Department of Justice, 1980).

32. C.E. Frazier, "Preadjudicatory Detention," in *Juvenile Justice: Policies, Programs, and Services,* ed. A. R. Roberts (Chicago: Dorsey Press, 1989).

33. C. E. Frazier and D. M. Bishop, "The Pretrial Detention of Juveniles and Its Impact on Case Dispositions," *Journal of Criminal Law and Criminology* 76 (1985): 301–321.

34. Frazier, "Preadjudicatory Detention."

35. For a review of the literature, see Burrell et al., *Crowding in Juvenile Detention Centers.*

36. M. Wordes and S. M. Jones, "Trends in Juvenile Detention and Steps Toward Reform," *Crime and Delinquency* 44 (1998): 544–560.

37. J. J. Senna and L. J. Siegel, *Juvenile Law: Cases and Comments*, 2nd ed. (St. Paul, MN: West, 1992).

38. *Schall v. Martin,* 467 U.S. 253 (1984).

39. See Burrell et al., *Crowding in Juvenile Detention Centers*; I. M. Schwartz et al., "Juvenile Detention: The Hidden Closets Revisited," *Justice Quarterly* 4 (1987): 219–235.

40. J. H. Kramer and D. J. Steffensmeier, "The Differential Detention/Jailing of Juveniles: Comparison of Detention and Nondetention Courts," *Pepperdine Law Review* 5 (1978): 795–807; P. Lerman, "Discussion of Differential Selection of Juveniles for Detention," *Journal of Research on Crime and Delinquency* 14 (1977): 166–172; Pappenfort and Young, "Use of Secure Detention of Juveniles and Alternatives to Its Use"; J. E. Poulin et al., *Juveniles in Assessment Centers and Jails: An Analysis of State Variations During the Mid 1970's, Reports of the National Juvenile Justice Assessment Centers* (Washington, DC: Office of Juvenile Justice and Delinquency Prevention, 1980).

41. H. T. Rubin, *Juveniles in Justice: A Book of Readings* (Santa Monica, CA: Goodyear, 1980).

42. J. Maupin and L. Bond-Maupin, "Detention Decision-Making in a Predominately Hispanic Region: Rural and Non-Rural Differences," *Juvenile and Family Court Journal* 50 (1999):11–21.

43. Puzzanchera, *Juvenile Court Statistics 1999*.

44. For reviews of this research, see Chesney-Lind and Shelden, *Girls, Delinquency and Juvenile Justice*; B. R. McCarthy, "Preventive Detention and Pretrial Custody in Juvenile Court," *Journal of Criminal Justice* 15 (1987):185–200.

45. Chesney-Lind and Shelden, *Girls, Delinquency and Juvenile Justice*.

46. J. A. Butts et al., *Juvenile Court Statistics, 1994* (Washington, DC: Office of Juvenile Justice and Delinquency Prevention, 1996).

47. Chesney-Lind and Shelden, *Girls, Delinquency and Juvenile Justice*; S. Hoyt and D. G. Scherer, "Female Juvenile Delinquency: Misunderstood by the Juvenile Justice System, Neglected by Social Science," *Law and Human Behavior* 22 (1998): 81–107.

48. Frazier and Bishop, "The Pretrial Detention of Juveniles and Its Impact on Case Dispositions"; J. C. Henretta, et al., "The Effect of Prior Case Outcomes on Juvenile Justice Decision Making," *Social Forces* 65 (1986): 542–562.

49. Parent et al., *Conditions of Confinement*.

50. Parent et al., *Conditions of Confinement*.

51. D. W. Roush, "Juvenile Detention Programming," in *Juvenile Delinquency: A Justice Perspective*, 3rd ed., ed. R.A. Weisheit and R.G. Culbertson (Prospect Heights, IL: Waveland, 1995).

52. P. M. Harrison and J. C. Karberg, "Prison and Jail Inmates at Midyear 2003." *Bureau of Justice Statistics Bulletin* (Washington, DC: U.S. Department of Justice, 2003).

53. Sickmund, "Juveniles in Corrections."

54. C. B. Renfrew, quoted in I. M. Schwartz, *(In)Justice for Juveniles: Rethinking the Best Interests of the Child* (Lexington, MA: Lexington Books, 1989), 66–67.

55. Senna and Siegel, *Juvenile Law*.

56. H. T. Rubin, *Juvenile Justice. Policies, Practices, and Programs* (Kingston, NJ: Civic Research Institute, 2003).

57. Snyder and Sickmund, *Juvenile Offenders and Victims: A National Report*.

58. B. Krisberg and J. F. Austin, *Reinventing Juvenile Justice* (Newbury Park, CA: Sage, 1993), 177.

59. Rubin, *Juvenile Justice. Policies, Practices, and Programs*.

60. I. M. Schwartz, *(In)Justice for Juveniles*.

61. Harrison and Karberg, "Prison and Jail Inmates at Midyear 2003."

62. R. Lotz et al., *Juvenile Delinquency and Juvenile Justice* (New York: Random House, 1985).

63. Schwartz, *(In)Justice for Juveniles*.

64. M. Chesney-Lind, "Girls in Jail," *Crime and Delinquency* 34 (1988): 150–168.

65. President's Commission on Law Enforcement and Administration of Justice, *Task Force Report: Corrections* (Washington, DC: U.S. Government Printing Office, 1967), 122.

66. Community Research Associates, *Juvenile Suicides in Adult Jails: Findings from a National Survey of Juveniles in Secure Detention Facilities* (Washington, DC: U.S. Department of Justice, 1983); M. Flaherty, "The National Incidence of Juvenile Suicide in Adult Jails and Juvenile Detention Facilities," *Suicide and Life-Threatening Behavior* 13, no. 2 (1983): 85–94; J. M. Memory, "Juvenile Suicides in Secure Detention Facilities: Correction of Published Rates," *Death Studies* 13 (1989): 455–463.

67. Community Research Associates, *Juvenile Suicides in Adult Jails*.

68. Memory, "Juvenile Suicides in Secure Detention Facilities."

69. A. Browne and D. Finkelhor, "Impact of Child Sexual Abuse: A Review of Research," *Psychological Bulletin* 99 (1986): 66–77.

70. Chesney-Lind and Shelden, *Girls, Delinquency and Juvenile Justice*.

71. Percentage calculated from Office of Juvenile Justice and Delinquency Prevention, *Census of Juveniles in Residential Placement Databook, Offense Profile of Committed Residents by Sex and Race/Ethnicity for United States*, 2001 available online at http://ojjdp.ncjrs.org/ojstatbb/cjrp/asp/Offense_Committed.asp.

72. Sickmund, "Juveniles in Corrections."

73. Parent et al., *Conditions of Confinement*.

74. Snyder and Sickmund, *Juvenile Offenders and Victims: A National Report*.

75. Parent et al., *Conditions of Confinement*.

76. R. J. Gemignani, "Juvenile Correctional Education: A Time for Change," *OJJDP Update on Research* (Washington, DC: Office of Juvenile Justice and Delinquency Prevention, 1994).

77. Parent et al., *Conditions of Confinement*.

78. Parent et al., *Conditions of Confinement*.

79. American Correctional Association, *Standards for Juvenile Training Schools* (Laurel, MD: American Correctional Association, 1988, 1991).

80. Parent et al., *Conditions of Confinement*.

81. L. J. Siegel et al., *Juvenile Delinquency, Theory, Practice, and Law*, 5th ed. (Belmont, CA: Wadsworth/Thompson, 2003).

82. C. Bartollas and S. J. Miller, *Juvenile Justice in America*, 4th ed. (Upper Saddle River, NJ: Pearson/Prentice Hall, 2005).

83. M. Soler, *Representing the Child Client* (New York: Matthew Bender, 1990).

84. American Correctional Association, *Standards for Juvenile Training Schools*.

85. Parent et al., *Conditions of Confinement*.

86. V. Leiter, "Special Analysis of Data from the OJJDP Conditions of Confinement Study," cited in Snyder and Sickmund, *Juvenile Offenders and Victims: A National Report*, 169.

87. G. A. Wasserman et al., "Assessing the Mental Health Status of Youth in Juvenile Justice Settings," *Juvenile Justice Bulletin* (Washington, DC: Office of Juvenile Justice and Delinquency Prevention, 2004).

88. M. Sickmund, "Juvenile Residential Facility Census, 2000 Selected Findings" *Juvenile Offenders and Victims National Report Series* (Washington, DC: Office of Juvenile Justice and Delinquency Prevention, 2002).

89. Sickmund, "Juvenile Residential Facility Census, 2000 Selected Findings"; Snyder and Sickmund, *Juvenile Offenders and Victims: A National Report*.

90. Data extracted from the Office of Juvenile Justice and Delinquency Prevention, *Census of Juveniles in Residential Placement Databook*.

91. Snyder and Sickmund, *Juvenile Offenders and Victims: A National Report*, 172.

92. Snyder and Sickmund, *Juvenile Offenders and Victims: A National Report*.

93. Parent et al., *Conditions of Confinement*.

94. Bartollas and Miller, *Juvenile Justice in America*.

95. J. Austin et al., *Juveniles Taken into Custody, Fiscal Year 1993* (Washington, DC: Office of Juvenile Justice and Delinquency Prevention, 1995).

96. Parent et al., *Conditions of Confinement*.

97. J. W. Rogers and G. L. Mays, *Juvenile Delinquency and Juvenile Justice* (New York: John Wiley, 1987).

98. Rogers and Mays, *Juvenile Delinquency and Juvenile Justice*, 435.

99. A. Binder et al., *Juvenile Delinquency*, 2nd ed. (Cincinnati, OH: Anderson, 1997).

100. B. B. Bourque et al., "Boot Camps for Juvenile Offenders: An Implementation Evaluation of Three Demonstration Programs," *National Institute of Justice Research in Brief* (Washington, DC: National Institute of Justice, 1996).

101. Binder et al., *Juvenile Delinquency*.

102. Bourque et al., *Boot Camps for Juvenile Offenders*.

103. Bourque et al., *Boot Camps for Juvenile Offenders*.

104. Bourque et al., *Boot Camps for Juvenile Offenders*.

105. D. L. MacKenzie et al., "A National Study Comparing the Environments of Boot Camps with Traditional Facilities for Juvenile Offenders," *National Institute of Justice Research in Brief* (Washington, DC: National Institute of Justice, 2001).

106. Sickmund, "Juveniles in Corrections."

107. Austin et al., *Juveniles Taken into Custody, Fiscal Year 1993*; Sickmund, "Juveniles in Corrections."

108. Schwartz, *(In)Justice for Juveniles*.

109. Sickmund, "Juveniles in Corrections."
110. Patricia Guttridge, cited in Schwartz, *(In)Justice for Juveniles*, 131.
111. Schwartz, *(In)Justice for Juveniles*, 131.
112. Schwartz, *(In)Justice for Juveniles*.
113. A. M. Durham, "Origins of Interest in the Privatization of Punishment: The Nineteenth and Twentieth Century American Experience," *Criminology* 27 (1989): 107–139.
114. Office of Juvenile Justice and Delinquency Prevention, "Effective Programs for Serious, Violent, and Chronic Juvenile Offenders" (national teleconference, Eastern Kentucky University, Training Resources Center, October 1995).
115. I. M. Schwartz et al., "The 'Hidden' System of Juvenile Control," *Crime and Delinquency* 30 (1984):371–385.
116. Schwartz, *(In)Justice for Juveniles*; Shichor and Bartollas, "Private and Public Juvenile Placements."
117. B. D. Lurie, cited in Schwartz, *(In)Justice for Juveniles*, 139.
118. Schwartz, *(In)Justice for Juveniles*.
119. It is likely that this figure overestimates the percentage increase in the juvenile population due to improvements in data collection. However, there is no doubt that the incarceration of youths for illegal behaviors has become more popular. See T. J. Flanagan and K. Maguire, *Sourcebook of Criminal Justice Statistics*, 1989 (Washington, DC: U.S. Government Printing Office, 1990), 559, Table 6.6; J. Moone, "Juveniles in Private Facilities, 1991–1995," *Fact Sheet no. 64* (Washington, DC: Office of Juvenile Justice and Delinquency Prevention, 1997); J. Moone, "States at a Glance: Juveniles in Public Facilities, 1995," *Fact Sheet no. 69* (Washington, DC: Office of Juvenile Justice and Delinquency Prevention, 1997); Sickmund, "Juveniles in Corrections."
120. Data from *Census of Juveniles in Residential Placement Databook*. Retrieved January 21, 2005, from http://OJJDP.ncjrs.org/ojstatbb/cjrp/asp/State_Offense.asp
121. B. Krisberg et al., *National Juvenile Custody Trends*, 1978–1989 (Washington, DC: Office of Juvenile Justice and Delinquency Prevention, 1992); Sickmund, "Juveniles in Corrections."
122. Data from *Census of Juveniles in Residential Placement Databook*. Retrieved January 21, 2005, from http://OJJDP.ncjrs.org/ojstatbb/cjrp/asp/Offense_Race.asp; http://OJJDP.ncjrs.org/ojstatbb/ezapop/asp/specific.asp. Re Sickmund, "Juveniles in Corrections."
123. C. E. Pope et al., "Disproportionate Minority Confinement: A Review of the Research Literature from 1989 through 2001." *OJJDP Juvenile Justice Bulletin* (Washington, DC: Office of Juvenile Justice and Delinquency Prevention, 2002).
124. Krisberg et al., *National Juvenile Custody Trends*, 1978–1989.
125. S. Miller, "Post-Institutional Adjustment of 443 Consecutive TICO Releases" (Ph.D. diss., Ohio State University, 1971).
126. I. J. Goldman, *Arrest and Reinstitutionalization after Release from State Schools and Other Facilities of the New York State Division of Youth: Three Studies of Youths Released January 1971 through March 1973* (Albany, NY: New York State Youth Division, 1974).
127. S. Lab and J. T. Whitehead, "A Meta-Analysis of Juvenile Correctional Treatment," *Journal of Research in Crime and Delinquency* 26 (1989): 276–295; R. Martinson, "What Works? Questions and Answers about Prison Reform," *The Public Interest*, no. 35, 1974, 22–54; W. E. Wright and M. C. Dixon, "Community Prevention and Treatment of Delinquency," *Journal of Research in Crime and Delinquency* 14 (1977): 35–67.
128. C. J. Garrett, "Effects of Residential Treatment on Adjudicated Delinquents," *Journal of Research in Crime and Delinquency* 22 (1985): 287–308.
129. M. W. Lipsey and David B. Wilson, "Effective Intervention for Serious Juvenile Offenders," in *Serious and Violent Juvenile Offenders, Risk Factors and Successful Interventions*, eds. R. Loeber and D. P. Farrington (Thousand Oaks, CA: Sage, 1998). Also see R. J. Lundman, *Prevention and Control of Juvenile Delinquency*, 3rd. ed. (New York: Oxford University Press, 2001), Chapter 10.
130. B. Krisberg, *Juvenile Justice: Improving the Quality of Care* (San Francisco: National Council on Crime and Delinquency, 1992); Lipsey and Wilson, "Effective Intervention for Serious Juvenile Offenders"; T. Palmer, "Programmatic and Nonprogrammatic Aspects of Successful Intervention," in *Choosing Correctional Options That Work: Defining the Demand and Evaluating the Supply*, ed. A. T. Harland (Thousand Oaks, CA: Sage, 1996).
131. Parent et al., *Conditions of Confinement*.
132. Sickmund, "Juveniles in Corrections."
133. Elrod et al., "Toward an Understanding of State Operated Juvenile Institutions."
134. C. Bartollas et al., *Juvenile Victimization: The Institutional Paradox* (New York: Wiley, 1976).

135. C. Sieverdes and C. Bartollas, "Security Level and Adjustment Patterns in Juvenile Institutions," *Journal of Criminal Justice* 14 (1986): 135–145.

136. C. M. Bartollas et al., "The Exploitation Matrix in a Juvenile Institution," *International Journal of Criminology and Penology* 4 (1976): 257–270; J. G. Miller, *Last One over the Wall: The Massachusetts Experiment in Closing Reform Schools* (Columbus, OH: Ohio State University Press, 1991).

137. Bartollas et al., "The Exploitation Matrix in a Juvenile Institution."

138. Bartollas and Miller, *Juvenile Justice in America*.

139. A. M. Propper, "Make-Believe Families and Homosexuality among Imprisoned Girls," *Criminology* 20 (1982): 127–138.

140. R. Giallombardo, *The Social World of Imprisoned Girls* (New York: Wiley, 1974).

141. Miller, *Last One over the Wall*; E. D. Poole and R. M. Regoli, "Violence in Juvenile Institutions," *Criminology* 21 (1983): 213–232; Sieverdes and Bartollas, "Security Level and Adjustment Patterns in Juvenile Institutions."

142. Poole and Regoli, "Violence in Juvenile Institutions."

143. Sieverdes and Bartollas, "Security Level and Adjustment Patterns in Juvenile Institutions."

144. R. Dembo and M. Dertke, "Work Environment Correlates of Staff Stress in a Youth Detention Facility," *Criminal Justice and Behavior* 13 (1986): 328–344.

145. R. Snarr, *Introduction to Corrections*, 3rd ed. (Dubuque, IA: Wm. C. Brown, 1996).

146. D. M. Altschuler et al., "Reintegration, Supervised Release, and Intensive Aftercare," *OJJDP Juvenile Justice Bulletin* (Washington, DC: Office of Juvenile Justice and Delinquency Prevention, 1999).

147. *Heryford v. Parker*, 396 F2d 393 (10th cir. 1968).

148. See Mich. Court Rule 5.943 (E)(2).

149. W. E. Thornton and L. Voight, *Delinquency and Justice*, 3rd ed. (New York: McGraw-Hill, 1992), 430.

150. *Holt v. Sarver*, 309 F Supp 362 (E.D. Ark 1970).

151. Thornton and Voight, *Delinquency and Justice*.

152. Mich. Comp. Law Ann. 712A.18(1)(e); Mich. Comp. Law Ann. 712A.18(c)(3); Mich. Comp. Law Ann. 712A.18d; Mich. Court Rule 5.944(c)(1)-(4).

153. Thornton and Voight, *Delinquency and Justice*.

154. *C.J.W. by and through L.W. v. State*, 853 P.2d 4 (Kan. 1993).

155. R. V. Del Carmen et al., *Briefs of Leading Cases in Juvenile Justice* (Cincinnati, OH: Anderson, 1998).

156. Sickmund, "Juveniles in Corrections."

The Status Offender in Juvenile Justice

Chapter Objectives

After studying this chapter, you should be able to:

- Articulate reasons supporting state intervention with status offenders
- Articulate reasons against state intervention with status offenders
- Describe how the concept of *parens patriae* has encouraged the involvement of juvenile courts in status offense cases
- Give examples of status offenses and indicate how these offenses differ from delinquent or criminal offenses
- Discuss research on the differences and similarities between status offenders and youths involved in criminal behavior
- Discuss the role that the Juvenile Justice and Delinquency Prevention Act of 1974 and subsequent amendments to that act have played in the deinstitutionalization of status offenders
- Describe the mechanisms used by different jurisdictions to incarcerate status offenders and discuss whether the incarceration of status offenders is good public policy
- Discuss different approaches that might be used with status offenders, including the strengths and weaknesses of each approach

Chapter Outline

- Introduction
- Arguments for and Against Juvenile Court Involvement in the Lives of Status Offenders
- The Legal Basis for Juvenile Court Jurisdiction over Status Offenders
- Juvenile Court Responses to Status Offenders: A Historical Perspective

■ Introduction

status offender A youth who comes within the jurisdiction of a juvenile court for a status offense rather than a delinquent or criminal offense. Alternative terms include *children in need of supervision* (*CINS* or *CHINS*), *undisciplined children*, *incorrigible youths*, and *persons in need of supervision* (*PINS*).[80]

parens patriae The doctrine that the state has a legal obligation to provide for the care, protection, treatment, and rehabilitation of children if the children's parents are unable to do so.

status offense "Any offense committed by a juvenile that would not be a crime if committed by an adult, according to the statutes or ordinances of the jurisdiction in which the offense was committed, and which is specifically applicable to juveniles because of their status as a juvenile."[81] Status offenses consist of two types: (1) acts and behaviors only possible for juveniles, such as incorrigibility, running away, and truancy, and (2) acts and behaviors that are legal for adults but are illegal for juveniles, such as smoking cigarettes, drinking alcohol, and sexual intimacy.

Since the development of the juvenile court, incorrigible youths, runaways, and truants (i.e., **status offenders**) have often been treated the same as youths who have engaged in criminal behaviors. Historically, the *parens patriae* philosophy of the juvenile courts has supported juvenile court intervention in a broad range of youthful behaviors. Indeed, many parents and other adults have demanded that juvenile courts exert their authority over problem youths even when they have avoided criminal behavior. From these parents' perspective, juvenile courts have an obligation to support parental and adult authority over children. Other people feel that juvenile court involvement is necessary in order to protect status offenders who place themselves at risk of harm. Still others believe that, if left unchecked, minor forms of child misbehavior will escalate into more serious forms of law violation. From their point of view, juvenile court involvement is necessary to protect the community from potential delinquents. Because juvenile courts are also political institutions that are sensitive to community demands for assistance, they often feel obliged to intervene in the lives of troubled children and their families.

Status Offenses Can Only Be Committed by Minors

A **status offense** is "any offense committed by a juvenile that would not be a crime if committed by an adult, according to the statutes or ordinances of the jurisdiction in which the offense was committed, and which is specifically applicable to juveniles because of their status as a juvenile."[1]

Despite the compelling nature of the arguments favoring juvenile court intervention in status offense cases, there are good reasons to question the ability of juvenile courts to effectively respond to these cases. Many status offense cases are the products of ongoing family dysfunction, abuse, neglect, and ineffective parenting, and these cases present a number of complicated issues to juvenile courts, which often lack the resources or expertise to intervene in these matters in a helpful way. Indeed, it is possible to question the ability of formal legal institutions to resolve the interpersonal, social, and psychological problems that often underlie status offense behaviors. At best, juvenile courts can play a role, along with other social agencies, in the handling of youths who defy their parents, run away from home, and skip school. To further complicate matters, the resources needed to deal with many of the problems that underlie status offense behaviors may not exist, may not be present in the community, may be difficult to access, or may be difficult to coordinate, which leads to incomplete, ineffective, and sometimes counterproductive responses. As a result, some juvenile courts have recognized their limitations in dealing with these cases and avoid them even when legal intervention is possible.

Should juvenile courts intervene in the lives of status offenders? If so, what actions represent appropriate and realistic responses? These are critical questions that continue to be debated within communities and juvenile courts around the country. This chapter reviews a variety of arguments for and against juvenile court intervention in the lives of status offenders, examines the legal basis for such intervention, and presents a historical overview of court responses. It also examines trends in the processing of status offenders, legal issues related to court processing, and research on the similarities and differences between status offenders and criminal offenders. It concludes with a discussion of effective responses to status offense cases.

■ Arguments For and Against Juvenile Court Involvement in the Lives of Status Offenders

As noted earlier, the juvenile court's role in dealing with status offenders is a matter of considerable controversy. Indeed, one of the most challenging issues that has confronted juvenile justice personnel in recent years is whether the juvenile court should "maintain, restrict, or abandon jurisdiction over status offenders."[2] Today there are two opposing camps whose positions are based on different assumptions regarding the degree of similarity between delinquents' behavior and needs and status offenders' behavior and needs.[3] Those who favor continued court intervention in the lives of status offenders present a number of arguments in support of their cause. These arguments include the following:

- Many status offenders have "special needs,"[4] and discontinuing court intervention will deny them services that they require and would not otherwise receive.[5] If the court does not intervene, no one else will, and thus very real problems faced by children, their families, and others in the community will be ignored.
- The treatment needs of status offenders are often more pronounced and require more professional and financial resources than those experienced by many delinquent offenders. Juvenile courts can assist status offenders by seeing that the necessary range of services are provided.

- Status offenses often escalate into more serious offenses[6] and are "predictive of a criminal career."[7] Behaviors such as failing to go to school, failing to obey one's parents, and running away from home can easily lead to criminal behaviors such as burglary and robbery if the court does not intervene.
- A failure of the court to intervene could result in youths' injuring themselves, because many status offense behaviors put children at risk for victimization, injury, and even death. Running away and incorrigibility may place youths in situations where they turn to drugs, alcohol, prostitution, and other high-risk behaviors.
- Court oversight of community services is often necessary to ensure that youths receive appropriate treatment and that adequate services are provided to children within the community.[8] Court orders are frequently necessary in order to "encourage" social service agencies to provide needed services to families and youths and to provide services in a timely manner.
- Because voluntary agencies do not afford youths the full range of due process protections, youths may be coerced into receiving treatment. Court oversight of the treatment process makes coercive treatment less likely.[9]
- If juvenile courts relinquish jurisdiction over status offenders, compulsory school laws would be undermined, because there would be no agency with the authority to enforce these laws.[10] Moreover, everyone in the community, children and adults alike, will bear the costs of youths who fail to get an adequate education. Juvenile court jurisdiction over truancy cases is necessary in order to uphold compulsory school laws.
- Children have a responsibility to obey their parents, and juvenile courts have a responsibility to support lawful parental authority.[11] Parents would have no effective way to deal with an unruly or troubled child without support from the local juvenile court.
- If juvenile courts surrendered their authority over status offense cases, they would, in effect, be abandoning their historical mission—protecting and serving the best interests of children.[12] They would also take a step toward becoming mini-adult courts.

The opponents of maintaining juvenile court jurisdiction over status offenders claim that the above arguments are based on a number of questionable assumptions.[13] They contend that juvenile court jurisdiction is inappropriate, frequently ineffective, and often harmful, and recommend strict limitations on or outright abandonment of juvenile court jurisdiction over status offenders (abandonment is often called **decriminalization**). Their arguments include the following:

- Family or juvenile codes dealing with status offenses are often vague, expose youths to **cruel and unusual punishment**, and fail to adequately provide youths with **equal protection under the law**, thus raising serious questions about their constitutionality.[14] Furthermore, juvenile courts have often failed to protect the constitutional rights of status offenders.[15]
- Jurisdiction over status offenses is based on faulty assumptions that these behaviors will escalate into more serious delinquent behaviors.[16]
- There is a lack of evidence that juvenile court intervention benefits status offenders.[17] Instead of relying on the juvenile courts, which often lack the

decriminalization
The process of removing proscriptions from legal codes (in the context of juvenile justice, the process of removing proscriptions against status offenses from juvenile codes).

cruel and unusual punishment Punishment handed down by a court that is too harsh for the offense committed. The Eighth Amendment of the U.S. Constitution explicitly prohibits the imposition of cruel and unusual punishment.

equal protection under the law The constitutional requirement that all laws be applied to all persons fairly and without regard to race, creed, color, national origin, sex, or age. The Fourteenth Amendment of the U.S. Constitution prohibits denial of equal protection of the laws to anyone.

necessary resources and expertise to intervene effectively, we should recognize that other social agencies are better equipped to meet the needs of these youths.

- The lack of juvenile court success in dealing with status offense cases often results in employment of the courts' ultimate constraint—institutionalization.[18] Use of the court's full coercive powers is inappropriate in cases where youths have not violated criminal laws.

- Youths learn disrespect for the law when they confront the reality that they are held to higher standards of behavior than adults, who are not required to go to school, stay at home, or agree with their parents.[19]

- Processing of status offense cases leads to negative labeling, which encourages additional acting out behaviors.[20]

- Most status offense behaviors are common but transitory behaviors that revolve around intrafamily problems that are not suited to formal court intervention.[21] Moreover, courts often weaken parental responsibility by being too willing to intervene in family relationships.[22]

- Court intervention often removes responsibility from schools and other social agencies to develop and implement out-of-court solutions to family and community problems.[23] Indeed, because the juvenile courts deal with these problems and often promise solutions, other agencies find it easy to avoid responsibility for these cases.[24]

- Status offense laws are unequally applied against poor children and females. Children of wealthy parents are able to avoid juvenile court because their parents can afford to purchase private services for them, including inpatient mental health and substance abuse treatment. However, poor children are subject to court jurisdiction because these services are not similarly available. Also, females are more likely to be processed by courts and detained as runaways.[25] As a result, minority youths and females are at greater risk of being treated like delinquent offenders and subject to punitive court action.

- Courts place an undue burden on children by holding them responsible for their behavior, but fail to hold parents responsible for what they contribute to the problem.[26] In many instances, parents, guardians, school personnel, and others play a significant role in the development of those behaviors that are subject to court intervention.

As the preceding material makes clear, there are a variety of arguments for and against juvenile court intrusion into the lives of status offenders. Regardless of the pros and cons, in most jurisdictions juvenile courts do have legal authority over status offense cases. This authority is part of a long tradition of juvenile court intervention in a broad range of situations involving youths. The following section examines the legal basis for juvenile court jurisdiction over status offense cases.

■ The Legal Basis for Juvenile Court Jurisdiction over Status Offenders

The legal basis for juvenile court jurisdiction over status offenders can be traced back to the development of the legal concept of *parens patriae* and its elaboration in the legislation that established juvenile courts in the United States. As noted in Chapter 5, the first juvenile courts were designed not only to respond to the criminal behaviors of the young, but also, in keeping with their *parens patriae* focus, to deal with troublesome children—children viewed as likely to become criminals without appropriate intervention.[27] Indeed, early juvenile courts were vested with the authority to deal with a wide range of youth behaviors and conditions, including criminal law violations, nuisance behaviors, and neglect by parents.[28]

Juvenile Codes Cover a Wide Range of Behaviors

The intent of the early reformers who worked for the establishment of juvenile courts was to avoid formal legal distinctions between delinquent, dependent, and neglected youths.[29] Consequently, statutory definitions of delinquency included acts that would be criminal if committed by an adult; behaviors that were in violation of local town, city, and county ordinances; and violations of vaguely worded laws prohibiting, for example, "vicious or immoral behavior," "incorrigibility," "profane or indecent behavior," "growing up in idleness," and "living with any vicious or disreputable person."[30]

Underlying the concept of *parens patriae* and status offense statutes is the belief that children occupy a unique position in our society. As noted in Chapter 5, juveniles are recognized by the law as having certain constitutional rights that are similar, although not identical, to those afforded adults in like circumstances. For example, the U.S. Supreme Court mandated, in the *Gault* (1967) decision, that juvenile offenders be given most of the due process protections possessed by adults—the right to assistance from counsel, the right to reasonable notice, the right to confront witnesses, and freedom from self-incrimination.[31] However, in *McKeiver v. Pennsylvania* (1971), the Court ruled that juveniles do not have a right to a jury trial. In part, this decision was based on the desire of the Court to maintain the informal nature of the juvenile courts and to avoid turning juvenile court adjudications into an adversarial process. In addition, the Court ruled that it was more important to allow the juvenile courts to continue to serve children's best interests—a goal not advanced by requiring these courts to have jury trials.[32] The *McKeiver* ruling not only refused to grant jury trials to juveniles, but also reinforced the idea that children enjoy a special status in society that is inconsistent with granting them all of the due process protections afforded adults.

State statutes also recognize the special status of children. For example, children (minors) cannot enter into binding contracts, possess real or personal property, marry, or seek employment except under special circumstances. From a legal standpoint, childhood is defined purely in terms of age. Yet underlying any age-related definition of childhood is the idea that persons under a certain age lack the life experience and maturity to act in their own best interests. Indeed, due to their age and limited experience, they could easily be victimized by unscrupulous adults. Consequently, the law recognizes the need to protect children in most legal or business relationships.

According to the concept of *parens patriae*, the state, in certain circumstances, has a duty and an obligation to protect children from adults, including their parents. Consequently, the state has established a variety of laws that spell out the obligations that parents, teachers, and others have toward children. These laws also permit state intervention in the lives of children and families when adults do not fulfill their obligations to youths. In return, the state has imposed special legal obligations on children. For example, state legal codes often require children to

- Obey the reasonable and lawful rules of parents, guardians, or custodians
- Attend school regularly and obey school rules
- Not run away or absent themselves from home without permission
- Obey curfews

Although it can be argued that the legal obligations imposed on children are reasonable and within the state's legitimate interest because they ensure that families are supported and children disciplined, any violation of the laws and ordinances that spell out these legal obligations is considered a status offense. These offenses consist of two types: (1) acts and behaviors only possible for juveniles, such as incorrigibility, running away, and truancy, and (2) acts and behaviors that are legal for adults but are illegal for juveniles, such as smoking cigarettes, drinking alcohol, and sexual intimacy.[33] In short, a status offense is an offense that can be committed only by children because of their "status" as minors. Moreover, these offenses are not inconsequential and can result in a child coming under the jurisdiction of a juvenile court.

Although many states have developed special designations for status offenders, such as children in need of supervision (CHINS or CINS) or persons in

Different Terms Are Used to Refer to Status Offenders Around the Country

Although *status offender* is the most common term used to refer to those youths who come within the jurisdiction of juvenile courts for status offenses, a variety of other names have been developed to identify these youths. In the District of Columbia they are referred to as children in need of supervision (CINS or **CHINS**); in North Carolina, they are termed undisciplined children; in Michigan, they are termed incorrigible youths; in New York State, they are persons in

CHINS Children in need of supervision (also abbreviated as CINS). This term is used to refer to status offenders in some states.

PINS Persons in need of supervision. This term is used to refer to status offenders in some states.

need of supervision (**PINS**).[34] These designations have been created in an effort to recognize the unique status of children and to differentiate status offenders from youths who engage in criminal behavior.

need of supervision (PINS), these labels often present a misleading picture of the "typical" status offender. For example, many children designated as ungovernable youths are, in fact, neglected or abused children who are forced onto the streets because of physical or sexual abuse or unlivable home conditions.[35] Using the power of the juvenile court to force this group of so-called status offenders to return home and obey their parents would hardly serve their best interests. In reality, many children who become status offenders come from neglectful or dysfunctional homes in which the parents or responsible adult have not cultivated the appropriate adult-child relationship and fundamental respect has broken down. Many children who have experienced domestic violence at a young age become incorrigible when they reach adolescence. In these cases, the children have lost respect for both parents and resort to making their own rules for behavior. They refuse to accept their father's authority, as he is often the abuser, and they know they cannot count on him. They initially look to their mother for protection, but she cannot protect herself, let alone them, so they soon learn not to rely on her either. At this point, incorrigibility is almost inevitable without outside intervention.

■ Juvenile Court Responses to Status Offenders: A Historical Perspective

Concerns about problem children are hardly new. Throughout recorded history, young people have engaged in behaviors that have been troublesome to adults. For example, in the Bible we find the following passage:

> If a man has a stubborn and rebellious son, who will not listen to them even when they chastise him, then his father and mother shall lay hands on him and bring him before the sheiks of his town at the local gateway, telling the sheiks of his town, "This son of ours is a stubborn and rebellious fellow who will not obey our orders; he is a spendthrift and a drunkard." Whereupon, all his fellow citizens shall stone him to death. So shall you eradicate evil from you, and all Israel shall hear and fear.[36]

MYTH VS. REALITY — Each Generation of Adults Has Been Concerned with Youths' Behavior

Myth—Concern over children's refusal to bow to parental authority is a recent problem.

Reality—Adult concern about children's unwillingness to adhere to adult wishes goes back to antiquity.

The legislation that established the juvenile courts codified a similar attitude to that of the previous Biblical passage. Indeed, concern over the behavior of youths—particularly poor children, who were seen as making up a population of potential criminals—played a prominent role in the development of such legislation.[37] The laws spelled out a variety of legal responses to a number of troublesome behaviors, such as incorrigibility, truancy, idleness, the use of profane language, living with disreputable persons, and immoral behavior, and they were treated as an ancillary means by which parents and other adults might exert proper control over the young so that they would develop into moral, disciplined, and productive adults.[38] Moreover, early laws vested the authority to enforce these statutes in the juvenile court, which, operating on the basis of *parens patriae*, was supposed to "act on behalf of the child and to provide care and protection equivalent to that of a parent."[39] The result was a court that had the authority to exert control over children—control viewed as critical by the early child savers, because adult courts had no authority to deal with behaviors that were not criminal.[40]

In their efforts to control problem and wayward children, juvenile courts have historically employed a variety of strategies, including warning youths, ordering them on probation, placing them out, and institutionalizing them in a variety of residential placements. These strategies were thought to be justified by the courts' mission: to support families, control children, help youths lead productive and moral lives, and protect the community.

Despite a lengthy history of responding to status offenses, many juvenile courts have come to recognize that their efforts to control and prevent such behaviors have not always been effective. For one thing, the resources and therapeutic techniques available to juvenile courts are often inadequate for responding to children's needs, and today more of these resources are being devoted to dealing with violent offenders. Moreover, many courts have come to the conclusion that efforts to control status offenders have often produced more harm than good. Unfortunately, status offenders have frequently been treated exactly like youths who have committed criminal offenses. In essence, these young offenders have been victimized by the juvenile courts, despite the courts' legal obligation to look out for their best interests.

By the early 1970s, some changes in the juvenile courts' response to status offenders were beginning to be seen in some jurisdictions. In 1972, Massachusetts,

deinstitutionalization
The removal of offenders from institutional settings. In this chapter, the term is used to refer to the removal of status offenders from detention centers and other secure institutional settings.

under the leadership of Jerome Miller, the commissioner of the Department of Youth Services, began to **deinstitutionalize** the juvenile offender population and moved toward a community-based system of corrections. Not only were status offenders released from institutions, but so were most youths who were committed to state care for criminal offenses.[41] In addition, the Juvenile Justice and Delinquency Prevention Act, which called for the nationwide deinstitutionalization of status offenders, was introduced in Congress that same year and was signed into law in 1974.[42]

According to the Juvenile Justice and Delinquency Prevention Act, within two years of the passage of the act, states were required to refrain from placing in a detention or correctional facility any youth who had committed an offense that would not be a crime if committed by an adult. Subsequent amendments to the act and clarification of the deinstitutionalization mandate confirmed Congress's intent to remove status offenders from juvenile correctional facilities and adult jails. The amendments required states to make sufficient progress in their deinstitutionalization efforts or risk losing their eligibility to receive federal formula grant funds. In addition, West Virginia and Tennessee abolished the institutionalization of status offenders as a result of lawsuits brought in those states, whereas other states, such as Alabama, New Mexico, New York, Delaware, Louisiana, and Washington, passed legislation supporting deinstitutionalization.[43]

MYTH VS. REALITY   Status Offenders Can Be Detained

Myth—The passage of the Juvenile Justice and Delinquency Prevention Act of 1974, as well as the passage of state laws in some jurisdictions, has eliminated the incarceration of status offenders in the United States.

Reality—Today, significant progress has been made in the deinstitutionalization of status offenders. However, the 1980 amendments to the Juvenile Justice and Delinquency Prevention Act make it possible for courts to detain juveniles initially referred to courts as status offenders. For example, the amendments allow juvenile court judges to detain status offenders for a limited period of time if they have violated a court order. The amendments also gave states additional time to comply with the law.[44] As a result, some 6.6% of the 334,000 status offense cases referred to juvenile courts in 1991 were detained at some point between referral to court and case disposition.[45]

Although the Juvenile Justice and Delinquency Prevention Act has led to a reduction in the detention and institutionalization of status offenders, a number of tactics are used at the local level to circumvent the intent of the act. These tactics include referring status offenders to a secure mental health facility, creating or waiting for delinquency jurisdiction, developing "semi-secure" facilities that only house status offenders, and using the court's power of contempt to detain a youth for failure to obey a lawful court order rather than for the status offense. This latter method was used with approval by the family court in South Carolina in the case of *In re Darlene C.* In its ruling on this case, the Supreme Court of South Carolina stated the following:

> *The issue is whether a juvenile who commits criminal contempt by running away in violation of a court order may be given a disposition reserved for delinquents who have committed offenses which would be crimes if committed by an adult. We conclude that, under the most egregious of circumstances as we have here, family courts may exercise their contempt power in such a manner that a status offender will be incarcerated in a secure facility.*[46]

The "most egregious circumstances" included the fact that the juvenile was 16, was female, and was a chronic runaway and school truant. The South Carolina Supreme Court supported the use of the contempt powers of the family court as a mechanism for circumventing federal restrictions regarding the treatment of status offenders. Indeed, the use of contempt powers to control status offenders allows most jurisdictions to incarcerate status offenders by elevating them to delinquency status.[47] However, in a few states, this approach to status offenders has been rejected. In the case of *W.M. v. State of Indiana*, the Indiana Court of Appeals stated that contempt of court "should not be used to accomplish the goal that the juvenile court in the instant case attempted. It is meant to allow the juvenile courts a means of keeping order in their courtrooms. . . . In light of the statutory prohibition of incarcerating juveniles whose acts are not crimes for adults, [contempt] cannot be used to incarcerate them when the underlying act is a status offense."[48]

These cases jointly indicate the continuing conflict within juvenile justice over the best ways to respond to status offenders. Many courts are unwilling to abandon their authority over status offenders and wish to retain their power to institutionalize these youths when they see fit. Nevertheless, some jurisdictions have recognized a need to curb the ability of courts to incarcerate youths who have not committed criminal offenses.

Although the deinstitutionalization of status offenders has not been completely successful, considerable progress has been made in efforts to remove these youths from juvenile and adult correctional facilities. However successful the movement to deinstitutionalize status offenders has been, though, it has not necessarily prepared courts or communities to deal more effectively with the status offender population. Significant numbers of status offenders continue to come to the attention of school personnel, law enforcement officers, and juvenile courts each year. The questions still remain as to who should respond to the problems presented by this population and which responses are most effective.

■ Status Offenders and Delinquents: Is There a Difference?

Arguments that status offense behaviors will escalate to become more serious offenses, that such offenses leave youths vulnerable to victimization, and that status offenders have special needs that can be best met by juvenile court intervention are often used to support continued court involvement in status offense cases. As the North Carolina Supreme Court ruled in the case of *In the Matter of Walker*, "In seeking solutions which provide in each case for the protection, treatment, rehabilitation, and correction of the child, it is impellingly relevant to the achievement of the state's objective that distinctions be made between undisciplined children, on the one hand, and delinquent children on the other. *The one may need protection while the other needs correction*"[49] (emphasis added).

This ruling clearly reflects the state's desire to assist children, and recognizes that status offenders and those involved in criminal behaviors may have different needs. If there are differences, they could be used as a justification for court intervention if status offenders have special needs or need to be protected by the court from youths involved in criminal activity or from adults in their lives. If there are no differences or if status offenses lead naturally to more serious delinquent activity, this too could serve as a justification for court intervention (the public needs protection from "potential criminals").

The extent of the differences between status offenders and youths involved in criminal behaviors is a matter of some debate. The majority of studies on this issue conducted before 1980 reported few differences between these two groups.[50] According to these studies, juvenile offense patterns represent a **cafeteria-style approach to delinquency** where youths engage in a variety of delinquent (status and criminal law-violating) behaviors over time.[51] This finding supports the strategy of treating status and criminal offenders in the same way.

In contrast, several other studies have reported that some status offenders are distinctly different than other types of delinquent youths and should be treated differently. In one early effort to shed light on this debate, Charles Thomas studied court records on juveniles who appeared for the first time in two Virginia juvenile courts between 1970 and 1974. Of these youths, 50.3% were charged with misdemeanors, 22.3% were charged with felonies, and 27.3% were charged with status offenses. In addition, he collected follow-up data on these youths to see if they returned to the court on subsequent charges.[52]

Thomas found that most of the first-time referrals were nonrecidivists during the follow-up period. However, he found that those charged with status offenses were more likely than youths charged with misdemeanors or felonies to engage in further delinquency. He also found that 12% of felony recidivists were referred a second time for a status offense, and 17% of misdemeanor recidivists were referred a second time for a status offense. Among youth initially referred for a status offense, 40% were referred a second time for a status offense and 60% were referred a second time for a criminal offense. Consequently, Thomas concluded that status offenders were not unlike other types of offenders who appear before the juvenile court. Moreover, he found little evidence that court processing of

cafeteria-style approach to delinquency The tendency of a youth to engage in a variety of status and criminal law-violating behaviors over time.

status offenders led to their subsequent involvement with the court for more serious offenses.[53]

Similar findings regarding the similarities between status offenders and youths who commit criminal offenses were produced in a study by Howard Snyder of the National Center for Juvenile Justice. In this study, Snyder analyzed the court careers of over 69,000 youths referred to juvenile courts in Utah and in Maricopa County (Phoenix), Arizona, and found that more than half of all youths referred to the juvenile court for a status offense were also referred for a criminal offense. Moreover, over one-quarter of the youths referred to the juvenile court at some time for a delinquent offense also had a referral for a status offense. Snyder found little evidence of offense specialization. He did find that youths charged with status offenses were less likely to escalate to more serious delinquent behaviors, especially violent behaviors, than youths who did not have status offenses in their offense careers.[54]

Somewhat different findings, however, resulted from a study conducted by Steven Clarke, who examined differences between types of juvenile offenders in Philadelphia. Clarke found that, among 3,475 males who were born in Philadelphia in 1945 and had acquired a police record before their 18th birthday, 23.4% were initially involved with the police for a status offense. Moreover, when Clarke compared these youths to those who were first arrested for a criminal offense, he found the following: (1) those first arrested for a criminal offense became recidivists at twice the rate of status offenders (61% of the criminal offenders and 30% of the status offenders had at least two offenses), and (2) 21% of the criminal offenders but only 10% of the status offenders were chronic offenders (i.e., had five or more arrests). In addition, he found that these patterns were the same for white and nonwhite males.[55]

Similar findings have been reported by Thomas Kelly. After examining the offense patterns of over 2,000 juveniles referred to the juvenile court in Wayne County (Detroit), Michigan, over a five-year period, Kelly found that status offenders and criminal offenders often have different offense careers. He found that status offenders were no more likely to engage in subsequent offending than those charged with a criminal offense—a finding at odds with the research done by Thomas. He also found that status offenders who returned to court often returned for a more serious offense than the one they had committed the first time, though typically for a less serious offense than other delinquent youths. These results indicate that juvenile court processing, rather than assisting status offenders, appears to encourage recidivism. Kelly concluded that status offenders are distinct from other delinquent offenders and should be removed from juvenile court jurisdiction.[56]

In another study, which examined 863 first referrals to the Clarke County (Las Vegas) juvenile court in 1980, Randall Shelden, John Horvath, and Sharon Tracy found differences in the offense careers of different types of status offenders. For example, they found little evidence that status offenders specialize in such behaviors or that the majority of youths who are initially referred to juvenile courts become serious offenders. However, they also found that, among status offenders who became recidivists, females were more likely to have a subsequent referral for a status offense than males. In addition, they found that status offenders who were referred to the juvenile court for running away or being unmanageable were less likely to be referred a subsequent time for a criminal offense than those petitioned for truancy, curfew infractions, or liquor law violations.[57]

Research by Solomon Kobrin and his associates, who studied status offenders in eight federal programs that focused on the deinstitutionalization of status offenders, also found that status offenders are not a homogeneous group. These researchers concluded that status offenders consist of three relatively distinct groups: status offenders who display little inclination to engage in more serious delinquent acts; status offenders whose records indicate a preponderance of delinquent (criminal) acts; and youths who had one citation for a status offense but had no prior or subsequent arrests for any offense (the largest group of the three).[58]

Overall, these findings indicate that many youths involved in status offense behaviors are similar (and in many respects identical) to youths who engage in delinquent or criminal offenses. Yet, there is also a population of youths who tend to be exclusively involved in status offense behaviors, even though chronic status offenders are rare. Consequently, status offenders are more appropriately seen as constituting a diverse category of juvenile offenders. Moreover, many of these youths require different responses than those that are typically used for youths involved in criminal offenses.

Box 13-1

Interview: Linda Vogler, Status Offender Caseworker—Family Court

Q: What is your educational background?

A: I have a B.A. in social work from Western Michigan University in 1979.

Q: What is your employment history, and do you have any special training with status offenders?

A: I have worked with status offenders and diversion cases since I was hired at the court in August of 1979. My training has been on the job. Prior to my court employment, I was in the clerical staff for the FBI in 1973. From 1974–1977, I was on active duty in the Navy doing a variety of jobs from investigator to plane captain. I left the Navy as an E4. While at Western Michigan University, I worked at the Calhoun County Juvenile Home.

Q: How big is your caseload, and what percentage of it is status offenders?

A: At any one time I have between 32 and 38 kids on my caseload. Virtually all of them are status offenders, but 50% may have delinquency charges.

Q: What special work or programming do you do with status offenders?

A: We focus on two fundamental concepts: rewarding kids for positive behavior and, thus, enhancing their self-esteem; and we hold kids accountable when they make poor choices. We contract with many of the kids, and we have a number of activities: art, sports, camps, modeling, and so on, that we get the kids involved in. These activities improve kids' self-esteem, and the close supervision over them aids in holding them accountable.

Q: What parts of your job do you like best?

A: I really enjoy the personal contact with the kids, the ability to interact with them one on one and, hopefully, have a positive effect on their decision making and attitudes.

Q: What parts of your job do you like the least?

A: Sometimes working in the court "system" can be frustrating because status offenders are at the bottom of the intervention ladder. It's hard when I see kids with needs that cannot be met due to the fact that they are not "bad" enough. In reality, we don't want them to get "bad" enough; but, without help, they might get there!

Q: What special skills or abilities do you believe you bring to the job that help you in working with these difficult and troubled youth?

A: First of all, I'm good at listening to these kids and communicating with them at their level. This really helps me identify their needs. I have a good grasp of the available community resources, which I can then match to their needs.

Q: What are your major day-to-day challenges?

A: Dealing with school systems who often take a zero tolerance attitude and can be rigid in their approach to kids. We need to be more creative by working together and looking at individual children's situations. My second biggest challenge is parents who don't discipline their children, who don't hold them accountable. If kids don't learn accountability and responsibility at home, it's difficult for society to teach it to them. Even formal court action and incarceration don't always help kids to understand that they must take individual responsibility for their actions. Parents who fail in this regard fail big time!

Q: Who do you report to?

A: The director of intake.

Q: What is your compensation?

A: About $43,000 a year.

Q: What qualities or skills would you recommend that a young person interested in the field of juvenile justice in general and in working with status offenders in particular should possess?

A: First, you must like kids as people and not be in any way judgmental. Leave that to the judges and referees. You must be creative in looking at alternatives to help kids and find creative ways of holding them accountable. You must like people in general and be systems wise. You need to understand how the court system interconnects with the welfare system, the school systems, and other agencies.

Q: What is your "wish list" for effective intervention with status offenders?

A: More than anything, our community needs two group homes: one for boys and one for girls. They need a place where they can feel safe, get a timeout, and get supportive counseling to address home and school problems; but they should not be intermingled with criminal offenders.

Q: Do you believe that courts should be involved with status offenders?

A: Yes, I believe that courts should be involved with status offenders. Early intervention can prevent later serious delinquency in my experience. Kids also need to know that they will be ultimately responsible to someone, and sometimes that someone needs to be the court. When they won't listen to their parents or me, they still need to listen.

Q: Is there anything else that you want our readers to know?

A: I like my job, even after all this time. Teenage status offenders are quite a challenge, and every day brings a new one. The successes are important and lead to life-long friendships.

Whether or not status offenders constitute a unique group of juvenile offenders, there are some practical and philosophical problems with using the coercive power of the court to intervene in their lives. In many instances, youths before the court for status offenses are responding to disruptive and unhealthy family relationships. Given the nature of these problems, it is reasonable to question the ability of the juvenile court to resolve them.

MYTH VS. REALITY — Juvenile Courts Have Limited Resources

Myth—Juvenile courts can use their enormous power to coerce families into solving their problems.

Reality—Juvenile courts possess limited methods and resources for resolving family problems. Although juvenile courts may play a key supportive role in helping families resolve their problems and can act as service brokers, they themselves are not in a position to provide the types of interventions that families often require.

Ultimately, a juvenile court must rely on its coercive powers as a legal institution and the expertise of other community agencies when it tackles family problems. Also, the efforts of a juvenile court to effect change in a family may feel the same to the child whether they are made under the guise of protection or correction. From the child's perspective, removal from a disruptive family environment and placement in detention, foster care, or some other out-of-home placement may still be perceived as punishment regardless of the court's reasons for removal. By using the power of the court to intervene in their lives and attempt to change behavior by sanctions, the state, at least philosophically, treats both delinquents and status offenders the same. Although the Juvenile Justice and Delinquency Prevention Act has discouraged the incarceration of status offenders, nothing prohibits the states from using a range of responses to treat status offenders, including probation, foster care placement, and institutionalization. In these circumstances, the distinction between status offenders and delinquent youths delineated in the *Walker* case becomes a distinction with no difference. Unfortunately, many status offenders are still treated like youths charged with criminal violations.

Attempts by juvenile courts to intervene in status offense cases, while clearly within their legal authority, are understandable given the *parens patriae focus* of the courts. Indeed, many juvenile court decision makers would like to help families that are facing problems. As noted in earlier chapters, youths who come to the attention of juvenile justice agencies often face a variety of problems. They are typically poor and have been unsuccessful in school, and many come from neglectful, abusive, and dysfunctional homes. Not surprisingly, many of these youths are in conflict with their parents and display complete disregard for parental and adult authority.

Conflict between children and their parents often leads to involvement with the juvenile court under four conditions. First, parents may seek relief from the court in their effort to establish parental authority. Second, other authority figures, such as school personnel and police, often come into contact with these youths and seek assistance or punishment from the court. Third, when these youths are out on the streets with no visible means of support, they often resort to criminal activities ranging from prostitution to robbery, which may lead to juvenile court involvement. Fourth, the juvenile court serves as the social service agency of last resort for the poor, who often lack other resources (e.g., counseling and treatment services) that might help them resolve their problems. The juvenile court becomes the default agency within the community and is forced to try to resolve a range of family and community problems that occur in poor neighborhoods. However, the court, lacking sufficient resources and expertise, will often be unable to resolve these problems, which sets the stage for more coercive responses to status offenders.

■ Factors Influencing Juvenile Justice Responses to Status Offenders

Under the legal philosophy of *parens patriae*, which has traditionally guided juvenile justice practice, there has been little reason to distinguish between status and criminal (delinquent) offenders. Moreover, any distinction that juvenile courts have made between these two groups has typically had no discernible effect on juvenile court practice. Given the lack of concern with due process for children, juvenile courts have often intervened without giving much thought to how the youths it deals with came to court. The result has frequently been similar treatment for status and delinquent offenders.

Despite the history of similar treatment, juvenile courts and legislatures have begun to take a much different approach to the status offender in some jurisdictions. Perhaps the most important factor motivating change in this area is the pressure on juvenile courts to deal with increasing numbers of serious juvenile offenders. In response to this pressure, courts have directed resources away from status offenders, and many courts no longer accept parental or school complaints as the sole basis for court intervention.

Another factor influencing the courts' handling of status offenders is the Juvenile Justice and Delinquency Prevention Act. Under the act, states have been encouraged to divert status offenders from the juvenile justice process. They have also been encouraged to refrain from placing them in detention or correctional facilities or risk losing JJDP Act grant funds.[59] As a result, almost every state has enacted laws designed to implement the mandates of the Juvenile Justice and Delinquency Prevention Act, such as the deinstitutionalization of status offenders and the removal of status offenders from adult jails.[60] Similarly, many juvenile courts have begun to examine a range of diversionary responses to status offenders.

Still another factor is the recognition by many juvenile justice practitioners, including judges, that juvenile courts lack the resources and expertise to successfully intervene in many status offense cases. This recognition has led courts to call for other community agencies to play a more prominent role in handling child and family problems within the community. Although courts understand that some status offenders may eventually become involved in more serious types of delinquent activity, many courts refuse to assume formal jurisdiction over them until they do. The tendency to avoid intervention in the lives of status offenders has no doubt been partly fostered by the advocacy of various groups such as the American Bar Association Juvenile Justice Standards Project, which has even proposed the elimination of general juvenile court jurisdiction over status offenders.[61]

■ Effective Interventions with the Status Offender

Juvenile justice agencies, including juvenile courts, have supported a wide range of responses to status offenders. Historically, these responses have ranged from institutionalization to avoidance of involvement in status offense cases. To a large extent, the range of responses reflects the ambivalence juvenile justice agencies display toward these often difficult cases. Although more juvenile courts might be inclined to intervene in these cases if they were confident of success, they must confront the reality that intervention in status offense cases is often ineffective. Indeed, in some instances, juvenile justice intervention is actually harmful.

Although juvenile justice agency involvement in status offense cases can hardly be called an unmitigated success, there are some indications that effective policies geared toward status offenders are possible. One policy that has received some support is **decriminalization**,[62] in particular, the removal of legal prohibitions against status offenses from juvenile codes. The effect of decriminalization is to keep status offense cases from falling under juvenile court jurisdiction. This approach would have some obvious advantages for juvenile justice agencies. By removing such cases from the juvenile justice process, concerns about ineffective or harmful juvenile justice responses would be alleviated.

Those who favor continuing juvenile court involvement in status offense cases contend that the effect of discontinuing court involvement would be to leave the real problems that confront families and communities around the country unaddressed. However, advocates of continued intervention in status offense cases do not always agree on how much intervention is appropriate. Some support more limited forms of juvenile justice intervention, such as diversion.

Programs designed to divert status offenders from the formal juvenile justice process appear to have met with some success in a number of jurisdictions. For example, a program developed in Sacramento, California, that was intended to divert status offenders from detention and to avoid subsequent referrals to the juvenile court has been found to be moderately successful.[63] This program, called the Sacramento County 601 Diversion Project, began operation in 1970. It focused on status offenders who were referred to the juvenile court by police, parents, and school officials. These youths were typically white, poor, female, under 15 years of age, and were referred to the program because of family problems.

decriminalization
The process of removing proscriptions from legal codes (in the context of juvenile justice, the process of removing proscriptions against status offenses from juvenile codes).

When a youth was referred to the court, specially trained probation officers would contact the juvenile's parents and request that they come to Juvenile Hall for an immediate counseling session. When the parents arrived at Juvenile Hall, a probation officer read the juvenile his or her Miranda warning and explained that participation was voluntary. The officer indicated, however, that if the juvenile did not wish to participate in an immediate counseling session, his or her case would be referred to the court. The officer added that if the juvenile agreed to participate in a counseling session, he or she could return for as many as five sessions. In most cases, the parents and the juvenile agreed to participate. Once the juvenile waived his or her rights, a counseling session began in which the probation officer used family intervention techniques. The counseling techniques were intended to get the family to look at the problem as a family problem that should be addressed by the entire family rather than by simply focusing attention on the youth.

The Sacramento County 601 Diversion Project was evaluated using an experimental design. The results indicated that the project was able to significantly reduce the number of youths being detained and petitioned to court. In addition, a follow-up of youths 12 months after their involvement in the project found that 46% of those who had gone through the project had another court referral and that 22% of these youths were referred to the court for a criminal offense. In comparison, 54% of youths handled by regular probation officers were subsequently referred to court, and almost 30% of those were referred for a criminal offense.[64]

Given the apparent success of the Sacramento County 601 Diversion Project, other jurisdictions in California developed similar programs and began to evaluate them in the mid-1970s. These programs were similar to the Sacramento project in a number of ways. For example, they substituted short-term family and individual counseling for detention and referral to juvenile court. However, there were also some important differences. For example, more of the youths involved in these projects were male, Hispanic, and African American, and a number of these youths were diverted after committing criminal offenses. The evaluators of these projects employed quasi-experimental designs and found that those youths who were diverted were arrested slightly less often during a six-month follow-up period than similar youths who were not diverted.[65]

In addition to the above interventions, programs designed to improve the family and social adjustment of "antisocial" youths and status offenders have shown some promise.[66] For example, programs designed to train parents and teachers to employ child management techniques at school and in the home have resulted in reductions in problem behaviors within educational and family settings.[67] One approach that emphasizes working with the child's family in an effort to reduce problem behaviors is multisystemic therapy. In this approach, counselors attempt to collaborate with the family in developing treatment goals and plans intended to help the family meet those goals. Treatment may also involve working directly with uncooperative family members, teachers, school administrators, and others in an effort to actively involve them in treatment. An important aspect of multisystemic therapy is that it focuses on identifying strengths and problem areas within various social systems, such as the family, school, workplace, and community, and encouraging the family to take action to resolve problems rather than just talking about them.[68] Although this approach has been used effectively with serious juvenile offenders, it appears to offer considerable promise for use with status offenders as well.[69]

■ Legal Issues

A fundamental question in juvenile justice is whether juvenile courts should exercise jurisdiction over juveniles who are only status offenders. As mentioned earlier, court involvement in the lives of status offenders is clearly justified under the doctrine of *parens patriae*, but is it appropriate or even feasible for the juvenile courts to take responsibility for all "troubled children"? Of course, some children who are chronically disobedient to parents, teachers, and other authority figures and who are continually running away from home will engage in more serious forms of delinquency over time and will become victims of crime. Moreover, because juvenile courts serve as social service agencies of last resort for many poor people, they will continue to feel pressure to respond to these cases.

There are, however, strong arguments against formal court involvement. Certainly one of the most compelling is that juvenile courts should focus their time and limited resources on the most seriously delinquent youths in order to protect the community. Devoting scant court resources, such as caseworkers' time or limited bed space, to deal with status offenders who are at most only dangerous to themselves squanders limited resources. Furthermore, juvenile court efforts to resolve the problems underlying status offenses are often ineffectual. In order to be truly effective, the courts need a range of resources and therapeutic techniques that are often unavailable to them. In view of these introductory arguments, what the courts have said about the legal issues involving status offenders needs to be explored. We will begin by examining the 1972 North Carolina Supreme Court case *In the Matter of Walker*.[70]

In the Matter of Walker

Facts

On August 2, 1971, Mrs. Katherine Walker, the mother of Valerie Lenise Walker, filed a petition alleging that Valerie was an "undisciplined" child. The petition alleged that Valerie was under 16 years of age and regularly disobedient to her parents, in that she came and went without permission, kept late hours, associated with persons objected to by her parents, and went places where her parents told her not to go. Mr. and Mrs. Walker had seven children, both parents worked, and Valerie and the children were left home during working hours. Valerie was often not home when her parents got home from work. At the time of the filing of the petition, Valerie was 14 years old. A hearing was held on the petition. Valerie and her mother were present, along with a court counselor. Valerie did not have an attorney. Valerie was found to be an "undisciplined" child and placed on probation.

On September 21, 1971, Ann Jones, the court counselor (and apparently a probation officer), filed another petition concerning Valerie. This petition asked that the case be reconsidered, apparently because of problems at school. At this hearing, Valerie was represented by a public defender, who moved to vacate the original findings due to Valerie's not having had an attorney. That motion was denied, as was a motion that status offense jurisdiction violates the Equal Protection Clause of the Fourteenth Amendment to the U.S. Constitution. In addition, after the hearing, the judge determined that Valerie was now a delinquent child and committed her to the North Carolina Board of Juvenile Correction for out-of-home placement.

Court Decision

There were two issues raised before the court: (1) Did Valerie have a constitutional right to an attorney at the initial hearing that resulted in her coming under the court's jurisdiction? (2) Does the status of being an "undisciplined" child violate the Equal Protection Clause because it subjects the child to probation and the potential risk of removal from home and incarceration when the child has committed no criminal offense (whereas adults are subject to these sanctions only after a criminal conviction)?

In addressing the first issue, the court said,

> *Appellant would have this court go further than* Gault *requires. She argues for the right to counsel at the hearing of an "undisciplined child" petition on the theory that such a hearing is a critical stage in the juvenile process since it subjects the child to the risk of probation and since a violation of probation means the child is "delinquent" and subject to commitment. In such fashion, appellant seeks to engraft upon the juvenile process the "critical stage" test used by the United States Supreme Court in determining the scope of the Sixth Amendment right to counsel in criminal prosecutions. We find no authority for such engraftment. Whatever may be the proper classification for a juvenile proceeding in which the child is alleged to be undisciplined, it is certainly not a criminal prosecution within the meaning of the Sixth Amendment which guarantees the assistance of counsel.*[71]

In addressing the equal protection argument, the court said,

> *The purpose of the Juvenile Court Act is not for the punishment of offenders but for the salvation of children. . . . The Act treats delinquent children not as criminals, but as wards, and undertakes . . . to give them the control and environment that may lead to their reformation, and enable them to become law-abiding and useful citizens. . . . The state must exercise its power as* parens patriae *to protect and provide for the comfort and well-being of such of its citizens as by reason of infancy . . . are unable to take care of themselves."*[72]

The court went on to say that

> . . . *the classification here challenged is based on differences between adults and children; and there are so many valid distinctions that the basis for challenge seems shallow. These differences are reasonably related to the purposes of the Act. . . . [It] is our view that the desire of the state to exercise its authority as* parens patriae *and provide for the care of children supplies a "compelling rational" justification for the classification.*[73]

Although the *Walker* case is over 20 years old, it clearly and forcefully reiterates the *parens patriae* philosophy of the juvenile court. The North Carolina Supreme Court chose to maintain the distinction between the juvenile court and adult court and, consequently, to limit juveniles' right to due process protections in status offense cases. In many respects, the Walker case reflects the continuing debate over the legal rights of children, a debate also reflected in the case of *State ex rel, Harris v. Calendine.*[74]

State ex rel, Harris v. Calendine

Facts

On April 9, 1976, Gilbert Harris, a 15-year-old, was petitioned to the juvenile court for irregular school attendance. At the time of the petition, he had missed 50 days of school. He and his mother and stepfather were summoned to appear in the Calhoun County (West Virginia) Juvenile Court, and a hearing was held on May 17, 1976. At that hearing, Gilbert appeared with his mother and an attorney. He did not contest the allegations and was adjudicated a delinquent child and committed to an industrial school for boys. The legal action that brought this case to appeal was based upon a petition for **habeas corpus** due to his being incarcerated for almost one year after his 16th birthday, which was in July 1976, and in view of the fact that at 16 school attendance was no longer required in West Virginia.

Court Decision

The habeas corpus action was based on a broad allegation that the West Virginia Juvenile Code was inherently unconstitutional because it combined status offenders and criminal offenders. The Supreme Court of Appeals of West Virginia did not determine the statute to be unconstitutional, but set forth parameters that limited the application of parts of the law. The court began by saying, "The statutes under consideration, in the absence of guidelines for their application, fail to meet the equal protection, substantive due process, and the cruel and unusual punishment standards because they permit the classification and treatment of status offenders in the same manner as criminal offenders."[75]

The court further said,

> *We are, however, concerned with incarceration of children for status offenses. . . . The Legislature has vested the juvenile court with jurisdiction over children who commit these status offenses so that the court may enforce order, safety, morality, and family discipline within the community. The intention of the law is laudable; however,*

habeas corpus Literally, that you have the body. The term refers to a writ or petition to a court requesting review of a person's incarceration to determine whether it is proper.

the means employed to accomplish these ends are unconstitutional insofar as they result in the commitment of status offenders to secure prison-like facilities which also house children guilty of criminal conduct, or needlessly subject status offenders to the degradation and physical abuse of incarceration.[76]

The court ended by stating,

Finally, it should be noted that status offender legislation discriminates invidiously against females. . . . Our society tends to condemn female promiscuity more severely than male promiscuity, and this tendency may explain why females often are unfairly classified and treated as status offenders. . . . Furthermore the court finds no rational connection between the legitimate legislative purpose of enforcing family discipline, protecting children, and protecting society from uncontrolled children and the means by which the state is permitted to accomplish these purposes, namely incarceration of children in secure, prison-like facilities. It is generally recognized that the greatest colleges for crime are prisons and reform schools.[77]

In the Calendine ruling, the West Virginia Supreme Court of Appeals recognized that status offenders were as likely to be harmed by treatment as helped. Removal from home and family was to be a last resort, for use after all other alternatives were exhausted; and even then placement could not be in a facility housing delinquent youths. Importantly, the court required the state to view status offenders as a separate group from delinquents and provide for them accordingly.

Although there are good arguments for limiting the types of responses available to juvenile courts or even removing status offenders from juvenile court jurisdiction, such options leave unaddressed the very real problems that children and families experience. Clearly, some families are seriously troubled and need assistance. Likewise, some status offenders are seriously troubled and need help. Moreover, leaving decisions for assistance up to the family may not be feasible, nor is it always in the best interests of the child or the community. Often courts wait until a status offender has committed a delinquent offense before actually intervening, yet intervention at an earlier stage may have prevented the child from engaging in criminal behavior and obtaining a delinquent or criminal record.

A similar issue concerns children who suffer from some form of mental disability that causes them to act out in delinquent ways or behave like a status offender. Most mental health services for children must be obtained for them by their parents and are in large part voluntary. But what if parents do not seek such help, and what if youths resist their parents' efforts to provide assistance? Should mentally challenged children be compelled to get treatment? In reality, many never get the help they need until they commit a delinquent act and are brought to the attention of the local juvenile court. In many communities, services for children who experience serious mental disturbance are severely limited or nonexistent. With seriously ill children, effective care often waits for juvenile court intervention.

■ Chapter Summary

As noted in Chapter 3, many families face obstacles that challenge their ability to provide a healthy environment for their children. This, of course, is not a new problem. Indeed, one of the primary reasons for the development of the juvenile court was to establish a legal mechanism for protecting children and controlling a host of predelinquent behaviors, such as incorrigibility, running away from home, and failing to attend school.[78] The result has been a court that has two often conflicting roles. On the one hand, the juvenile court is expected to control youths and protect the community, which often requires the court to take coercive measures. On the other hand, the needs of youths often require the court to act as a nurturing and helpful parent.[79]

Historically, juvenile court intervention into the lives of status offenders has been supported for a variety of reasons. Considerable pressure has been placed on juvenile courts to support parental authority, help families deal with vexing problems, and check juvenile behaviors that are believed to be precursors to more serious criminality. The responses of juvenile courts have ranged from inaction to detention and institutionalization. Not all of the interventions employed by juvenile courts have been effective, and indeed some have produced questionable results. Some have even been patently harmful. Part of the explanation for this is that the juvenile courts have not always recognized that some status offenders have different needs than youths who have committed criminal offenses.

In an effort to restrain some of the more harmful responses to status offenders, Congress passed the Juvenile Justice and Delinquency Prevention Act of 1974, which, along with subsequent amendments to that act, discouraged states from incarcerating status offenders. As a result of this federal effort, considerable progress in the deinstitutionalization of status offenders has been made in some states. Despite this progress, however, status offenders are still incarcerated in some jurisdictions.

To a large extent, the continuing involvement of juvenile courts in status offense cases reflects the historical conflict inherent in the *parens patriae* doctrine upon which the juvenile courts were founded. Juvenile courts often feel an obligation to control or help children and their families, but they must confront the reality that they do not always possess the resources or expertise needed to intervene effectively. In some jurisdictions, the realization that juvenile courts are sometimes ill equipped to deal with status offense cases has led to policies preventing court involvement in these cases. In other jurisdictions, juvenile courts continue to intervene, and there is evidence that effective interventions for some troubled families exist. The key for a juvenile court is to determine to what extent it can play a role in helping families and children, and at the same time avoid reliance on coercion and incarceration in those cases where its efforts to help fail.

KEY CONCEPTS

cafeteria-style approach to delinquency The tendency of a youth to engage in a variety of status and criminal law-violating behaviors over time.

CHINS Children in need of supervision (also abbreviated as CINS). This term is used to refer to status offenders in some states.

cruel and unusual punishment Punishment handed down by a court that is too harsh for the offense committed. The Eighth Amendment of the U.S. Constitution explicitly prohibits the imposition of cruel and unusual punishment.

decriminalization The process of removing proscriptions from legal codes (in the context of juvenile justice, the process of removing proscriptions against status offenses from juvenile codes).

deinstitutionalization The removal of offenders from institutional settings. In this chapter, the term is used to refer to the removal of status offenders from detention centers and other secure institutional settings.

equal protection under the law The constitutional requirement that all laws be applied to all persons fairly and without regard to race, creed, color, national origin, sex, or age. The Fourteenth Amendment of the U.S. Constitution prohibits denial of equal protection of the laws to anyone.

habeas corpus Literally, that you have the body. The term refers to a writ or petition to a court requesting review of a person's incarceration to determine whether it is proper.

parens patriae. The doctrine that the state has a legal obligation to provide for the care, protection, treatment, and rehabilitation of children if the children's parents are unable to do so.

PINS Persons in need of supervision. This term is used to refer to status offenders in some states.

status offender A youth who comes within the jurisdiction of a juvenile court for a status offense rather than a delinquent or criminal offense. Alternative terms include *children in need of supervision (CINS or CHINS)*, *undisciplined children*, *incorrigible youths*, and *persons in need of supervision (PINS)*.[80]

status offense "Any offense committed by a juvenile that would not be a crime if committed by an adult, according to the statutes or ordinances of the jurisdiction in which the offense was committed, and which is specifically applicable to juveniles because of their status as a juvenile."[81] Status offenses consist of two types: (1) acts and behaviors only possible for juveniles, such as incorrigibility, running away, and truancy, and (2) acts and behaviors that are legal for adults but are illegal for juveniles, such as smoking cigarettes, drinking alcohol, and sexual intimacy.

REVIEW QUESTIONS

1. Should juvenile courts exercise jurisdiction and authority over status offenders?

_____

_____

_____

_____

_____

2. Does society have a "compelling interest" in the control of status offenders?

_____

_____

_____

_____

_____

3. Should status offenders be detained for status offenses? For running away from home?

_____

_____

_____

_____

_____

4. Are status offenders similar enough to juvenile delinquents that they should be handled by the same system?

_____

_____

_____

_____

_____

5. How does early court intervention at the status offending stage affect a youth's chances of becoming a delinquent?

_____

_____

_____

_____

_____

6. Could juvenile court intervention contribute to the eventual delinquency of status offenders?

_____

_____

_____

_____

_____

7. Should status offenders be afforded the same legal rights at juvenile court hearings as delinquent offenders?

_____

_____

_____

_____

_____

8. What alternative options are available to the community for addressing the issue of status offenses?

_____

_____

_____

_____

_____

9. What are the arguments for and against the incarceration of status offenders?

_____

_____

_____

_____

_____

10. Is it true that exposing status offenders to delinquent youths will cause the status offenders to become delinquents?

_____

_____

_____

_____

_____

11. Where do mentally ill children fit in the spectrum between status offenders and delinquents?

_____

_____

_____

_____

_____

12. How should we handle youths when this status offending or delinquent acting out is the result of mental illness?

_____

_____

_____

_____

_____

ADDITIONAL READINGS

Gendreau, P. and B. Ross. 1979. "Effective Correctional Treatment: Bibliotherapy for Cynics." *Crime and Delinquency* 25: 463–489.

Kelly, T. M. 1983. "Status Offenders Can Be Different: A Comparative Study of Delinquent Careers." *Crime and Delinquency* 29: 365–380.

Kobrin, S. et al. 1980. "Offense Patterns of Status Offenders." In *Critical Issues in Juvenile Delinquency*, ed. D. Schichor and D. Kelly. Lexington, MA: Lexington Books.

Logan, C. H. and S. P. Rausch. 1985. "Why Deinstitutionalizing Status Offenders Is Pointless." *Crime and Delinquency* 31: 501–517.

Lundman, R. J. 2001. *Prevention and Control of Juvenile Delinquency*, 3rd ed. New York: Oxford University Press, 125–148.

Shelden, R. G. et al. 1989. "Do Status Offenders Get Worse? Some Clarification on the Question of Escalation." 35: 202–216.

NOTES

1. Council of State Governments, *Juvenile Facilities: Functional Criteria* (Lexington, KY: Council of State Governments, 1977), 41.

2. J. G. Weis et al., *Jurisdiction and the Elusive Status Offender: Comparison of Involvement in Delinquent Behavior and Status Offenses.* Reports of the National Assessment Centers (Washington, DC: Office of Juvenile Justice and Delinquency Prevention, 1980), vii.

3. Weis et al., *Jurisdiction and the Elusive Status Offender*.

4. Weis et al., *Jurisdiction and the Elusive Status Offender*.

5. H.T. Rubin, *Juvenile Justice Policy, Practice and Law*, 2nd ed. (New York: Random House, 1985).

6. Rubin, *Juvenile Justice Policy, Practice and Law*.

7. Weis et al., *Jurisdiction and the Elusive Status Offender*, vii.

8. Rubin, *Juvenile Justice Policy, Practice and Law*.

9. Rubin, *Juvenile Justice Policy, Practice and Law*.

10. Rubin, *Juvenile Justice Policy, Practice and Law*.

11. Rubin, *Juvenile Justice Policy, Practice and Law*.

12. Rubin, *Juvenile Justice Policy, Practice and Law*.

13. A. R. Gough, "Beyond-Control Youth in the Juvenile Court—The Climate for Change," in *Juveniles in Justice: A Book of Readings*, ed. H. T. Rubin (Santa Monica, CA: Goodyear, 1980).

14. C. W. Thomas, "Are Status Offenders Really Different? A Comparative and Longitudinal Assessment," 22 (1976): 438–455.

15. Rubin, *Juvenile Justice Policy, Practice and Law*.

16. Gough, "Beyond-Control Youth in the Juvenile Court"; Weis et al., *Jurisdiction and the Elusive Status Offender*.

17. Gough, "Beyond-Control Youth in the Juvenile Court."

18. Rubin, *Juvenile Justice Policy, Practice and Law*.

19. Rubin, *Juvenile Justice Policy, Practice and Law*.

20. Thomas, "Are Status Offenders Really Different?"; Rubin, *Juvenile Justice Policy, Practice and Law*.

21. Gough, "Beyond-Control Youth in the Juvenile Court."

22. Rubin, *Juvenile Justice Policy, Practice and Law*.

23. Rubin, *Juvenile Justice Policy, Practice and Law*.

24. D. L. Bazelon, "Beyond Control of the Juvenile Court," *Juvenile Court Journal* (Summer 1970): 44.

25. Rubin, *Juvenile Justice Policy, Practice and Law*.

26. Rubin, *Juvenile Justice Policy, Practice and Law*.

27. A.M. Platt, *The Child Savers: The Invention of Delinquency*, 2nd ed. (Chicago: University of Chicago Press, 1977).

28. Rubin, *Juvenile Justice Policy, Practice and Law*.

29. J. Handler, "The Juvenile Court and the Adversary System: Problems of Function and Form," *Wisconsin Law Review* (1965): 9.

30. Platt, *The Child Savers*, 138.

31. J. J. Senna and L. J. Siegel, *Juvenile Law: Cases and Comments*, 2nd ed. (St. Paul, MN: West, 1992).

32. See, for example, T. J. Bernard, *The Cycle of Juvenile Justice* (New York: Oxford University Press, 1992).

33. Weis et al., *Jurisdiction and the Elusive Status Offender*.

34. North Carolina General Statutes, Chapter 7A-289.6 (1988); D.C. Code 1981, Sec. 16-2301 (8) (A) (i)-(iii); N.Y. Family Court Act, Sec 712 (b) (McKinney Supp 1978); Michigan Compiled Laws 712A.2a (1), (2), (3) & (4).

35. Notes, "Ungovernability: The Unjustifiable Jurisdiction," *Yale Law Journal* 83 (1974): 1383.

36. Deuteronomy 21:18–21.

37. Platt, *The Child Savers*.

38. Platt, *The Child Savers*.

39. Senna and Siegel, *Juvenile Law*, 1.

40. See, for example, Bernard, *The Cycle of Juvenile Justice*.

41. J. G. Miller, *Last One over the Wall: The Massachusetts Experiment in Closing Reform Schools* (Columbus, OH: Ohio State University Press, 1991).

42. G.A. Holden and R.A. Kapler, "Deinstitutionalizing Status Offenders: A Record of Progress," *Juvenile Justice* (Office of Juvenile Justice and Delinquency Prevention) 2, no. 2 (1995): 3–10.

43. Holden and Kapler, "Deinstitutionalizing Status Offenders"; C. H. Logan and S. P. Rausch, "Why Deinstitutionalizing Status Offenders Is Pointless," *Crime and Delinquency* 31 (1985): 501–517.

44. Holden and Kapler, "Deinstitutionalizing Status Offenders."

45. J. Austin et al., *Juveniles Taken into Custody, Fiscal Year 1993* (Washington, DC: Office of Juvenile Justice and Delinquency Prevention, 1995).

46. *In re Darlene C.*, 278 SC 644, 301 SE 2nd 136 (1983).

47. R. H. Mnookin and D. K. Weisberg, *Child, Family and State: Problems and Materials on Children and the Law*, 3rd ed. (Boston: Little, Brown, 1995).

48. *W.M. v. State of Indiana*, 437 NE 2nd 1028 (1982).

49. *In the Matter of Walker*, 282 NC 28, 191 SE 2nd 702 (1992).

50. S. Kobrin et al., "Offense Patterns of Status Offenders," in *Critical Issues in Juvenile Delinquency*, ed. D. Schichor and D. Kelly (Lexington, MA: Lexington Books, 1980); M. L. Erickson, "Some Empirical Questions Concerning the Current Revolution in Juvenile Justice," in *The Future of Childhood and Juvenile Justice*, ed. L. T. Empey (Charlottesville, VA: University of Virginia Press, 1979); M. W. Klein, *Street Gangs and Street Workers* (Englewood Cliffs, NJ: Prentice-Hall, 1971); C. Thomas, "Are Status Offenders Really Different?" 438–455.

51. Kobrin et al., "Offense Patterns of Status Offenders."

52. Thomas, "Are Status Offenders Really Different?"

53. Thomas, "Are Status Offenders Really Different?"

54. H. Snyder, *Court Careers of Juvenile Offenders* (Washington, DC: Office of Juvenile Justice and Delinquency Prevention, 1988).

55. S. H. Clarke, "Some Implications for North Carolina of Recent Research in Juvenile Delinquency," *Journal of Research in Crime and Delinquency* 12 (1975): 51–60.

56. Kelly, T. M. "Status Offenders Can Be Different: A Comparative Study of Delinquent Careers," 29 (1983): 365–380.

57. R. G. Shelden et al., "Do Status Offenders Get Worse? Some Clarification on the Question of Escalation," *Crime and Delinquency* 35 (1989): 202–216.

58. Kobrin et al., "Offense Patterns of Status Offenders."

59. Holden and Kapler, "Deinstitutionalizing Status Offenders."

60. J. W. Brown, "Beyond the Mandates," *Juvenile Justice* (Office of Juvenile Justice and Delinquency Prevention) 2, no. 2 (1995): 22–24.

61. American Bar Association, *Juvenile Justice Standards for Noncriminal Misbehavior*, 7–8 (1982); Mnookin and Weisberg, *Child, Family, and State*, 942.

62. S. P. Rausch and C. H. Logan, "Diversion from Juvenile Court: Panacea or Pandora's Box?" in *Evaluating Juvenile Justice*, ed. J.R. Kluegel (Beverly Hills, CA: Sage, 1983).

63. For a review of the Sacramento 601 Diversion Project, see R.J. Lundman, *Prevention and Control of Juvenile Delinquency*, 3rd ed. (New York: Oxford University Press, 1993).

64. R. Baron and F. Feeney, *Juvenile Diversion through Family Counseling* (Washington, DC: U.S. Government Printing Office, 1976).

65. T. Palmer et al., *The Evaluation of Juvenile Diversion Projects: Final Report* (Sacramento, CA: Division of Research, California Youth Authority, 1978).

66. T. Palmer, *The Re-Emergence of Correctional Intervention* (Newbury Park, CA: Sage, 1992).

67. P. Gendreau and B. Ross, "Effective Correctional Treatment: Bibliotherapy for Cynics," *Crime and Delinquency* 25 (1979): 463–489.

68. S. W. Henggeler and C. M. Borduin, *Family Therapy and Beyond: A Multisystemic Approach to Treating the Behavior Problems of Children and Adolescents* (Pacific Grove, CA: Brooks/Cole, 1990).

69. For an example of the use of multisystemic therapy with serious juvenile offenders, see S. W. Henggler et al., "Family Preservation Using Multisystemic Therapy: An Effective Alternative to Incarcerating Serious Juvenile Offenders," *Journal of Consulting and Clinical Psychology* 60 (1992): 953–961.

70. *In the Matter of Walker*, 282 NC 28, 191 SE 2nd 708 (1972).

71. *In the Matter of Walker*, 282 NC 28, 191 SE 2nd 709 (1972).

72. *In the Matter of Walker*, 282 NC 28, 191 SE 2nd 702 (1972).

73. *In the Matter of Walker*, 282 NC 28, 191 SE 2nd 702 (1972).

74. *State ex rel, Harris v. Calendine*, 233 SE 2nd 318 (1977).

75. *State ex rel, Harris v. Calendine*, 233 SE 2nd 318 (1977).

76. *State ex rel, Harris v. Calendine*, 233 SE 2nd 324 (1977).

77. *State ex rel, Harris v. Calendine*, 233 SE 2nd 325 (1977).

78. Platt, *The Child Savers*.

79. C. P. Smith et al., *A Preliminary National Assessment of the Status Offender and the Juvenile Justice System: Role Conflicts, Constraints, and Information Gaps, Reports of the National Juvenile Justice Assessment Centers* (Washington, DC: Office of Juvenile Justice and Delinquency Prevention, 1980).

80. North Carolina General Statutes, Chapter 7A-289.6 (1988); D.C. Code 1981, Sec. 16-2301 (8) (A) (i)-(iii); N.Y. Family Court Act, Sec 712 (b) (McKinney Supp 1978); Michigan Compiled Laws, 712A.2a (1), (2), (3) & (4).

81. Council of State Governments, Juvenile Facilities.

Juvenile Justice and the Serious, Chronic, or Violent Juvenile Offender

Chapter Objectives

After studying this chapter, you should be able to:

- Define violent juvenile offender, serious juvenile offender, and chronic juvenile offender and distinguish among these kinds of youths
- Describe the methods and outcomes of the Philadelphia cohort studies and the role they played in understanding the chronic juvenile offender
- Describe the extent of juvenile violence in the United States
- Describe the relationship between gender and violent juvenile crime
- Describe the relationship between race and violent juvenile crime
- Describe recent trends in violent juvenile offending
- Explain how the cultural, political, economic, community, peer group, family, and individual context of youth influences juvenile violence
- Describe the characteristics of recent violence prevention efforts initiated around the United States
- Describe different types of juvenile justice responses to violent juvenile behavior

Chapter Outline

- Introduction
- Defining the Violent Juvenile Offender
- Trends in Violent Juvenile Offending
- Explaining Violent Juvenile Crime

- Responses to Violent Juvenile Crime
- Legal Issues
- Chapter Summary
- Key Concepts
- Review Questions
- Additional Readings
- Notes

Introduction

Reports of serious or violent crimes committed by juveniles are an all-too-common feature of contemporary urban life. In cities like Los Angeles, Chicago, Houston, Detroit, and New York City,[1] reports of gang-related violence and other serious forms of juvenile and adult crime are regular features of nightly news broadcasts and local newspapers. In many instances, the stories of violence have become routine. Occasionally a sensational event will occur, like the "Central Park jogger" incident, where more than 30 teenagers, on a night of "wilding," allegedly attacked anyone they saw, including a 26-year-old female investment banker who was beaten with bricks and a lead pipe before being gang raped and left for dead. In December 1997, a 14-year-old high school student opened fire on classmates at a Paducah, Kentucky, high school, killing three and injuring five others. In Littleton, Colorado, two youths killed 13 people at Columbine High School. Although events such as these are uncommon, they undoubtedly symbolized what many perceive to be a stark reality—that no place is immune from violent juvenile crime. Moreover, the events that unfolded in Paducah and Columbine may have sparked even deeper fears in others—that violence is not relegated to urban minority youths, and that perpetrators and victims can be white and affluent. Indeed, reports by the media have regularly, even nightly, documented what some have referred to as an epidemic of teenage murder and violence. As former Secretary of Health and Human Services Louis Sullivan noted in a 1991 article published in *U.S. News and World Report*, "During every 100 hours on our streets we lose more young men than were killed in 100 hours of ground war in the Persian Gulf."[2] The sobering reality is that there is a significant amount of violent crime committed by youths in the United States.

Serious and violent juvenile crime is a significant problem in the United States, and constant media reminders of juvenile violence ensure that the problem is brought to the attention of the public. Although juvenile violence is a problem, serious violence is more of a problem in certain communities than others, and overall levels of violent juvenile crime have actually been decreasing since 1994.[3] Violent juvenile crime trend projections made during the early 1990s indicated that if present trends continued, juvenile arrests for violent juvenile crime would double by the year 2010.[4] More recently, other researchers, such as James Fox of Northeastern University, have predicted an increase in violent juvenile crime as the number of poor teenagers who lack adequate parental supervision and guidance increases in the coming years.[5] Yet, these predictions have not proven to be correct. This is not to say that the present downward trend in violent juvenile crime will continue. Some increase in juvenile crime, including violent juvenile crime, may be seen in the near future depending on cultural, economic and political trends.[6]

This chapter examines the social and individual contexts of violent juvenile crime and the juvenile justice response to violent offenses. In addition, it considers several critical issues that are important for understanding violent juvenile crime and the juvenile justice response to it. For example, it defines the serious violent offender, explores the extent of juvenile violence in the United States, identifies recent trends in violent juvenile crime, and discusses recent policy proposals for reducing the levels of violent juvenile crime in the United States. The chapter ends by examining several important legal issues related to the processing of <u>violent juvenile offenders</u>.

■ Defining the Violent Juvenile Offender

A variety of terms are used to refer to youths who engage in serious or violent juvenile crime. Indeed, it is not uncommon to hear terms like *violent offender*, *chronic offender*, or *serious offender* used to denote youths who have committed offenses that result in bodily injury to others. Although these terms are used indiscriminately by the media and members of the public, they are not synonymous.

A <u>serious offense</u> can be directed at a person or at property. Juveniles do commit crimes such as murder, rape, and aggravated assault, which are clear examples of serious offenses involving violence. However, youths may also engage in serious offenses that do not cause bodily harm, but do considerable damage to or result in significant loss of property. Moreover, the victims (the property owners) may suffer significant financial and psychological harm as a result of these offenses.

A *chronic offender* engages in repeated acts of delinquency. These acts need not be seriously harmful or violent. Although many chronic offenders do engage in serious and violent juvenile crime, some chronic status offenders merely skip school, fail to obey their parents, or run away from home regularly. Such behaviors are not violent, nor do they represent a significant threat to others, although youths who run away may place themselves at risk for victimization. Similarly, some youths may engage in a number of less serious forms of delinquency, such as shoplifting or minor assaultive behavior. These forms of delinquency are not terribly serious as far as crime goes, although assaultive behavior is clearly violent in nature, even though no serious bodily harm results.

Research on chronic offending has been done since at least the 1930s, when Sheldon and Eleanor Glueck of Harvard University began conducting research on delinquent careers.[7] Their research led them to the conclusion that early involvement in delinquent behavior, as well as a variety of personal and family factors such as body type, personality, intelligence, the quality of discipline provided by parents, and emotional ties to parents, were good predictors of delinquent careers.[8] The term <u>chronic delinquent offender</u> was popularized slightly later, by Marvin

violent juvenile offender A juvenile who has committed a violent criminal offense against a person, such as rape, murder, aggravated assault, or robbery.

serious offense An offense that results in significant personal injury, property loss, or damage.

chronic delinquent offender A juvenile who, over time, accumulates a number of delinquency charges and convictions. The related term *chronic recidivist* was used by Wolfgang and colleagues to refer to juvenile offenders who had five or more police contacts.

cohort study A study that looks at a group of people born at a specific time in a particular location (a cohort).

Wolfgang, Robert Figlio, and Thorsten Sellin, as part of their landmark publication *Delinquency in a Birth Cohort* (see Chapter 2). After studying a **cohort** of 9,945 boys born in Philadelphia in 1945, Wolfgang and his associates discovered that about one-third of the youths in the cohort (35%) had some police contact from the time they were born until their 18th birthday. They also found that about half (54%) of those youths had more than one police contact, and these youths were further categorized into two groups. Members of one of the groups were called *nonchronic recidivists*. These were youths who had more than one police contact but less than five. Members of the other group were called *chronic recidivists*. These were youths who had five or more arrests. Chronic recidivists, who comprised a mere 6% of the entire birth cohort, were responsible for a disproportionate share of all delinquent activity committed by the youths in the study. For example, chronic recidivists accounted for 52% of all arrests attributable to the cohort. An even more striking finding was that this 6% accounted for 71% of the homicides, 73% of the rapes, 82% of the robberies, and 69% of the aggravated assaults committed by members of the cohort.[9]

Similar findings were produced in a subsequent study by Wolfgang, Figlio, and Paul Tracy. These researchers looked at a cohort of 27,160 males and females born in Philadelphia in 1958. Again they found that chronic recidivists accounted for a disproportionate share of crime, including serious violent crime, attributable to the cohort. For example, of the males in the cohort, 7.5% were classified as chronic recidivists (i.e., they had five or more arrests). The chronic recidivists accounted for 61% of all arrests of cohort members, as well as 61% of the homicides, 76% of the rapes, 73% of the robberies, and 65% of the aggravated assaults.[10] Research conducted by Lyle Shannon, which looked at cohorts of youths born in Racine, Wisconsin, in 1942, 1949, and 1955, also found that a minority of youths were responsible for a disproportionate percentage of delinquent behavior.[11]

Franklyn Dunford and Delbert Elliott analyzed data from the National Youth Survey (NYS), a large-scale study that employs self-reports of delinquent behavior, to understand criminal careers. Their research indicated that previous studies probably underestimated the actual size of the chronic offending population because they relied solely on **official data**. Dunford and Elliott defined serious career offenders as youths who had committed at least three felonies, such as aggravated or sexual assault, gang fighting, car theft, or strong-arm robbery, two or more years in a row. Their results indicated that only 24% of serious career offenders had ever been arrested.[12] Thus, it is likely that the chronic offender population is somewhat larger than the 6–8% of the juvenile population identified by Wolfgang and his colleagues, although the exact size of the population is not known.

official data Data collected by juvenile justice agencies such as the police, courts, and correctional agencies.

MYTH VS. REALITY

There Is a Small Population of Chronic Violent Offenders

Myth—Youths who in engage in chronic delinquent behavior also engage in violent crime.

Reality—There is a small population of chronic violent offenders. However, some youths are merely chronic status offenders or engage in minor forms of deliquency and are not involved in violent or serious criminality.

The violent offender is still another distinct type of offender. In the broad sense, violence "is any act, event, or condition resulting from human behavior that causes death, injury, or some other damage to the physical or psychological well-being of one or more individuals."[13] However, although violence can be thought of as including actions that produce only psychological trauma, the common conception of violence is that it necessarily involves face-to-face interactions that produce physical harm or present the threat of physical harm. This conception of violence is reflected in conventional definitions of violent crime, and consequently a range of behaviors, from homicide, robbery, and rape to simple assault, are considered **violent offenses**. Certainly, violent offenses can result in serious injury or even death. However, as the case of robbery illustrates, violent crimes do not always result in the physical injury of the victim, because many robberies are carried out without anyone being hurt. Furthermore, violent offenses are not always serious offenses. For example, a simple assault can result from a schoolyard conflict, but all participants might escape injury.

Data on violent juvenile crime arrests indicate that violent juvenile crime increased substantially between 1988 and 1994 after being quite stable for the 15 years prior to 1988, but has declined since then.[14] (A more detailed examination of violent juvenile crime trends is contained in the next section.) Although juvenile crime has been decreasing, for many it continues to be at an unacceptably high level, particularly in some communities. Moreover, the over-representation of minority youths as well as an increase in the proportion of violent offenses attributed to females in recent years raises important questions for policy makers and juvenile justice practitioners.[15]

Regardless of the trends in violent juvenile offending, it must be viewed within the total context of delinquent behavior. Even during the late 1980s and the first half of the 1990s, violent juvenile offending still accounted for only a very small proportion of juvenile arrests. For example, in 2003 only about 15% of arrests of persons under 18 years of age involved an Index violent offense.[16]

The relatively small size of the violent juvenile offender population has also been noted in research that has focused on the offense careers of violent delinquents. One early study was carried out by Donna Hamparian and associates at Ohio State University. This study, called the "Dangerous Offender Project," examined a cohort of 1,138 youths who were born in Franklin County (Columbus),

violent offense
An offense directed at the person or property of another that is intended to inflict injury on another and/or deprive them of their property. Violent offenses, however, do not always result in significant personal injury to the victim.[145]

Ohio, in 1956 and 1960 and had been arrested for at least one violent offense before reaching age 18. The researchers found that only about 2% of the youths in Columbus had been arrested for a violent offense, and those who were arrested for a serious violent offense made up only about .5% of the cohort. The researchers also found that, although members of the cohort averaged about four arrests before reaching age 18, very few of them were arrested for repetitive acts of violence.[17] A subsequent analysis of the same data noted that the chronic violent offender is rare, and that most violent juvenile crimes do not involve the use of weapons.[18]

MYTH VS. REALITY

Violent Crimes Do Not Always Result In Serious Injury to Victims

Myth—Youths who are arrested for violent crimes have engaged in actions causing serious physical injury to their victims.
Reality—Some youths who are arrested for violent offenses have not seriously harmed their victims.

Similar findings were brought to light by an analysis of the delinquent careers of more than 151,000 youths who turned 18 years of age between 1980 and 1995 in Maricopa County (Phoenix), Arizona. This research, conducted by Howard Snyder of the National Center for Juvenile Justice, found that 5 in 6 youths referred to the juvenile court for a violent offense were never referred again on a violent charge. Moreover, this same study found that the frequency with which juveniles engage in violence has not changed in recent years.[19] Although too many youths engage in violent behavior, the average violent offender today is no more violent than in the past.

Box 14-1

Interview: Sandy Losota, Victim Advocate and Witness Coordinator, Kalamazoo County Prosecutor's Office—Family Division

Q: What is your full job title, and for whom do you work?
A: I am the victim advocate and witness coordinator for the Family Division of the Kalamazoo County prosecutor's office. In Michigan, the prosecuting attorney has the primary responsibility, under the law, for enforcing victims' rights. My direct employer is the Kalamazoo County Prosecuting Attorney.
Q: What are your job responsibilities?
A: I have a number of job duties when it comes to victims; most of them are interrelated and fall under the broad category of advocacy/communica-

tion. I send out victim impact statements; notify victims of upcoming court dates; send out information to victims regarding restitution, reimbursement, and/or compensation; monitor all court cases involving victims and follow up on the collection of restitution; attend hearings, sometimes with victims and sometimes on their behalf; and I enter all victim information and data into the office computer system.

Q: What is your educational background?

A: I have a B.S. in criminal justice from Michigan State University in 1975.

Q: What is your employment history that has led up to your present position?

A: Upon graduation from college, I went to work in a private law office as a secretary. I then moved to the Kalamazoo County prosecutor's office, where I have held a number of positions. I have been the word processing supervisor; intake supervisor, screening warrant requests and typing the actual warrants from the police, as well as entering them into the computer. I have spent the last year and a half in my present position.

Q: What is your usual contact with victims of juvenile crime?

A: Very few victims come into court until a formal hearing takes place. My initial contact with them, sending out the victim impact statement, generates quite a few phone calls; but unless the matter gets on the formal court docket, I would not have in-person contact with them. Once having met them, my job is to shepherd them through the system.

Q: What do you see as the major concern of victims?

A: Virtually all the victims that I see want the juvenile offender punished. They even want this more than restitution in most cases. I am often told by them that their children could not get away with something similar without punishment, and they don't want the offender to get away with anything!

Q: What are the most difficult challenges you face in your job?

A: Not being able to hold juveniles accountable. What is difficult is that the court's focus seems to be primarily on rehabilitation, not punishment. Victims want punitive action. In this way, the court is out of touch with victims; but, in fairness, the court does a good job in communicating with victims and addressing their concerns when they are made known to the court.

Q: What do you find especially rewarding about your job?

A: Being able to satisfy victims; letting them know that the prosecuting attorney's office cares about them. I enjoy the special human touch that I can bring to these cases, and securing an actual restitution award is also very gratifying.

Q: What part of your job do you least care for?

A: It hurts me to have to tell victims that I cannot help them. Often the juvenile and his/her family have no money and restitution is not available. For people that may also be facing economic hardship, this can be very difficult for them, having a loss that is not their fault but for which they will receive nothing.

Q: How much do you get paid?

A: My salary is $38,000 a year.

Q: What advice would you give victims in dealing with their victimization?

A: Don't give up! Force the communication with those persons and agencies that are legally responsible for helping you. Try to find a person in a police department, prosecutor's office, or court that will listen to you. But most of all, keep trying. Victims deserve help!

Q: What would you tell someone who wanted to become involved in your line of work?

A: Learn about the court system, how police work, and what prosecutors do. You cannot help victims if you are unable to see how they "fit" into the overall system. It is also very important to know who the key people are in the local police departments, courts, and related agencies. Effective advocacy is done between people. Be nice to everyone, establish your credibility by always going the extra mile, and accumulate favors so that you can use them for your clients.

Q: What are the major challenges you see for victim advocacy in the future?

A: The overwhelming problem is, and will continue to be, too many victims and not enough advocates. There is just so much work to do and not enough people to do it.

self-report study
A study in which the participants report on their own behavior.

Although violent juvenile crime represents only a small proportion of the criminal behavior that occurs each year, research that has employed self-report methods indicates that youth involvement in violent offending is more widespread than is suggested by arrest data. One of the most comprehensive **self-report studies** to date is the Causes and Correlates of Delinquency Study, which consists of three longitudinal projects supported by the Office of Juvenile Justice and Delinquency Prevention (OJJDP). These projects—which are being conducted in Rochester, New York; Denver; and Pittsburgh—are based on samples of more than 4,000 youths. The research found that, before reaching their late teen years, approximately 40% of the males and between 16% and 32% of the females reported involvement in one or more serious violent offenses, such as aggravated assault, robbery, rape, and gang fighting. It also found, however, that serious violent behavior tends to be intermittent rather than part of a consistent pattern. For example, after more than 5 years of research in Denver, it was found that 42% of violent offenders were actively involved in violence during only one year out of five.[20] Furthermore, caution needs to be exercised in generalizing the findings to other localities. Youths in many other communities may be less involved in violence than youths in the three cities studied. In addition, some of the delinquent activity reported in these studies may be less serious than it appears. It is probably safe to say, however, that juvenile involvement in serious violent behavior is more widespread than arrest data suggest—although still not as common as some members of the public may believe.

Certainly, one type of offense that draws considerable public attention is homicide. Like many other types of violent offenses committed by juveniles, homicides by youths under age 18 increased from 1988 through 1994. However, even in urban areas with substantial increases in homicide rates, the actual numbers of homicides have been relatively small. In fact, most communities within the United States do not experience juvenile homicides.[21]

MYTH VS. REALITY

Serious Juvenile Violence Tends to Be Concentrated in Certain Communities in the U.S.

Myth—Serious violent juvenile crime is a significant problem across the country.

Reality—Rather than being spread evenly across communities within the United States, violent juvenile crime tends to be located in certain areas. For example, in 1997, only about one-third of the states had juvenile violent arrest rates above the national average. Moreover, only 14% of the 3,141 counties in the United States had Violent Crime Index rates above the national average.[22]

As noted earlier, serious juvenile offenders, chronic juvenile offenders, and violent juvenile offenders can and should be distinguished from each other. It is perhaps more important, however, to distinguish between the types of offenses committed, such as serious offenses and violent offenses, because juvenile offenders, even the small number who can be considered chronic violent offenders, typically engage in a range of illegal actions rather than specialize in a particular type.[23]

Gender and Violent Juvenile Offending

Arrest and self-report data indicate that males are much more likely to be involved in violent criminal activities than females. For example, in 1997, the male arrest rate for Index violent offenses was five times the female rate. However, a growing number of females have been arrested for violent offenses in recent years. Indeed, in 1981, the Index violent offense arrest rate for females was 12% of the male rate, but it had increased to 20% of the male rate by 1997.[24] Moreover, the number of females referred to juvenile court for person offenses increased by 107% between 1990 and 1999, compared to 42% for males.[25]

FYI

Girls' Involvement in Violence Has Been Increasing

Girls' share of arrests of persons under 18 years of age for crime index violent offenses increased in recent years, from 11% in 1983, to 15% in 1996, to more than 18% by 2003.[26]

Self-report studies also indicate that male juveniles are much more likely to engage in violent offenses than females. Although more recent self-report studies indicate that levels of offending for males and females are similar for a number of minor offenses, such as truancy, driving without a license, running away from home, and minor forms of theft, differences in violent offenses are pronounced. Indeed, self-report studies indicate that a number of violent offenses, such as gang fighting, carrying a hidden weapon, strong-arming students and others, aggravated assault, hitting students, and sexual assault, can be considered male-dominated offenses.[27]

Males Engage in Higher Levels of Serious Violence than Females

There is some indication that more females are engaging in violent crimes today than at any time in recent history. For example, an increasing number of females have been arrested in recent years for assaultive behaviors.[28] However, as Meda Chesney-Lind and Randall Shelden note in their book, *Girls, Delinquency and Juvenile Justice*, although there has been a general increase in juvenile arrests for violent crime over the past two decades, the greatest increases for violent crime have "been for relatively minor assaults and boys have clearly been in the lead. The male of the species still has a considerable lead in committing violent acts."[29]

Race and Violent Offending

As noted in Chapter 2, African American youths are disproportionately represented in UCR arrest data for index violent offenses. For example, in 2003, African American juveniles accounted for approximately 45% of all juvenile arrests for violent crimes.[30] In addition, more recent self-report studies indicate that black youths are more likely to report involvement in more serious crimes than white youths, although the data suggest that the differences in offending may not be as great as would be expected based on arrest data. To some extent, differences in the arrest of blacks and whites for involvement in violent offenses may be a product of police bias. However, they also seem to be partly a product of behavioral differences. In a recent review of the influence of race on justice entitled *The Color of Justice*, Samuel Walker, Cassia Spohn, and Miriam DeLone found that differences in arrest rates between blacks and whites for at least some offenses, such robbery and assault, are the product of differences in offending rather than selection bias or racism on the part of law enforcement agencies.[31]

Racial Differences in Violent Female Offending Are Small

Myth—Black females are much more delinquent than white females. Moreover, black girls engage in much higher levels of violent or "masculine" forms of delinquency.

Reality—As Meda Chesney-Lind and Randall Shelden note, "race differences in girls' offending are not as marked as some might expect. Especially questionable are notions that black girls are far more delinquent than their white counterparts and that their delinquency is far more 'masculine' in content. In fact, although there are some differences between black and white girls in delinquency content, the differences tend to be less marked as the girls mature. More to the point, black girls, like their white counterparts, are still quite likely to be arrested for traditionally female offenses."[32]

■ Trends in Violent Juvenile Offending

Several types of data sources can be used to understand recent trends in juvenile violence. These data sources include UCR arrest data, <u>clearance data</u>, and victimization data. Arrest trends for crime index violent offenses for persons under age 18 have changed since the late 1980s. Arrests of persons under age 18 for violent crimes increased from 1988 through 1994—a full 70%.[33] Between 1994 and 2003, however, they declined 48%.[34] The decline in arrests of youths for violent offenses during this period may represent a general trend toward the more stable pattern of violent juvenile arrests seen between the mid-1970s and 1988. Indeed, other than the period between 1988 and 1994, arrests of persons under 18 for violent offenses have been remarkably stable.[35] The overall trend in arrest rates of persons under age 18 for crime index violent offenses is shown in **Figure 14-1**.

clearance data Data on cleared offenses. A cleared offense is an offense for which at least one person has been arrested, charged with the commission of the offense, and referred to court.

Figure 14.1 Arrest rates per 100,000 youths 10–17 years of age (Index Violent Offenses, 1980–2003). *Source*: U.S. Bureau of the Census, *Current Regulation Reports* (Washington, DC: U.S. Government Printing Office, 1982, 1992); FBI, Crime in the United States, 1980–2003.

MYTH VS. REALITY

Arrests of Youths for Violent Offenses Have Been Generally Stable Over Time

Myth—Arrests of persons under age 18 for crime index violent offenses have increased steadily over time.

Reality—Arrests of persons under age 18 for crime index violent offenses were quite stable from 1975 through 1987. They did increase from 1988 to 1994, but they have been declining since that time.[36]

Another way to investigate trends in violent juvenile crime is to examine violent crime clearance data (i.e., data on the crimes cleared or solved by law enforcement agencies). Law enforcement agencies clear an offense "when at least one person is arrested, charged with the commission of the offense, and turned over to the court for prosecution."[37] If two persons commit a robbery and both are arrested for the offense, the UCR counts this as two arrests, but clearance data would record this as one crime cleared or solved by an arrest. Thus, clearance data give a better picture of the amount of crime committed by juveniles.

When clearance trends for violent crimes are examined, they look much like the trends for violent arrests. However, the proportion of violent crimes cleared by the arrest of youths under age 18 indicates that juveniles accounted for a greater proportion of violent crime in 1994 and 1995 than at any time since the mid-

1970s. According to Melissa Sickmund, Howard Snyder, and Eileen Poe-Yamagata of the National Center for Juvenile Justice, "In both 1994 and 1995, 14% of violent crimes cleared by law enforcement were cleared by the arrest of a juvenile. Based on these statistics, juveniles were responsible for a greater proportion of the nation's violent crime problem in 1994 and 1995 than in any other year in the previous two decades."[38] Only in 1975 did juveniles account for almost as much violent crime. In that year, persons under age 18 accounted for 13% of violent crime clearances. However, the percentage of Index violent offenses cleared by the arrest of persons under 18 years of age has been declining since the peak years of 1994 and 1995, although it is still at roughly 1975 levels. In other words, juveniles continue to account for a significant portion of violent offense arrests. The percentages of violent crimes cleared by the arrest of a juvenile since 1975 are shown in **Figure 14-2**.

Figure 14.2 Percent of Index Violent Offenses cleared by arrest of persons 10–17 years of age, 1975–2003. *Source:* FBI, Crime in the United States, reports for 1975–2003.

Additional information on violent juvenile crime can be gleaned from data collected as part of the National Crime Victimization Survey (NCVS), which is a **victimization study**. As noted in Chapter 2, the NCVS asks a nationally representative sample of persons 12 years or older about the characteristics of offenders. According to the NCVS, the number of violent victimizations dropped 20% from 1994 to 1995, the largest drop in violent victimizations since the NCVS was initiated nearly 25 years ago.[39] Nevertheless, by 1996, juveniles were at greater risk of violent victimization than any other group.[40]

Arrest, clearance, and victimization data on violent juvenile crime present a conflicting picture. Recent arrest and victimization data indicate that violent juvenile crime rates have been decreasing.[41] However, some evidence indicates a larger number of juveniles are being arrested for repeat violent offenses in some jurisdictions.[42] Moreover, clearance data indicate that youths today continue to be responsible for a significant proportion of violent crime, although there is some indication that violent juvenile crime has declined in recent years.

victimization study
A study in which the participants are asked to report their victimization experiences.

It must be borne in mind, however, that some communities experience far more violent juvenile crime than others, and that the high levels of violent crime experienced in the United States are primarily the result of adult, not juvenile, violence. Indeed, in 2003, 88% of crime index violent crimes cleared by arrest were committed by adults.[43]

MYTH VS. REALITY Adults Commit Disproportionately More Homicides than Juveniles

Myth—Today juveniles account for a disproportionate share of all homicides committed in the United States.
Reality—Juveniles are underrepresented in homicide clearance data. In 2003, persons under 18 years of age accounted for only 5.2% of all homicides cleared by arrest.[44]

■ Explaining Violent Juvenile Crime

There is not one single cause of juvenile violence. Like other types of illegal behavior, violent juvenile crime is best understood by considering the interrelationships among the cultural, political, economic, community, peer group, family, and individual context within which it occurs (see Chapter 3 for a discussion of this context). Also like other types of illegal behavior, it has been explained using a variety of theories. Although it is not possible to discuss thoroughly every explanation of violent juvenile behavior that has been proposed, some of the more recent explanations are presented in the following sections.

The Cultural Context of Violent Juvenile Crime

Violence has always played a large role in the cultural make-up of the West and of the United States in particular. Even before the founding of the country, violence was seen as an acceptable way to subjugate and "civilize" native peoples. Later it was viewed as a primary means of achieving the nation's "manifest destiny," of protecting corporate interests from strikers, and of protecting the country's international interests. Indeed, the history of the United States has been filled with ongoing conflict between a variety of groups, including, but hardly limited to, capitalists and workers, men and women, minorities and majorities, environmentalists and corporate executives, civil rights activists and segregationists, anti-war protestors and pro-war supporters, and the young and the old. Because people are more connected to the present than the past, many people feel that violence has reached an unprecedented level today. In reality, violence has always been a regular feature of American culture.

Perhaps the pervasiveness of violence is best understood by considering the television viewing habits of youths and the content of TV programs. Systematic viewing of television begins at 2.5 years of age and remains at a high level through-out the teen years. It has been estimated that, on average, preschool children watch 27.8 hours of TV per week, children 6 to 11 watch 24.3 hours per week, and teenagers watch 23 hours per week.[45] Furthermore, a considerable amount of the content of television programming involves aggression, whether it is Saturday morning cartoons or prime-time dramas. Indeed, some research indicates that 80% of television programs contain some degree of violence, and it is estimated that, by the time a child's secondary education is completed, he or she will have witnessed 13,000 violent deaths on television and seen more than 200,000 violent acts.[46]

A number of research studies have found a relationship between exposure to TV violence and aggression among children.[47] In addition, the American Psychological Association and the National Institute of Mental Health have gone on record in support of the evidence showing a link between media violence and **interpersonal violence**.[48] Nonetheless, a direct causal link has not been firmly established. Perhaps the best that can be said at this point is that the literature generally portrays media violence as a reinforcer for youths who have been exposed to a number of violent models. Of course, TV violence is only one type of media violence to which children are exposed. Numerous violent models are also found in movies, comic books, and video games.

interpersonal violence Face-to-face interaction that produces physical harm or presents a threat of physical harm.

Poor Youths Watch More TV than Affluent Youths

Approximately 98% of households in the United States have a television set, and many of these have multiple sets. Moreover, television viewing occupies more of a child's waking time than any other nonschool activity. However, time spent in front of the television is not evenly distributed across the youth popu-lation. For example, the heaviest viewers of television are poor youths, because they often lack alternative leisure time opportunities.[49]

Violence is also a part of many other American institutions. A considerable amount of violence is found in sports, in schools, in communities, and in fami-lies (an issue examined later in this chapter). The effect of pervasive violence on youths is not completely understood. Nevertheless, some researchers maintain that regular exposure to violence on the part of others in the environment and in the media, particularly when such violence is rewarded, encourages violent *modeling* on the part of youths.[50] Others believe that regular exposure to violence leads to a more negative and cynical view of the world, while still others main-tain that youths may actually develop *biological adaptations* that make them more susceptible to aggressive responses to the violence they experience.

The Political Economy of Violent Juvenile Crime

As noted in Chapter 3, the *political economy* of a society at a particular point in time plays a critical role in the production of delinquency, because it determines the ways in which economic and political resources are developed, managed, and distributed, which, in turn, affect the ability of other socializing institutions (e.g., families, schools, and communities) to meet people's needs.[51] For example, political and economic decisions made by corporations and government representatives influence the distribution of jobs, the pay and benefits earned by workers, and the conditions in which people work. Similarly, political and economic decisions influence the quality of schools and the education that children receive as well as the types and quality of the TV programs that they view. Obviously, the decisions made by government and businesses have a profound effect on the quality of life experienced by all Americans.

Although there are exceptions, it is clear that violent juvenile offenders typically come from those groups who are the least advantaged—minorities and the poor. Self-report studies indicate that lower-class youths report more involvement in serious and violent juvenile crimes than middle-class youths.[52] Although poverty affects large numbers of youths of all races, it is particularly prevalent among minority youths. According to the Annual Demographic Survey conducted by the Bureau of Labor Statistics and the Bureau of the Census, 17.6% of all persons under 18 lived in poverty in 2003. However, this figure was 34.1% for African American children and 29.7% for Hispanic children.[53] Consequently, it is not surprising that UCR, self-report, and juvenile court data indicate that African American youths are overrepresented among the perpetrators of violent juvenile crime.[54]

The Community and Peer Group Contexts of Violent Juvenile Crime

According to a number of different theoretical explanations of delinquency, community factors play an important role in delinquent behavior. Research indicates that delinquency rates appear to be influenced by the economic and social conditions found in communities. Today, approximately 13 million youths live in poverty, and this number increased by almost 1 million youths between 2002 and 2003 after showing some decline during the last half of the 1990s.[55] Although poverty is a problem for youths of all racial groups, it is especially problematic for African American and Hispanic youths. Levels of poverty are important, because the economic situation in many communities is believed to contribute to urban violence.[56]

A major problem in inner-city areas, which often have predominately minority populations, is that many youths no longer have access to entry-level jobs in factories and shops. These jobs either have disappeared as the economy has become more service oriented or have moved to other locations, often in foreign countries, where wages are low. The jobs that do remain are typically low paying and offer few opportunities for advancement. Because of the lack of legitimate options for earning income, many youths in urban and other poor areas turn to drug dealing, theft, and violence in order to make money and handle the strains of urban life.[57]

Importantly, **ecological approaches** to delinquency (i.e., attempts to understand delinquency that focus on the geographic distribution of delinquent behaviors and the factors related to this distribution) maintain that community characteristics are related to delinquency in important ways. For example, factors such as population density, poverty, job loss, income inequality, transience, dilapidation, the existence of deviant subcultures, the mixed use of property (i.e., interspersion of residential and commercial land), and a lack of participation by residents in community organizations have been found to be associated with increases in delinquency.[58]

ecological approach to delinquency An approach to understanding delinquency that focuses on the geographic distribution of delinquent behaviors and the factors that are related to this distribution.

A number of researchers that have examined high-delinquency areas have found that factors such as those mentioned above have a significant impact on the social organization of the community. For example, research indicates that communities characterized by conditions such as poverty, population density, mixed use of land, transience, and dilapidation experience **social disorganization** (i.e., an inability to exert effective social control over community members). Research on children in disorganized communities indicates that they have less involvement with conventional social institutions. As a result, such youths are more vulnerable to participation in delinquent peer groups and are more likely to engage in interpersonal aggression.[59] Moreover, the lack of employment opportunities for youths and adults in poor communities is associated with high levels of predatory crime. In these communities, the scarcity of jobs tends to destabilize families, and children from these families are more likely to use violence and aggression as a means of dealing with their limited options.[60]

social disorganization The state of a community or society characterized by a lack of social control over community members.

Youths who live in socially disorganized communities experience a variety of stresses that have been found to be related to delinquency.[61] Studies of inner-city youths indicate that substantial numbers have witnessed violence within their environment. For example, surveys of middle and high school students in high crime areas in Chicago found that as many as 35% had witnessed a stabbing, 39% had witnessed a shooting, and 24% had seen someone killed. In the majority of these cases, the children indicated that they knew the victims, and 47% of the victims were characterized as friends, family members, classmates, or neighbors. Other studies have produced similar findings.[62] Importantly, children's exposure to violence can have an adverse effect on their mental health. One problem experienced by many youths exposed to violence is **posttraumatic stress disorder** (an affliction common among combat veterans). In addition, youths exposed to violence can have difficulty forming close personal relationships, believe that they will not reach adulthood, suffer from lowered self-esteem, perform poorly in school, and display aggression toward others.[63]

posttraumatic stress disorder A disorder where a significant traumatic event is reexperienced by the individual and produces intense fear, anxiety, helplessness, and avoidance of stimuli associated with the trauma.

Youths in socially disorganized neighborhoods also make various social adaptations to their environment. These adaptations can include carrying weapons for protection and joining delinquent peer groups such as gangs. For example, a study of 835 male inmates in correctional facilities in California, New Jersey, Illinois, and Louisiana found that the major reason youths carried guns was for self-protection.[64] Indeed, weapon carrying is not uncommon, particularly in some communities. In a 2003 survey of a nationally representative sample of students in grades 9 through 12, 6.1% reported carrying a weapon at least once in the previous month. Moreover, 6.1% also reported carrying a weapon on school grounds at least once in the 30 days prior to the survey.[65] The ready accessibility of guns is another factor that has contributed to high levels of violent crime, including homicide, in recent years.

The Number of Persons Under 18 Arrested for Murder Has Been Decreasing

The number of gun homicides committed by juveniles doubled from 1980 to 1994.[66] However, by 2000 juvenile arrests for murder were lower than they had been during any year in the previous 20 years.

The social and economic conditions in many communities have also been factors leading to the proliferation of youth gangs. According to Pamela Irving Jackson, the development of gangs in many areas is a natural consequence of the shift from a manufacturing economy, with many high-paying jobs, to a low-wage service economy. As well-paying manufacturing jobs leave communities, youths face diminished economic opportunities. Under these conditions, family stability is eroded and gangs become attractive alternatives for many youths.[67]

MYTH VS. REALITY
Gangs Can Be Found in Various Geographic Locations

Myth—Youth gangs are only a problem in large urban areas.
Reality—According to the 2002 *National Youth Gang Survey*, 38% of suburban county law enforcement agencies, 27% of small city agencies, and 12% of rural agencies reported gang problems in 2002.[68]

Youths join gangs, like they develop other affiliations, for a variety of reasons. In some instances, gang involvement is part of a cultural family tradition passed from one generation to the next. For many youths, a gang serves as a substitute family, where they find a sense of belonging and status, or as a social outlet, where they can meet with friends, have fun, and "hang out."[69] For others, gang membership may be seen as a practical way to protect themselves in a hostile environment—a means of achieving at least marginal safety within the confines of their neighborhood.[70] Or a gang may seem to offer a way of achieving economic rewards not available through more conventional means.[71] For youths with serious psychopathologies, a gang may provide an environment where their enjoyment of violence is rewarded and respected.[72]

Not All Gang Members Are Heavily Involved in Violence

Myth—Today's gangs are much more oriented toward violence than gangs in the 1950s or 1960s.

Reality—It is true that gang violence has increased in severity. There is evidence that some gangs recruit members who own guns, and many other youths obtain firearms after they join a gang.[73] As a result, many gang members possess firearms, including automatic weapons, which increases the likelihood of serious injury or death. However, it is a mistake to assume that all gang members are involved in violence or that all gangs are engaged in violent activity. In fact, some gangs engage in relatively little serious or violent delinquent behavior. In a study of gangs in Chicago, San Diego, and Los Angeles, Jeffrey Fagan found that some gangs are more oriented toward social activities than delinquency.[74]

Gang membership is important because it, like involvement in other types of groups, plays an important role in shaping the behavior of many youths. For example, in one study of juvenile gang members in Rochester, New York, researchers found that many youths who joined gangs were no more delinquent before they joined than nongang youths. Once they became involved in the gangs, however, their rate of delinquency increased, particularly their involvement in crimes against persons.[75] Indeed, violent criminal activities are encouraged in some gangs because group norms call for violent responses to those who insult gang members, encroach on their territory, or present a threat to their illegal activities.[76] Furthermore, violence may be condoned as a tactic for settling internal gang disputes, as a means of intimidating neighborhood residents, or as part of a gang initiation. In addition, violence may help strengthen gang solidarity and may serve as an outlet for more aggressive members to exercise their skills and develop and/or maintain reputations.[77]

Gang Membership Encourages Deliquency

One important study that examined gang member involvement in delinquency, the Rochester Youth Development Study, found that 30% of a representative sample of seventh and eighth graders in Rochester, New York, indicated they belonged to a gang. The study revealed that gang youths were disproportionately involved in juvenile crime. Although gang members accounted for only 30% of the sample population, over a 4-year period they accounted for 65% of all delinquent acts and 69% of all violent acts committed by youths in the sample.[78]

The Family Context of Juvenile Violence

The family should be understood as only one part of the mosaic of social institutions that influence children. It does, however, play a critical role in child development. Consequently, it is not surprising that research indicates that the family also plays an important role in delinquency.[79] Indeed, a number of family-related factors, such as family conflict and breakup, **child neglect** and abuse, inconsistent discipline, parental deviance, parental economic status, family structure, and a lack of emotional ties between parents and children, have been found to be related to delinquent behavior.[80] Unfortunately, many children's lives are influenced by a combination of these factors. For example, a child may live with a single unemployed parent who has a serious substance abuse problem. This same child may have siblings who are involved in criminal activities, and the child may be a victim of abuse and neglect. Moreover, as noted in Chapter 3, the family itself is part of an even larger social context that influences the quality of family life. Indeed, the wider political and economic contexts within which families are located determine the types of jobs available to adults and youths, the pay associated with those jobs, the health benefits that are available, the quality of schools that youths attend, and the types of recreational and leisure activities that can be easily accessed by family members.

child neglect Failure of a caregiver to provide for the physical, mental, or educational needs of a child.

Youths Have a High Rate of Violent Victimization

Children and teenagers suffer more victimization than adults in almost every category of violent crime, including physical abuse, sibling assaults, bullying, sexual abuse, and rape.[81] In 2003, the rate of violent victimization of youths from 12 to 15 years of age was higher than those of any other age group except for those between 16 and 19.[82]

One factor that has been found to be closely related to juvenile aggression is the quality of the interactions among family members. Research indicates that violent interactions within the family encourage subsequent aggression and violence, not only within the family, but also in other social contexts. After looking at the discipline practices of parents, Murray Straus found a strong relationship between physical punishment within the family and later aggression.[83] Physical punishment not only may serve as a model of aggressive behavior for children, but also may weaken the bond between the parent and child, lower the child's self-esteem, result in self-labeling as a "bad kid," and undermine the child's sense of justice.

The difference between spanking and **child physical abuse** is one of degree, not kind. Moreover, the precise point at which spanking and other types of physical punishment become abuse is difficult to determine. Consequently, there is a lack of agreement among social scientists and others as to what constitutes physical abuse. Nevertheless, physical abuse is generally defined as actions that result in a nonaccidental injury to a child by a caregiver.[84]

Although it is not possible to provide an accurate estimate of the extent of child physical abuse and neglect in the United States, the available evidence indicates that it is a significant social problem. Indeed, in 1999 state agencies handled some 3 million cases of alleged abuse and neglect, and 800,000 of these cases were substantiated.[85] The actual number may even be higher, because many instances of abuse and neglect go unreported. Moreover, child welfare agencies often lack the necessary resources to adequately investigate each alleged incident of abuse or neglect.[86]

Although even mild forms of violence against children may encourage aggression, it is clear that physical abuse is strongly related to a variety of childhood adjustment problems, including violent juvenile crime.[87] The victimization of children can have a profound influence on their development.[88] Obviously, physical abuse can result in physical injury and even death. It can also lead to emotional and psychological trauma, which can affect the individual throughout his or her life. Research indicates that problems such as chronic anxiety, lowered self-esteem, depression, school problems, aggressive peer relations, pregnancy, and emotional and mental health problems are associated with child abuse.[89] Furthermore, children who have been abused are three times more likely than their peers to become substance abusers.[90]

There also appears to be a strong relationship between child abuse and later delinquency and adult criminality. In one study that examined the relationship between child abuse and subsequent offending, Cathy Spatz Widom found that child abuse increases the odds of future juvenile and adult criminality by 40%. Widom followed the offending careers of 908 youths who had been reported as victims of abuse and neglect between 1967 and 1971, and she compared them to a control group of 667 youths who had not been maltreated. She found that 29% of those who had histories of abuse or neglect had adult criminal records, compared with 21% for the control group. Widom also found that age, race, and gender affected the likelihood that a youth with a history of abuse would become involved in criminal activity. For example, her analysis indicated that older black males who had been abused were more likely to become involved in adult criminality

child physical abuse
Actions by a caregiver that result in a nonaccidental injury to a child.

than other youths. Indeed, almost 67% of this group had adult criminal records, compared with only 4% of white females with no history of abuse. According to Widom, being abused or neglected as a child increased the likelihood of having a juvenile arrest by 53%, it increased the likelihood of being arrested as an adult by 38%, and it increased the likelihood of being arrested for a violent crime by 38%. Indeed, children who had been the victims of physical abuse had a violent crime arrest rate that was twice that of the controls. Interestingly, Widom discovered that children who had been neglected also had a higher rate of violent arrests than the controls.[91] Subsequent research by Widom indicates that neglect is as damaging to children as physical abuse.[92]

A study that examined a representative sample of 1,000 seventh and eighth graders in Rochester, New York, also found a strong link between child maltreatment and violent delinquency. The researchers interviewed the caretakers of sample subjects and collected data from police, Department of Social Service, school, and other agency records at different points in time. They found that children from disadvantaged families and those where at least one biological parent was missing were more likely to experience maltreatment. They also found that maltreated youths (i.e., youths who had been the victim of some form of abuse or neglect) were significantly more likely to become involved in delinquency, including violent delinquency, than other youths.[93]

Homicide Is a Leading Cause of Death Among Youths

Homicide is the fourth leading cause of death for children between the ages of 1 and 4, and it is the third leading cause of death for those youths aged 5 to 14.[94]

family violence Violence that occurs between persons who are living together as a family unit. Often this violence, which can be emotional as well as physical, takes place in the family dwelling.

Child abuse is only one type of violence found in American families. Significant numbers of families also experience spouse abuse, sibling–sibling abuse, and abuse of a parent by a child. Indeed, it is estimated that between 2 and 6 million women each year are abused by their partners, and battery is the major cause of injury to women in the United States.[95] Although little research has been done on violence between siblings, such violence is another form of **family violence**,[96] and it is one that likely goes undetected by authorities. The abuse of a parent by a child is still another type of family violence. In an analysis of family violence, Richard Gelles found that 3 out of every 100 teenagers aged 15 to 17 had kicked, bit, punched, hit with an object, beat up, threatened, or used a gun or knife against a parent.[97] Moreover, it is estimated that between 700,000 and 1 million elderly persons may be abused each year.[98]

Clearly, being the victim of abuse or neglect is strongly associated with a variety of youth adjustment problems as well as violent behavior. However, there is evidence that witnessing violence also leads to a variety of developmental problems among children.[99]

Multiassaultive Families

Research indicates that at least 7% of all intact families can be classified as **multiassaultive families**. These families are characterized by ongoing patterns of physical aggression among their members. Research indicates that the patterns of assault that are learned in these families are carried into other contexts as well. For example, males in these families are five times more likely to have assaulted a nonfamily member than males from nonassaultive families.[100]

multiassaultive family
A family whose members exhibit ongoing patterns of physical aggression.

The Individual Context of Juvenile Violence

Although it is important to understand the political, economic, family, community, and peer group contexts of juvenile violence, there is also an individual dimension that must be considered. Clearly, individuals do make decisions to act, although these decisions are not always carefully thought out. Sometimes youths plan criminal activities and take precautions to avoid detection; however, in other instances, delinquent activity is best viewed as a type of "pick-up game." In other words, it is the product of the coming together of individuals at a certain time, in a certain place, and under certain conditions.[101] Moreover, although there is no evidence that the majority of youths who engage in delinquent activities, including violent criminal activities, are suffering from an identifiable biological or psychological pathology, there is evidence that a small proportion of violent juveniles have a biological and/or psychological abnormality that influences their behavior.

A number of researchers have maintained that there is a genetic or biological basis for delinquent behavior. However, even these researchers often argue that a combination of environmental and biological factors cause delinquency. They tend to posit that biological and genetic factors are important because they influence central nervous system (i.e., brain) functioning in ways that lead to antisocial and criminal behaviors. In their view, how the brain responds to environmental stimuli is determined by one's biology.

A number of approaches have been taken by researchers to uncover a biological or genetic basis for criminality. Some researchers have conducted twin studies that investigate similarities in delinquent behavior between identical and fraternal twins. These studies have typically found that identical twins are more similar in their behavior than fraternal twins, indicating there may be a biological basis for some criminality.[102] However, research has also shown that identical twins are more likely to be treated alike, so these studies have been unable to disentangle the unique influences of the environment and genetics on delinquent behavior.

In an effort to overcome the methodological problems of twin studies, other researchers have studied children who have been removed from their biological parents. The goal of their research is to compare the level of criminality of the children, that of their biological parents, and that of their adoptive parents. If the level of criminality of the adopted children is like that of their biological parents and not like that of their adoptive parents, this would indicate that a genetic basis for criminality exists. The results of these studies, like the results of the twin studies, indicate that genetics possibly plays some role in criminality.[103] However, those studies that have been large enough to examine violent and property crimes separately have found that genetic factors seem to play a larger role in property crimes than in violent crimes. As Adrian Raine notes, "This seems counterintuitive since one would imagine that violent, more serious offending, being more extreme, would be more 'hard wired' and have a higher heritability than property offending, which one could imagine is more driven by social and economic factors."[104]

Another approach to investigating delinquency focuses on intelligence and its relationship to delinquent behavior. Some research has found a weak to moderate inverse relationship between IQ test scores and delinquency (i.e., as IQ test scores decrease to the low normal range, delinquency increases).[105] Other research has failed to find a strong relationship between IQ scores and delinquency.[106] The debate on the relationship of IQ to delinquency rests on assumptions about what IQ tests actually measure. Some maintain that intelligence is a heritable trait that can be measured by standardized IQ tests. These researchers believe that a strong inverse relationship between IQ scores and delinquency is evidence that innate intelligence is inversely related to delinquent behavior. However, others maintain that IQ tests are culturally biased and that they do not measure innate intelligence but rather cultural and educational background.

There Is Considerable Debate Over the Meaning of the Relationship Between IQ and Delinquency

Some researchers suggest that the relationship between IQ and delinquency is indirect. According to their view, low IQ is related to school failure, which in turn leads to delinquent behavior. There is indeed considerable evidence of a strong link between school failure and delinquency. However, it is important to recognize that this view is still predicated on the assumption that IQ tests measure innate intelligence. If they do not, then it may be more accurate to state that a deprived cultural and educational background is related to school failure, which in turn is related to delinquency.

In addition to the previously mentioned studies, a variety of other efforts have been made to uncover a biological basis for criminal behavior. For example, studies have been done positing a relationship between delinquency and factors such as body build; temperament; neurological dysfunctions, such as attention deficit hyperactivity disorder (ADHD), other learning disabilities, and minimal brain dysfunction (MBD); and hormonal imbalances. Moreover, some neurological dysfunctions, such as MBD and learning disabilities, have been linked to impulsive and aggressive behaviors among youth.[107] So have hormonal imbalances.[108] This general approach has been used to explain what Terrie Moffitt calls **life-course-persistent (LCP) delinquents**—individuals who begin involvement in antisocial behaviors at a very early age and engage in criminal activity of increasing seriousness, including violent criminal activity, well into their adult years. According to Moffitt, many of these youths display minor neuropsychological problems during their childhoods, including difficult temperaments, attention deficit hyperactivity disorders, and learning difficulties in school. When these youths come into contact with mental health professionals during childhood, they are often diagnosed as conduct disordered. Contacts made during their adult years often result in a diagnosis of antisocial personality disorder.[109]

life-course-persistent (LCP) delinquent
A person who engages in antisocial behavior at a very early age; gradually becomes involved in increasingly more serious criminal activity; including violent criminal activity, and continues to commit criminal offenses well into his or her adult years.

Some researchers focus on psychological abnormalities, while admitting that these abnormalities could well have a biological basis. For example, some think that youths with psychopathic personalities suffer from biological defects that influence their behavior. Psychopaths typically are unreliable; fail to develop strong attachments to others; repeatedly engage in antisocial, often violent behavior; and lack remorse. Some researchers suggest that psychopaths suffer from a defect in the autonomic nervous system (ANS) that leads to low arousal. Low arousal, in turn, causes the individuals to act out in an effort to receive stimulation. However, because of low arousal, they have difficulty in learning from their behaviors.[110] Finally, some researchers focus on the personality, and these tend to think that antisocial or criminal personalities are the product of emotionally disturbed parents, a lack of parental affection, parental rejection, and inconsistent discipline.[111]

The literature on violent juvenile crime does contain some evidence indicating a link between a number of biological and psychological factors and violent offending. However, there are several weaknesses with the strategy of looking for such links. First, the relationships found between biological and/or psychological abnormalities and delinquency are often weak. This is particularly true for the studies that are more rigorous and have employed sound methodologies.[112] Second, some of the supposed links between biological and/or psychological abnormalities are counterintuitive, such as the link between biological factors and property crime.[113] Third, existing research indicates that only a small proportion of juvenile crime, including violent juvenile crime, is the product of biological or psychological abnormalities. This does not mean that biological or psychological factors are not important. It seems reasonable to believe that they play a major role in a small proportion of the violent crimes committed by juveniles each year. However, many violent juvenile crimes are committed by youths who appear to suffer from no biological or psychological abnormality. Instead, violence by these youths, in some cases at least, seems like a normal and expected response to the family, community, peer, political, economic, and cultural contexts within which they live.

■ Responses to Violent Juvenile Crime

Violent juvenile crime represents a significant problem in many communities around the country. Not surprisingly, a variety of responses have been implemented, from programs designed to prevent violent juvenile crime to legislation making it easier to try juveniles in adult courts. Nearly every state has enacted legislation changing the way that juvenile justice operates, and many of the reforms have focused specifically on the violent juvenile offender.

Violence Prevention

In an effort to take a proactive approach to violent juvenile crime, some states, as well as some local communities, have implemented violence prevention initiatives, in some cases with the support of the federal government through the Office of Juvenile Justice and Delinquency Prevention (OJJDP). In recent years, OJJDP has been in the process of developing what it refers to as a "Comprehensive Strategy for Serious, Violent, and Chronic Offenders." This strategy is predicated on a **risk-focused approach to delinquency prevention** that attempts to identify and then address those factors that contribute to delinquent behavior. These include:

risk-focused delinquency prevention An approach to delinquency prevention in which factors that contribute to delinquent behavior are identified and addressed.

1. Individual characteristics such as alienation, rebelliousness, and lack of bonding to society
2. Family influences such as parental conflict, child abuse, and family history of problem behavior (substance abuse, criminality, teen pregnancy, and school dropouts)
3. School experiences such as early academic failure and lack of commitment to school;
4. Peer group influences such as friends who engage in problem behavior (minor criminality, gangs, and violence)
5. Neighborhood and community factors such as economic deprivation, high rates of substance abuse and crime, and low neighborhood attachment[114]

By encouraging states and communities to develop effective responses to these factors, OJJDP hopes to play a role in reducing levels of violent juvenile crime.

States have also taken steps to support violence prevention initiatives. For example, the Illinois Violence Prevention Act of 1995 established the Illinois Violence Prevention Authority, which is designed to coordinate statewide violence prevention programs, raise funds for state and community organizations involved in comprehensive violence prevention efforts, and provide assistance to communities and organizations attempting to develop, implement, and evaluate violence prevention strategies. The act also established a violence prevention fund, which is a repository for appropriations and grants from federal, state, and private sources earmarked for violence prevention efforts. Missouri and Oregon have completed comprehensive juvenile justice reforms that offer tax breaks for individuals and organizations that become involved in delinquency prevention activities. Missouri's

Youth Opportunities and Violence Prevention Act provides tax credits for individuals or businesses that make monetary or physical contributions to programs that provide new or expanded educational opportunities for youths. The Oregon program provides tax incentives for programs that hire at-risk youths. In Mississippi, legislation has been enacted that is intended to foster collaboration between schools, families, and local agencies involved in youth development activities and in programs that attempt to provide cost-effective responses to youth-related problems before they become more serious.[115]

In addition to state-level support for delinquency and juvenile violence prevention efforts, a number of local communities have developed and implemented their own programs. In some instances, these programs have been state or federally supported; in other instances they have been "locally grown." For example, a number of communities in Maryland have implemented late-night basketball programs, and these are believed to be associated with a 60% decrease in drug-related crimes in that state.[116] In Allegheny County, Pennsylvania, a countywide comprehensive antiviolence effort was launched in the early 1990s, after a dramatic increase in violent juvenile crime between 1989 and 1993. This effort can be best characterized as a large-scale attempt to respond to youth violence through the collaboration of law enforcement agencies, other government agencies, juvenile justice agencies, educational institutions, the religious community, grass-roots community activists, the media, health and human services, and community representatives. Among the accomplishments to date are reductions in turf warfare between agencies, better coordination of law enforcement efforts, the development of over 20 family support centers, the establishment of youth sports leagues in 12 communities, a summer youth jobs program, and neighborhood antiviolence initiatives. According to 1995 arrest data, total juvenile arrests had declined from 1994 levels by 13%, and arrests of juveniles for violent offenses had dropped by 30%. When these decreases are compared with a 2% statewide decline in the number of total juvenile arrests and 9% statewide decline in the number of violent crime arrests, Allegheny County's efforts appear to have paid off.[117]

Another approach taken by some communities has been to implement curfews as a way of reducing juvenile violence. Although curfews have been in existence around the country since the 1890s, they have been receiving renewed attention in recent years. Although there are differences in curfew laws, they typically require juveniles to be at home between 11 p.m. and 6 a.m., with exceptions for weekends and during the summer.[118] Despite their growing popularity, curfew laws do have opponents. First, some of these ordinances have been challenged on constitutional grounds. The American Civil Liberties Union (ACLU) has challenged curfew laws because it feels they violate the rights of children and parents to free speech and association. Others maintain that these laws give police undue powers to detain and question children without probable cause. Still others are concerned that these laws may act to widen the net of social control by subjecting a new category of children, often minority children, to police and court intervention.[119] There is also the question of whether curfews are effective in reducing delinquency. A few evaluations have found that curfews appear to play a significant role in reducing juvenile crime, including violent juvenile crime. Other studies indicate that curfews have little impact on delinquency or produce a **displacement effect** (i.e., juvenile crime decreases during curfew hours but

displacement effect
Moving an activity of a certain kind to another location or another time. Curfews can have the effect of causing a decrease in juvenile crime during curfew hours and an offsetting increase in juvenile crime during other times of the day.

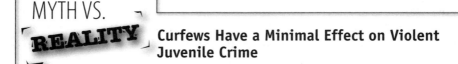

MYTH VS. REALITY   Curfews Have a Minimal Effect on Violent Juvenile Crime

Myth—Curfew laws reduce levels of violent juvenile crime because most of this crime occurs late in the evening or very early in the morning.

Reality—Curfews may have little effect on violent juvenile crime because, in many communities, most such crime occurs during the afternoon and early evening. A study of FBI crime data from eight states found that on school days the peak time for violent juvenile crime was 3 p.m. (just after school), and on weekends it was 11 p.m.[121]

increases during other times of the day).[120]

Gang prevention and intervention efforts have also been implemented in some communities in order to reduce levels of juvenile violence. One such prevention program is the *Gang Resistance Education and Training (GREAT)* program. This program targets sixth and seventh graders who may be exposed to gangs. GREAT consists of a curriculum that stresses ways to resist peer pressure, improve self-esteem, and avoid gang influences and violence. The program curriculum is taught by uniformed police officers within a school setting. However, one evaluation of this program indicated that it was only moderately successful at changing youths' attitudes about gangs.[122] However, a more recent evaluation found that it had no significant impact on youths' gang participation, although participation in GREAT was associated with less victimization, less risk-taking behavior, more positive attitudes toward the police, and more negative views of gangs.[123]

Other community approaches to dealing with gangs range from assigning youth workers to gangs (in an effort to redirect youth gangs and their members to more conventional activities) to providing intensive supervision to gang members who are on probation for minor offenses. In addition, grass-roots programs have been developed in a number of communities. For example, in Philadelphia, Sister Falaka Fattah, after learning that her 16-year-old son had become involved in gang activity, invited gang members to live in her home, which she named the House of Umoja (umoja means "unity" in Swahili). Her idea was to provide a refuge and supportive network for gang members, who, in turn, were required to avoid illegal activity and participate in activities within the home. These activities included performing regular household chores, participating in an early morning conference at which they set goals for the day, and attending school. Over time the program spread to include other homes in the community; connections were established with the juvenile courts, which began to place youths there for care; and additional assistance was provided by various sources of support.[124]

One other approach to dealing with serious and violent juvenile offenders that has begun to receive considerable attention is multisystemic therapy (MST), which is being pioneered by Scott Henggeler at the Medical University of South

Carolina. One of the goals of this approach is to address the factors that lead to problem behaviors among youths. Consequently, MST focuses attention on the individual, family, school, peer group, and community context of delinquency. Indeed, at the heart of this approach is the recognition that the family is a key institution in a child's life. Direct services are provided by a therapist who has a caseload of four to six families and provides approximately 60 hours of service over approximately 4 months. In addition, the direct service provider is supported by a supervisor. Services consist of individualized interventions that address family, school, peer, or other issues related to antisocial behavior. Evaluations of MST programs indicate that this approach has considerable potential. For example, one study that compared a randomly selected group who received MST services and a control group who received regular services showed that the former had significantly fewer arrests (.87 arrests on average, versus 1.52) over a 59-week postreferral follow-up. In another study, violent offenders who received MST services had a four-year **recidivism** rate of 22%, compared with a rate of 72% for youths receiving individual therapy and a rate of 87% for youths who refused to participate in the program.[125]

recidivist A person who engages in further criminal activity following arrest, court conviction, or release from incarceration.

There are also a number of programs that have been found to have considerable potential in efforts to prevent violence. One good example of these programs is nurse visitation programs such as the ones operated in Montreal (Montreal Home Visitation Study) and in the Appalachian region of New York State (University of Rochester Home Visitation Program). An essential feature of these programs is the provision of a range of health, infant care, and child development and supportive services to working class or low income mothers. Both of these programs were associated with a variety of positive outcomes such as fewer child injuries, the development of a more positive home environment, improved parenting skills, and lower levels of child abuse.[126] Similar outcomes have also been found in programs that provide a range of educational, health, safety, and support services to pregnant women and to their preschool children.[127]

In addition to programs for preschool youths, several programs for school-aged children have received favorable evaluations. One of the more promising of these programs is the Seattle Social Development Project (SSDP). The SSDP is a comprehensive approach to violence prevention that uses a variety of interventions designed to strengthen youths' bonds to their family, school, and community groups by providing them with opportunities for active involvement, and giving them the skills to participate successfully. This is accomplished through a variety of interventions including classroom management and interactive and cooperative instructional practices, along with parent training. Interviews with youths 6 years after program completion revealed that students who had successfully completed the program were less likely to report involvement in violent behavior, to engage in heavy drinking, to participate in sexual intercourse, and had greater school achievement than controls.[128]

Juvenile Justice Responses to Violent Juvenile Crime

As a result of the decline in juvenile arrests in recent years, the number of juvenile court cases, including cases involving violent juvenile crime, has also decreased. In 1999, juvenile courts handled an estimated 387,100 offenses classified as person offenses.[129] As a response to concerns about juvenile crime, particularly violent juvenile crime, state legislatures have enacted new laws designed to change the ways that juvenile courts respond to violent juvenile offenders. In addition, juvenile courts themselves have developed new policies and programs in an effort to deal with those youths who are charged with violent offenses. In the period between 1992 and 1995, legislatures in 47 states passed laws that toughened state responses to juvenile crime.[130] What is significant about these changes is that they represent a significant change in the philosophy of the juvenile court. Indeed, the basic rationale behind the reforms was to "punish, hold accountable, and incarcerate for longer periods of time those juveniles who, by history or instant offense, passed a tolerable threshold of tolerated juvenile law violating behavior."[131] In effect, the recent wave of legislation has pushed the juvenile courts away from their traditional *parens patriae* philosophy toward a more punitive criminal court philosophy.

Recent state legislative changes have been directed at five basic areas of juvenile court and juvenile justice operation: (1) jurisdictional authority, (2) sentencing authority, (3) confidentiality, (4) victims' rights, and (5) correctional programming.[132] For example, 41 states passed laws that removed juvenile court jurisdiction over certain categories of juvenile offenders charged with violent offenses and gave jurisdiction over these cases to criminal courts. Twenty-five states expanded the sentencing options available to juvenile and criminal courts with respect to juvenile offenders. Forty states passed legislation that made juvenile court records and proceedings more open. In 22 states, laws were passed that gave victims an increased role in the juvenile justice process. In response to these legislative changes, juvenile correctional administrators have felt increasing pressure to develop new programs for serious and violent offenders.[133]

One of the most significant changes in juvenile justice in recent years has been the passage of legislation that makes it easier to try some juveniles as adults. As noted in Chapter 9, each state has a mechanism—typically called waiver, transfer, certification, or remand—by which juveniles may be tried in adult courts. Between 1992 and 1995, 41 states passed laws that make it easier to try certain juveniles in criminal courts. In 34 states, *statutory exclusion laws* were enacted. These laws remove certain juveniles from juvenile court jurisdiction if they meet specific age and offense criteria. In addition, 11 states lowered the age at which a juvenile can be transferred to adult courts, 10 states expanded the list of offenses for which a juvenile could be waived, 9 states passed **presumptive waiver** statutes requiring certain youths to be waived unless they can prove they are suitable for juvenile justice services, and 2 states passed legislation that allows some youths to be transferred if they have prior records. Moreover, 3 states lowered the upper age limit for original juvenile court jurisdiction, thereby bringing larger numbers of youths under adult court jurisdiction.[134] The result of these changes is to remove increasing numbers of youths from juvenile court jurisdiction and place them under the control of the criminal courts. However, research in Florida comparing the recidivism rate for youths transferred to criminal courts and the rate for a matched group of youths handled by juvenile courts found that waiver to adult court tended to increase recidivism.[135] In fact, there is little evidence that the transfer of juveniles to adult courts helps to meet their needs or protects community safety.

presumptive waiver
A statute-mandated waiver of a youth to criminal court that can be countermanded if the youth is able to prove he or she is suitable for juvenile justice services.

There Are Various Mechanisms by Which Youths Can Be Transferred to Adult Courts

Overall, the number of cases being judicially waived to adult court has been decreasing in recent years.[136] However, because there are a number of mechanisms whereby juveniles can be transferred to adult court, the actual number of juveniles subjected to adult court jurisdiction is difficult to determine, and it may be increasing.

A number of states, as well as individual juvenile courts, have made substantial changes in the types of dispositions and sentencing options employed in juvenile cases. Between 1992 and 1995, 16 states passed statutes requiring mandatory sentences for certain types of violent or serious offenses. During this same period, 12 states expanded the dispositional age limit (usually to 21), which allows courts in those states to maintain jurisdiction over juvenile offenders for a longer period of time. In addition, between 1992 and 1995, 16 states passed laws that make **blended sentences** possible. Blended sentences allow courts to impose juvenile and/or adult correctional sanctions on some categories of offenders, such as youths designated as "aggravated juvenile offenders" or "youthful offenders." For example, in New Mexico, certain youths can be sentenced to an adult or a juvenile correctional program by a juvenile court. In Colorado, Massachusetts, Rhode Island, South Carolina, and Texas, the juvenile court may make a disposition that remains in force beyond the age of its extended jurisdiction, at which time the youth may be transferred to the adult correctional system. In California, Colorado, Florida, Idaho, Michigan, and Virginia, youths tried in criminal courts may be sentenced to an adult or a juvenile correctional program.[137] The result of allowing blended sentences has been to expand the range of dispositional options available in juvenile cases, give juvenile courts control over youths for longer periods, and increase criminal court involvement in a wider range of juvenile cases.

Although a "get tough" approach to juvenile crime is seen as good politics by many legislators and prosecutors, it often produces the opposite of what its supporters claim. This is nowhere better illustrated than in Cook County (Chicago), Illinois, when Richard M. Daly was the state's attorney. In the late 1980s, his "political" response to the rise of teenage violence was to prosecute 97% of all cases forwarded to him by the police. This effort, although good "copy" for the media, had several detrimental effects:

- Almost 70% of the cases were dropped as not prosecutable due to witness problems and lack of necessary information. As a result, many youths lost respect for a system that made threats that it couldn't enforce.

blended sentence
A sentence handed down by a juvenile court that encompasses juvenile and adult correctional sanctions. Blended sentences are allowed in some jurisdictions for certain categories of offenders, such as those designated as "aggravated juvenile offenders" or "youthful offenders."

- The system got "clogged" with kids who did not pose a serious threat to the community. As a result, precious and limited resources were expended on nonserious cases, while many serious cases failed to receive the necessary attention.
- Because many courts had limited resources, probation got overused, caseloads got too high, and effective supervision over probationers became impossible, contributing further to the disrespect felt by juveniles toward the juvenile justice process.[138]

Although court practices in recent years have shifted more toward accountability, punishment, and a concern for community safety and away from rehabilitation,[139] a variety of promising correctional programs have begun to be developed for violent and serious juvenile offenders. Many of these interventions are community based. As noted earlier, MST appears to be a very promising community-based intervention that can be employed with many youths charged with violent offenses. There is some indication that intensive supervision programs can be as effective as placement in state training schools for some violent offenders. In an evaluation of three home-based intensive supervision programs in Wayne County, Michigan, William Barton and Jeffrey Butts found that intensive community-based programs were as effective as state placement for some serious and violent juvenile offenders, as well as being more cost-effective.[140]

Day treatment programs represent another community-based option that can be used effectively with some violent offenders. The Bethesda Day Treatment Center Program in West Milton, Pennsylvania, provides a range of services to youths, including intensive supervision, counseling, study skills development, cultural enrichment, life and job skills training, and coordination of other services intended to meet youths' needs. Preliminary results suggest that many youths could benefit from this and similar programs.[141]

Institutional placement is another viable option for dealing with violent youths. Offering the most promise are small, secure facilities that have good leadership and well-trained staff. One such program is the Florida Environmental Institute (FEI), also known as "The Last Chance Ranch." This small program, which is located in a remote area of the Everglades, accepts some of Florida's most violent and serious offenders. It is operated by a staff who focus on education, hard work, developing caring relationships with youths, and providing quality aftercare to youths once they return to the community. Despite the large percentage of violent and serious offenders among the participants, evaluations of the "ranch" indicate that only about 30% of youths who go through this program become recidivists, compared with 70% to 75% of youths who go through other institutional programs.[142]

For youths who are placed in out-of-home settings, intensive aftercare is essential to helping them make a successful transition to the community and preventing recidivism. Unfortunately, aftercare has been given little attention within juvenile justice, despite the fact that it is widely recognized as an important component of rehabilitation. To some extent, the lack of attention given to aftercare programs is being remedied in some jurisdictions. Moreover, there is a developing body of research that is spelling out the key ingredients of successful aftercare programs. This research indicates that effective programs begin planning for release while youths are incarcerated; provide a structured transition to the community, involving institutional and aftercare staff; and provide long-term reintegrative services.[143] There is some evidence that aftercare programs that are well implemented and that provide intensive supervision reduce the probability of rearrest. Importantly, good aftercare programs can shorten the time youths spend in institutions, thus saving precious resources without endangering public safety.[144]

Clearly, a number of promising programs are being developed for violent offenders. These programs tend to offer a number of potential benefits: (1) they address problems related to youths' involvement in delinquency, (2) they are associated with reductions in delinquency, (3) they protect community safety, and (4) they are cost effective. Despite the development of these programs, however, juvenile corrections still faces a number of challenging problems. Too many juvenile corrections programs still lack any of the above listed benefits, and consequently fail to achieve their basic mission, which is to assist youths and protect community safety.

■ Legal Issues

The problem of violent juvenile crime has led to a reexamination of the philosophical underpinnings of juvenile justice. Are violent offenses less serious because they are committed by juveniles? Because of their age, do juveniles really lack the capacity of adults to control their behavior? If, in fact, they lack the same capacity, should they not be treated less severely? These questions need to be systematically raised and discussed. In trying to reach meaningful answers to these questions, careful consideration needs to be given to the fact that not all juveniles are alike. They vary not only in terms of age, but also in terms of mental capacity, level of maturity, emotional development, and social circumstances.

Even if the issue of juveniles' responsibility for their actions can be resolved, the appropriate response to juvenile offenses remains an open question. Historically, the official focus of the juvenile court has been on treatment rather than punishment. Nevertheless, punishment and treatment often go hand in hand. Being placed in an institution for purposes of treatment may seem a lot like punishment. Moreover, placement in an institutional setting for an indeterminate time period may result in the juvenile doing more time than a comparable adult defendant. This possible disproportionateness of punishment is an issue that needs to be addressed by juvenile courts and state legislatures.

The final legal issue concerns due process protection for juveniles. Regardless of how competent an attorney is, do 13- and 14-year-olds really understand what they are facing when they are transferred to criminal court? Is it absolutely certain that they are competent to stand trial as adults? Given that juveniles are not likely to be reformed in adult prisons, does society's interest in retribution outweigh its *parens patriae* obligations? How much due process protection will be enough in cases where juveniles are tried as adults? Is it necessary to ensure that they understand what is happening, or are additional due process procedures only going to make matters more confusing?

■ Chapter Summary

Violent juvenile crime has received considerable attention in recent years. As a result, a number of significant changes in state laws and juvenile court practices have been implemented. These changes signal an important shift away from the traditional juvenile court focus on treatment toward a new focus on punishment and accountability. This shift is evident in legislation that is intended to remove many youths from juvenile court jurisdiction by lowering the age limits of juvenile court jurisdiction, by making it easier to transfer youths to criminal courts, by increasing the time that juvenile courts may maintain jurisdiction over youths, and by increasing the severity of sanctions imposed by juvenile courts. The result of such legislation is that increasing numbers of youths are falling under the jurisdiction of juvenile and adult courts for longer periods of time. Although "getting tough" on juvenile crime continues to be a politically popular stance, the reduction in crime promised by "get tough" advocates remains unrealized.

Several problems presently confront the juvenile justice process in its efforts to effectively deal with violent juvenile crime. First, in order to reduce violent juvenile crime, juvenile justice agencies and local community leaders must be willing to commit the resources to address the social, political, economic, community, family, and individual contexts of youth violence. Second, strong prevention and early intervention programs need to be developed and implemented. Third, each of the components of the juvenile justice process (police, courts, and corrections) must be given the resources and support it needs to accomplish its mission. Fourth, state agencies, local communities, and juvenile courts need to be actively involved in developing and implementing a wider range of effective options for violent offenders.

Despite the lack of effective interventions for violent juvenile offenders in many jurisdictions, a growing body of research indicates that viable programs for these youths can be developed, including violence prevention programs and community-based and institutional programs that treat existing violent offenders. Indeed, the development of cost-effective programs that address the factors that contribute to violent behavior holds some promise as a way of dealing with juvenile violence while safeguarding the ideal of the juvenile court—to serve the best interests of children and simultaneously protect public safety.

KEY CONCEPTS

blended sentence A sentence handed down by a juvenile court that encompasses juvenile and adult correctional sanctions. Blended sentences are allowed in some jurisdictions for certain categories of offenders, such as those designated as "aggravated juvenile offenders" or "youthful offenders."

child neglect Failure of a caregiver to provide for the physical, mental, or educational needs of a child.

child physical abuse Actions by a caregiver that result in a nonaccidental injury to a child.

chronic delinquent offender A juvenile who, over time, accumulates a number of delinquency charges and convictions. The related term *chronic recidivist* was used by Wolfgang and colleagues to refer to juvenile offenders who had five or more police contacts.

clearance data Data on cleared offenses. A cleared offense is an offense for which at least one person has been arrested, charged with the commission of the offense, and referred to court.

cohort study A study that looks at a group of people born at a specific time in a particular location (a cohort).

displacement effect Moving an activity of a certain kind to another location or another time. Curfews can have the effect of causing a decrease in juvenile crime during curfew hours and an offsetting increase in juvenile crime during other times of the day.

family violence Violence that occurs between persons who are living together as a family unit. Often this violence, which can be emotional as well as physical, takes place in the family dwelling.

ecological approach to delinquency An approach to understanding delinquency that focuses on the geographic distribution of delinquent behaviors and the factors that are related to this distribution.

interpersonal violence Face-to-face interaction that produces physical harm or presents a threat of physical harm.

life-course-persistent (LCP) delinquent A person who engages in antisocial behavior at a very early age; gradually becomes involved in increasingly more serious criminal activity, including violent criminal activity; and continues to commit criminal offenses well into his or her adult years.

multiassaultive family A family whose members exhibit ongoing patterns of physical aggression.

official data Data collected by juvenile justice agencies such as the police, courts, and correctional agencies.

posttraumatic stress disorder A disorder where a significant traumatic event is reexperienced by the individual and produces intense fear, anxiety, helplessness, and avoidance of stimuli associated with the trauma.

presumptive waiver A statute-mandated waiver of a youth to criminal court that can be countermanded if the youth is able to prove he or she is suitable for juvenile justice services.

recidivist A person who engages in further criminal activity following arrest, court conviction, or release from incarceration.

risk-focused delinquency prevention An approach to delinquency prevention in which factors that contribute to delinquent behavior are identified and addressed.

self-report study A study in which the participants report on their own behavior.

serious offense An offense that results in significant personal injury, property loss, or damage.

social disorganization The state of a community or society characterized by a lack of social control over community members.

victimization study A study in which the participants are asked to report their victimization experiences.

violent juvenile offender A juvenile who has committed a violent criminal offense against a person, such as rape, murder, aggravated assault, or robbery.

violent offense An offense directed at the person or property of another that is intended to inflict injury on another and/or deprive them of their property. Violent offenses, however, do not always result in significant personal injury to the victim.[145]

REVIEW QUESTIONS

1. What are the differences between chronic, serious, and violent juvenile offenders?

_____

_____

_____

_____

_____

2. Many people feel that violent crime is much worse now than at any time in our country's history. Is this true?

_____

_____

_____

_____

_____

3. How much violence is committed by chronic juvenile offenders?

_____

_____

_____

_____

_____

4. In what ways has violent juvenile crime changed in recent years?

_____

_____

_____

_____

_____

5. In what ways should society respond to the serious offender and the chronic offender?

_____

_____

_____

_____

_____

6. What is the relationship between gender and violent juvenile crime?

_____

_____

_____

_____

_____

7. What is the relationship between race and violent juvenile crime?

_____

_____

_____

_____

_____

8. In what ways do cultural, political, economic, community, peer, family, and individual factors influence juvenile violence?

_____

_____

_____

_____

_____

9. What factors need to be considered in the development of violence prevention initiatives for youths?

_____

_____

_____

_____

_____

10. What types of youth violence prevention programs have been developed in recent years?

_____

_____

_____

_____

_____

11. What are recent trends in the juvenile justice response to chronic and serious offenders?

_____

_____

_____

_____

_____

12. What types of programs appear to hold the most promise in the battle to reduce violent juvenile crime?

_____

_____

_____

_____

_____

13. Should more juveniles be transferred to adult courts? Why or why not?

_____

_____

_____

_____

_____

Chapter Resources

ADDITIONAL READINGS

Coordinating Council on Juvenile Justice and Delinquency Prevention. 1996. *Combating Violence and Delinquency: The National Juvenile Justice Action Plan*. Washington, DC: Office of Juvenile Justice and Delinquency Prevention.

Hamparian, D. M. et al. 1985. *The Young Criminal Years of the Violent Few*. Washington, DC: Office of Juvenile Justice and Delinquency Prevention.

Henggeler, S. W. 1997. "Treating Serious Anti-Social Behavior in Youth: The MST Approach." *OJJDP Juvenile Justice Bulletin*. Washington, DC: Office of Juvenile Justice and Delinquency Prevention.

Howell, J. C. et al., eds., 1995. *A Sourcebook, Serious, Violent, and Chronic Juvenile Offenders*. Thousand Oaks, CA: Sage.

Klein, M. W. et al. 1995. *The Modern Gang Reader*. Los Angeles: Roxbury.

Loeber, R. and D. P. Farrington, eds., 1998. *Serious and Violent Juvenile Offenders, Risk Factors and Successful Interventions*. Thousand Oaks, CA: Sage.

National Criminal Justice Association. 1997. *Juvenile Justice Reform Initiatives in the States: 1994–1996*. Washington, DC: Office of Juvenile Justice and Delinquency Prevention.

Office of Juvenile Justice and Delinquency Prevention. 1993. *Comprehensive Strategy for Serious, Violent, and Chronic Juvenile Offenders, Program Summary*. Washington, DC: Office of Juvenile Justice and Delinquency Prevention.

Widom, C. S. 1996. "The Cycle of Violence Revisited." *Research Review*. Washington, DC: National Institute of Justice.

NOTES

1. M. Sickmund et al., *Juvenile Offenders and Victims: 1997 Update on Violence* (Washington, DC: Office of Juvenile Justice and Delinquency Prevention, 1997).
2. G. Witkin et al., "Kids Who Kill," *U.S. News and World Report,* 18 April 1991: 26–32.
3. H. N. Snyder and M. Sickmund, *Juvenile Offenders and Victims: 1999 National Report* (Washington, DC: Office of Juvenile Justice and Delinquency Prevention, 1999); also see Federal Bureau of Investigation, *Crime in the United States*, 2003 (Washington, DC: U.S. Department of Justice, 2004).
4. H. N. Snyder and M. Sickmund, *Juvenile Offenders and Victims: A Focus on Violence* (Washington, DC: Office of Juvenile Justice and Delinquency Prevention, 1995).
5. J. A. Fox, *Trends in Juvenile Violence: A Report to the United States Attorney General on Current and Future Rates of Juvenile Offending* (Washington, DC: Bureau of Justice Statistics, 1996).
6. See M. L. Benson and G. L. Fox, "When Violence Hits Home: How Economics and Neighborhood Play a Role," *NIJ Research in Brief* (Washington, DC: National Institute of Justice, 2004).
7. J. Laub and R. Sampson, "The Sutherland-Glueck Debate: On the Sociology of Criminological Knowledge," *American Journal of Sociology* 96 (1991): 1402–1440.
8. S. Glueck and E. Glueck, *Predicting Delinquency and Crime* (Cambridge, MA: Harvard University Press, 1967.
9. M. Wolfgang et al., *Delinquency in a Birth Cohort* (Chicago: University of Chicago Press, 1972).
10. P. Tracy et al., *Delinquency in Two Birth Cohorts: Executive Summary* (Washington, DC: U.S. Department of Justice, 1985).
11. L. Shannon, *Assessing the Relationship of Adult Criminal Careers to Juvenile Careers: A Summary* (Washington, DC: U.S. Department of Justice, 1982).
12. F. Dunford and D. Elliott, "Identifying Career Offenders Using Self-Report Data," *Journal of Research in Crime and Delinquency* 21 (1984): 57–86.
13. R. J. Michalowski, *Order, Law, and Crime: An Introduction to Criminology* (New York: Random House, 1985), 277–278.
14. Snyder and Sickmund, *Juvenile Offenders and Victims: 1999 National Report*.
15. Snyder and Sickmund, *Juvenile Offenders and Victims: 1999 National Report*.
16. Federal Bureau of Investigation, *Uniform Crime Reports for the United States, 2003*. Retrieved December 29, 2004, from http://www.fbi.gov/ucr/cius_03/pdf/03sec4.pdf.
17. D. M. Hamparian et al., *The Violent Few* (Lexington, MA: Lexington Books, 1978).
18. D. M.Hamparian et al., *The Young Criminal Years of the Violent Few* (Washington, DC: Office of Juvenile Justice and Delinquency Prevention, 1985).

19. Howard N. Snyder, cited in Sickmund et al., *Juvenile Offenders and Victims: 1997 Update on Violence*, 24.

20. B. T. Kelley et al., "Epidemiology of Serious Violence," *OJJDP Juvenile Justice Bulletin* (Washington, DC: Office of Juvenile Justice and Deliquency Prevention, June 1997).

21. Snyder and Sickmund, *Juvenile Offenders and Victims: 1999 National Report*.

22. Snyder and Sickmund, *Juvenile Offenders and Victims: 1999 National Report*.

23. R. Loeber et al., "Serious and Violent Juvenile Offenders," in *Serious and Violent Juvenile Offenders, Risk Factors and Successful Interventions*, eds. R. Loeber and D. Farrington (Thousand Oaks, CA: Sage, 1998); Hamparian et al., *The Young Criminal Years of the Violent Few*; H. N. Snyder and M. Sickmund, *Juvenile Offenders and Victims: A National Report* (Washington, DC: Office of Juvenile Justice and Delinquency Prevention, 1995).

24. Snyder and Sickmund, *Juvenile Offenders and Victims: 1999 National Report*.

25. C. Puzzanchera et al., *Juvenile Court Statistics 1999* (Washington, DC: Office of Juvenile Justice and Delinquency Prevention, 2003).

26. Federal Bureau of Investigation, *Crime in the United States*, 1996; E. Poe-Yamagata and J. A. Butts, *Female Offenders in the Juvenile Justice System, Statistics Summary* (Washington, DC: Office of Juvenile Justice and Delinquency Prevention, 1996); Federal Bureau of Investigation, *Uniform Crime Reports for the United States, 2003*.

27. M. Chesney-Lind and R. G. Shelden, *Girls, Delinquency, and Juvenile Justice*, 3rd ed. (Belmont, CA: Thomson/Wadsworth, 2004).

28. Chesney-Lind and Shelden, *Girls, Delinquency, and Juvenile Justice*.

29. Chesney-Lind and Shelden, *Girls, Delinquency, and Juvenile Justice*, 15.

30. Federal Bureau of Investigation, *Uniform Crime Reports for the United States, 2003*.

31. S. Walker et al., *The Color of Justice* (Belmont, CA: Wadsworth, 1996).

32. Chesney-Lind and Shelden, *Girls, Delinquency, and Juvenile Justice*, 28.

33. Sickmund et al., *Juvenile Offenders and Victims: 1997 Update on Violence*.

34. Calculations based on data from the Federal Bureau of Investigation, *Crime in the United States*, reports for the years 1994 through 2003.

35. See P. J. Cook and J. H. Laub, "The (Surprising) Stability of Youth Crime Rates," *Journal of Quantitative Criminology* 2 (1986): 265–277; D. W. Osgood et al., "Time Trends and Age Trends in Arrests and Self-Reported Illegal Behavior," *Criminology* 27 (1989): 389–417; Snyder and Sickmund, *Juvenile Offenders and Victims: A National Report*.

36. FBI, *Uniform Crime Reports for the United States*, 1980–2003; Snyder and Sickmund, *Juvenile Offenders and Victims: 1999 National Report*.

37. Federal Bureau of Investigation, *Uniform Crime Reports for the United States, 1996*, 203.

38. Sickmund et al., *Juvenile Offenders and Victims: 1997 Update on Violence*, 19.

39. Sickmund et al., *Juvenile Offenders and Victims: 1997 Update on Violence*.

40. Snyder and Sickmund, *Juvenile Offenders and Victims: 1999 National Report*.

41. Sickmund et al., *Juvenile Offenders and Victims: 1997 Update on Violence*; Snyder and Sickmund, *Juvenile Offenders and Victims: 1999 National Report*.

42. Howard N. Snyder, cited in Sickmund et al., *Juvenile Offenders and Victims: 1997 Update on Violence*, 24.

43. Federal Bureau of Investigation, *Crime in the United States, 2003*. Retrieved December 30, 2004 from http://www.fbi.gov/ucr/cius_03/pdf/03sec3.pdf.

44. Federal Bureau of Investigation, *Crime in the United States, 2003*.

45. D. Anderson et al., "Television Viewing at Home: Age Trends in Visual Attention Time with TV," *Child Development* 57 (1986): 1024–1033.

46. G. Gerbner et al., "The Demonstration of Power: Violence Profile No. 10," *Journal of Communication* 29 (1979): 177–196; A. C. Huston et al., *Big World, Small Screen: The Role of Television in American Society* (Lincoln, NE: University of Nebraska Press, 1992).

47. L. Friedrich-Cofer and A. Huston, "Television Violence and Aggression: The Debate Continues," *Psychological Bulletin* 100 (1986): 364–371; B. Bushman and C. Anderson, "Media Violence and the American Public," *American Psychologist* 56 (2001): 477–489; J. Johnson et al., "Television Viewing and Aggressive Behavior During Adolescence and Adulthood," *Science* 295 (2002): 2468–2471.

48. American Psychological Association, *Violence on TV: Social Issue Release from the Board of Social and Ethical Responsibility for Psychology* (Washington, DC: American Psychological Association, 1985); National Institute of Mental Health, *Television and Behavior: Ten Years of Scientific Progress and Implications for the Eighties* (Washington, DC: U.S. Government Printing Office, 1982).

Chapter Resources

49. E. Donnerstein et al., "The Mass Media and Youth Aggression," in *Reason to Hope: A Psychological Perspective on Violence and Youth*, ed. L. D. Eron et al. (Washington, DC: American Psychological Association, 1994); R. Kuby and M. Csikszentmihalyi, *Television and the Quality of Life: How Viewing Shapes Everyday Experience* (Hillsdale, NJ: Erlbaum, 1990).

50. A. Bandura and R. Walters, *Social Learning and Personality Development* (New York: Holt, Rinehart & Winston, 1963); B. Carlson, "Children's Beliefs about Punishment," *American Journal of Orthopsychiatry* 56 (1986): 308–312; D. Perry et al., "Cognitive Social Learning Mediators of Aggression," *Child Development* 57 (1986): 700–711.

51. An excellent examination of the relationship between the political economic context and delinquency can be found in H. Schwendinger and J. S. Schwendinger, *Adolescent Subcultures and Delinquency* (New York: Praeger, 1985).

52. D. Elliott and S. Ageton, "Reconciling Race and Class Differences in Self-Reported and Official Estimates of Delinquency," *American Sociological Review* 45 (1980): 95–110.

53. U.S. Census Bureau, *Annual Demographic Survey, March Supplement*. Retrieved December 30, 2004, from http://ferret.bls.census.gov/macro/032004/pov/new01_100.htm.

54. Federal Bureau of Investigation, *Crime in the United States, 2003*; D. Huizinga and D. Elliott, "Juvenile Offenders: Prevalence, Offender Incidence, and Arrest Rates by Race," *Crime and Delinquency* 33 (1987): 206–223; Snyder and Sickmund, *Juvenile Offenders and Victims: 1999 National Report*. Walker et al., *The Color of Justice*.

55. C. DeNavis-Walt et al., *Income, Poverty and Health Insurance Coverage in the United States: 2003*. U.S. Census Bureau, Current Population Reports, P60-226 (Washington, DC: U.S. Government Printing Office, 2003).

56. L. Carroll and P. I. Jackson, "Inequality, Opportunity, and Crime Rates in Central Cities," *Criminology* 21 (1983): 178-194; National Research Council, *Common Destiny: Blacks and American Society* (Washington, DC: National Research Council, 1989).

57. W .J. Wilson, "Studying Inner-City Social Dislocations: The Challenge of Public Agenda Research," *American Sociological Review* 56 (1991): 1–14.

58. J. Blau and P. Blau, "The Cost of Inequality: Metropolitan Structure and Violent Crime," *American Sociological Review* 147 (1982): 114–129; R. Block, "Community Environment and Violent Crime," *Criminology* 17 (1982): 46–57; R. McGahey, "Economic Conditions, Organization, and Urban Crime," in *Communities and Crime*, ed. A. Reiss and M. Tonry (Chicago: University of Chicago Press, 1986); R. Rosenfeld, "Urban Crime Rates: Effects of Inequality, Welfare Dependency, Region and Race," in *The Social Ecology of Crime*, ed. J. Byrne and R. Sampson (New York: Springer-Verlag, 1985); R. Sampson, "Structural Sources of Variation in Race-Age-Specific Rates of Offending across Major U.S. Cities," *Criminology* 23 (1985): 647–673; L. Scheurman and S. Kobrin, "Community Careers in Crime," in *Communities and Crime*, ed. A. Reiss and M. Tonry (Chicago: University of Chicago Press, 1986); C. R. Shaw and H. D. McKay, *Juvenile Delinquency and Urban Areas* (Chicago: University of Chicago Press, 1972); O. Simcha-Fagan and J. Schwartz, "Neighborhood and Delinquency: An Assessment of Contextual Effects," *Criminology* 24 (1986): 667–703; R. Stark, "Deviant Places: A Theory of the Ecology of Crime," *Criminology* 25 (1987): 893–909.

59. D. Gottfredson et al., "Social Area Influences on Delinquency: A Multilevel Analysis," *Journal of Research in Crime and Delinquency* 28 (1991): 197–206.

60. Benson and Fox, "When Violence Hits Home: How Economics and Neighborhood Play a Role"; R. McGahey, "Economic Conditions, Organization, and Urban Crime," in *Communities and Crime*, ed. A. Reiss and M. Tonry (Chicago: University of Chicago Press, 1986).

61. For a discussion of the relationship between strain and delinquent behavior, see R. Agnew, "A Revised Strain Theory of Delinquency," *Social Forces* 64 (1985): 151–167; R. Agnew, "Foundation for a General Strain Theory of Crime and Delinquency," *Criminology* 30 (1992): 47–87.

62. C. C. Bell and E. J. Jenkins, "Traumatic Stress and Children," *Journal of Health Care for the Poor and Underserved* 2, no. 1 (1991): 175–184.

63. Bell and Jenkins, "Traumatic Stress and Children."

64. J. F. Sheley and J. D. Wright, "Motivations for Gun Possession and Carrying among Serious Juvenile Offenders," *Behavioral Sciences and the Law* 11 (1993): 375–388.

65. J. A. Grunbaum, "Youth Risk Behavior Surveillance-United States, 2003," *Morbidity and Mortality Weekly Report* 53 (Atlanta, GA: Centers for Disease Control and Prevention, 2004).

66. A. Blumstein, "Violence by Young People: Why the Deadly Nexus?" *U.S. Department of Justice National Institute of Justice Journal*, August 1995, 2–9.

67. P. I. Jackson, "Crime, Youth Gangs, and Urban Transition: The Social Dislocations of Postindustrial Economic Development," *Justice Quarterly* 8 (1991): 379–397.

68. A. Egley and A. K. Major, "Highlights of the 2002 National Youth Gang Survey." *OJJDP Fact Sheet* (Washington, DC: Office of Juvenile Justice and Delinquency Prevention, 2004).

69. W. B. Miller, "Lower Class Culture as a Generating Milieu of Gang Delinquency," *Journal of Social Issues* 14 (1958): 5–19; J. Moore, *Going Down to the Barrio* (Philadelphia: Temple University Press, 1991); I. Spergel, *The Youth Gang Problem: A Community Approach* (New York: Oxford University Press, 1995); F. Thrasher, *The Gang* (Chicago: University of Chicago Press, 1927); J. D. Vigil, *Barrio Gangs* (Austin, TX: Texas University Press, 1988).

70. Spergel, *The Youth Gang Problem*.

71. Jackson, "Crime, Youth Gangs, and Urban Transition: The Social Dislocations of Postindustrial Economic Development"; M. Sullivan, *Getting Paid: Youth Crime and Work in the Inner City* (Ithaca, NY: Cornell University Press, 1989).

72. M. Klein, *The American Street Gang, Its Nature, Prevalence and Control* (New York: Oxford University Press, 1995); L. Yablonsky, *The Violent Gang* (Baltimore: Penguin, 1966).

73. B. Bjerregaard and A. Lizotte, "Gun Ownership and Gang Membership," *Journal of Criminal Law and Criminology* 86 (1995): 37–53.

74. J. Fagan, "The Social Organization of Drug Use and Drug Dealing among Urban Gangs," *Criminology* 27 (1989): 633–669.

75. T. Thornberry et al., "The Role of Juvenile Gangs in Facilitating Delinquent Behavior," *Journal of Research in Crime and Delinquency* 30 (1993): 55–87.

76. W. Sanders, *Gangbangs and Drive-bys: Grounded Culture and Juvenile Gang Violence* (New York: Aldine de Gruyter, 1994).

77. A. Campbell, "Female Participation in Gangs," in *Gangs in America*, ed. C. R. Huff (Beverly Hills, CA: Sage, 1990); Sanders, *Gangbangs and Drive-bys*; F. L. Strodtbeck and J. F. Short, Jr., "The Response of Gang Leaders to Status Threats: An Observation on Group Process and Delinquent Behavior," *American Journal of Sociology* 68 (1983): 571–578.

78. T. P. Thornberry and J. H. Burch II, "Gang Members and Delinquent Behavior," *OJJDP Juvenile Justice Bulletin*, (Washington, DC: Office of Juvenile Justice and Deliquency Prevention, June 1997).

79. W. Gove and R. D. Crutchfield, "The Family and Delinquency," *Sociological Quarterly* 23 (1982): 301–319; R. L. Simmons et al., *Families, Delinquency, and Crime* (Los Angeles, CA: Roxbury, 2004).

80. S. M. Dornbysch et al., "Single Parents, Extended Households, and the Control of Adolescents," *Child Development* 56 (1985): 326–341; D. P. Farrington, "Early Predictors of Adolescent Aggression and Violence," *Violence and Victims* 4 (1989): 79–100; S. Hengeller, *Delinquency in Adolescence* (Newbury Park, CA: Sage, 1989); G. R. Patterson, *Coercive Family Process* (Eugene, OR: Castalia Press, 1982); P. H. Tolan and R. P. Lorion, "Multivariate Approaches to the Identification of Delinquency Proneness in Males," *American Journal of Community Psychology* 16 (1988): 547–561; L. E. Wells and J. H. Rankin, "Families and Delinquency: A Meta-analysis of the Impact of Broken Homes," *Social Problems* 38 (1991): 71–90.

81. L. D. Eron and R. G. Slaby, "Introduction," in *Reason to Hope: A Psychological Perspective on Violence and Youth*, ed. L. D. Eron et al. (Washington, DC: American Psychological Association, 1994).

82. S. M. Catalano, "Criminal Victimization, 2003," *Bureau of Justice Statistics, National Crime Survey* (Washington, DC: U.S. Department of Justice, 2004).

83. E. T. Gershoff, "Corporal Punishment by Parents and Associated Child Behaviors and Experiences: A Meta-analytic and Theoretical Review," *Psychological Bulletin* 128 (2002): 539–579; M. Straus, "Discipline and Deviance: Physical Punishment of Children and Violence and Other Crime and Adulthood," *Social Problems* 38 (1991): 133–154.

84. M. Clement, *The Juvenile Justice System: Law and Process* (Boston: Butterworth-Heinemann, 1997).

85. R. Chalk et al., "The Multiple Dimensions of Child Abuse and Neglect: New Insights into an Old Problem," *Child Trends Research Brief* (Washington, DC: Child Trends, 2002).

86. Chalk et al., "The Multiple Dimensions of Child Abuse and Neglect."

87. Simmons et al., *Families, Delinquency, and Crime*; Straus, "Discipline and Deviance."

88. S. Boney-McCoy and D. Funkelhor, "Psychological Sequelae of Violent Victimization in a National Youth Sample," *Journal of Consulting and Clinical Psychology* 63 (1995): 726–736; Chalk et al., "The Multiple Dimensions of Child Abuse and Neglect."

89. Boney-McCoy and Funkelhor, "Psychological Sequelae of Violent Victimization in a National Youth Sample"; J. Briere, *Child Abuse Trauma* (Newbury Park, CA: Sage, 1992); Chalk et al., "The Multiple Dimensions of Child Abuse and Neglect"; B. T. Kelley et al., "In the Wake of Childhood Maltreatment," *OJJDP Juvenile Justice Bulletin* (Washington, DC: OJJDP, August 1997).

Chapter Resources

90. Chalk et al., "The Multiple Dimensions of Child Abuse and Neglect"; D. Finkelhor and J. Dziuba-Leatherman, "Victimization of Children," *American Psychologist* 49 (1994): 173–183; K. E. Scott, "Childhood Sexual Abuse: Impact on a Community's Mental Health Status," *Child Abuse and Neglect* 16 (1992): 285–295.

91. C. S. Widom, *The Cycle of Violence* (Washington, DC: National Institute of Justice, 1992).

92. C. S. Widom, *The Cycle of Violence Revisited*, research review (Washington, DC: National Institute of Justice, 1996).

93. Kelley et al., "In the Wake of Childhood Maltreatment."

94. Snyder and Sickmund, "Juvenile Offenders and Victims; 1999 National Report"; also see J. Dawson and P. Langan, *Murder in Families, Bureau of Justice Statistics Special Report* (Washington, DC: Bureau of Justice Statistics, 1994).

95. A. Browne, *When Battered Women Kill* (New York: The Free Press, 1987).

96. L. Ohlin and M. Tonry, "Family Violence in Perspective," in *Family Violence*, vol. 11, ed. L. Ohlin and M. Tonry (Chicago: University of Chicago Press, 1989).

97. R. J. Gelles, "Domestic Criminal Violence," in *Criminal Violence*, ed. M. E. Wolfgang and N. A. Weiner (Beverly Hills, CA: Sage, 1982).

98. K. Pillemer and D. Finkelhor, "The Prevalence of Elder Abuse," *The Gerontologist* 28 (1988): 51–57.

99. E. L. Feindler and J. V. Becker, cited in Eron et al., eds., *Reason to Hope*.

100. G. T. Hotaling and M. A. Straus, "Intrafamily Violence, and Crime and Violence Outside the Family," in *Family Violence*, vol. 11, ed. L. Ohlin and M. Tonry (Chicago: University of Chicago Press, 1989).

101. M. Gold, *Delinquent Behavior in an American City* (Monterey, CA: Brooks/Cole, 1970).

102. A. Raine, *The Psychopathology of Crime: Criminal Behavior as a Clinical Disorder* (San Diego, CA: Academic Press, 1993).

103. B. Hutchings and S. A. Mednick, "Criminality in Adoptees and Their Adoptive and Biological Parents: A Pilot Study," in *Biosocial Bases of Criminal Behavior*, ed. S. A. Mednick and K. O. Christiansen (New York: Gardner Press, 1977).

104. Raine, *The Psychopathology of Crime*, 66.

105. R. A. Gordon, "SES versus IQ in the Race-IQ Delinquency Model," *International Journal of Sociology and Social Policy* 7 (1987): 29–96; T. Hirschi and M. Hindelang, "Intelligence and Delinquency: A Revisionist Review," *American Sociological Review* 42 (1977): 471–586; J. Q. Wilson and R. Herrnstein, *Crime and Human Nature* (New York: Simon & Schuster, 1985).

106. S. Menard and B. Morse, "A Structuralist Critique of the IQ-Delinquency Hypothesis: Theory and Evidence," *American Journal of Sociology* 89 (1984): 1347–1378.

107. T. Moffitt, "The Neuropsychology of Conduct Disorder" (Madison, WI: University of Wisconsin, mimeo).

108. D. Fishbein, "Biological Perspectives in Criminology," *Criminology* 28 (1990): 27–72; C. M. Buchanan et al., "Are Adolescents the Victims of Raging Hormones? Evidence for Activational Effects of Hormones on Moods and Behavior at Adolescence," *Psychological Bulletin* 111 (1992): 62–107.

109. T. Moffitt, "Adolescence-Limited and Life-Course-Persistent Antisocial Behavior: A Developmental Taxonomy," *Psychological Review* 100 (1993): 674–701.

110. Mednick and Christiansen, eds., *Biosocial Bases of Criminal Behavior*.

111. S. Rathus, *Psychology* (New York: Holt, Rinehart & Winston, 1996).

112. G. D. Walters, "A Meta-Analysis of the Gene-Crime Relationship," *Criminology* 30 (1992): 595–613.

113. Raine, *The Psychopathology of Crime*.

114. Office of Juvenile Justice and Delinquency Prevention, *Comprehensive Strategy for Serious, Violent, and Chronic Juvenile Offenders* (Washington, DC: Office of Juvenile Justice and Delinquency Prevention, 1993), 13.

115. National Criminal Justice Association, *Juvenile Justice Reform Initiatives in the States, 1994–1996* (Washington, DC: Office of Juvenile Justice and Delinquency Prevention, 1993).

116. National Criminal Justice Association, *Juvenile Justice Reform Initiatives in the States, 1994–1996*.

117. H. M. Hsia, "Allegheny County, PA: Mobilizing to Reduce Juvenile Crime," *OJJDP Juvenile Justice Bulletin*, (Washington, DC: OJJDP, June 1997).

118. National Criminal Justice Association, *Juvenile Justice Reform Initiatives in the States, 1994–1996*.

119. W. Ruefle and K. M. Reynolds, "Curfews and Delinquency in Major American Cities," *Crime and Delinquency* 41 (1995): 347–363.

120. D. Hirschel et al., "Juvenile Curfews and Race: A Cautionary Note," *Criminal Justice Policy Review* 12 (2001): 197–214; M. A. Males, "Vernon, Connecticut's Juvenile Curfew: The Circumstances of Youths Cited and Effects on Crime," *Criminal Justice Policy Review* 11 (2000): 254–267; Ruefle and Reynolds, "Curfews and Delinquency in Major American Cities."

121. Sickmund et al., *Juvenile Offenders and Victims: 1997 Update on Violence*.

122. D. Palumbo and J. Ferguson, "Evaluating Gang Resistance Education and Training (GREAT): Is the Impact the Same as the Drug Abuse Resistance Education (DARE)?" *Evaluation Review* 19 (1995): 597–619.

123. F. Esbensen et al., "How Great Is G.R.E.A.T? Results from a Longitudinal Quasi-Experimental Design," *Criminology and Public Policy* 1 (2001): 87–118.

124. A. Binder et al., *Juvenile Delinquency: Historical, Cultural, and Legal Perspectives*, 2nd ed. (Cincinnati, OH: Anderson, 1997).

125. S. W. Henggeler, "Treating Serious Anti-Social Behavior in Youth: The MST Approach," *OJJDP Juvenile Justice Bulletin* (Washington DC: OJJDP, May 1997).

126. C. C. P. Larson, "Efficacy of Prenatal and Postpartum Home Visits on Child Health and Development," *Pediatrics* 66 (1980): 191–197; Olds et al., "Improving the Life-Course Development of Socially Disadvantaged Mothers: A Random Trial of Nurse Home Visitation," *American Journal of Public Health* 78 (1988): 1436–1445.

127. J. R. Berrueta-Clement et al., *Changed Lives: The Effects of the Perry Preschool Program on Youths through Age 19* (Ypsilanti, MI: High/Scope, 1984); D. L. Johnson and T. Walker, "Primary Prevention of Behavior Problems in Mexican-American Children," *American Journal of Community Psychology* 15 (1987): 375–385; J. R. Lally et al., "The Syracuse University Family Development Research Project: Long-Range Impact of an Early Intervention with Low-Income Children and their Families," in *Annual Advances in Applied Developmental Psychology*, ed. D. R. Powell (Norwood, NJ: Ablex Publishing Corporation, 1988).

128. G. A. Wasserman and L. S. Miller, "The Prevention of Serious and Violent Juvenile Offending," in *Serious and Violent Juvenile Offenders, Risk Factors and Successful Interventions*, ed. R. Loeber and D. P. Farrington (Thousand Oaks, CA: Sage, 1998).

129. Puzzanchera et al., "Juvenile Court Statistics 1999."

130. Sickmund et al., *Juvenile Offenders and Victims: 1997 Update on Violence*.

131. P. R. Torbet, *State Responses to Serious and Violent Juvenile Crime* (Washington, DC: Office of Juvenile Justice and Delinquency Prevention, 1996), cited in Sickmund et al., *Juvenile Offenders and Victims: 1997 Update on Violence*, 28.

132. Torbet et al., *State Responses to Serious and Violent Juvenile Crime*, cited in Sickmund et al., *Juvenile Offenders and Victims: 1997 Update on Violence*, 28.

133. Torbet et al., *State Responses to Serious and Violent Juvenile Crime*, cited in Sickmund et al., *Juvenile Offenders and Victims: 1997 Update on Violence*, 28.

134. Sickmund et al., *Juvenile Offenders and Victims: 1997 Update on Violence*.

135. L. Winner et al., "The Transfer of Juveniles to Criminal Court: Reexamining Recidivism over the Long Term," *Crime and Delinquency* 43 (1997): 548–563.

136. Puzzanchera et al., "Juvenile Court Statistics 1999."

137. Sickmund et al., *Juvenile Offenders and Victims: 1997 Update on Violence*.

138. F. Butterfield, "With Juvenile Courts in Chaos, Critics Propose Their Demise," *New York Times*, 21 July 1997, A-13.

139. Sickmund et al., *Juvenile Offenders and Victims: 1997 Update on Violence*.

140. W. H. Barton and J. A. Butts, "Viable Options: Intensive Supervision Programs for Juvenile Delinquents," *Crime and Delinquency* 36 (1990): 238–256.

141. Coordinating Council on Juvenile Justice and Delinquency Prevention, *Combating Violence and Delinquency: The National Juvenile Justice Action Plan* (Washington, DC: Office of Juvenile Justice and Delinquency Prevention, 1996).

142. J. C. Howell, ed., *Guide for Implementing the Comprehensive Strategy for Serious, Violent, and Chronic Juvenile Offenders* (Washington, DC: Office of Juvenile Justice and Delinquency Prevention, 1995).

143. D. M. Altschuler and T. L. Armstrong, "Aftercare Not Afterthought: Testing the IAP Model," in *Juvenile Justice* (Washington, DC: Office of Juvenile Justice and Delinquency Prevention, 1996), 15–22.

144. J. A. Fagan, "Treatment and Reintegration of Violent Juvenile Offenders: Experimental Results," *Justice Quarterly* 7 (1990): 233–263.

145. M. H. Beers et al. *The Merck Manual of Diagnosis and Therapy*, 17th ed. (West Point, PA: Merck and Company, 1999).

Chapter Resources

Present Conditions and Future Directions in Juvenile Justice

15

- The Future of Juvenile Justice
- Problems That Will Continue to Influence America's Youth, Families, Communities, and Juvenile Courts
- Legal Issues
- Chapter Summary
- Key Concepts
- Review Questions
- Additional Readings
- Notes

■ Introduction

Since the development of the first juvenile court in Cook County, Illinois, over 100 years ago, juvenile justice has undergone a number of changes. Some of these changes represent fundamental differences in how juvenile justice agencies view juvenile crime, how they are organized, and how they respond to youth problems. In other respects, however, changes in juvenile justice are more cosmetic than real. Indeed, it can be argued that juvenile justice has changed very little over time. Today, many youths involved in the juvenile justice process and the people who work with them face many of the same problems that were confronted by juvenile justice clients and practitioners over 100 years ago. Moreover, given the present state of juvenile justice practice, it appears that many of the problems that have historically confronted juvenile justice will continue to affect the quality of services received by juvenile justice clients as well as the working conditions within juvenile justice agencies. Yet, although a number of problems continue to challenge juvenile justice professionals and the children they work with, there are hopeful signs for the future. This chapter examines some of the major changes that have occurred in juvenile justice since the first juvenile court was established. In addition, it explores problems that presently affect the quality of juvenile justice practice and predicts what can be expected to occur in juvenile justice in the early part of the 21st century.

■ Changes in the Quality of Juvenile Justice Practice

Juvenile justice has changed in a number of ways in the last 100 years. Indeed, changes in juvenile law, the structure and operation of juvenile courts, and the quality and quantity of services offered to youths and their families have occurred in this period. Many of the changes have undoubtedly improved the ability of juvenile justice agencies to meet the needs of youth, families, and the communities they serve. Other changes, however, have had little or no positive effect on children, families, or community safety.

Early Reforms: The Development of the Juvenile Court

The early juvenile courts were developed in response to a particular set of problems confronting urban citizens during the late 1800s. One of these problems was the large and ever-growing number of neglected, dependent, and wayward youths, many of whom were also engaged in illegal behavior. From the perspective of many city dwellers, these children—more than a few the offspring of immigrants who did not speak English—were felt to constitute a "dangerous class" that threatened the established order.[1] As Julia Lathrop, a noted 19th-century **child saver** stated,

> *There are at the present moment in the state of Illinois, especially in the city of Chicago, thousands of children in need of active intervention for their preservation from physical, mental, and moral destruction. Such intervention is demanded, not only by sympathetic consideration for their well-being, but also in the name of the commonwealth, for the preservation of the State. If the child is the material out of which men and women are made, the neglected child is the material out of which paupers and criminals are made.[2]*

Although wealthy citizens in developing urban areas desired to control the increasing number of problem children, they confronted several major obstacles. First, criminal courts had no jurisdiction over youths unless they engaged in criminal activity. Second, the existing criminal justice system was ineffective at dealing with these youths, because it punished youth offenders either too severely or not at all.[3] For example, when young children were brought before criminal courts, the outcomes were often less than satisfactory. In many instances, young children were placed in adult jails and prisons, despite the fact that it was generally recognized that placing children with hardened criminals was hardly an effective way to protect children or community safety. The judges who felt this way tended to balk at jailing children for minor criminal behaviors. As a result, many citizens began to see adult courts as an ineffective mechanism for controlling problem youths.

In addition to criminal courts, a number of other mechanisms were established throughout the 1800s to deal with problem youths, including houses of refuge, the placing out system, probation, training schools, and orphanages. None of these mechanisms, however, proved to be particularly effective. Furthermore, the Illinois Supreme Court, in *People v. Turner* (1870), ruled that a child could not be institutionalized without having committed an offense. The court noted that the institutionalization and punishment of youths who had not committed a criminal offense was a violation of basic **due process** protections.[4] Although this case did not affect efforts to control youths outside of Illinois, it prevented the institutionalization of youths who had not committed crimes in that state and required that, prior to commitment, youths be given due process protection. In effect, this case stopped the long-standing practice of institutionalizing problem youths who had not been charged with a crime—a practice that many people favored because criminal courts could not legally control these children.

child savers Affluent 19th-century reformers, who were concerned about controlling and training youths so that they could become productive, law-abiding citizens. The child savers played a major role in the establishment of the juvenile courts.

due process An established course of proceedings designed to protect individual rights and liberties and to ensure that people are treated with fairness before the law. Although juveniles have most of the due process protections afforded adults in the criminal courts, they do not have all of these rights. Moreover, because of the informality found in many juvenile courts, legal protections that are theoretically available may in fact be unavailable.

The inability of criminal courts to effectively deal with a large population of wayward youths, along with the Illinois Supreme Court's decision in *Turner*, presented a major obstacle to child savers in Chicago, who were interested in controlling and reforming problem children. To circumvent the *Turner* decision and to reestablish control over wayward children, the child savers invented a new legal institution designed to control and reform troublesome youths. This new legal institution was the juvenile court.[5]

Once the first juvenile court was established in Chicago, the idea of a court designed especially for children became so appealing that by 1920 all but three states had developed juvenile courts.[6] These courts were designed to be different than adult courts. They were essentially **chancery** (or civil) **courts** founded on the legal concept of ***parens patriae*** (i.e., the idea that the state must act as protector of children).[7] The goals of the new courts were, at least in theory, to serve the best interests of children and to protect the community. What distinguished the juvenile courts from adult criminal courts was their concern for the interests of the children who came before the bench. As Judge Julian Mack of the Cook County (Chicago) Juvenile Court wrote in 1909, "The problem for the determination by the judge is not, Has this boy or girl committed a specific wrong, but What is he, how has he become what he is, and what had best be done in his interest and in the interest of the state to save him from a downward career."[8] As this statement makes clear, the focus of the juvenile courts was on training and reform as opposed to punishment.

In order to facilitate their efforts to reform children, the new courts were given broad discretionary powers over a range of youths—not only youths involved in criminal behaviors, but also neglected, dependent, and other types of wayward or unfortunate youths. They were also soon given expanded powers to intervene in the relationships between parents and their children.[9] To allow them to handle child and family problems, juvenile court procedures were designed to be flexible and **informal**, another characteristic that distinguished juvenile courts from criminal courts. Importantly, due process protections were felt to hinder the ability of the juvenile court to serve the best interests of children. As a result, juvenile court hearings were typically held in an office, representation by counsel was rare, no records of hearings were made, specific charges were avoided in favor of general charges of delinquency, and witnesses and complainants were rarely present.

The particular approach taken by juvenile courts in exercising their powers was influenced by reformers' beliefs about the causes of delinquency. The reformers who were instrumental in the establishment of the juvenile courts thought that several factors contributed to delinquent behavior, including biological and mental defects as well as poverty and inappropriate parenting.[10] However, even poverty and inappropriate parenting were seen as types of individual pathologies rather than the products of the political and economic conditions within which people lived. Thus, the responses to youth crime and waywardness developed by juvenile courts were focused on controlling, treating, and reforming the individual rather than on changing harmful social conditions. As Barry Krisberg and James Austin note in their book *Reinventing Juvenile Justice*,

chancery court English court concerned primarily with property rights. Over time these courts became more involved in the welfare of families and children, and served as a model for the development of the juvenile court.

parens patriae A legal doctrine according to which the state has an obligation to protect and serve the interests of children.

informal juvenile court A court that places little emphasis on due process protections for youths. Because juvenile courts exist, in theory, to protect the best interests of children, legal protections for children are felt to be unnecessary within an informal court.

The thrust of Progressive Era reforms was to found a more perfect control system to restore social stability while guaranteeing the contin-ued hegemony of those with wealth and privilege. Reforms such as the juvenile court are ideologically significant because they preserved the notion that social problems (in this case delinquency, dependency, and neglect) could be dealt with on a case-by-case basis, rather than through broad-based efforts to redistribute wealth and power throughout society. The chief dilemma for advocates of the juvenile court was to develop an apparently apolitical or neutral system, while preserving differential treatment for various groups of children. The juvenile court at first lacked a core of functionaries who could supply the rationale for individualized care for wayward youth, but soon these needs were answered by the emergence of psychiatry, psychology, and criminology, as well as by the expanding profession of social work.[11]

The belief that youth problems were products of individual pathology led juvenile justice agencies to focus on control, treatment, and rehabilitation rather than other types of correctional responses, at least up until the late 1970s.

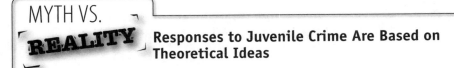

MYTH VS. REALITY — Responses to Juvenile Crime Are Based on Theoretical Ideas

Myth—Theories of delinquent behavior have no practical significance.
Reality—Every response to juvenile crime is based on some theory as to the causes of delinquency. For example, reformers who developed the juvenile court created responses to juvenile crime that were consonant with their under-standing of delinquent and problem behaviors. Similarly, contemporary responses to juvenile crime are also based on theories of delinquent behavior.

The development of the juvenile court represented a significant development in efforts to control and reform children. It became the legal mechanism by which power over children and their parents could be exercised. However, the court itself, regardless of its reform and treatment orientation, had little influence on other institutions that had been established to deal with troublesome youths. The development of the juvenile court did nothing to change, for example, the often appalling conditions found in the facilities—the houses of refuge, reform schools, and jails and prisons—that typically housed delinquent youths. Moreover, the informal procedures of the juvenile court lent themselves to abuses by judges, who had few checks on their power.

The Legal Reform Years: The Supreme Court's Focus on the Juvenile Court, 1966–1971

Until the mid-1960s, juvenile court operations continued to be characterized by the same informality found in the first juvenile courts. Despite this informality (or perhaps because of it), juvenile courts did provide assistance to many children and their families. By the mid-1960s, however, juvenile court practices began to be more carefully scrutinized by the U.S. Supreme Court. Beginning with the case of *Kent v. United States* (1966), the Supreme Court made it clear that it was going to take a closer look at the due process protections afforded youths in the juvenile court.[12] In subsequent decisions, such as those in the landmark cases *In re Gault* (1967)[13] and *In re Winship* (1970),[14] the Supreme Court expanded the due process rights of children appearing before a juvenile court. As a result of these cases, much of the traditional informality of the courts was replaced with procedural requirements designed to give children enhanced legal protections. By 1971, youths had the right to counsel, the right to adequate and timely written notice, the right to confront and cross-examine witnesses, and the privilege against self-incrimination in instances where they were subject to interrogation.

In *In re Gault*, the Supreme Court Addresses the Informal Nature of Juvenile Court

The following excerpts from the *Gault* decision clearly demonstrate the U.S. Supreme Court's concern with the informality that characterized juvenile court up until that time. In the opinion, the majority justices noted the following:

> *The early reformers were appalled by adult procedures and penalties, and by the fact that children could be given long prison sentences and mixed in jails with hardened criminals. They were profoundly convinced that society's duty to children could not be confined by the concept of justice alone. . . . The rules of criminal procedure were therefore altogether inapplicable. . . . [However] Juvenile Court history has again demonstrated that unbridled discretion, however benevolently motivated, is frequently a poor substitute for principle and procedure. . . . The absence of procedural rules based upon constitutional principle has not always produced fair, efficient, and effective procedures. Departures from established principles of due process have frequently resulted not in enlightened procedure, but in arbitrariness. . . . Under our Constitution, the condition of being a boy does not justify a kangaroo court.*[15]

Closer scrutiny of juvenile court practice by the Supreme Court led to more formality in juvenile courts and increased concern with the due process rights of juveniles. These Supreme Court cases can be seen as important milestones in the development of the juvenile court. However, in another important case, *McKeiver v. Pennsylvania* (1971), the Supreme Court ruled that juveniles do *not* have a constitutional right to a jury trial.[16] This case indicated that the Supreme Court was not willing to grant juveniles all of the due process protections given adults in criminal courts. In the majority opinion, the justices argued that jury trials would make the juvenile courts indistinguishable from criminal courts and would create a more adversarial type of proceeding, to the disadvantage of children.[17]

The *McKeiver* case gave notice that the Supreme Court was not willing to completely remove the informality that has been the hallmark of juvenile courts. Overall, the cases heard by the Supreme Court between 1966 and 1971 that dealt with juvenile court practices indicated that juvenile courts should be more concerned about due process rights, and limited the blatant informality common in juvenile courts up until that time. Yet, the Supreme Court also indicated that it continued to support some degree of informality, which it believed was necessary if the juvenile courts were to accomplish their mission of serving the best interests of children and protecting the community.

The "Get Tough" Movement in Juvenile Justice

The period between 1966 and 1971, during which the Supreme Court focused attention on the need for due process protections in the juvenile courts, was immediately followed by another important development in juvenile justice history—the appearance of the <u>"get tough" movement</u>. The "get tough" movement is part of a conservative reform movement that began to dominate the national debate over juvenile crime by the late 1970s[18] and that continues to play a significant role in juvenile justice policy development. According to advocates of a "get tough" approach to juvenile crime, the primary obligation of juvenile justice agencies is to protect citizens from youthful offenders. Conservatives have called for a number of changes in the juvenile justice process, particularly in how juvenile justice agencies deal with chronic, serious, and violent juvenile offenders. These changes include developing a more adversarial model of juvenile justice, strengthening the position and improving the effectiveness of the prosecuting attorney in the juvenile justice process, developing mandatory sentencing guidelines, opening access to juvenile court records and procedures, and developing mechanisms for identifying potentially violent or chronic offenders.[19] From a correctional standpoint, "get tough" reformers are concerned with accountability, deterrence, and punishment as opposed to treatment and rehabilitation.[20]

The "get tough" movement has pushed legislators to make a number of substantive changes in the operation of juvenile justice. Initially, a number of states implemented reforms intended to fine-tune juvenile justice operations. More recently, there has been more of a focus on implementing comprehensive juvenile justice reforms.[21] As a result, a number of states have given prosecutors a more prominent role in the juvenile justice process, penalties for juvenile offenders have been stiffened, mandatory sentencing guidelines have been established,

"get tough" movement
A legal reform movement that began during the late 1970s. The goals of the movement included developing a more punitive juvenile court orientation or abolishing the juvenile courts so that more severe sanctions could be leveled against juvenile offenders.

new mechanisms for transferring juveniles to criminal courts have been created and existing procedures simplified, and increasing numbers of youths have been incarcerated in juvenile and adult correctional facilities.[22] In short, many juvenile courts have become more concerned with protecting the community and with punishing offenders than with their traditional mission—serving the best interests of children.[23]

The Resiliency of Treatment and Rehabilitation

rehabilitation The process of restoring an individual's ability to play a constructive role in society through education, training, or therapy.

aftercare Services delivered to youths after they are released from a residential setting and while they are still under the jurisdiction of the state or juvenile court.

Despite the reforms just mentioned, treatment and <u>rehabilitation</u> continue to play a significant role in juvenile justice. Today, a wider range of treatment and rehabilitation programs exist for juvenile offenders than at any other time in our nation's history. To begin with, there is a wide variety of community-based programs, institutions, and interventions, such as diversion programs, probation, intensive supervision programs, multisystemic therapy, house arrest, group homes, shelter facilities, day treatment programs, treatment foster care, restitution, mediation, and community service programs, outdoor experiential programs, intensive <u>aftercare</u> programs, and short-term substance abuse treatment programs. In addition, a number of residential treatment programs for youths exist around the country, some of which provide highly specialized treatment for chronic, serious, and violent juvenile offenders. Moreover, other programs are presently being developed.

MYTH VS.
REALITY Most Citizens Favor the Treatment of Juvenile Offenders

Myth—Most Americans believe that the goal of juvenile courts should be the punishment of youths who violate the law.
Reality—Research on public attitudes toward crime and justice has consistently found that most Americans believe that treatment and rehabilitation are the appropriate goals of juvenile justice.[24]

There continues to be widespread support for rehabilitation and treatment among both the general public and those who work in juvenile justice. Public opinion surveys have consistently found high levels of support for the treatment and rehabilitation of juvenile offenders. For example, a national survey conducted in the 1980s found that 73% of the respondents felt that treatment and rehabilitation, rather than punishment, should be the primary mission of the juvenile courts. Similar findings are evident in polls conducted in California, Illinois, and Michigan in the 1980s and early 1990s and in more recent national polls.[25] In a 1995 national poll conducted as part of the survey research program at Sam Houston State University, approximately 48% of respondents indicated that the most important purposes to be achieved in sentencing juveniles were the training, education, and counseling of offenders. In comparison, only 30% indicated that the purpose of juvenile court sentencing should be to give offenders the punishment that they deserved.[26] Similar support for corrective treatment of juvenile offenders has also been found in state-level surveys and among judges and correctional administrators.[27] In a study that examined a sample of Florida judges' support for the treatment ethic, Gordon Bazemore and Lynette Feder found that the majority of juvenile division judges supported rehabilitation, although these judges felt that crime control was also important.[28] Similarly, a survey of chief administrators of state-operated secure correctional facilities in 1992 found that 84% of the respondents indicated that treatment and rehabilitation characterized the correctional philosophy employed at their facility.[29]

Juvenile justice practice has changed in a number of ways over the last 100 years. The informality of earlier juvenile courts has been mitigated by the inclusion of formal procedures intended to ensure due process. Moreover, the *parens patriae* philosophy, with its focus on serving the best interests of the child, has been supplanted in many instances by a punishment and deterrence orientation. Yet despite these changes, many juvenile courts continue to retain a great deal of informality. Although there has been a clear movement toward punishment, rehabilitation continues to be a key ingredient in many court and correctional programs. The following section examines how the opposing orientations—informality and rehabilitation versus formality and punishment—influence the operation of contemporary juvenile justice practice.

■ Contemporary Problems and Issues in Juvenile Justice

Today, juvenile justice faces a number of problems and issues, some of which call into question the continued existence of separate courts for juveniles. Can the juvenile courts protect the due process rights of children? Are juvenile corrections programs able to rehabilitate youths and protect the community? Should the juvenile courts be abolished? These are some of the important questions that presently confront those who work in the field of juvenile justice.

Problems Confronting Juvenile Courts

The problems currently confronting the juvenile courts are not new. In fact, they have existed since the development of a separate justice process for children. Nevertheless, the ability of the juvenile courts to successfully deal with these problems may well determine their very future. These problems include a failure to protect the due process rights of children, a lack of appropriate responses to status offenders, the processing of many youths who could be diverted from juvenile justice agencies, the lack of coordinated resources for responding to problem youths, and the failure to recognize the limits of juvenile justice in solving youth problems.

Juvenile courts have often failed to protect the due process rights of children. An ongoing conflict continues between the state's caretaking responsibilities and its obligation to ensure that families and children are treated fairly. The state is being increasingly called upon to intervene in the lives of serious delinquents and exercise "control" over them. If juvenile courts are going to continue to incarcerate larger numbers of youths and transfer more youths to adult courts, they must take children's due process protections ever more seriously to ensure that justice is done.

Although Supreme Court rulings have changed juvenile court practices, it is debatable how fundamental these changes have been.[30] Overall, juvenile courts have become more cognizant of the need to protect juveniles' due process rights. Nevertheless, juvenile courts are still characterized by considerable informality. In reality, juveniles often lack adequate representation (or any representation at all) in juvenile proceedings.[31] As noted in Chapter 10, courts must do more than pay passing attention to due process if it is to become an integral part of juvenile court operations. In order to receive the benefits of due process, youths and their parents or guardians must know what their rights are and feel comfortable in exercising those rights. However, there is evidence that many of those who come before juvenile courts neither understand their rights nor feel comfortable in exercising them.[32]

Since the inception of the juvenile courts, their procedural informality has been seen as one of their strengths. However, this alleged strength has also been a major weakness. When juveniles and their parents do not understand their rights, do not feel comfortable exercising those rights, or do not have adequate representation, they run the risk of being harmed rather than helped by the juvenile courts.

The inability or unwillingness of juvenile courts to adequately protect the due process rights of juveniles and parents has led some to call for the abolishment of traditional juvenile courts. One proponent of abolishment is Professor Barry Feld of the University of Minnesota Law School. Feld maintains that juveniles have not received adequate protection of their rights within the juvenile court, nor is he optimistic that such protection will be afforded juveniles in the future. He argues that juveniles would be better served in criminal courts, where there is more emphasis on due process protections. Because of their age, Feld maintains that juveniles deserve special due process protections. For example, he argues that juveniles should automatically receive assistance of counsel when they are taken into custody and should be allowed to consult with counsel before the right to counsel can be waived or before any interrogation takes place.[33]

Feld also believes, however, that before adult courts should be allowed to handle younger defendants, they need to be changed. One important change would be to give youths a "discount" in recognition of their diminished culpability. According to Feld, "Criminal courts can provide shorter sentences for reduced culpability with fractional reductions of adult sentences in the form of an explicit 'youth discount.' For example, a 14-year-old might receive 33% of the adult penalty, a 16-year-old 66%, and an 18-year-old the adult penalty, as is presently the case."[34]

From Feld's point of view, a **youth discount** would remove many of the inconsistencies and injustices presently found in juvenile court practices. For example, currently a youth who is one day shy of his or her 18th birthday is tried as a juvenile, while a youth who has just turned 18 is tried as an adult. In addition, by trying juveniles in adult courts, costly and time-consuming waiver hearings would be eliminated.[35]

youth discount A stratagem for taking into account the supposed diminished culpability of youths (based on their immaturity) in the sentencing process. The use of youth discounts would generally result in shorter sentences for juveniles than for adults who had committed identical offenses.

Unlike Feld, who is worried about issues of justice, some juvenile court critics are essentially concerned with holding youths accountable when they break the law. These critics maintain that the criminal courts are better able to punish juvenile offenders than are the juvenile courts, which have traditionally paid too much attention to the needs of the youth and ignored the needs of victims and the community.

Another problem facing juvenile courts is the lack of appropriate responses to status offenders. The juvenile court was established, in part, as a legal mechanism for controlling and reforming status offenders. However, the ability of juvenile courts to control these youths or to "fix" the problems that contribute to their behavior is often limited. Historically, they have never had the resources they need to deal effectively with the large number of youths and families who need assistance of various kinds. Moreover, they may not be particularly well equipped to respond to problems such as truancy, running away from home, and incorrigibility. They have often resorted to coercion and incarceration in dealing with these problems, even though the wisdom of such treatment is suspect.

Recognizing the potentially harmful effects of the incarceration of status offenders, Congress passed the Juvenile Justice and Delinquency Prevention Act of 1974, which encouraged states to remove status offenders from secure detention and correctional facilities and to separate juvenile and adult offenders. In addition, the federal government encouraged the development of diversionary programs as a way of responding to status offenders. However, neither of these approaches has been completely successful. Although the number of status offenders detained in secure facilities has decreased since the passage of the Juvenile Justice and Delinquency Prevention Act, status offenders continue to be detained in secure facilities in some jurisdictions. Moreover, diversionary programs, which are used for many status offenders, also face a number of problems, as noted below.

There is a tendency to process many youths who could instead be diverted from juvenile justice. Status offenders are obvious candidates for diversion, but diversion could reasonably be used for many youths who commit nonserious criminal offenses. Because of the pervasive nature of delinquency, juvenile justice agencies cannot possibly respond to each act of delinquency that comes to their attention. Moreover, even if they could respond to each act, the appropriateness of doing so would be questionable for at least three reasons. First, juvenile justice agencies lack the technologies and resources to deal effectively with each case of delinquent behavior. As noted in Chapter 13, although juvenile courts can play a role in helping youths and families respond to many problems that are brought to their attention, the courts themselves, as well as other juvenile justice agencies, are limited in their ability to respond to many of the problems that underlie delinquent behavior. Second, there is evidence that juvenile justice intervention can encourage additional delinquency rather than prevent it.[36] Third, most youths who commit delinquent acts eventually desist from delinquent behavior even though they are not identified and processed by juvenile justice agencies. Consequently, it is important that juvenile justice agencies divert cases from the juvenile justice process when it is possible and appropriate. Unfortunately, it is not always easy to agree about which cases are appropriate for diversion.

Diversion, of course, has its own problems, and the development of diversion programs needs to be approached with caution. As noted in Chapter 7, diversion programs have the potential to increase the costs associated with case processing, they can be stigmatizing for those involved, they can be coercive and deny youths and parents basic due process rights, and they can widen the net of social control by increasing the number of youths under correctional supervision.[37] All of these outcomes are at odds with the typical goals of diversion programs.

Many communities lack a coordinated set of responses that allow courts to deal with youth problems effectively. Juvenile courts have historically been responsible for dealing with the needs of a diverse population of youths. Indeed, the youths who come to the attention of juvenile courts vary in a number of ways, such as in their level of psychological, emotional, and physical maturity, and their economic, family, and social background. Consequently, it is important that juvenile courts have access to a <u>**continuum of quality services**</u> that they can draw on. Courts in many jurisdictions, however, have relatively few resources. In some communities juvenile court judges and other juvenile justice professionals have taken an active role in developing and coordinating services for youth, but in others juvenile court judges and juvenile justice professionals have instead been content to maintain the status quo. Juvenile courts and other juvenile justice agencies often fail to recognize that they possess limited technologies and resources for responding to youth problems. They play an important role in dealing with delinquency, of course, but without the support of other institutions, they are unable to substantially change the behavior of many youth offenders. Although the juvenile courts possess considerable power, the ability to influence behavior ultimately rests on coercion, which has limited utility in reforming behavior. The courts therefore need to develop strong linkages with other community agencies and to determine when it is appropriate to use coercive responses, when to use noncoercive responses, and when to avoid responses altogether.

continuum of quality services A range of effective services consisting of prevention and correctional responses for deliquent youth.

MYTH VS. REALITY
The Court's Ability to Solve Family Problems Is Limited

Myth—Courts are the most effective institutions for controlling problem youths.
Reality—Courts can play an important role in helping youths and families deal with their problems. However, courts themselves often have limited resources for assisting families. As a result, they often rely on their coercive powers to solve problems, which limits their effectiveness in many instances.

Problems Confronting Juvenile Corrections

Since the development of the first specialized correctional institutions for youths in the 1800s, treatment and rehabilitation have been primary goals of juvenile corrections—in word if not in deed. Indeed, by the 1960s and into the 1970s, rehabilitation dominated progressive correctional thinking. However, in the late 1960s and mid-1970s, several literature reviews appeared that raised questions about the effectiveness of many correctional interventions.[38] In what is undoubtedly the most often cited of these reviews, Robert Martinson stated that "with few and isolated exceptions, the rehabilitative efforts that have been reported so far have had no appreciable effect on recidivism."[39]

Research that questioned the effectiveness of juvenile correctional programs proved to be powerful ammunition for conservatives who wished to "get tough" with juvenile offenders. Moreover, it struck a chord with liberals who recognized that correctional programs were, in many cases, designed to punish rather than rehabilitate youths. Doubts about the effectiveness of correctional programs were supported by additional research done during the 1980s. For example, in an extensive review of evaluations of juvenile correctional programs published in professional journals from 1975 to 1984, John Whitehead and Steven Lab concluded that these studies indicated that existing correctional treatments had "little effect on recidivism."[40]

Certainly, a number of questions have been raised about the utility of many correctional programs for juvenile offenders. Yet there continues to be strong evidence that some programs are effective in assisting youths and protecting the community. Indeed, during the 1970s and 1980s, a period in which a number of articles appeared questioning the effectiveness of correctional programs, other studies and reviews indicated that some programs did work.[41] For example, a <u>meta-analysis</u> of studies of correctional programs completed between 1960 and 1983 by Carol Garrett found that there were few negative effects of treatment. Instead, she reported that treatment of delinquents in institutional and community settings generally was associated with positive effects, even though in many instances these effects were not large. Potentially promising types of interventions uncovered by Garrett included some that employed a cognitive-behavioral approach or incorporated a family therapy component.[42] Moreover, articles that questioned the effectiveness of correctional programs in general acknowledged that some programs did reduce recidivism. Even Robert Martinson's research indicated that some programs may work.[43]

Interestingly, although there was a definite shift in many juvenile courts toward deterrence, punishment, and protecting community safety during the 1990s, there has also been increasing evidence of the viability of treatment and rehabilitation in juvenile corrections.[44] There is good empirical evidence that a number of community-based and institutional juvenile correctional programs serve the best interests of youths and are effective at reducing recidivism, even in the case of serious and violent juvenile offenders.[45] Moreover, our understanding of the characteristics of effective correctional programs and our knowledge of which types of interventions are most effective with which types of youths have improved considerably in recent years.[46]

meta-analysis A statistical technique that makes possible statements about the effects of interventions across a number of different studies.

Although good programs do exist, there are many other programs that fail to provide quality services to youth, are poorly managed, have poorly trained and paid staff, and do little to protect community safety. Some jurisdictions have what might be called "file drawer" probation where overworked probation officers are responsible for large caseloads that preclude effective supervision or assistance to clients. Parole or aftercare programs are often similarly overburdened. Moreover, the quality of care and services provided in many group homes, foster homes, shelter-care facilities, detention centers, and institutional placements is far less than youths need or deserve.

Many correctional personnel lack the education, training, pay, and motivation needed to perform their jobs well. Part of the explanation for this is that there is a dearth of basic and follow-up training programs for correctional staff. Although some state systems require basic training and certification of juvenile justice personnel and provide good training programs for this purpose, ongoing training of staff is rare. It is also possible to find staff who are in their positions because of who they know, not what they know.[47] One result is the extensive use, in many correctional programs, of "boob tube" therapy—the relegation of youths to watching television most of the time.[48]

People's penchant for fads and panaceas in responding to juvenile crime needs to be recognized. One problem that confronts juvenile corrections is that those in policy-making positions often promote faddish and simplistic responses to juvenile crime that have little or no correctional utility. In fact, such approaches may actually be counterproductive. Nevertheless, faddish approaches are often easily "sold" as effective because they have intuitive appeal. Unfortunately, intuitive appeal is not always associated with true effectiveness. According to criminologist James Finckenauer, many people are looking for simple and inexpensive solutions to the problems of crime and delinquency.[49] Thus, responses like the Scared Straight! program, in which hardened adult offenders attempt to frighten youths away from a life of crime, or the passage of tougher laws come to be seen, at least for a short period, as panaceas. Many people want to believe that such approaches to delinquency work. Moreover, programs like Scared Straight! are much less costly than attempting to respond to the problems faced by families and the lack of good educational, employment, and recreational opportunities for youths. However, as research by Finckenauer has shown, youths who went through the Scared Straight! program actually engaged in more new offenses than juveniles in a control group.[50] Furthermore, the passage of tougher laws appears to have had no appreciable effect on delinquency. In fact, violent juvenile crime actually increased after states began to "get tough" on juvenile crime.

MYTH VS. REALITY

Commonsense Approaches to Deliquency Are Not Always Effective

Myth—If we just use "common sense," we can solve the delinquency problem.
Reality—A problem with commonsense solutions to crime is that seemingly good ideas do not always translate into effective interventions. If they did, the "delinquency problem" would have been solved long ago. In reality, delinquency is a complex phenomenon that is not amenable to simple solutions, and programs need to be carefully evaluated assess their effectiveness.

Perhaps the most favored recent panacea is the juvenile offender boot camp.[51] Despite the growing popularity of such boot camps, however, research on their effectiveness at reducing recidivism has not produced encouraging results. For example, an evaluation of three federally funded boot camp programs in Cleveland, Denver, and Mobile indicated that youths who participated in these programs were no less likely to relapse than a comparable group of youths who did not attend boot camps. Indeed, in Cleveland, the percentage of recidivists among the boot camp participants was much higher (72%) than for the control group (50%). In Denver and Mobile, the percentages of boot camp participants and controls who relapsed were roughly comparable.[52] Thus, there is no evidence that these programs have a significant impact on recidivism. Boot camp participants in Cleveland and Mobile did improve their academic skills,[53] but these types of improvements are found in a variety of correctional programs.

Although boot camps may have considerable appeal, especially to those with prior military experience, the theoretical basis for using military-style training to rehabilitate youth offenders appears to be suspect. As Merry Morash and Lila Rucker noted in a critique of boot camps, "The very idea of using physically and verbally aggressive tactics in an effort to 'train' people to act in a prosocial manner is fraught with contradiction. The idea rests on the assumption that forceful control is to be valued. The other unstated assumption is that alternative methods of promoting prosocial behavior, such as development of empathy or a stake in conformity (e.g., through employment), are not equally valued."[54]

Another serious problem is the lack of attention given to the needs of female juvenile offenders. This problem is not new, but it is becoming more severe because of the increased number of girls who are entering the juvenile justice process.[55] Women face a number of unique problems within our society, and these problems are even more difficult for young girls. Physical abuse, sexual abuse, neglect, and exploitation are often found in the backgrounds of female juvenile offenders,[56] but few resources have been available for girls who face these problems. As a result, girls have often been forced to flee from abusive situations, but in leaving these situations, they sometimes find themselves victimized again by juvenile justice agencies, which either force them back home or incarcerate them. Those placed in juvenile or adult correctional facilities might then suffer further abuse, neglect, and exploitation—this time at the hands of those who are charged with caring for their needs.[57]

If the needs of female offenders are to be met, policy makers and those responsible for implementing programs must begin to take seriously the reality that gender does influence one's life experiences. As Meda Chesney-Lind and Randal Shelden note, "Girls, in short, experience a childhood and adolescence heavily colored by their gender. It is simply not possible to discuss their problems, their delinquency, and what they encounter in the juvenile justice system without considering gender in all its dimensions."[58] Recognition of the importance of gender is required for the development of the types of interventions that can help girls deal with issues such as sexual and physical abuse, neglect, exploitation, pregnancy, parenting, substance abuse, and family dysfunction. It is also required for the development of the types of aftercare and support services that females need. For example, some girls do not have homes or families that want them or can care for their needs. Consequently, it is important that support networks be developed for girls, as well as housing and employment options that will help them live successfully on their own.[59]

Careful attention needs to be given to *disproportionate minority confinement (DMC)* as well as the over-representation of minority youths at each stage of the juvenile justice process.[60] The reality is that black youths are more likely to be referred to juvenile court than white youths, and once referred, to have their cases result in a formal petition being filed than youths of other races. They are also more likely to be detained. Moreover, although black youths are less likely than white youths to be adjudicated, when they are adjudicated, they are more likely to placed out of their home than white youths.[61]

Many Minority Groups Are Over-Represented in Juvenile Justice

Although Latino, Native American, Asian, and Pacific Island youths do not appear to be over-represented in the juvenile justice process when national-level data are examined, this is *not* the case when data for individual states and jurisdictions is examined.[62]

Of course, the disproportionate over-representation of minority youths in juvenile justice could be the result of their disproportionate involvement in crime. Research that has examined the extent to which incarceration rates reflect crime rates has found little support for this hypothesis. In a large-scale study of the relationship between offense behavior and incarceration, David Huizinga and Delbert Elliott drew the following conclusion: "A summary of the findings would suggest that differences in incarceration rates among racial groups cannot be explained by differences in offense behavior among these groups. . . . There is some indication of differential arrest rates for serious crimes among the racial groups, but further investigation of the relationship of race to arrest and juvenile justice system processing is required if reasons underlying differences in incarceration are to be more fully understood."[63]

A similar conclusion was reached in a study of juvenile incarceration in California conducted by the National Council on Crime and Delinquency. As the researchers noted, "In broad terms, this analysis unveils a picture of persistent differential treatment for some minority groups after having accounted for prereferral factors such as offense and prior record. This leads one to the observation that some ethnic disparities in detention and sentencing outcomes are limited to African-Americans and cannot be fully explained by the juvenile justice attributes of that group."[64] These findings raise serious concerns, because they indicate that, at least in some jurisdictions, institutional racism or racial bias influences decisions about which youths should be incarcerated and results in the disproportionate incarceration of minority youths.

There will, of course, be instances when youths will require out-of-home placement. Consequently, it is imperative that a continuum of services be available for youths who, for at least a short time, or for periods of the day, require more structured placements than are available in their own homes. This continuum of services should consist of primarily community-based programs such as day treatment, foster care, and group home placement, as well as small secure treatment programs for serious juvenile offenders. Unfortunately, too many youths are placed in large locked facilities that provide little effective treatment and almost no aftercare upon their release. Moreover, many community-based programs lack the staff and resources to provide quality services to youths and families. Fortunately, however, there are good community-based and secure residential programs that do more than warehouse youths, and these programs can serve as models for effective correctional intervention.

Strong aftercare programs must also be developed in order for correctional programs to reduce recidivism. Certainly one of the most significant problems facing juvenile corrections today is the need to develop sound aftercare programs for youths placed in residential settings. Unfortunately, aftercare has too often been an afterthought in juvenile corrections. In far too many instances, youths have been placed in institutional settings, then released without careful preparation and planning. Once in the community again, they have often received very little in the way of services, because those responsible for aftercare have large caseloads that preclude meaningful supervision or assistance.

The transition of a youth from a residential setting to the community often requires considerable adjustment on the part of the youth, the parents or guardian, and others in the community, such as teachers, neighbors, and victims. Moreover, in most cases, the environment to which the youth is returning has played a significant role in his or her delinquent behavior. Indeed, chaotic, neglectful, and/or abusive family situations; poor schools; and a scarcity of conventional economic and leisure activities for children are critical problems that need to be addressed. In addition, youths often face a number of system barriers that prevent them from becoming immediately involved in positive structured activities within the community.[65] Consequently, it is important that steps are taken to prepare youths to deal effectively with their environment and to support their efforts to avoid delinquent behavior.

In an effort to support the development of effective aftercare services for youth, the Office of Juvenile Justice and Delinquency Prevention began the Intensive Community-based Aftercare Programs (IAP) initiative in 1988, under the direction of David Altschuler of Johns Hopkins University and Troy Armstrong of California State University, Sacramento. The IAP model, as developed thus far, focuses attention on three areas: (1) planning and preparation for release during placement, (2) implementing a structured transition that requires the participation of institutional and aftercare staff prior to and following community reentry, and (3) engaging in long-term reintegrative activities that provide adequate levels of services to clients and ensure social control.[66] The IAP model objectives include

- Preparing juveniles for progressively increased responsibility and freedom in the community
- Facilitating interaction and involvement between juveniles and the community
- Working with offenders and targeted community support systems (families, peers, schools, employers) on the qualities needed for constructive interactions that advance the juvenile's reintegration into the community
- Developing new resources and support as needed
- Monitoring and testing juveniles and the community on their capacity to deal with each other productively[67]

The IAP model represents a comprehensive approach to aftercare. Unfortunately, in far too many instances, youths do not receive the types of aftercare services they need and deserve.

Box 15-1

Interview: George Page, Southern Regional Administrator, Kentucky Baptist Homes for Children

Age: 57
Education: MSW, University of Tennessee
Salary: $59,800

Q: You have been involved in youth-serving organizations for a number of years. Could you say something about how long you have worked with juvenile offenders and in what capacity?

A: I have been working with delinquent youth for 29 years. I began my career with the Kentucky Department of Child Welfare in 1969 as Assistant Superintendent at the Frenchburg Boys Center. I also served as the Superintendent of Woodsbend Boys Camp and as Superintendent of the Lake Cumberland Boys Camp. From 1991 through 1994, I was the Assistant Director of Treatment for the Kentucky Department of Social Services, Residential Services. I was responsible for 49 different state programs including residential, day treatment, group home, and clinical programs. I presently serve as the Director, Southern Region, Kentucky Baptist Homes for Children.

Q: What are your present job responsibilities with KBHC?

A: My primary responsibilities include developing and overseeing a variety of youth-serving programs in 27 Kentucky counties, which include an 18-bed residential/shelter program, a 24-bed wilderness treatment program, 31 foster care homes, a Family Preservation and Reunification program, and an Evening Reporting Center. I am also responsible for developing and monitoring an annual budget of approximately $3.2 million, fund raising, coordinating the building of new facilities, and overseeing staff development and training.

Q: What are the biggest challenges you face in your work?

A: Probably the biggest challenge we face is finding appropriate placements for youth when they leave our residential programs. In an effort to deal with this we have been developing more treatment foster homes for youth to help them make the transition from placement back to the community. Another challenge I face is developing an annual budget that allocates funds in such a way that the best possible services are delivered to youth in our care, and then staying within that budget.

Q: Over the course of your career, what changes have you seen in how we deal with delinquent youth?

A: Kids today seem to have more emotional and chemical dependency issues. In Kentucky, basically, we are still doing the same thing today that we were doing in 1965 with small residential programs. There are, however, a few more options today than in the past, such as day treatment and family preservation and reunification. Also, more private child care agencies are willing to take delinquents than in the past and treatment planning is more sophisticated today.

Q: Are there any things that you don't particularly like about your job?

A: I don't like dealing with allegations of abuse by staff. Fortunately, this does not occur often.

Q: What do you like about this type of work?

A: I enjoy the challenges of this job. It requires a great deal of creativity to meet the needs of the children and families that we deal with, but I feel that what we do really does make a difference. It's fun.

Q: What qualities does a person need to successfully work with youth in an institutional setting?

A: To be good in this work you have to have two ears that work, lots of patience, a great deal of empathy, a creative mind, lots of energy, and a strong faith. You must also sincerely like kids and have an unquestionable belief that kids can and want to change. Finally, you can't give up on kids.

Juveniles' Right to Treatment in Correctional Institutions

The most frequently used justification for placing juvenile offenders in correctional institutions is that they can receive treatment. Therefore, many people maintain that juvenile correctional institutions have an obligation to provide meaningful treatment. In fact, some attorneys argue that juveniles have a legal right to treatment. As noted earlier, meaningful treatment is often lacking in residential placements.

The legal right of children to receive humane and appropriate treatment in juvenile correctional institutions has been the focus of a number of lawsuits brought on behalf of incarcerated youths since the early 1970s. The first of these cases was *Inmates of the Boys' Training School v. Affleck* (1972).[68] This case brought to light the inhumane conditions that existed in some juvenile institutions, such as the use of solitary confinement, the use of strip cells, and a lack of educational opportunities. The decision of the lower court in this case was that conditions in the institutions violated the prohibition against cruel and unusual punishment spelled out in the U.S. Constitution.[69]

Inhumane conditions have also been pointed out in several subsequent cases. For example, in the 1974 case of *Nelson v. Hayne*, the Seventh Circuit Court of Appeals in Indiana ruled that the use of unnecessary and excessive corporal punishment, the use of tranquilizing drugs to control behavior, and the failure to provide rehabilitative treatment to youths violated the youths' Eighth and Fourteenth Amendment rights.[70] This case was the first appellate court decision to affirm juveniles' right to rehabilitative treatment. In another important 1974 case, *Morales v. Turman*, the court found that the Texas Youth Council violated the Eighth Amendment.[71] Problems addressed by the court included the use of tear gas, the segregation and solitary confinement of residents, enforced silence, and the use of excessive and unnecessary corporal punishments, among others. The court found that youths were denied their right to treatment and indicated that the state has an obligation to rehabilitate delinquent youths.[72]

Although these cases remedied some abuses in juvenile correctional facilities, such problems continue to emerge. In a 1987 case, *Gary H. v. Hegstrom*, a federal judge found that isolation punishments employed at the McClaren School for Boys in Oregon were inappropriate and that juveniles were being denied their right to treatment.[73] More recently, in a 1995 South Carolina case, *Alexander v. Boyd*, the plaintiffs argued that four institutional facilities violated youths' constitutional rights in a number of ways, including these: (1) they relied on lock-up units and CS gas (a potent form of tear gas) for handling discipline problems; (2) they placed youths in padlocked cells in violation of fire safety regulations; (3) they served youths food containing cockroaches and other foreign substances; (4) they failed to provide good medical care; (5) they failed to provide adequate living space; and (6) they failed to provide minimally adequate programming. In its ruling, the district court noted that the purpose of the facilities was to rehabilitate juvenile offenders, and existing practices, in failing to do this, violated constitutional and statutory requirements. As a result, the court ordered that an acceptable plan be developed in order to remedy existing deficiencies.[74]

These cases document continued denial of quality care and treatment in juvenile correctional facilities. Importantly, they have helped rectify many problematic situations in institutions around the country. However, the federal courts have, to date, dealt with these problems on a case-by-case basis and have refused to make sweeping rulings that would affect a wide range of institutions around the country. Moreover, they have refused to make mandatory current professional association standards for the care of youths.

Besides lawsuits brought on behalf of individual youths in institutions or even groups of youths, a mechanism that can be used to protect the rights of youths in institutional settings is the *Civil Rights of Institutionalizied Persons Act (CRIPA)*. This act was authorized in 1980, and it gives the Civil Rights Division within the U.S. Department of Justice the power to take legal action against states or local governments for "violating the civil rights of persons institutionalized in publicly operated facilities."[75] Moreover, the act has been used on a number of occasions to initiate investigations regarding the treatment of children in various juvenile correctional facilities.

The privatization of juvenile corrections needs to be carefully scrutinized. The number of youths admitted to public juvenile correctional institutions has been growing dramatically in recent years, but less dramatically than the number of youths incarcerated in private facilities. For example, from 1991 through 1999, public facility admissions grew 29% while private facility admissions grew 57%.[76] Similarly, the number of delinquent youths placed in private facilities increased 60% between 1991 and 1999, whereas the number of youths placed in public facilities increased 36%.[77]

There are many privately operated juvenile correctional facilities around the country. Some of these facilities contain the most innovative and effective programs for juvenile offenders in the country. Consequently, these institutions represent an important element in the continuum of care for youths committed to residential settings. Nevertheless, it is a mistake to assume that, just because a program is privately operated, it provides quality care. Indeed, there is considerable evidence that some privately operated programs fail to provide quality care to their clients.[78] For example, in one study that examined private facilities in Southern California, David Shichor and Clemens Bartollas found that many agencies that place children in private facilities do not have the staff to adequately monitor the quality of services provided and that many staff in private facilities lacked appropriate qualifications. In addition, they found that the desire of some institutions "to make money from kids" resulted in lowering the number and quality of staff in some programs.[79] A potential problem with private facilities is that the desire to make a profit or keep the program operating acts as an incentive to cut corners and acts as a disincentive to help the residents complete treatment in a timely manner.

Placing juveniles in adult jails is harmful. As noted in Chapter 12, the incarceration of juveniles in adult jails creates a number of problems for both juveniles and jail administrators. Many jails have difficulty meeting the needs of adult inmates, much less the needs of incarcerated juveniles. Sadly, there is a long history of abuse of juveniles in these facilities. Moreover, there is clear evidence that placing juveniles in adult jails places them at risk for self-destructive behaviors.

Jails Are Not Appropriate Placements for Juveniles

The suicide rate for juveniles in adult jails is estimated to be 4.6 times higher than the rate for juveniles in the general population and 7.7 times higher than the rate for youths in juvenile detention facilities. Moreover, there is some evidence that it may even be higher.[80]

In recognition of the problems associated with the jailing of juveniles, the federal government has encouraged states to remove juveniles from jails through passage of the Juvenile Justice and Delinquency Prevention Act. Despite the encouragement, jail removal has not been completely successful. Although progress has been made in some jurisdictions, juveniles are still being incarcerated in adult jails. In fact, there is some pressure from those who argue for a "get-tough" approach to juvenile crime to place more juveniles in adult facilities.[81] As a result of such pressure, a failure by the states to develop juvenile detention facilities, and an unwillingness to develop appropriate alternatives to the jailing of juveniles, the juvenile population in adult jails is once again on the rise in a number of jurisdictions.

The lack of adequate program development and evaluation hinders the development of effective juvenile corrections programs. Few juvenile corrections programs engage in a systematic and ongoing process of evaluating the extent to which program interventions are being implemented as intended and achieving their stated goals and objectives. As a result, many program managers have little information to help them modify and improve the services they provide to youths, and they have little or no information about their programs' effects on recidivism. Many program managers in juvenile corrections would like to have good information, but they face several obstacles in trying to collect it. For one thing, they often lack the necessary expertise to collect and analyze information about program quality and effectiveness. For another, they rarely have adequate budgets for contracting with outside organizations that have the expertise to assist them. Then, there are those program managers who seem to have little interest in collecting data and scrutinizing program operations. The end result is an appalling lack of good information about the quality and effectiveness of services provided to youths placed in juvenile correctional programs.

The Future of Juvenile Justice

Predicting the future is fraught with problems. Nevertheless, if present circumstances are any indication of future trends, then it is possible to make some reasonably educated guesses about what juvenile justice will look like in the near future and the challenges that it will face.

Perhaps the most important issue confronting juvenile justice is whether the juvenile courts should continue to exist. Some have called for the elimination of the juvenile courts on the basis that they are either unwilling or unable to protect the due process rights of juveniles. Critics of the juvenile courts, such as law professor Barry Feld, maintain that the *parens patriae* focus of the juvenile courts have led to court practices that deny juveniles the same due process protections afforded adults, even though the juvenile courts today operate much like adult courts, and juveniles may face the same sanctions as adults. According to Feld, this means that youths receive neither treatment nor justice in the juvenile courts.[82]

Other critics of the juvenile courts claim that they represent an ineffective response to delinquent behavior, especially serious or violent delinquent behavior. According to these critics, the criminal courts are better equipped to deal with chronic, serious, and violent juvenile offenders.

Some juvenile justice experts agree with these critics to an extent—that is, they believe that adult and serious juvenile offenders are not vastly different—but they argue that the juvenile court model should be extended to the criminal courts. Travis Hirschi and Michael Gottfredson, for example, maintain that juvenile and adult offenders have many similarities and that many adult crimes are neither more nor less serious than many juvenile offenses. In addition, they argue that juvenile and adult criminality is the product of low self-control and that adults are as amenable to reform as juveniles. Consequently, they call for one criminal justice system, but one that is constructed along the lines of the *juvenile justice process*. From their point of view, such a system not only would be more cost efficient, but also would provide more consistent responses to offenders and possibly focus more resources on the prevention of criminal behavior.[83]

The debate over the future of juvenile justice will undoubtedly continue for some time. However, in the near future, it is unlikely that the juvenile courts will be abolished, and it is equally unlikely that the criminal courts will be reconstructed according to the juvenile court model. What is probable is that in some quarters considerable energy will be directed toward protecting the status quo, and in others the push will continue to reform juvenile justice practice.[84] Because juvenile courts are political institutions, judges, being elected officials in many cases, are often hesitant to take risks or act in ways that can be perceived as soft on crime. Hence, juvenile courts, like other courts, tend to be rather conservative institutions that approach change cautiously.

Critics on the reform end of the scale include Barry Krisberg and James Austin, who are in favor of "reinventing" juvenile justice. The kind of reinvention they have in mind involves four steps. First, a public health approach to serious juvenile crime would be taken. This approach would place greater emphasis on the environmental factors that contribute to delinquency and on community organizing and self-help strategies in dealing with youth crime. Second, knowledge about child development would be incorporated into juvenile justice policies and procedures. Third, the juvenile justice process would be altered so as to fully protect youths' legal rights. Fourth, "the whole child in his or her family and community context" would be treated by developing coordinated and comprehensive approaches to youth problems.[85]

Other advocates of reform, such as Preston Elrod and Daryl Kelley, have also called for a critical approach to juvenile justice. Elrod and Kelley argue that neither liberal nor conservative reforms are capable of effectively meeting the needs of youths or effectively responding to juvenile crime. Consequently, they call for a different approach, which includes the development of cooperative relationships between social researchers and communities, including the juvenile justice community, in an effort to devise responses to youth problems that are more humane, more effective, and at the same time sensitive to the rights of children.[86] Elrod contends that traditional liberal and conservative approaches to juvenile crime share similar flaws that invariably will produce ineffective juvenile justice policies. For example, one major flaw is the failure of traditional approaches to address the cultural, economic, political, and social factors that contribute to delinquent behavior. Elrod further argues that traditional approaches have employed a very narrow range of methodologies in their efforts to understand delinquency and develop correctional responses. Consequently, he calls for juvenile justice responses that (1) focus attention on the needs of youths and families, (2) work to integrate offenders into the community, (3) limit juvenile justice involvement when possible, (4) protect the rights of children, and (5) are regularly evaluated to determine their effectiveness.[87]

Some change will undoubtedly occur in the future due to the perceived ineffectiveness of existing juvenile justice policies. Moreover, it is possible that some communities will implement some fundamentally different responses to juvenile crime. One response likely to receive attention is the so-called **balanced approach to juvenile justice**. The balanced approach is based on recognition that three important community needs exist: the need to sanction crime, the need to rehabilitate offenders, and the need to ensure public safety. As a result, the balanced approach emphasizes the following three goals for juvenile justice:

1. *Accountability*. The goal is for offenders to take responsibility for their crimes and the harm caused to victims by making amends and restoring losses. Moreover, accountability requires that victims and offenders be involved in the sanctioning process, when feasible.

2. *Competency*. The goal is to require that youths measurably improve those skills that will allow them to function as productive adults (successful rehabilitation). These skills include various educational, social, work, and civic skills.

3. *Public safety*. The goal is for communities to invest more heavily in crime prevention, which includes ensuring that youth offenders placed in institutions spend large amounts of time preparing for their return to the community and that they develop connections with community resources that will help make their return successful.[88]

Because of its focus on the needs of offenders, victims, and the community at large, the balanced approach to juvenile justice will, perhaps, serve as an important bridge to more humane and productive ways of responding to juvenile crime while simultaneously protecting community safety. One thing is for sure, however: Juvenile courts will continue to handle a variety of problems presented by a diverse and growing population of youths over the next decade.

balanced approach to juvenile justice An approach based on the recognition that a community has three needs: to sanction crime, to rehabilitate offenders, and to ensure public safety. The balanced approach emphasizes three corresponding goals for juvenile justice: accountability, competency, and public safety.

■ Problems That Will Continue to Influence America's Youths, Families, Communities, and Juvenile Courts

One of the more obvious, but also insidious, problems that many young people face is the breakup of their families. Moreover, the single parents (usually the mothers) who are left with the primary responsibility for raising these children will confront a number of difficulties. Many of these parents will struggle financially to provide the necessities for their children, who themselves will often suffer hardship. In addition, these single parents will have difficulty supervising their children, because working will take them out of the home. Without a significant adult present, these children will be exposed to a variety of questionable influences.

In other families, youths will be exposed to violence and in many instances will be the victims of domestic conflict. Many children will learn that it is "okay" for men and women to hit one another, and that using violence is an acceptable way to resolve interpersonal disputes. Sadly, significant numbers of children will live in homes where their educational, psychological, and emotional needs will not be met. The scars associated with their treatment will not be visible, but the resulting psychological and social impairment will play an important role in how these youths adapt to their environment. Because children learn how to be parents from their own experiences, harmful parenting practices will be repeated from generation to generation unless the cycle is broken by effective intervention. In the future, youths will continue to face important decisions about sexual relationships, drug use, education, and careers. Many youths will choose drugs as a form of self-medication, to escape from difficult family situations or as a way of dealing with other problems. Along with this drug use will come addiction, serious health problems, and even death. The emotional retardation that can result from serious drug use will take years to overcome. When youths use drugs to keep from feeling painful emotions—emotions that people need to learn about as part of maturing—they fail to benefit from necessary life experiences. Remaining in this drug-induced emotional immaturity can have effects that will carry over into adulthood and impair their ability to parent their own children. Drug use will also bring many young people into contact with individuals involved in a range of illegal activities and will place them in situations where there is an increased probability of violence and arrest.

Many youths will have to deal with the consequences of unprotected sex and unwanted pregnancies. Sexual encounters carry with them a number of serious risks. Sexually transmitted diseases and the emotional entanglements that intimacy can bring will take their toll on many youths. Romantic relationships must be based upon more than physical attraction; when they are not, they are likely to go bad within a short time. In addition, the burden of unwanted or unplanned pregnancies resulting from failed relationships will continue to fall primarily on young girls and their parents or guardians. Having a child is the most important decision that two people can make together, regardless of age or marital status. Many children, however, will continue to have children of their own but without the skills and maturity necessary to provide them the love and nurturing they need.

Many young people will attend schools that possess few resources to meet their varied educational needs. Still others will be exposed to messages, often subtle but sometimes explicit, that they are not capable learners. Far too many youths will find learning boring and school irrelevant. They will likely experience school failure and face diminished prospects for well-paying and meaningful employment. As regards employment prospects, the developing global economy promises tremendous opportunities for some, but there is a real possibility that many others will find themselves competing with low-paid workers in foreign lands who are employed by corporations that have more loyalty to profit than to their employees and communities. Still others will find their lives changed as a result of the trend toward corporate downsizing. Companies are getting "leaner and meaner," eliminating blue-collar and white-collar jobs across the board and requiring the remaining employees to do more and varied tasks.

Economic and political arrangements will ensure that poverty will be a major problem confronting many Americans, both young and old. Consequently, large numbers of individuals will lack access to legal, medical, or educational resources, not to mention the kinds of social services that could help them deal with family or child problems. In these circumstances, many juvenile courts will continue to serve as the only available social service agencies for large numbers of citizens, yet it is doubtful that they will have the resources or expertise to deal with all of their clients' problems.

Gangs will continue to be a serious problem, especially in urban areas. They will continue to serve as substitute families for many youths. For its members, a gang provides a sense of belonging, a kind of status, and an identity, valuable commodities in situations characterized by uncertainty and instability.

Clearly, juvenile justice will face a number of serious challenges throughout the 21st century; however, these problems should not be viewed as insurmountable. In fact, there is considerable room for optimism. For example, the extent of public and juvenile justice support for rehabilitation is an encouraging sign. Although there has been some movement toward making the juvenile court more punitive, it should be remembered that there has always been a punitive side to juvenile justice. Nevertheless, rehabilitation continues to have many adherents in juvenile justice, and we believe that support for it will become even more pronounced in the future.

We believe several factors will increase the amount of attention paid to the treatment of juvenile offenders in the future. First, juvenile justice will become increasingly more professional in its approach to youths, partly because juvenile justice personnel will be better educated and better trained. Second, delinquency will increasingly be seen as a community problem that requires a coordinated and comprehensive community response. This coordinated response will lead to more efforts directed toward delinquency prevention and to the recognition that effective correctional responses are based on the existence of sound preventive mechanisms. To a large extent, the development of more coordinated and comprehensive responses to youths will be driven by the failure of punitive responses to solve juvenile crime and by the realization that they are not cost-effective. Third, the continued professionalization of juvenile justice will lead to more emphasis on program development and evaluation, which in turn will help juvenile justice agencies deal more effectively with their clients. Fourth, continued program development and evaluation will expand our knowledge about what works and for whom, which will increase the ability of juvenile justice to provide effective treatment to youths.

■ Legal Issues

Three significant legal issues will have an ever-increasing impact on juvenile delinquents, their families, and the juvenile justice system. The first is the development of the *family court* and the resulting change in process, procedure, and attitude toward dealing with youths that is associated with the family court model. The second issue is the recognition that *therapeutic jurisprudence* and *restorative justice* are concepts that need to be integrated into the juvenile justice system. Third, but not at all least, is the lack of access to justice faced by many families and children. The reality is that the costs of private or court-appointed legal representation continue to increase, and this jeopardizes the availability and quality of attorneys for juveniles and their families.

The idea of a "family court" where all legal matters involving families can be taken care of (i.e., divorce, custody, child support, visitation, delinquency, and neglect) is not a new idea but one that is gaining increasing acceptance. Combining courts with similar jurisdictions to handle both domestic and delinquency matters can often save money in the long run due to staff reductions and increased efficiency. This is possible because deliquency, child protection, and other domestic relations cases revolve around common issues and require similar types of services.

One of several key characteristics that need to be found in a family court for it to be effective is that the court needs to be outcome-focused without ignoring due process. Services need to be available to treat the multitude of issues that children and families bring to the court. Without the ability to effectively intervene and encourage change, the delinquent and his or her family will continue in their old, dysfunctional patterns of behavior. Second, the family problems need to be identified as early as legally possible, but no later than disposition or sentencing. To accomplish this, the court must have a knowledgeable and experienced intake department so that assessments can be made early on. Access to psychological and substance abuse services is also important in developing an accurate picture of the family. A mechanism has to be in place that forces the extended family, parents, siblings, and influencing relatives to cooperate with assessment and issue identification. Focusing dispositional orders on the juvenile alone is a waste of time. Children are much more likely, on a day-to-day basis, to be influenced by family and family associations than by a judge, referee, or probation officer. Third, the creation of this type of problem identification–centered court requires dedicated leadership from the judiciary. Judges elected to or assigned to the family court must want to be there, and hopefully are not looking to use it as a "stepping stone" to a higher or more publicly visible court. These judges must also be dedicated to staying in family court so that over time they can develop the necessary expertise to be fully effective.

Understanding the subtle connections between delinquency and child neglect and abuse and delinquency and domestic violence requires experience that can be gained only by dealing with the individual cases. These same judges need to be active in their communities in advocating for services and must be creative in utilizing both the public and the private sector. One of the catch phrases of the family court is "one judge; one family." This means that by handling all matters involving the family, from divorce to delinquency, the judge can develop an understanding of each family's individual dynamics and not need to be educated anew for each case. The only drawback to this approach is that it can be emotionally draining and frustrating to see the same family repeatedly, not to mention the second and third generations of that family.

A family court judge must be able to maintain empathy for families that are repeatedly before the court, while at the same time holding all of them accountable. This intersection of accountability and compassion is an important stop on the road map of therapeutic jurisprudence and restorative justice. **Therapeutic jurisprudence** is defined as ". . . the use of social science to study the extent to which a legal rule or practice promotes the psychological and physical well-being of the people it affects."[89] In simple terms, it means that when the court interacts with those persons that come before it, it needs to be aware of both the psychological and legal impact it can have on those people and make certain that the impact is a positive one. It involves the use of court orders directing people to make positive life changes after ensuring that there is a supportive system to help those changes occur. It also involves a willingness to use the power of the court (e.g., contempt and detention) to confront persons with the immediate need to change or face court-imposed sanctions.[91]

A common example of courts engaging in therapeutic jurisprudence is *drug courts*. These courts focus on identifying risk factors associated with the individuals' illegal activity (i.e., drug use) and then dealing with these risk factors. Those involved in drug courts are encouraged by consequences, jail, or prison, to make behavioral changes, and are treated in a supportive and affirming manner by the judge and court personnel rather than in a harsh, more punitive fashion.

One of the keys to effective intervention is accountability. Accountability has two parts—the individual's accepting of responsibility for his or her acts and understanding the impact on the community and the victim, if any, and the court's willingness to hold the person accountable. The court can obviously play a key role in the second part of accountability, holding the person accountable by available sanctions. But just as important, if not more important, is the role the court can play in helping the juvenile to accept responsibility for his or her behavior. If the juvenile enters a plea, the court can refuse to allow a no contest plea, forcing the juvenile to admit in open court, in front of his or her parents and possibly the victim, to what he or she did. Without an up-front acknowledgement of wrongdoing or misbehavior, it is virtually impossible to convince a person that there is some behavior that needs to be changed. Constant reminders of that prior inappropriate behavior may also be necessary to continue the momentum necessary for change.

therapeutic jurisprudence The use of the legal process to promote the psychological and physical well-being of the people it serves.

Including victims in the dispositional process is essential in many instances. Youths involved in illegal behaviors need to be able to understand the impact of their actions on others. They need to know and understand that more than just a law was broken. The peace of their community was violated, and real people were hurt. This can be accomplished in a number of ways:

- Victim impact statements available to the court
- Requirements for restitution by the juvenile and the parent, guardian, or custodian responsible for supervision when the criminal act occurred
- Community service restitution
- Group or offender educational programming
- Face-to-face victim offender mediation
- Focus on restoring a balance in the community

One of the key components of restorative justice is the idea that the offender must also be "restored." The offender must be helped back into balance with his or her community, the victim, and the family. By problem solving to determine the causes of the crime and by addressing those causes, the juvenile offender can "leave" the system in a better position than when he or she entered it. Indeed, the victim, the community, and the offender are all integral players in the idea of restorative justice. "Paying back" both the victim and the community restores the offender as a productive citizen, as well as hopefully increases their knowledge about their community and their understanding of the impact of their behavior on that community.

The third legal issue that needs to be discussed is access to justice. This issue has two parts, one involving the problems surrounding the actual litigant–court interaction and the other involving the access to legal representation. In order for courts to be accessible, they need to be in a physical location that can be conveniently accessed by all. They need to be on a bus line, and have adequate parking and handicap access. As much as possible they should provide as many services in one place as possible so that people who have limited means and difficulties with transportation do not have to unnecessarily spend precious time and money to get to court. Court staff need to be knowledgeable and helpful when citizens have difficulties or need assistance with legal issues. An intake system needs to be available so that they are directed to the correct person or place. Court staff need to understand that courts are created for the citizenry, not for the employees! Making people feel that the courts are available to help them resolve their legal issues is too often an overlooked part of the "access to justice" issue.

The second aspect of the "access to justice" issue is the availability of attorneys for litigants, including juveniles who cannot afford to retain counsel. The U.S. Supreme Court in the case of *Gideon v. Wainwright*, 372 US 335 (1963) held that: ". . . in our adversary system of criminal justice, any person haled into court, who is too poor to hire a lawyer, cannot be assured a fair trial unless counsel is provided for him."[91] At first, the *Gideon* precedent was limited to felony offenses, but now it applies to all criminal offenses (see *Alabama v. Shelton*, 535 US 654, 2002).

The legacy of *In re Gault*, *Gideon*, and *Shelton* would seem to be undeniable— that juveniles charged with any delinquent offense are entitled to counsel. Counsel for minors would seem to be even more of a necessity than for adults in similar circumstances because of the lack of life experience and recognized immaturity. However, the promise of the case law is often unfulfilled. Since the finding of appointed counsel for indigent adult defendants and juvenile offenders is often left to local funding units, the compensation rates for attorneys can vary widely. Equal protection would seem to require equal compensation, but that is not always the case. The great majority of persons arrested in the United States cannot afford an attorney and must rely on the government to provide one. For youths the rate is arguably higher because the presumption is that *all* juveniles are indigent, since by law they cannot own property. Currently 46 states provide some funding for indigent defendants. Of those, 24 provide 100% of the public defense costs. Only 4 states place the responsibility on the local funding unit, which causes the reimbursements to fluctuate widely based upon the economic health of the individual counties. Although most people recognize that criminal defendants and juvenile offenders deserve representation, it is hard to convince politicians that they should spend more than the bare minimum on "criminals" who they believe are probably guilty anyway.

Who provides legal representation may also vary. It may be members of the private bar who are paid an hourly rate. That rate may be limited to "in court time," so that legal research and case preparation are not compensated. Some jurisdictions use public defenders who are salaried attorneys rather than paid by the hour or by the case. Often public defenders are overwhelmed by case loads and cannot give each defendant or juvenile offender the time needed. If the attorney is paid by the case (e.g., $500 for a simple or class 1 felony), then there is more incentive to plea bargain than go to trial. All of the compensation methods have flaws that affect the quality of legal representation and thus the legitimate access to justice for the accused.

Problem-solving family courts, where families can go to have their problems addressed, are clearly a goal for the future. These courts must be litigant friendly, easy to access, and all in one location. They should be staffed by judges and court personnel who treat everyone with civility, dignity, and respect; who are focused on protecting rights and identifying problems; and who believe in accountability and a person's ability to change. Through these "new" courts the historical purpose of the juvenile court may be realized. In such courts, children would be given a more realistic opportunity to overcome youthful mistakes so that they can become contributing adult citizens.

■ Chapter Summary

The purpose of this chapter was to (1) examine historical changes in the ways that juvenile courts have dealt with problem youths, (2) review problems that currently confront contemporary juvenile justice agencies, and (3) consider issues that are likely to confront juvenile justice in the near future. Clearly, juvenile justice practice has changed over time. Although juvenile courts were developed as informal chancery courts and had little concern for due process protections, U.S. Supreme Court decisions during the mid-1960s and early 1970s required juvenile courts to take due process more seriously. Nevertheless, the Supreme Court balked at giving children all of the due process protections afforded adults. As a result, juvenile courts continue to be more informal than criminal courts, causing some critics to call for their reform or abolishment.

One type of reform effort that began in the late 1970s was the "get tough" movement. This movement was led by reformers who urged the abandonment of the juvenile court's traditional *parens patriae* philosophy, with its focus on serving the best interests of children. Instead of the traditional *parens patriae* approach to juvenile justice, "get tough" reformers sought to substitute a more punitive philosophy, and some even called for the abolishment of the juvenile court. The "get tough" movement continues to influence the development of juvenile justice policy today, although rehabilitation and treatment still maintain a strong foothold in juvenile justice, at least in rhetoric if not always in actual practice.

Today, the juvenile courts, despite numerous operational changes, continue to be plagued by many problems of long standing: the failure to protect the due process rights of children, a lack of appropriate responses to status offenders, the processing of many youths who could be diverted from juvenile justice, the lack of coordinated resources for responding to problem youths, and the failure to recognize the limits of juvenile justice in solving youth problems. Juvenile corrections also continues to face significant problems: the existence of ineffective programs; inadequately trained, paid, and motivated staff; a penchant for relying on simplistic panaceas; a paucity of interventions that address the special needs of girls; the disproportionate processing and incarceration of minority youths; a shortage of good aftercare programs for youths; a failure to give sufficient attention to youths' right to treatment; a reliance on the privatization of juvenile corrections; the continued jailing of juveniles; and the lack of sound **program development and evaluation**.

Perhaps the most important issue currently facing juvenile justice is whether the juvenile courts should continue to exist. As noted earlier, some have called for the elimination of the juvenile courts, arguing that they are either unwilling or unable to protect the due process rights of youths. Still others hold that they do not constitute an effective response to juvenile crime, particularly serious juvenile crime. There are those who argue that the juvenile court needs fundamental change, whereas others argue that adult courts should be patterned after the juvenile court model.

Although the future will continue to see considerable debate and conflict over the juvenile courts, it seems likely that they will continue to exist with some changes. One possible change is a move toward a more balanced approach to juvenile justice, one that stresses accountability for crimes, helping offenders develop competencies to enable them to lead productive lives, and protecting public safety. Another possible change is toward more formality in juvenile court operations.

No matter what the changes in their operation, juvenile courts will continue to serve as primary social agencies responsible for dealing with a variety of problems that confront children, families, schools, and communities. These problems include family dissolution; child abuse and neglect; drug abuse; poor educational, recreational, and employment opportunities; and poverty. The good news is that juvenile justice will become more professional, it will employ better educated and

program development and evaluation
A systematic process of collecting and using data to determine program effectiveness, with the goal of making changes that will improve the level of effectiveness.

trained personnel, more emphasis will be placed on preventing delinquency, and the linkage between prevention and effective corrections will become clearer. Moreover, more attention will be given to program development and evaluation, which will expand our knowledge of effective juvenile justice interventions.

There are those who doubt the continued existence of juvenile justice. Our belief is that, on the contrary, juvenile justice is likely to become even more vital in the near future. In any case, one thing is certain: The future of juvenile justice ultimately depends on its ability to deliver on its promise of serving the best interests of children and protecting the community.

KEY CONCEPTS

aftercare Services delivered to youths after they are released from a residential setting and while they are still under the jurisdiction of the state or juvenile court.

balanced approach to juvenile justice An approach based on the recognition that a community has three needs: to sanction crime, to rehabilitate offenders, and to ensure public safety. The balanced approach emphasizes three corresponding goals for juvenile justice: accountability, competency, and public safety.

chancery court English court concerned primarily with property rights. Over time these courts became more involved in the welfare of families and children, and served as a model for the development of the juvenile court.

child savers Affluent 19th-century reformers, who were concerned about controlling and training youths so that they could become productive, law-abiding citizens. The child savers played a major role in the establishment of the juvenile courts.

continuum of quality services A range of effective services consisting of prevention and correctional responses for deliquent youth.

due process An established course of proceedings designed to protect individual rights and liberties and to ensure that people are treated with fairness before the law. Although juveniles have most of the due process protections afforded adults in the criminal courts, they do not have all of these rights. Moreover, because of the informality found in many juvenile courts, legal protections that are theoretically available may in fact be unavailable.

"get tough" movement A legal reform movement that began during the late 1970s. The goals of the movement included developing a more punitive juvenile court orientation or abolishing the juvenile courts so that more severe sanctions could be leveled against juvenile offenders.

informal juvenile court A court that places little emphasis on due process protections for youths. Because juvenile courts exist, in theory, to protect the best interests of children, legal protections for children are felt to be unnecessary within an informal court.

meta-analysis A statistical technique that makes possible statements about the effects of interventions across a number of different studies.

parens patriae A legal doctrine according to which the state has an obligation to protect and serve the interests of children.

program development and evaluation A systematic process of collecting and using data to determine program effectiveness, with the goal of making changes that will improve the level of effectiveness.

rehabilitation The process of restoring an individual's ability to play a constructive role in society through education, training, or therapy.

therapeutic jurisprudence The use of the legal process to promote the psychological and physical well-being of the people it serves.

youth discount A stratagem for taking into account the supposed diminished culpability of youths (based on their immaturity) in the sentencing process. The use of youth discounts would generally result in shorter sentences for juveniles than for adults who had committed identical offenses.

REVIEW QUESTIONS

1. In what ways have juvenile justice practices changed since the development of the first juvenile courts?

_____

_____

_____

_____

_____

2. What were the concerns of early child savers, and how did these concerns lead to the development of the juvenile courts?

_____

_____

_____

_____

_____

3. What was the position taken by the U.S. Supreme Court during the mid-1960s and early 1970s toward the informal character of the juvenile courts?

_____

_____

_____

_____

_____

4. What is the "get tough" movement, and how has it affected juvenile justice practice?

_____

_____

_____

_____

_____

5. How has rehabilitation fared in the face of the "get tough" movement?

_____

_____

_____

_____

_____

6. What are some of the most important problems that presently face the juvenile courts?

_____

_____

_____

_____

_____

7. What are some of the most important problems that presently face juvenile corrections?

_____

_____

_____

_____

_____

8. What are the essential components of the Intensive Aftercare Program (IAP) model?

_____

_____

_____

_____

_____

9. What arguments support the abolishment of the juvenile courts?

_____

_____

_____

_____

_____

10. What arguments support the continuation of the juvenile courts?

_____

_____

_____

_____

_____

11. What are the primary assumptions of the balanced approach to juvenile justice? Do you agree with the authors' prediction that it will receive substantial attention in the future?

_____

_____

_____

_____

_____

12. What problems will continue to confront the juvenile courts? How will these problems affect the juvenile courts?

_____

_____

_____

_____

_____

ADDITIONAL READINGS

Altschuler, D. M. and T. L. Armstrong. 1996. "Aftercare Not Afterthought: Testing the IAP Model." *Juvenile Justice* 3 (Washington, DC: Office of Juvenile Justice and Delinquency Prevention): 15–22.

Antonowicz, D. H. and R. R. Ross. 1994. "Essential Components of Successful Rehabilitation Programs for Offenders." *International Journal of Offender Therapy and Comparative Criminology* 38: 97–104.

Bazemore, G. and S. E. Day. 1996. "Restoring the Balance: Juvenile and Community Justice." *Juvenile Justice* 3 (Washington, DC: Office of Juvenile Justice and Delinquency Prevention): 3–14.

Bernard, T. J. 1992. *The Cycle of Juvenile Justice*. New York: Oxford University Press.

Chesney-Lind, M. and R. G. Shelden. 2004. *Girls, Delinquency, and Juvenile Justice*. 3rd ed. Belmont, CA: Thomson/Wadsworth.

Elrod, P. and D. Kelley. 1995. "The Ideological Context of Changing Juvenile Justice." *Journal of Sociology and Social Welfare* 22: 57–75.

Feld, B.C. 1993. "Juvenile (In)Justice and the Criminal Court Alternative." *Crime and Delinquency* 39:403–424.

Gendreau, P. and C. Goggin. 2000. "Correctional Treatment: Accomplishments and Realities." In *Correctional Counseling and Rehabilitation*, 4th ed., eds. P. Van Voorhis et al. Cincinnati, OH: Anderson.

Krisberg, B. 2005. *Juvenile Justice, Redeeming Our Children*. Thousand Oaks, CA: Sage.

Krisberg, B. and J. F. Austin. 1993. *Reinventing Juvenile Justice*. Newbury Park, CA: Sage.

Platt, A. M. 1977. *The Child Savers: The Invention of Delinquency*. 2nd ed. Chicago: University of Chicago Press.

Puritz, P. et al. 1995. *A Call for Justice: An Assessment of Access to Counsel and Quality of Representation in Delinquency Proceedings*. Washington, DC: American Bar Association.

Ryerson, E. 1978. *The Best Laid Plans: America's Juvenile Court Experiment*. New York: Hill and Wang.

NOTES

1. A. M. Platt, *The Child Savers: The Invention of Delinquency*, 2nd ed. (Chicago: University of Chicago Press, 1977).

2. Julia Lathrop, quoted in R. M. Mennel, *Thorns and Thistles* (Hanover, NH: University Press of New England, 1973), 129.

3. T. J. Bernard, *The Cycle of Juvenile Justice* (New York: Oxford University Press, 1992).

4. *People v. Turner*, 55 Ill. 280 (1870).

5. Bernard, *The Cycle of Juvenile Justice*.

6. E. Ryerson, *The Best Laid Plans: America's Juvenile Court Experiment* (New York: Hill and Wang, 1978).

7. Bernard, *The Cycle of Juvenile Justice*.

8. J. M. Mack, "The Juvenile Court," *Harvard Law Review* 23 (1909): 104.

9. Ryerson, *The Best Laid Plans*.

10. Bernard, *The Cycle of Juvenile Justice*; B. Krisberg and J. F. Austin, *Reinventing Juvenile Justice* (Newbury Park, CA: Sage, 1993).

11. Krisberg and Austin, *Reinventing Juvenile Justice*, 31–32.

12. *Kent v. United States*, 383 U.S. 541; 86 S. Ct. 1045; 16 L. Ed. 2d 84 (1966).

13. *In re Gault*, 387 U.S.1; 87 S. Ct. 1428; 18 L. Ed.2d 527 (1967).

14. *In re Winship*, 397 U.S. 358; 90 S. Ct. 1068; 25 L. Ed. 2d 368 (1970).

15. *In re Gault*, 387 U.S. at 15, 18, 19, and 28 (1967).

16. *McKeiver v. Pennsylvania*, 403 U.S. 528; 91 S. Ct. 1976; 29 L. Ed. 2d 647 (1971).

17. Bernard, *The Cycle of Juvenile Justice*.

18. Krisberg and Austin, *Reinventing Juvenile Justice*.

19. P. Elrod and D. Kelley, "The Ideological Context of Changing Juvenile Justice," *Journal of Sociology and Social Welfare* 22 (1995): 57–75.

20. Krisberg and Austin, *Reinventing Juvenile Justice*.

21. B. Krisberg et al., "The Watershed of Juvenile Justice Reform," *Crime and Delinquency* 32 (1986): 5–38; National Criminal Justice Association, *Juvenile Justice Reform Initiatives in the States, 1994–1996* (Washington, DC: Office of Juvenile Justice and Delinquency Prevention, 1997); I. M. Schwartz, *(In)Justice for Juveniles* (Lexington, MA: Lexington Books, 1989).

22. Krisberg and Austin, *Reinventing Juvenile Justice* ; H.T. Rubin, "The Emerging Prosecutor Dominance of the Juvenile Court Intake Process," *Crime and Delinquency* 26 (1980): 299–318; M. Sickmund et al., *Juvenile Offenders and Victims: 1997 Update on Violence* (Washington, DC: Office of Juvenile Justice and Delinquency Prevention, 1997).

23. B. C. Feld, "Juvenile (In)Justice and the Criminal Court Alternative," *Crime and Delinquency* 39 (1993): 403–424.

24. F .T. Cullen et al., "Is Child Saving Dead? Attitudes toward Juvenile Rehabilitation in Illinois," *Journal of Criminal Justice* 11 (1983): 1–13; Krisberg and Austin, *Reinventing Juvenile Justice*; K. Maguire and A.L. Pastore, eds., *Sourcebook of Criminal Justice Statistics, 1995* (Washington, DC: U.S. Government Printing Office, 1996).

25. Cullen et al., "Is Child Saving Dead?"; Krisberg and Austin, *Reinventing Juvenile Justice*.

26. Maguire and Pastore, eds., *Sourcebook of Criminal Justice Statistics, 1995*.

27. G. Bazemore and L. Feder, "Rehabilitation in the New Juvenile Court: Do Judges Support the Treatment Ethic?" *American Journal of Criminal Justice* 21 (1997): 181–212; Cullen et al., "Is Child Saving Dead?"; M. M. Moon et al., "Is Child Saving Dead? Public Support for Juvenile Rehabilitation," *Crime and Delinquency* 46 (2000): 38–60.

28. Bazemore and Feder, "Rehabilitation in the New Juvenile Court."

29. P. Elrod et al., "Toward an Understanding of State Operated Juvenile Institutions: Preliminary Descriptive Analysis of a Survey of Chief Correctional Administrators" (paper presented at the annual meeting of the American Society of Criminology, New Orleans, November 1992).

30. Bernard, *The Cycle of Juvenile Justice*.

31. B. C. Feld, "*In re Gault* Revisited: A Cross-State Comparison of the Right to Counsel in Juvenile Court," *Crime and Delinquency* 34 (1988): 393–424; P. Puritz et al., *A Call for Justice: An Assessment of Access to Counsel and Quality of Representation in Delinquency Proceedings* (Washington, DC: American Bar Association, 1995); J. B. Sanborn, Jr., "Remnants of Parens Patriae in the Adjudicatory Hearing: Is a Fair Trial Possible in the Juvenile Court?" *Crime and Delinquency* 40 (1994): 599–615.

32. M. A. Bortner, *Inside a Juvenile Court: The Tarnished Ideal of Individualized Justice* (New York: New York University Press, 1982); Puritz et al., *A Call for Justice*.

33. Feld, "Juvenile (In)Justice and the Criminal Court Alternative"; B. C. Feld, *Justice for Children: The Right to Counsel and the Juvenile Courts* (Boston: Northeastern University Press, 1993).

34. Feld, "Juvenile (In)Justice and the Criminal Court Alternative," 418.

35. Feld, "Juvenile (In)Justice and the Criminal Court Alternative."

36. E. M. Schur, *Radical Nonintervention: Rethinking the Delinquency Problem* (Englewood Cliffs, NJ: Prentice-Hall, 1973).

37. B. Bullington et al., "A Critique of Diversionary Juvenile Justice," *Crime and Delinquency* 24 (1978): 59–71; F. D. Dunford, *National Evaluation of Diversion Projects* (Washington, DC: Office of Juvenile Justice and Delinquency Prevention, 1981); T. Palmer and R. V. Lewis, *An Evaluation of Juvenile Diversion* (Cambridge, MA: Oelgeschlager, Gunn & Hain, 1980).

38. W. C. Bailey, "Correctional Outcome: An Evaluation of 100 Reports," *Journal of Criminal Law, Criminology, and Police Science* 57 (1966): 153–160; R. Martinson, "What Works? Questions and Answers about Prison Reform," *Public Interest* 35 (1974): 22–54; J. Robinson and G. Smith, "The Effectiveness of Correctional Programs," *Crime and Delinquency* 17 (1971): 67–80; W. E. Wright and M. C. Dixon, "Community Prevention and Treatment of Delinquency," *Journal of Research in Crime and Delinquency* 14 (1977): 35–67.

39. Martinson, "What Works?," 23.

40. J. T. Whitehead and S. P. Lab, "A Meta-Analysis of Juvenile Correctional Treatment," *Journal of Research in Crime and Delinquency* 26 (1989): 276–295.

41. P. Gendreau and B. Ross, "Effective Correctional Treatment: Bibliotherapy for Cynics," *Crime and Delinquency* 25 (1979): 463–489; P. Gendreau and R. R. Ross, "Correctional Treatment: Some Recommendations for Effective Intervention," *Juvenile and Family Court Journal* 34 (1983): 31–39; L. Genevie et al., "How Effective Is Correctional Intervention?" *Social Policy* 16 (1986): 52–57.

42. C. J. Garrett, "Effects of Residential Treatment on Adjudicated Delinquents: A Meta-Analysis," *Journal of Research in Crime and Delinquency* 22 (1985): 287–308.

43. T. Palmer, "Martinson Revisited," *Journal of Research in Crime and Delinquency* 12 (1975): 133–152.

44. See, for example, T. Palmer, *The Re-emergence of Correctional Intervention* (Newbury Park, CA: Sage, 1992).

45. G. R. Grissom and W. L. Dubnov, *Without Locks and Bars: Reforming Our Reform Schools* (New York: Praeger, 1989); S. W. Henggeler, "Treating Serious Anti-Social Behavior in Youth," *OJJDP Juvenile Justice Bulletin*, May 1997; J. C. Howell, ed., *Guide for Implementing the Comprehensive Strategy for Serious, Violent, and Chronic Juvenile Offenders* (Washington, DC: Office of Juvenile Justice and Delinquency Prevention, 1995); A. L. Schneider, "Restitution and Recidivism Rates of Juvenile Offenders: Results from Four Experimental Studies," *Criminology* 24 (1995): 533–551.

46. D. H. Antonowicz and R. R. Ross, "Essential Components of Successful Rehabilitation Programs for Offenders," *International Journal of Offender Therapy and Comparative Criminology* 38 (1994): 97–104; P. Gendreau and C. Goggin, "Correctional Treatment: Accomplishments and Realities," in *Correctional Counseling and Rehabilitation*, 4th ed., ed. P. Van Voorhis et al. (Cincinnati, OH: Anderson, 2000).

47. See, for example, J. G. Miller, *Last One over the Wall: The Massachusetts Experiment in Closing Reform Schools* (Columbus, OH: Ohio State University Press, 1991).

48. See, for example, C. E. Silberman, *Criminal Violence, Criminal Justice* (New York: Random House, 1978).

49. J. O. Finckenauer, *Scared Straight! and the Panacea Phenomenon* (Englewood Cliffs, NJ: Prentice-Hall, 1982).

50. Finckenauer, *Scared Straight! and the Panacea Phenomenon*.

51. M. Peters et al., *Boot Camps for Juvenile Offenders: Program Summary* (Washington, DC: Office of Juvenile Justice and Delinquency Prevention, 1997).

52. Peters et al., *Boot Camps for Juvenile Offenders*.

53. Peters et al., *Boot Camps for Juvenile Offenders*.

54. M. Morash and L. Rucker, "A Critical Idea of Boot Camp as a Correctional Reform," *Crime and Delinquency* 36 (1990): 204–222.

55. C. Puzzanchera et al., *Juvenile Court Statistics 1999* (Washington, DC: Office of Juvenile Justice and Delinquency Prevention, 2003).

56. M. Chesney-Lind and R. G. Shelden, *Girls, Delinquency, and Juvenile Justice*, 3rd ed. (Belmont, CA: Thomson/Wadsworth, 2004).

57. M. Chesney-Lind, "Girls in Jail," *Crime and Delinquency* 34 (1988): 150–168; Schwartz, *(In)Justice for Juveniles*.

58. M. Chesney-Lind and R. G. Shelden, *Girls, Delinquency, and Juvenile Justice*, 2nd ed. (Belmont, CA: West/Wadsworth, 1998): 241–242.

59. Chesney-Lind and Shelden, *Girls, Delinquency, and Juvenile Justice*, 3rd ed.

60. Krisberg and Austin, *Reinventing Juvenile Justice*.

61. Puzzanchera et al., *Juvenile Court Statistics 1999*.

62. See B. Krisberg, *Juvenile Justice, Redeeming Our Children* (Thousand Oaks, CA: Sage, 2005).

63. D. Huizinga and D.S. Elliott, "Juvenile Offenders: Prevalence, Offender Incidence, and Arrest Rates by Race," *Crime and Delinquency* 33 (1987): 206–223.

64. Krisberg and Austin, *Reinventing Juvenile Justice*, 127.

65. C. Lehman, "Oregon Initiative for Reintegrating Adjudicated Youth," *Reaching Today's Youth* 2 (1997): 65–68.

66. D. M. Altschuler and T. L. Armstrong, "Aftercare Not Afterthought: Testing the IAP Model," *Juvenile Justice* 3 (1996): 15–22.

67. Altschuler and Armstrong, "Aftercare Not Afterthought: Testing the IAP Model," 16.

68. *Inmates of Boys' Training School v. Affleck*, 356 F. Supp. 1354 (D.R.I. 1972).

69. R. V. del Carmen et al., *Briefs of Leading Cases in Juvenile Justice* (Cincinnati, OH: Anderson, 1998).

70. *Nelson v. Hayne*, 491 F. 2d 353 (7th Cir. 1974).

71. *Morales v. Turman*, 383 F. Supp. 53 (E.D. Tex. 1974).

72. G. Hemmens et al., *Significant Cases in Juvenile Justice* (Los Angeles, CA: Roxbury, 2004).

73. *Gary H. v. Hegstrom*, 831 F. 2d 1430 (1987).

74. *Alexander v. Boyd*, 876 F. Supp. 773 (1995).

75. P. Puritz and M. A. Scali, *Beyond the Walls, Improving the Conditions of Confinement for Youth in Custody* (Washington, DC: Office of Juvenile Justice and Delinquency Prevention, 1998), 1.

76. Austin et al., *Juveniles Taken into Custody, Fiscal Year 1993*.

77. M. Sickmund, "Juveniles in Corrections" *Juvenile Offenders and Victims, National Report Series* (Washington, DC: Office of Juvenile Justice and Delinquency Prevention, 2001).

78. Schwartz, *(In)Justice for Juveniles*.

79. D. Shichor and C. Bartollas, "Private and Public Juvenile Placements: Is There a Difference?" *Crime and Delinquency* 36 (1990): 286–299.

80. Community Research Associates, "Juvenile Suicides in Adult Jails: Findings from a National Survey of Juveniles in Secure Detention Facilities," in *Juvenile Delinquency: A Justice Perspective*, ed. R. G. Culbertson (Prospect Heights, IL: Waveland, 1985); J. M. Memory, "Juvenile Suicides in Secure Detention Facilities: Correction of Published Rates," *Death Studies* 13 (1989).

81. Schwartz, *(In)Justice for Juveniles*, 18, 81–82.

82. Feld, "Juvenile (In)Justice and the Criminal Court Alternative"; B. Feld, "The Punitive Juvenile Court and the Quality of Procedural Justice: Disjunctions between Rhetoric and Reality," *Crime and Delinquency* 36 (1990): 443–466.

83. T. Hirschi and M. Gottfredson, "Rethinking the Juvenile Justice System," *Crime and Delinquency* 39 (1993): 262–271.

84. See, for example, Krisberg and Austin, *Reinventing Juvenile Justice*.

85. Krisberg and Austin, *Reinventing Juvenile Justice*, 186.

86. Elrod and Kelley, "The Ideological Context of Changing Juvenile Justice."

87. P. Elrod, "The Similarities in Conservative and Liberal Juvenile Justice Policies," in *Cutting the Edge: Current Perspectives in Radical/Critical Criminology and Criminal Justice*, ed. J.I. Ross (Westport, CT: Praeger, 1998).

88. G. Bazemore and S. E. Day, "Restoring the Balance: Juvenile and Community Justice," *Juvenile Justice* 3 (1996).

89. C. Slobogen, "Therapeutic Jurisprudence: Five Dilemmas to Ponder," *Psychology, Public Policy and Law* (1995): 193, 196

91. P. F. Hora, et al., "Therapeutic Jurisprudence and the Drug Treatment Court Movement: Revolutionizing the Criminal Justice System's Response to Drug Abuse and Crime in America," *Notre Dame Law Review*, 24 (1999): 439–537.

92. *Gideon v. Wainwright*, 372 U.S. 335 (1963).

Working in Juvenile Justice

Chapter Objectives

After studying this chapter, you should be able to:

- Describe those factors that contribute to conflict between juvenile justice practitioners

- Describe how factors such as race, ethnicity, gender, social class, education, governmental association, and age can influence relationships between juvenile justice practitioners and their clients

- Describe how the offenses that youths commit can influence the relationship between juvenile justice practitioners and their clients

- Explain why coercive power makes juvenile justice agencies different than other social welfare agencies

- Describe the ways in which clients may be harmed through the exercise of coercive and persuasive force

- Explain why due process is important in juvenile justice, and indicate the four elements that guide practitioners in their efforts to ensure that juvenile justice clients are treated fairly and justly

- Explain what is meant by the presumption of innocence, and describe why it is important to ensure the fair treatment of youths and families

- Describe why conflicts of interest or the appearance of these conflicts should be avoided

- Explain the importance of maintaining confidentiality, and describe steps that practitioners can take to ensure confidentiality

- Describe how different types of bias can adversely influence juvenile justice practice

- Explain how economics can lead to ethical problems in juvenile justice

- Describe how ethics affect the quality of juvenile justice practice
- Describe the basic job functions of intake officers, psychologists, substance abuse counselors, court administrators, judges, and casework and child care staff
- Describe the required skills for effective juvenile justice practice

Chapter Outline

- Introduction
- Dealing with Conflict in Juvenile Justice
- Ethics in Juvenile Justice
- Critical Positions in Juvenile Justice
- Required Skills for Effective Juvenile Justice Practice
- Chapter Summary
- Key Concepts
- Review Questions
- Additional Readings
- Notes

■ Introduction

The purpose of this chapter is to review important jobs that are performed in juvenile justice and to examine the skills needed for effective juvenile justice practice. Although descriptions of the essential functions of many of the important positions found in juvenile justice are provided throughout this text, the objective of this chapter is to examine practical and professional requisites for successfully negotiating the everyday work of juvenile justice and to provide additional information on some of the important positions found in juvenile justice agencies. The first order of business, however, is to explore two topics that receive far too little attention in the classroom, the courtroom, and the community: the importance of dealing with conflicts of interest in juvenile justice and the role of **ethics** in juvenile justice practice. Our belief is that how people go about their jobs in juvenile justice is as important as what is achieved—that the "end" result does not always justify the "means," and that the process is an important part of the result.

ethics The principles of right conduct. These principles determine which actions are right and which are wrong.

■ Dealing with Conflict in Juvenile Justice

Very few people seek work in juvenile justice without a sincere desire to help youths and their families. The process of helping others, however, is often complicated by a number of factors. People who work in juvenile justice must, at times, deal with conflict between their own personal goals and those of others with whom they have contact, including coworkers. In addition, cultural and social factors, as well as a juvenile justice practitioner's gender, race, age, and association with governmental power (e.g., the power of the police, courts, and correctional institutions), also complicate the practitioner's dealings with others. How the practitioner handles these issues, which is colored by his or her individual and professional values, largely determines how effective the practitioner is in his or her job and how much satisfaction he or she receives from doing it.

Conflicts between Juvenile Justice Practitioners

As we noted in Chapter 1, *conflict* is an inherent feature of juvenile justice practice. In fact, the juvenile justice process can be best conceptualized as a mechanism for resolving conflicts. Most of the conflicts are not terribly serious, and these can often (though not always) be resolved with relatively little expenditure of time and other resources. Some conflicts, on the other hand, involve serious criminal behaviors and result in serious property damage, personal injury, and even death. These conflicts usually require considerable expenditure of time and resources.

MYTH VS. REALITY

Status Offenses and Minor Deliquent Offenses May Require the Expenditure of Large Amounts of Resources

Myth—Status offenses and other minor types of delinquent behaviors require very little expenditure of time and energy on the part of juvenile justice practitioners.

Reality—Many status offense cases and those involving minor delinquent behaviors can be quickly and effectively resolved. Many others, however, require considerable expenditure of time and other resources. For example, status offense cases that result from long-standing family dysfunction can be very difficult and time-consuming to mediate.

Because juvenile justice is essentially a *conflict resolution process*, disagreement over the best way to handle particular cases is a normal part of everyday juvenile justice practice. Such disagreement is found in interactions that take place between treatment providers and probation officers, between law enforcement officers and intake personnel, between judges and probation staff, and between defense and prosecuting attorneys, among others. For example, a probation officer may recommend that a youth be placed on probation in his or her home, believing that this is the best disposition for the youth and the community. However, the judge hearing the case may decide, in agreement with the prosecuting attorney, that a residential placement is in the best interests of the youth and the community. In another instance, a police officer called to the scene of a minor altercation between two youths may think that a formal complaint is unnecessary but will file one anyway because the complainant demands that an arrest be made.

Conflicts such as those just described are common in juvenile justice because those involved—whether victims, offenders, witnesses, or juvenile justice professionals—do not always look at things from the same perspective. The police officer in the previous example may view the altercation as a typical disagreement between two youths that could be effectively dealt with by parental intervention. From the perspective of one of the participants, however, it may look like an assault that should be handled by the legal authorities. The other participant may see it as a matter of self-defense. The important thing to note about cases that come to the attention of juvenile justice practitioners is that rarely will all parties involved in a case look at it in the same way.

Conflicts also arise within juvenile justice practice because different practitioners have different roles and responsibilities. Indeed, effective operation of the juvenile justice process often depends on conflicting perspectives. The most obvious example of this is the conflict that occurs as a result of the different roles and responsibilities of defense and prosecuting attorneys. However, there are some role differences between any two juvenile justice positions. As a result, probation officers and treatment providers, judges and probation officers, or police officers and intake workers should not be expected to agree as to how cases should be handled. A probation officer may have a rather different picture of a client than does a judge because of the amount of time the probation officer has spent with the youth, the family, and school officials trying to develop a good understanding of the case. Similarly, a probation officer, who sees the youth in the community, may have a very different view than a detention worker, who will have only interacted with the youth in a secure setting. This is not to say that one perspective is best. It is simply to acknowledge that differences in perspectives can be expected.

Of course, there is also some overlap in the responsibilities of those who work in juvenile justice. Everyone, regardless of position, is obligated to act with **integrity** in carrying out his or her responsibilities. Thus, the goal of the prosecuting attorney is not to win cases at any cost. The prosecuting attorney should not ignore evidence of a defendant's innocence in order to claim another successful prosecution. Nor should a probation officer omit information regarding a client's violations of probation in order to increase the chances that the judge will agree with the probation officer's **dispositional recommendation**. Regardless of the fact that these individuals have different roles and responsibilities, they both have an obligation to act with integrity and to uphold the basic legal and moral principles upon which juvenile justice is founded.

Even when individuals act with integrity, however, different outcomes are possible. In the example in which the judge and the probation officer disagree about the proper placement of a juvenile offender, each may have considered the options carefully and become fully convinced of the correctness of his or her opinion. The decision by the judge to place a youth in a residential setting despite a different recommendation by the probation officer may be because the judge feels that a residential placement is the best placement for the youth and the community, which is exactly what drove the probation officer to recommend placement at home. Because of the judge's position of authority, the judge's wishes will prevail, but this is not to say that the judge's opinion is the only one that a

integrity The quality of acting in a moral, sincere, and honest way.

dispositional recommendation A recommendation made to the court about how a case should be handled.

reasonable person could have arrived at. Clearly, two different people acting with integrity may make different recommendations about a case. Juvenile justice decision making, it should be clear, is fraught with conflicts. Moreover, these conflicts can arise between people who are acting with integrity and who are following what they believe to be the most appropriate course of action. The question then arises as to how these conflicts should be dealt with. There is no simple answer, except to say that juvenile justice practitioners have an obligation to act with professional integrity.

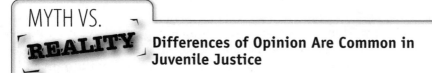

MYTH VS. REALITY Differences of Opinion Are Common in Juvenile Justice

Myth—It is necessary for everyone to agree about the best way to handle a case.

Reality—It is important that juvenile justice practitioners recognize that honest, decent individuals acting with integrity do not always reach the same conclusions. Because individuals are called upon to provide honest and candid appraisals of issues that arise in juvenile justice practice, disagreements are to be expected. When disagreements between persons acting with integrity occur, they must be respected.

Of course, respecting another's differing opinion is usually more palatable when one is confident that the other person is well meaning, competent, and honest. If the person is felt to lack these qualities, respecting his or her opinion can be difficult. Unfortunately, juvenile justice, like other social domains, contains some people who are neither competent nor honest. Indeed, the history of juvenile justice, like the history of criminal justice, is replete with individuals who harmed youths in their care and themselves engaged in unprofessional and even illegal behavior.

Malevolence is probably a less widespread problem than incompetence. Fortunately, decisions based on sound logic and strong evidence will usually prevail over those predicated on faulty arguments. Moreover, incompetence is often easily revealed, although not always easily dealt with. In some situations, individuals are doing the best that they can given their level of education and training. In others, individuals' performance is unacceptable despite their education and training. Dealing with these situations is never easy. One solution is to terminate the individuals in question, but they could also be moved to jobs where they are less likely to cause problems. Or they could be given additional training and be required to improve their performance or risk termination. Ideally, those in supervisory positions will deal directly with these cases, but, of course, sometimes it is the supervisor who is the problem.

Conflicts Between Juvenile Justice Practitioners and Others

The interactions between youths who have engaged in illegal behaviors (or behaviors that are alleged to be unlawful) and social control agents such as police officers, court personnel, probation officers, and other corrections workers are often conflictive. Even relationships between defendants and their defense attorneys can be problematic, especially if there is a lack of trust and poor communication. Similarly, relationships between probation officers and their clients, and between treatment program staff and their clients, can be difficult to establish and maintain, even though probation and treatment personnel believe their role is to help youths.

One obstacle to working with others is the difficulty people have in communicating with those who are different—those who come from different backgrounds and have different life experiences. This obstacle is prevalent within juvenile justice, because juvenile justice practitioners usually differ in a number of ways from the youths they work with. Juvenile justice practitioners are often of a different race, ethnic group, gender, or social class than their clients, and they are generally more educated and older. And even when these differences do not apply, there is one important distinguishing feature: juvenile justice practitioners are representatives of the juvenile justice process. As a result, they likely have a different perspective of the world and its possibilities than their clients, they frequently speak a language filled with professional jargon, and they typically live in different communities. Not surprisingly, communication between practitioners and their clients is often problematic.

Another obstacle to working with youths and their significant others is the difficulty of developing trust. Many Americans have the perception that government should not be trusted—a perception enhanced by the media emphasis on the lack of personal accountability and ethical commitment of persons in government. Moreover, at least some of those who come into contact with social welfare agency employees have not been helped but instead have encountered "red tape," condescension, excuses, and sometimes insults and abuse. Unfortunately, the phrase "I'm from the government, and I'm here to help" is often interpreted as "I'm from the government and I'm here to help myself!"

Of course, there are good reasons for youths not to trust juvenile justice practitioners. After all, those who work in juvenile justice, from police to judges to parole officers, are representatives of "the authorities." For many youths and their parents, such individuals are associated with hassles, arrests, violence, and loss of freedom. And in some cases, of course, youths are involved in activities they wish to hide from juvenile justice practitioners.

Juvenile justice practitioners face another problem in dealing with youth offenders. Some of these youths have engaged in actions that caused serious physical or psychological trauma to their victims, or even resulted in the death of another human being. Confronting the anguish of the victims can be difficult, but it is imperative that juvenile justice practitioners be able to effectively communicate with their clients and provide quality services congruent with their job responsibilities. Even when a youth has engaged in behaviors that a practitioner finds abhorrent, the practitioner must conscientiously perform his or her duties.

This does not mean that the practitioner has to condone the offender's behavior. It simply means that the practitioner needs to take his or her responsibilities seriously in all cases, not just in those cases where it is easy.

As the previous discussion indicates, juvenile justice practitioners confront a number of conflicts and difficulties in performing their jobs. Many of these conflicts and problems can be considered "normal" aspects of juvenile justice practice, and practitioners must successfully handle these conflicts in ways that do not impede their ability to carry out their professional duties (and feel a sense of accomplishment at the same time). Therefore, the main question is not how to avoid such conflicts, but how to successfully negotiate these conflicts. Part of the answer is that successful, effective, and professional practice is based on integrity and honesty. Another part of the answer is that such practice also depends on adherence to some important *ethical principles*. These principles are spelled out in the next section.

■ Ethics in Juvenile Justice

Juvenile justice practice always involves—sometimes subtly and at other times overtly—the exercise of **power**. Police officers, probation officers, detention personnel, attorneys, and others who work in juvenile justice are in a position to influence youths and their significant others in innumerable ways. Indeed, decisions made by juvenile justice personnel can result in orders to attend counseling, intrusion into individual and family privacy, loss of freedom, and termination of parental rights. Although the juvenile court is, in many respects, a social service agency that attempts to provide a range of helpful services to its clients, it differs from other types of social service agencies in one important respect: the court possesses *coercive power* that can be used to compel its clients to follow its directives.[1]

power The ability to influence how other people think or act.

Coercive force is a type of coercive power and involves the use of physical means or violence in order to influence an individual's behavior. Coercive force is often associated with law enforcement officers, but it is also employed by other juvenile justice personnel who have policing functions, such as probation officers, parole officers, and correctional and treatment staff. The use of coercive force must be taken seriously for at least five reasons. First, the use of coercive force on a juvenile can affect the safety of the juvenile justice agent, the juvenile, and others. Second, the use of coercive force threatens the liberty interests of the individual, which are constitutionally protected. Third, the use of coercion and violence against another human being threatens the sanctity of the individual. Fourth, the use of coercive force can lead to negative perceptions of juvenile justice on the part of the juvenile and others. Fifth, the use of coercive force denies the juvenile the ability to make free decisions.

coercive force Physical force threatened or applied with the intention of influencing an individual's behavior.

Although the use of coercive force is the most blatant example of the power wielded by juvenile justice practitioners, **coercive persuasion** is another type of coercive power, and constitutes a more subtle and more pervasive use of power. Coercive persuasion involves the use of one's position to influence the decisions and behaviors of others. For example, juvenile justice clients may be well aware

coercive persuasion The use of one's position to influence the decisions and behaviors of others.

that probation officers can revoke their probation unless they comply with the officers' requests. Similarly, a youth and his or her parents may agree to participate in a diversion program when given the not so subtle hint that failure to do so will result in a court hearing and the possibility of more severe sanctions. Also, many of the same problems associated with the use of coercive force can occur in instances of coercive persuasion. These include threats to the liberty interests of the individual, the development of negative perceptions of juvenile justice, and the denial of human agency.

Coercive Force and Persuasion Must Be Used Wisely

A number of potentially harmful consequences are associated with the use of coercive force and persuasion. Nevertheless, coercive force and persuasion are sometimes necessary. For example, both coercive force and persuasion may be used to protect the safety of clients, practitioners, or others; to assist clients in meeting legitimate treatment goals; or to accomplish other legitimate court goals. The problem here, of course, is that people will frequently disagree as to what constitutes client or practitioner safety, or legitimate treatment or court goals.

Because of the power they could eventually wield, those who embark on careers in juvenile justice must clearly understand the potentially harmful consequences associated with their positions. They must also act in ways that are consonant with the underlying ethical principles upon which juvenile justice practice is based. What are these principles? Perhaps one way to answer this question is to think of courts as institutions for dispensing justice. People who work in juvenile justice must ensure that people are treated in ways that are just, fair, and right. Because their actions are constrained by certain procedural requirements, it is helpful, as a way of approaching the topic of ethical principles, to consider these requirements. Among the most important are the following:

- Ensuring due process
- Maintaining the presumption of innocence
- Avoiding conflicts of interest
- Maintaining confidentiality
- Avoiding bias

Ensuring Due Process

due process An established course of proceedings designed to protect individual rights and liberties.

<u>Due process</u> is the backbone of our judicial system. A fundamental tenet of due process is that the end does not justify the means. Consequently, those involved in the practice of juvenile justice must realize that being "fair" requires a personal commitment to making the decision-making process understandable, predictable, and reliable. Only people do justice, but a fair system or process can help ensure

that the person who determines the outcomes will be fair. Crucial to due process and fair decision making are the following:

- Reasonable notice of the issues to be addressed
- A realistic opportunity to have one's position considered by the decision maker
- A decision maker who is objective and unbiased
- A search for the truth

Every person working in juvenile justice must be aware of and understand these four essential elements of due process. *Reasonable notice* means that a person brought before a court must be forewarned of the charges. The notice must accurately describe the charges so that the individual has an opportunity to defend him- or herself effectively. Without reasonable notice, the individual would face a serious handicap in trying to protect him- or herself against governmental power.

Reasonable notice would be irrelevant if the individual lacked an opportunity to present his or her position to the decision maker and have it considered. First, the accused must be physically present so that he or she (or his or her counsel) can personally respond to the charges. Second, the accused must be able to present information on his or her behalf and challenge information that is contrary to his or her interests. Third, disputed issues must be identified and ruled on by a person who has the expertise to do so. Fourth, the decision maker must be able to hear and understand the position of the accused.

Having one's case fully considered depends on having a decision maker who

Being Heard Is Necessary for a Fair Hearing

Although it is important that a decision maker be able to hear and understand the arguments made by the accused, the decision maker does not have to agree with those arguments. Nevertheless, without the ability to consider different points of view, a fair hearing is impossible.

knows how the decision process works and how evidence or other information is used in making decisions, but at the same time is unbiased regarding the facts presented. When prejudiced or biased views are allowed to influence juvenile justice decision making, such decision making is no longer fair. Consequently, caseworkers and other persons working in the courts must guard against pre-adjudication contact with judges and referees so as not to influence decision making. Similarly, actions that may result in biased or prejudicial decisions made by other decision makers, such as intake officers and probation staff, should be avoided.

Reasonable notice, the opportunity to present one's case, and objectivity on the part of the decision maker increase the likelihood that the truth will emerge. Uncovering the truth is in fact one of the main goals of the juvenile justice process, and the search for the truth (i.e., what really happened as opposed to what the state would like to believe happened) is what sets our legal system apart. Juvenile justice practitioners should be fully aware that the search for the truth is fundamental to fairness and due process, and should constantly strive to see that the truth is presented.

Maintaining the Presumption of Innocence

presumption of innocence The presumption that an individual accused of a crime is innocent of that crime until proven guilty. In the U.S. legal system, the government has the burden of proving an accused person is guilty rather than the accused person having the burden of proving he or she is innocent.

Of all the legal principles that make up the foundation of American jurisprudence, the **presumption of innocence** is perhaps the most important. According to this principle, rather than an accused being responsible for proving his or her innocence, the state is responsible for proving the guilt of the accused. It is the belief in this principle that allows attorneys to represent the guilty with a clear conscience: They know that guilt must be proved by the state beyond a reasonable doubt, and that any person has the right to test the state's ability to prove the case regardless of his or her culpability.

The importance of the presumption of innocence is typically associated with the trial or adjudication stage of the juvenile justice process. Nevertheless, this principle is equally important during the preadjudicatory decision-making stages. For example, presumption of guilt at the intake stage can result in intrusive, coercive, and inappropriate treatment being forced on a youth and his or her family without the benefit of a fair hearing. Similarly, presumption that a youth is guilty of misbehavior within an institutional setting can also serve as the basis for coercive and inappropriate actions on the part of correctional staff. Therefore, it is critical that presumption of guilt be avoided at every stage of the juvenile justice process pending a careful assessment of the facts.

Avoiding Conflicts of Interest

conflict of interest A situation in which an individual has two interests that cannot jointly be satisfied. The term is frequently used to refer to cases in which an individual has a responsibility to act in a certain way but has other reasons, perhaps self-interested ones, for acting otherwise.

Conflict of interest is a problem because a person cannot faithfully serve two different masters when those masters have conflicting interests. Juvenile justice personnel need to be sensitive to this issue because many people today must have more than one job to make ends meet. Outside associations also can lead to conflicts of interest. Consider, for example, a juvenile court judge who teaches at a local community college part-time in the area of juvenile justice and delinquency. She teaches local law enforcement officers and probation officers. She receives remuneration from the college. One day a juvenile who is accused of breaking into the college and doing severe damage is brought in front of the judge. Would her association with the college as an employee affect her ability to judge that juvenile without bias? Even if she was certain she would and could be fair, there is present "an appearance" of a conflict, and that should be enough to cause the judge to recuse herself and have the matter heard by another judge.

Obviously, fairness is critical to the operation of juvenile justice. However, just as important as actual fairness is the *appearance of fairness*. Consequently, juvenile justice practitioners must be sensitive to how their friendships, relationships, and voluntary associations may lead to conflicts of interest. The best way and only way to ethically handle such a situation is to disclose the conflict and step away from the case.

Maintaining Confidentiality

Historically, juvenile court proceedings and juvenile court records have been closed to the public. The purpose of closed hearings and records was to protect youths from the stigmatization that might occur if information about them was to become known to the community. Although the secretive nature of juvenile justice proceedings may have helped youths in many instances, it prevented careful scrutiny of juvenile court operations and sometimes resulted in unfair and adverse responses to youths.

Over the past 15 years, a number of reforms have opened court operations to public view. Today, many juvenile courts are open to the public (i.e., the public can sit in court and observe the proceedings). In addition, information collected within juvenile justice agencies is becoming more readily available to outsiders for employment and licensing purposes.[2] This does not mean that juvenile justice employees are free to disclose intimate facts of juvenile cases to everyone. Indeed, many people still recognize that the stigmatization that may result from juvenile court involvement serves neither the interests of youths nor those of the community. As a result, individual jurisdictions have rules about what information is confidential and what can be released to interested members of the public. In Michigan, for example, the legal file and those documents that have been admitted into evidence are available for public view, including perusal by the media, whereas other documents that contain intimate social information about the youth and his or her family and associates are not available. Some courts, unlike those in Michigan, allow some members of the public to access both legal and social files for specific purposes. For instance, many courts have policies that allow those involved in scholarly research to access files normally denied members of the public as long as the research does not identify specific youths.

MYTH VS. REALITY — Can Criminal Courts Access Juvenile Court Records?

Myth—Criminal courts do not have access to juvenile court records when making sentencing decisions.

Reality—Every state allows for prosecutor and/or court access to juvenile court records at some stage of the judicial process. Also, almost every state allows criminal courts to consider a defendant's juvenile record for sentencing purposes.[3]

Another issue that has received considerable attention in recent years is the need to share information among different agencies who work with youths who are involved in the juvenile justice process or at risk of juvenile justice involvement. For example, probation agencies, schools, social welfare programs, and mental health agencies may desire to share information on youths in order to improve the effectiveness of their services and keep youths from juvenile justice involvement. However, state and federal laws may restrict the sharing of some information among these agencies. State laws may require parents or guardians to provide written consent before agency records can be released or before agency personnel can discuss a particular case. The disclosure of students' education records is limited by the **Family Educational Rights and Privacy Act** (FERPA), and some states have enacted similar laws.[4]

FERPA is a complicated law that gives a student's parents the right to inspect and review their child's educational records and to request amendments to those records, and it gives the parents considerable control over the release of information in those records. However, there are some exceptions. Educators can share information with other agencies or individuals based on their personal knowledge or observations, as long as that information does not come from educational records. Educators can also release information without prior consent if this information is given to other educators who have a legitimate educational interest in the information, if the release is for the purpose of complying with judicial orders or a lawfully issued subpoena, or if state statute allows for the disclosure of information to juvenile justice authorities in order to improve the ability of juvenile justice agencies to better meet the needs of specific youths involved in the juvenile justice process. In addition, FERPA allows "law enforcement units" within schools to collect and disseminate information to others outside the school, even the media, without parental consent.[5]

Sharing information between agencies is often necessary in order to provide effective and cost-efficient services to youths and their families. Yet, it is important that juvenile justice personnel recognize the potential deleterious effects that can result from sharing information when it is not legally mandated. For example, knowledge that a youth is involved in the juvenile justice process may be used against him or her. School and other agency personnel may refuse to give the youth the same opportunities given to others. The youth may be denied opportunities to participate in a variety of school-sponsored activities. He or she may be singled out for minor behaviors and suspended when similar behaviors by other youths would result in different responses. As a result, juvenile justice practitioners should be extremely careful about sharing information that is not legally mandated. When it is necessary to discuss information about clients, the following rules are helpful:

- Make certain that releases are appropriately signed before obtaining reports from counselors, therapists, psychologists, and so on.
- Make certain that when you request an order for any examination or for counseling, the person subject to that order knows that a report will be forwarded to the court.
- Remember that nothing constructive is ever gained by embarrassing someone.
- If there are sensitive issues in reports presented to juvenile justice decision makers, be sure to alert the decision maker, the youth, and the family prior

Family Educational Rights and Privacy Act A federal law that gives a student's parents the right to inspect and review their child's educational records, request amendments to those records, and exert considerable control over the release of those records.

to the time that the decision is made so that they will not be surprised if these sensitive issues are made public.

- Let the client know that as a juvenile justice employee, everything said to you must be made available to the court and that you will not be able to counsel the client as would a private counselor or therapist.
- Understand court policies and relevant laws about **confidentiality** in your jurisdiction and become familiar with the rules of agencies that you regularly deal with.
- Guard and respect other people's confidences as if they were your own.

confidentiality
The maintaining of information in strict privacy. Legal confidentiality requirements prohibit certain information from being made available to the general public.

New technology that is available to courts today can have a definite impact on the quality of justice received by individual juveniles. For example, many courts have the technology to arraign juveniles, advise them of their rights, and even conduct hearings by video rather than in person. Many states are experimenting with this procedure to save money and time. Juveniles could remain in the detention center or even at the police station or jail and not have to be transported to the court's location. Hearing officers would not have to leave their offices or courtrooms, and probation officers could remain at the court.

These experimental procedures do raise several legal and practical concerns, however:

- Does the impersonal viewing of the hearing officer by the juvenile lessen the potential impact of the proceeding?
- What about the rights of the parents to be present or participate in hearings? Can they meaningfully interact with their child and advise him or her through video technology?
- How does the juvenile's attorney counsel with the juvenile if not in their presence?

All of these questions and many others need to be answered by courts before implementing plans to use available technology.

Another by-product of increased technology is the potential impact it can have on issues of confidentiality. With the increased use of computers to hold information about youths, issues regarding limitations on the access to that information arise. Court files stored at a courthouse cannot be "hacked" into, but files stored on a computer can be compromised by enterprising individuals with computer skills. It is not beyond imagination that a delinquent youth with computer skills could gain entry into a court system and delete, destroy, or modify files.

On the other hand, the increased use of data transfer and the ability to share information can increase communication between courts and law enforcement agencies, giving the police more vital information much faster than before. Serious and careful data collection can aid in planning for the efficient use of available dollars and programming.

Avoiding Bias

bias A tendency, preference, or prejudice, especially one that inhibits objectivity.

Bias is defined as an inclination, a preconceived opinion, or a predisposition to decide a course or an issue in a certain way that renders an otherwise objective person unable to exercise impartiality.[6] Bias can be influenced by a number of factors—race or ethnic background, gender, geography, social class, or religion, to name a few. Furthermore, often two or more factors are combined. Bias is learned and is often strongly ingrained. Moreover, just because a person works for a juvenile justice agency does not mean that he or she is free of bias or prejudice. Consequently, it is imperative that court employees recognize any bias that they have and work to overcome it so that all persons can receive just and fair treatment at each stage of the juvenile justice process.

Racial and ethnic bias has a long history in this country, and the courts have unfortunately played a role in maintaining this bias. It is not, however, within the scope of this chapter to detail the struggle of racial and ethnic minorities for equality. The main point to be made here is that if justice is truly to be blind, a person's color or nationality should have nothing to do with his or her right to receive justice. Those who work in juvenile justice must face the unpleasant fact that minority youths are disproportionately represented in the juvenile justice process, particularly in juvenile courts and in secure juvenile correctional institutions.[7] They must also face the fact that the disproportionate involvement of minority youths in juvenile justice is, at least in some jurisdictions, the product of institutionalized bias among juvenile justice decision makers.[8] Despite stereotypical portrayals of inner-city minority youths as criminal offenders by the media and despite institutionalized responses to juvenile crime that focus attention on lower-class minority communities, juvenile justice decision makers must avoid making assumptions about youths based on the color of their skin.

The gender of youth offenders is also associated with differential treatment of youths within juvenile justice. As noted earlier, delinquent or incorrigible females do present a significant challenge to juvenile justice decision makers, and more young women are coming to the attention of the police and juvenile courts. Unfortunately, juvenile justice personnel often lack the resources or expertise to deal with their special needs and problems. As a result, girls have experienced considerable neglect, which has led to a lack of sound programming options and support systems that might help them cope with difficult home lives and avoid future juvenile justice involvement.

Not only have girls experienced *gender bias* within the juvenile justice process, but so have young women, including single mothers.[9] There is some evidence that youths from single-parent families may be at greater risk of juvenile justice processing than similar youths from two-parent homes.[10] Because the system deals with so many single parents, the vast majority of whom are women, juvenile justice decision makers must avoid biased attitudes toward them. Instead, each case must be examined on its own merits.

Even female juvenile justice staff sometimes suffer from the gender bias of their colleagues. Today, women make up a larger part of the professional personnel employed in juvenile justice than was true in the past. Female attorneys are far more numerous today than in the 1970s, 1980s, and 1990s. Similarly, many women serve as juvenile court judges, probation officers, and police officers and in various other positions within juvenile justice. Most importantly, these women have made significant contributions to the improvement of juvenile justice operations, which must be recognized and supported by everyone within juvenile justice.

Another type of bias that can influence juvenile justice operations is *religious bias*. The First Amendment to the U.S. Constitution guarantees that the government, which includes the courts, cannot support a particular religious philosophy or place any restriction on the free exercise of religious practice. From this constitutional pronouncement has come the commonly accepted idea that church and state should be kept separate. In fact, churches, synagogues, and mosques are not subject to the taxation power of the state because that power could be used to restrict or modify their practices and beliefs.

Courts encounter religion in any number of ways. Indeed, many juvenile justice employees hold religious beliefs that influence their ideas about how people should conduct their lives. However, such religious beliefs must never interfere with an individual's objective judgment in a case. In addition, religious beliefs partly determine the child-rearing practices of juvenile justice clients. Indeed, in some instances people use religion to justify child-rearing practices that run contrary to state laws. Moreover, religious beliefs may lead to a lack of cooperation with those who are charged with providing different types of services to youths. In situations like these, juvenile justice decision makers must attempt to determine what constitutes acceptable community or legal standards without violating the individual's constitutional right to practice his or her religion without interference from governmental authorities.

Since September 11, 2001, there is no question that many Americans have modified their views toward and relationships with individuals of Muslim faith and ethnic background. Whether juveniles of Middle Eastern background have been treated unfairly as a result of prejudice in the juvenile justice system is open to debate, but it is clear that such potential exists and must be recognized as bias and combated at every opportunity.

The last forms of bias that need to be discussed are *geographic* and *social class bias*. Some people are inclined to think that youths or families who reside on "the wrong side of town" or in certain communities are potentially more dangerous or less amenable to intervention. Like racial or gender bias, such thinking prevents juvenile justice practitioners from making objective decisions about cases. Geography or social class is neither a cure for nor a cause of delinquency. Where a person lives can influence peer and adult associations and the values a youth is exposed to, but location is not an indicator of criminality or receptivity to treatment. Even in the most crime-ridden communities, there are many people who avoid criminality and lead conventional and law-abiding lives.

Economics and Ethical Dilemmas in Juvenile Justice

Juvenile justice agencies, like other governmental agencies, are influenced to a considerable degree by what is often called the "golden rule" in government—those who have the gold, rule! By appropriating money for certain programs or to support particular policies and by withdrawing funding from other programs and policies, political decision makers have tremendous influence over the range of programs and resources that are available to juvenile justice practitioners. If officials at the local level believe that "getting tough" with juvenile offenders is an effective way to reduce juvenile crime, detention facility budgets may be increased to hire more staff or expand the number of facility beds. Moreover, a judge's views about what juvenile justice strategies are effective may influence the proposed spending of money even before the budget gets to the spending source.

Too often, dispositional alternatives that might help children do not exist because they lack funding support. As a result, juvenile justice practitioners may face an ethical dilemma when figuring out what to recommend for their clients. Do they recommend what is truly in their clients' "best interest" regardless of cost, or do they recommend only that which they believe is financially available? Faced with this dilemma, juvenile justice practitioners must attempt to provide for clients' needs while simultaneously being a good steward of public funds. They also need to be knowledgeable about and creative with other revenue sources, such as federal dollars, grant money, and private insurance funds, in order to best meet client needs.[11]

How juvenile justice practitioners deal with the ethical issues discussed in this section will determine the extent to which juvenile justice is able to effectively meet its two main goals: serving the interests of children and protecting the community. Our belief is that each person who works in juvenile justice should give considerable thought to these issues and should conduct him- or herself in a manner that meets the highest standards of professional conduct. To do otherwise would compromise the ability of juvenile justice to deal fairly and justly with children and their families.

◼ Critical Positions in Juvenile Justice

Throughout this text, the phrase "juvenile justice process" has been used to refer to the totality of decision-making stages at which juvenile justice practitioners make determinations about the appropriate ways of handling juvenile offenders. This term was chosen because, unlike "juvenile justice system," it does not misleadingly hide the considerable conflict that exists between and within juvenile justice agencies—conflict that often precludes practitioners from working toward common goals and objectives. As noted, some of this conflict is unavoidable, but it is also important to keep in mind that the effectiveness of juvenile justice agencies in helping children and families depends on cooperation between the individuals who work in these agencies.

This section examines some of the important positions that make up juvenile justice. These positions require different skills, but they each play a critical role in effective juvenile justice practice. Moreover, they each offer a tremendous opportunity for individuals who are looking for a challenging career that provides a chance to assist others, can contribute to the quality of community life, and offers many intrinsic rewards as well.

Intake Officers

Juvenile justice agencies, including juvenile courts, are under increasing pressure from many sides to deal effectively with juvenile crime. In response, some legislatures are limiting the power, discretion, and jurisdiction of the juvenile courts by lowering the age for waiver and expanding automatic waiver provisions. At the same time, local governments are demanding more accountability for each tax dollar spent. This often means that juvenile justice agencies are being called upon to do more with static or diminishing resources. Consequently, intake—the process of screening cases and determining which juveniles should enter the juvenile court system—is more critical today than ever before.

Although the intake process varies from jurisdiction to jurisdiction, there are some common intake practices across communities. For example, regardless of who does the screening—probation officers, referees, court-designated workers, or prosecutors—the screeners must be knowledgeable about the following:

- Viable diversion programs, including non-court alternative interventions such as counseling programs, drug treatment programs, mediation programs, and alternative educational programs
- The seriousness of the various offenses and whether they are appropriate for a court hearing
- Nonjudicially ordered placement options, such as runaway shelters and temporary foster care placements
- Mental health services for juveniles

Intake officers also need to communicate well with youths and their parents, law enforcement officers, judges and other hearing officers, social service agency staff, school personnel, and others. Moreover, many intake officers provide short-term crisis intervention services to children and their families in an effort to avoid formal court processing. Most of these areas of knowledge are self-explanatory, but one of them needs special emphasis. Because all courts have only limited docket or hearing time, it needs to be put to the most productive use possible. Spending court time trying cases that are poorly investigated and improperly charged is a major waste of time and resources. Consequently, intake screeners must have some legal knowledge to help them evaluate the legal merits of juvenile delinquency charges. They also need the authority to require the responsible police agency to conduct further investigation when needed.

Psychologists and Substance Abuse Counselors

Many courts employ or have access to the services of a psychologist who has expertise in diagnosing children and identifying family problems. Because many youths who come to the court exhibit "acting out" behaviors that are rooted in their own personal and family problems, it is important that steps be taken to identify and respond to these problems as early as possible.

Similarly, many courts use the services of substance abuse counselors. Youths who come to the attention of juvenile justice agencies are often affected in some way by substance abuse. In many instances, it is the youth who has a substance abuse problem. In many other instances, one or both parents, or another significant other, has the problem. Consequently, it is important that courts have access to individuals who have the expertise to conduct good substance abuse assessments and provide effective interventions when appropriate.

Box 15-1

Interview: Jim Kendrick, Assistant Professor, Western Michigan University and Director of Field Education for the Specialty Program in Alcohol and Drug Abuse (SPADA)

Education: B.A. (education); M.A. (sports administration); M.A. (clinical mental health); and a specialist's certificate in alcohol and drug abuse, all from WMU Salary: $60,000, WMU plus between $20,000 and $25,000 for contract and consultant work each year

Q: What is your present employment position, and what do you do?

A: I am presently an assistant professor in the College of Health and Human Services. I work in the School of Community Health Services as the Director of Field Education for the Specialty Program in Alcohol and Drug Abuse (SPADA). I teach classes in the SPADA program training students to become substance abuse therapists and counselors. I am also the clinic coordinator for the WMU substance abuse clinic, which is a community clinic treating persons with substance abuse problems. In the clinic, I do direct counseling with patients. I also coordinate the administration of the substance abuse treatment programs in correctional programs for southwestern Michigan. The broad category of correctional programs includes those in many juvenile facilities.

Q: What is your employment background that built up to your present position?

A: My first "field-related" job was from 1979 to 1981 as a delinquency caseworker for the Michigan Department of Social Services working with serious delinquent youth who had been committed to the state by the juvenile courts. From 1981 to 1984, I worked as a probation officer with the Kalamazoo County Juvenile Court and started the substance abuse assessment program for delinquent youth, which is mandatory for all delinquents now. From 1984 to 1985, I was director of a half-way house for young adult felony offenders, called the Twin County Community Probation Center. From 1985 to 1991, I worked in private practice as a clinical substance abuse and mental health therapist. Since 1991, I have held my present position at WMU.

Q: How did you begin your career that relates to juvenile justice?

A: Sports was my gateway to connecting with young people. I played all the major sports in high school—football, basketball, baseball, and ran track. I then played basketball for two years at Kellogg Community College in Battle Creek, Michigan, before going on to finish my degree at WMU. My interest in athletics, especially basketball, was a natural connector with many of the delinquent youth that I encountered working for DSS and the court.

Q: What aspects of your job do you enjoy most? What parts of your job don't you enjoy?

A: I enjoy three parts of my job a great deal. I enjoy direct therapy and the personal contact with children and families. I also enjoy creating treatment delivery systems and programs that can help deliver treatment to a larger community. I also enjoy training people in these larger systems. The thing that frustrates me continually is the institutional or bureaucratic resistance in agencies and the government. The constant concern with one's turf frustrates progress and treatment effectiveness too often. With the limited dollars available, however, some of this survival mentality has to be expected.

Q: What are the major challenges you see in the future for the juvenile justice system as it relates to your job?

A: I believe that the next 10 years will be crucial in determining whether a juvenile justice system independent of the adult criminal justice system will continue to exist. I genuinely fear that political reactions to some violent acts by juveniles will end the independent juvenile court system, and that will be a shame. I believe that children need to be treated differently from adults in most cases.

Q: What individual personal qualities do you believe a person must possess to successfully do your job?

A: Most important are honesty and personal integrity. Without these qualities, a person will not have any credibility professionally or in the community. From this personal integrity must come a solid foundation of values and beliefs. Unless a person is deeply rooted, the aberrant values they will encounter doing therapy can blow them away!

Q: What would you want to tell a person who is thinking about entering your field?

A: I would tell them that they need to be committed to a career that impacts the lives of children and families. This commitment must exceed their needs and desires for material goals. You can make a decent living in my field, but the career has some monetary limits, and they need to know that going in.

Court Administrators

The court administrator (the individual mainly responsible for the daily operation of the court) is playing an increasingly important and public role in juvenile courts. The court administrator is often responsible for the fiscal management of court programs and for presenting the court's position to other agencies and the public, and he or she may have a great deal of influence in determining the types of programs that are operated by the court. Moreover, as competition for tax dollars increases, the court administrator must have a great deal of political savvy; in particular, he or she must know where the local government funding power is and be able to sell the court, its programs, and its mission.

The court administrator must take a lead role in formulating and defining the court's mission. To do this, the court administrator needs a comprehensive understanding of the juvenile justice process, from intake to treatment. The court administrator needs to know how the court can effectively network with other social agencies in both the public and private sectors to effectively deliver therapy, treatment, and other programming to juveniles and their families. The court administrator also must know the applicable laws and court rules governing the individual court and must be familiar with employment law and good administrative practices. The court administrator is responsible for interviewing, hiring, reviewing, and terminating employees, so knowledge of these techniques, effective communication skills, and interpersonal skills are a must. Because of the diverse legal expertise required in the administrator position, many courts are requiring formal legal training and/or a law degree for this position.

Finally, the court administrator must be computer literate so that the court can join the information age in its handling of its files, process, and docket. The importance of this position cannot be overstated.

Box 15-2

Interview: Doug Slade, Administrator for the Family Division of the Circuit Court for Kalamazoo County

Q: How did you begin your career in juvenile justice, and when?

A: In 1966, while I was a police administration major at Michigan State, a probation officer's job came open in the Ingham County, Michigan, Juvenile Court. Although my career goal at that time was to work in industrial security, I was nearing graduation and my wife thought I should apply for the position. I did and was hired at the lofty salary of $6,300 per year. Two years later, I was appointed to a referee's position holding preliminary hearings, removal hearings, and doing other intake functions. In 1970, I helped persuade our judge that the court needed an intake unit, and he appointed me to run it. In 1975, I was appointed Probate Registrar so I went over to the Estate Division and worked there until 1977 when I was appointed court administrator as well. I held the combined positions from 1977 to 1987 when I resigned from the court.

I was the Director of Central Michigan Legal Aid from 1987 to 1993, serving eight counties in central Michigan. In 1993, I was hired to be the court administrator for the Kalamazoo County Juvenile Court. As a result of the legislatively mandated combination of the juvenile court and the "family jurisdiction" of the circuit court, I am now the administrator in charge of the Family Division of the Circuit Court.

Q: What is your educational background?

A: I did graduate with a B.A. in police administration from MSU in 1966. I immediately started graduate school in that field but did not finish. My area of study was juvenile programs, but my increasing court responsibilities interfered with my continuing. In 1971, I attended the summer session of the National Judicial College at Reno, Nevada, and started law school at Cooley School of Law in Lansing, Michigan, in 1984. I finished on an accelerated program earning my juris doctoris in December 1986, while working full time as court administrator. As soon as I graduated from Cooley, I began teaching there as an adjunct professor teaching probate practice. I've also attended and taught at too many workshops and training seminars to count!

Q: What is your present job, how much do you make, and what are your major responsibilities?

A: As the Court Administrator for the Family Division of the Circuit Court, my annual salary is $88,000 per year. My major responsibilities are as follows:

a. Leadership to court staff—not just supervising or managing behavior, my job requires vision and being proactive in solving the myriad of challenges the court faces.

b. Planning—both day to day and long term. The creation of the local family court is a planning enterprise I was intimately involved in.

c. Public relations for the court—along with, and in addition to, the judges, I have major responsibilities in dealing with the media, citizenry, agencies, the county government, and others.

d. Community collaboration—one important duty is to identify areas in which the court and the community can work together and support each other.

e. Budget and finance—taking care of the taxpayers' dollars and meeting the needs of the children and families that appear in front of the court are very important duties.

f. Personnel administration—dealing with the many court employees and unions.

g. Mentoring other administrators and managers—I have long believed that persons who have experience owe newcomers all the help they can give them.

Q: What do you like and/or dislike about your job?

A: I enjoy the impact that I can have in my position on community issues of social justice and social change. The responsibilities of leadership mean a lot to me, and I take them very seriously. At my professional level, the monetary benefits are substantial, and I do enjoy the financial rewards. I don't like dealing with negative people. There is always somebody who will see how something can't work or can't be accomplished. I don't like people who don't work hard, who don't do their jobs. They make it harder for everyone else who have to pick up what's left undone.

Q: What major challenges do you see in the future for the juvenile justice system?

A: The major challenge that I see is overcoming the media praise of the idea that kids and families are so bad and out of control that they should just be treated as adults. I believe that the very existence of the juvenile justice system is threatened by the media misrepresentation of the facts about juvenile crime and delinquency and the juvenile court's ability to intervene effectively. I'm afraid that the real independence of a separate court for juveniles is threatened.

Q: What personal qualities do you believe a person must possess to successfully do your job?

A: Let me see if I can count them off for you: patience, being proactive, good listener, ability to see other perspectives, internally motivated, strong moral principles, quick learner, good communication skills, and humble. By humble, I mean that I need to always remind myself that I don't have all the answers or always know the best way to solve problems.

Q: What do you think a young person who is thinking about entering the juvenile justice field should know?

A: Foremost, they need to understand why kids come into the system—not because they are somehow inherently bad kids, but because their families and their community have failed them. They need to have empathy for the kids.

> Q: Who are you responsible to?
> A: First, I am responsible to the taxpayers whose money pays my salary and supports the court. Second, I am responsible to the presiding judge of the court.
> Q: How many people do you supervise?
> A: I am responsible for 164 employees: 10 managers, 67 professionals, and 87 clerical staff.

Judges

No position in the juvenile court is more important than that of juvenile court judge. Juvenile court judges must wear many hats. First, they need to be well versed in the juvenile and family law statutes of their particular jurisdiction. In addition, because of the increasing concern over due process in juvenile courts and because of the increasing visibility of juvenile court proceedings, judges should be well versed in procedural requirements. Increasingly, juvenile court judges will need trial experience in order to effectively perform their jobs and avoid making errors that would open the way for costly and time-consuming appeals.

In addition to having the legal acumen to make correct decisions regarding matters of law, juvenile court judges must be able to communicate well with youths, their parents, and a range of professional staff. Cases involving children and families are often the most emotionally draining cases that can be heard in courts. Consequently, juvenile court judges must have a sincere dedication to helping families and children. They must also have the ability to advocate for resources and services needed to enable the juvenile court to do its job. They may spend considerable time meeting with community groups, community leaders, and politicians in addition to the time they spend on court cases. Therefore, juvenile court judges must have vision and an understanding of how the court works and the ability to effectively articulate this to others. Lastly, they need to fully understand how the court interfaces with other public and private agencies in order to meet the needs of children and families and to protect public safety.

Juvenile court and family court judges must increasingly be able to handle heavy dockets and deal with increasing workloads. Given these difficult economic times for many state and local governments, judges are having to do more work with the same or reduced staff. For example, in Michigan, since 1990 the number of abused and neglected children in out-of-home care has increased by 17%.[12] Michigan law requires 90-day review hearings for all children in care and termination of parental rights of parents if the children remain out of home for 364 consecutive days. The number of hearings involving these children has increased exponentially, decreasing the time available for dealing with delinquent juveniles. In addition to the docket pressure, the delinquency rate in Michigan has increased since 1990, requiring more court interventions and court hearings that place considerable strain on the docket.

Appointed Judicial Officers

Regardless of whether they are called referees, magistrates, administrative law judges, masters, or commissioners, appointed judicial officers play an important role in juvenile courts. Their role may grow in importance, in fact, because they cost far less than judges. Routinely, their salaries are less and they require fewer "trappings" of office than judges. Many judges have personal secretaries, court reporters, and bailiffs. It is rare for referees to have "personal" staff. Appointed judicial officers have another advantage: unlike many judges, they are not elected, and therefore they can focus more of their attention on hearing cases. Moreover, many of them are attorneys with expertise in juvenile and family law, which makes them an important legal asset to the court. Too often juvenile court judges see the juvenile court bench as a stepping-stone to a higher court. In contrast, judicial officers who apply for juvenile court positions often do so because they have an interest in helping families and children and are more likely to make their service a career, thus giving the court needed stability.

Casework and Child Care Staff

Because of the increasing emphasis on the court's role in protecting the community, those juvenile justice practitioners who have the responsibility for supervising youths, providing services to youths and families, and making placement recommendations play a critical role in juvenile justice. These practitioners include

- Probation and aftercare officers
- Detention home workers
- Detention facility administrators

Probation officers perform multiple roles. They must monitor clients' adherence to court orders (e.g., the rules of probation), they must be able to provide services to clients, they must have knowledge of effective services and make referrals to appropriate service agencies when necessary, and they must complete a variety of administrative tasks in a timely manner. Consequently, probation officers need to know how to access the court process in order to set up probation violation hearings and effectuate necessary changes in placement. Probation officers need to be part police officer, part court officer, part therapist, and part friend. (See Chapter 10 for a detailed description of the duties of probation officers.)

Detention facilities serve as short-term secure settings for a range of juvenile offenders, from youths who have committed serious crimes against persons and property to youths who have run away from another jurisdiction. They are also used for temporarily holding youths awaiting long-term placement and, increasingly, for youths serving determinate sentences handed down by juvenile courts. Their diverse population makes the job of detention facility administrators even more difficult. Detention facility administrators must ensure the physical safety of residents and staff and maintain order. They must also provide basic educational programs for youths and see that their physical and mental health needs are addressed. In addition, they are responsible for hiring, evaluating, training, and terminating staff; taking a lead role in staff and program development; and communicating with other agencies and the public regarding detention facility operations.

Box 15-3

Interview: John P. Dantis, MSW, Director of the Bernalillo County, New Mexico, Juvenile Detention Center

Q: What is your educational and employment background?

A: I have a bachelor's degree from the College of Santa Fe and an MSW from New Mexico Highlands University. I also spent four years in the Marine Corps with a tour of Vietnam in 1968. My professional work experience includes 8 years as a juvenile probation officer, 2 years as the coordinator of probation services for the state of New Mexico, and 11 years as the director of Child Haven, a youth facility for delinquents.

Q: How long have you been the detention center director for Bernalillo County? What is your pay?

A: I am the director of the center, and have also been involved in the management of the county jail. I make over $80,000.

Q: What aspects of your job do you like and dislike?

A: The most important part of the job to me is the personal contact with the juveniles and their families. The real ability to impact kids and their families in positive directions gives the job real meaning for me. The part of the job that is most difficult is working in a political arena where I have to compete for limited resources with other county departments.

Q: What do you think that a college student should know about your job?

A: To understand the reality of any juvenile detention center today, the student must understand that success—the safe and productive management of these juveniles—requires them to understand cultural issues, educational issues, community resources, and mental health issues for the juvenile. They must also understand that detention workers are not "jailers" but must be counselors, teachers, social workers, and even coaches.

Q: What are the major challenges that you face today and in the immediate future?

A: Managing the continual growth in population, and continuing to educate politicians and other community leaders that the system needs balances. The need for secure detention is real, but it is not the only answer. Community-based corrections are vital to our doing the job in managing delinquent youth. The community must understand the spectrum of correctional choices that needs to be available.

Q: What are the demographics of your detention population? How many youth can your facility hold, and what is the average daily census?

A: Our population's median age is 15–16 years old, although we have had youth as young as 8. Our ethnic breakdown at any given time is 60–65% Hispanic, 25% Caucasian, 10% Native American, and 5% African American. Our license is for 78 youth at any one time, but our population average is about 100 plus 30 to 35 youth on community release and monitoring.

Q: Who operates the detention center? Who funds it?

A: It is a county facility supported primarily by tax dollars from Bernalillo County. I report directly to the county manager.

Q: How many employees work at the detention center?

A: We employ 76 child care workers in two classifications who are responsible for the daily hands-on management of the youth. We have four program managers who are responsible for managing "clusters" of youth. Their responsibility is not 8:00 a.m. to 5:00 p.m., but they manage their clusters 24 hours a day. We have 16 to 18 teachers in the school, which has components for special education and emotionally impaired students, as well as a **GED [general education diploma] program**.

Q: What different types of programs for youth do you offer?

A: First of all, you need to understand that we are a preadjudicative facility. Our residents are presumed innocent, so there are legal limits on what we can force youth to do. Having said that, we offer programs in the following areas:

- Education
- Leisure activities and the management of free time
- Substance abuse
- Gang intervention
- Communication skills (with a "Toastmasters" program)
- Computer education
- Family planning and AIDS education

Q: What types of offenses are you seeing juveniles detained for?

A: Primarily youth in the facility are here for offenses against persons, not property offenses. We also receive a lot of habitual or repeat offenders. Many of our population have gang connections and suffer from substance abuse problems.

Q: What other serious issues are you seeing with these juveniles?

A: Too many of our residents are parents as well as delinquents. They need serious and relevant parenting education, including instruction in child development issues. We have, as noted before, a chronic overcrowding problem, with the political and financial implications that carries with it. Finally, because we do a good job managing the youth sent to us, there is always going to be pressure to become a sentencing alternative or residential facility. I don't see any of these pressures or issues going away soon.

Q: My last question: What are you most proud of in your accomplishments as the detention center director?

A: We are the only accredited juvenile detention facility in the state of New Mexico. This is a tribute to the hard work of our staff and the vision of our political leaders. Each staff person takes individual responsibility for the children under the nine principles of direct supervision. To help maintain the high level of care, each of our staff gets 96 hours of training a year with one-third of the staff in training each week.

general education diploma A high school equivalency diploma awarded to adults who have not graduated from high school but have passed an official test covering a number of academic subjects. Commonly referred to by its acronym GED.

Detention employees, like their supervisors, must understand how to deal with a diverse resident population. This population consists of youths who have committed a wide range of offenses and who come from varied backgrounds. A major responsibility of detention staff is to ensure the safety of each resident under their care and to make certain that no one escapes. They are responsible for ensuring the safety of other staff as well. Consequently, they must be capable of forceful physical interventions. At the same time they must understand that they are often dealing with children. Although responsible for the safety and security of residents, detention staff are also in a position to offer the residents guidance and basic counseling when appropriate. They must, therefore, have the skills to work effectively with children and must be able to communicate with caseworkers, treatment staff, and detention supervisors about the needs of the youths under their care.

Required Skills for Effective Juvenile Justice Practice

The above discussion of the critical jobs in juvenile justice should make evident some of the skills and knowledge needed to perform those jobs. First, everyone working for a juvenile justice agency must *understand the interrelations among the agencies* that make up the juvenile justice process. Each worker also must understand how his or her position interrelates with other positions within the same agency.

Given the importance of communication, *computer skills* are clearly essential. Courts are now using computers to store and track files, docket cases, perform routine paperwork, and do any number of other jobs. Computers enable the printing of court orders at the close of court hearings for service on those present. Judges, judicial officers, prosecutors, and attorneys who practice in the court are doing legal research by computer. As a result, law libraries will become less important in the future. In addition, more and more probation, detention, and treatment staff are using computerized systems to track cases, complete reports, communicate with other agencies, and monitor program effectiveness.

Interpersonal skills will continue to be the key to effective job performance. Juvenile justice practitioners need the ability to communicate well, both orally and in writing. Moreover, given the diverse population served by juvenile justice personnel, they must be able to relate to and get along with many different kinds of people. Being fluent in other languages besides English, particularly Spanish, is very important, as are good listening skills and the ability to work effectively with people from varied racial, cultural, and ethnic backgrounds.

Two final prerequisites for effective juvenile justice practice are a *desire* to work with troubled youths and their families and a *strong commitment to professional ethics*. These qualities serve as the foundation upon which effective practice rests. They can be the key to a career that is both professionally and personally rewarding.

Chapter Summary

This chapter began with an examination of the role of conflict within juvenile justice. As noted, conflict is an inherent characteristic of juvenile justice practice and occurs between juvenile justice professionals and between these professionals and their clients. Indeed, conflicts between juvenile justice practitioners are to be expected, because those who work in juvenile justice have different ideas about how juvenile crime should be dealt with and because they have different responsibilities. Such conflicts will invariably lead to tension and disagreements between juvenile justice practitioners, who therefore must learn how to negotiate disagreements effectively.

Although conflicts between juvenile justice practitioners are an inherent feature of juvenile justice practice, so too are conflicts between juvenile justice professionals and their clients. Differences in social class, education, gender, race, and orientation toward authority can lead to difficulties in working together, not to mention the behaviors and attitudes exhibited by some juvenile justice clients. Nevertheless, it is important that juvenile justice practitioners recognize the potential problems presented by working with those who are different from themselves, that they have skills that allow them to work with a diverse client population, and that they act with honesty and integrity at all times.

Indeed, upholding the ethical principles upon which juvenile justice is based is fundamental to effective juvenile justice practice. Because of the coercive powers of juvenile justice agencies, juvenile justice practitioners must act so as to ensure due process, maintain the presumption of innocence, avoid conflicts of interest, guard confidentiality when appropriate, and avoid bias. To the extent that juvenile justice practitioners adhere to these ethical principles, they will help ensure that juvenile justice agencies live up to the mission of dealing with children and their families fairly, looking after the best interests of children, and at the same time protecting the community.

As is true of any profession, juvenile justice is made up of a diverse group of individuals who perform a variety of interrelated functions. If these individuals understand how their positions and responsibilities are related to those of others, both in their own agency and elsewhere, they will be better able to do their jobs effectively and efficiently. Their performance, whether they are intake officers, substance abuse counselors, judges, probation officers, or detention workers, determines the quality of justice given to youths and their families. The work is challenging and demanding, but it provides tremendous benefits to the community as well as to those who decide on a career in this exciting field.

KEY CONCEPTS

bias A tendency, preference, or prejudice, especially one that inhibits objectivity.

coercive force Physical force threatened or applied with the intention of influencing an individual's behavior.

coercive persuasion The use of one's position to influence the decisions and behaviors of others.

confidentiality The maintaining of information in strict privacy. Legal confidentiality requirements prohibit certain information from being made available to the general public.

conflict of interest A situation in which an individual has two interests that cannot jointly be satisfied. The term is frequently used to refer to cases in which an individual has a responsibility to act in a certain way but has other reasons, perhaps self-interested ones, for acting otherwise.

dispositional recommendation A recommendation made to the court about how a case should be handled.

due process An established course of proceedings designed to protect individual rights and liberties.

ethics The principles of right conduct. These principles determine which actions are right and which are wrong.

Family Educational Rights and Privacy Act A federal law that gives a student's parents the right to inspect and review their child's educational records, request amendments to those records, and exert considerable control over the release of those records.

general education diploma A high school equivalency diploma awarded to adults who have not graduated from high school but have passed an official test covering a number of academic subjects. Commonly referred to by its acronym GED.

integrity The quality of acting in a moral, sincere, and honest way.

power The ability to influence how other people think or act.

presumption of innocence The presumption that an individual accused of a crime is innocent of that crime until proven guilty. In the U.S. legal system, the government has the burden of proving an accused person is guilty rather than the accused person having the burden of proving he or she is innocent.

REVIEW QUESTIONS

1. What factors lead to conflict between juvenile justice professionals?

_____

_____

_____

_____

_____

2. What factors lead to conflict between juvenile justice practitioners and their clients?

_____

_____

_____

_____

_____

3. What are the potential negative outcomes associated with the use of coercive force and coercive persuasion in juvenile justice?

_____

_____

_____

_____

_____

4. In what ways can racial, gender, religious, geographic, social class, or other types of bias affect fairness in the juvenile justice process?

_____

_____

_____

_____

_____

5. What is meant by a conflict of interest? Give an example of how a conflict of interest could affect fairness in the juvenile justice process.

_____

_____

_____

_____

_____

6. What is meant by the presumption of innocence? How is it fundamental to our system of justice?

_____

_____

_____

_____

_____

7. What steps should juvenile justice practitioners take to ensure their clients' right to due process?

_____

_____

_____

_____

_____

8. What steps should juvenile justice professionals take to ensure confidentiality in the juvenile justice process?

_____

_____

_____

_____

_____

9. What is the "golden rule" in government and politics, and how does it affect client services?

_____

_____

_____

_____

_____

10. If you entered the field of juvenile justice, what would you like to do? Why is the type of job you chose appealing to you?

_____

_____

_____

_____

_____

11. Why is it important for juvenile justice practitioners to be able to work effectively with a diverse population of youths?

_____

_____

_____

_____

_____

12. What types of skills are required for successful juvenile justice practice?

_____

_____

_____

_____

_____

ADDITIONAL READINGS

Braswell, M. et al. 2002. *Justice, Crime and Ethics.* 4th ed. Cincinnati, OH: Anderson.

DeLucia, R. C. and T. J. Doyle. 1998. *Career Planning in Criminal Justice.* 3rd ed. Cincinnati, OH: Anderson.

Gordon, G. R. et al., 2005. *Criminal Justice Internships: Theory into Practice.* 5th ed. Cincinnati, OH: Anderson.

Souryal, S. 2003. *Ethics in Criminal Justice: In Search of the Truth.* Cincinnati, OH: Anderson.

NOTES

1. T. J. Bernard, *The Cycle of Juvenile Justice* (New York: Oxford University Press, 1992).
2. SEARCH, *Privacy and Juvenile Justice Records: A Mid-Decade Status Report* (Washington, DC: Bureau of Justice Statistics, 1997).
3. N. Miller, "State Laws on Prosecutors' and Judges' Use of Juvenile Records," *National Institute of Justice Research in Brief* (Washington, DC: National Institute of Justice, 1995).
4. M. L. Medaris et al., *Sharing Information: A Guide to the Family Educational Rights and Privacy Act and Participation in Juvenile Justice Programs* (Washington, DC: Office of Juvenile Justice and Delinquency Prevention, 1997).
5. Medaris et al., *Sharing Information.*
6. H. C. Black, *Black's Law Dictionary*, 4th ed. (St. Paul, MN: West, 1968).
7. H. N. Snyder and M. Sickmund, *Juvenile Offenders and Victims: 1999 National Report* (Washington, DC: Office of Juvenile Justice and Delinquency Prevention, 1999); B. Krisberg and J. F. Austin, *Reinventing Juvenile Justice* (Newbury Park, CA: Sage, 1993).
8. Krisberg and Austin, *Reinventing Juvenile Justice.*
9. See, for example, R.E. Johnson, "Family Structure and Delinquency: General Patterns and Gender Differences," *Criminology* 24 (1986): 65–84.
10. T. Hirschi, *Causes of Delinquency* (Berkeley, CA: University of California Press, 1969); Johnson, "Family Structure and Delinquency."
11. For a good example of the struggles that juvenile justice practitioners encounter in order to meet clients' needs, see M.D. Jacobs, *Screwing the System and Making It Work* (Chicago: University of Chicago Press, 1990).
12. Child Welfare League of America, *The Link* 2 (Summer 2003), 1.

Chapter Resources

Index